Enzymology of Complex Alpha-Glucans

Editor

Felix Nitschke

Assistant Professor
University of Texas Southwestern Medical Center
Department of Pediatrics and Biochemistry
Dallas, TX, USA

CRC Press
Taylor & Francis Group
Boca Raton London New York

CRC Press is an imprint of the
Taylor & Francis Group, an **Informa** business

A SCIENCE PUBLISHERS BOOK

Cover images by courtesy of persons mentioned below:

SEM image of Arabidopsis thaliana starch granules: Christine Lancelon-Pin (CERMAV-CNRS, Grenoble, France). Fabrice Wattebled (UGSF, University of Lille, France).

TEM image of negatively stained rabbit liver glycogen particles: Jean-Luc Putaux (CERMAV-CNRS, Grenoble, France).

Schematic of glycogen molecules (left) and starch granules (right): Christophe Colleoni (CNRS, University of Lille, France).

First edition published 2021
by CRC Press
6000 Broken Sound Parkway NW, Suite 300, Boca Raton, FL 33487-2742

and by CRC Press
2 Park Square, Milton Park, Abingdon, Oxon, OX14 4RN

© 2021 Taylor & Francis Group, LLC

CRC Press is an imprint of Taylor & Francis Group, LLC

Library of Congress Cataloging-in-Publication Data
```
Names: Nitschke, Felix, 1983- editor.
Title: Enzymology of complex alpha-glucans / editor, Felix Nitschke.
Description: First edition. | Boca Raton : CRC Press, Taylor & Francis
    Group, 2021. | "A science publishers book." | Includes bibliographical
    references and index.
Identifiers: LCCN 2020053240 | ISBN 9781138505209 (hardcover)
Subjects: MESH: Glucans--chemistry | Glucans--metabolism | Glycogen Storage
    Disease
Classification: LCC QR92.G5 | NLM QU 83 | DDC 572/.566--dc23
LC record available at https://lccn.loc.gov/2020053240
```

ISBN: 978-1-138-50520-9 (hbk)
ISBN: 978-0-367-76305-3 (pbk)
ISBN: 978-1-315-14644-7 (ebk)

Typeset in Times New Roman
by Radiant Productions

Preface

Glycogen and starch are the most abundant storage carbohydrates in the living world. For several reasons, the two types of storage carbohydrates are chemically similar: they are essentially composed of glucosyl residues and there are only two types of inter-glucose bonds that link the glucosyl moieties, alpha-1,4- and alpha-1,6 linkages. In both glycogen and starch, more than 90% of these bonds are alpha-1,4-linkages; less than 10% are alpha-1,6-bonds which lead to chain branchings. In glycogen and starch, most of the enzyme activities that form or cleave these inter-glucose bonds are similar and evolutionary related. Despite these (bio)chemical similarities, the intracellular location of glycogen and starch is different: typically, glycogen resides in the cytosol of carbohydrate-storing cells. By contrast, starch is usually located in green or non-green plastids of plants. In animals, indirect mode of debranching is usually dominant whereas in starch degradation of plants direct debranching occurs, which leads to an extensive maltodextrin metabolism. Direct debranching also occurs in the prokaryotic glycogen metabolism. Furthermore, there is an important difference in physico-chemical features of glycogen and starch that are also clearly biochemically relevant: normally, glycogen is a size-limited and hydro-soluble polyglucan molecule. In mammals, hydro-insolubility of glycogen has been observed in association with aging or with genetic diseases that can be very severe. Starch, however, is normally made of hydro-insoluble particles, called granules. In principle, these particles possess an essentially unlimited capacity of growth. Furthermore, the internal structure of starch granules seems to be evolutionary conserved. As opposed to hydro-soluble glycogen where branching points seem to be homogeneously distributed, those in starch follow a clustered arrangement giving rise to areas with largely unbranched chains forming double helices that deter chain hydration and hence promote hydro-insolubility. This allows the cells of a plant to synthesize a highly dense form of storage carbohydrates.

It is generally accepted that one isoform of glycogen synthase and the branching enzyme are sufficient to synthesize glycogen molecules from a glycogen primer, such as autoglycosylated glycogenin. Nevertheless, glycogen metabolism is highly regulated and both glycogen content and its physiological role differ between organs, tissues and cell types. The existence of various glycogen storage disorders emphasizes the importance of a functional glycogen metabolism even in tissues such as brain that have been shown to contain much less glycogen than, for instance, liver or muscle. In plants, metabolism of starch appears to be complex as there are two types of starch, assimilatory and reserve starch, which differ in metabolism and turnover. Assimilatory starch is generally located in chloroplasts and is synthesized in close relation to ongoing photosynthesis. By contrast, reserve starch located in non-green plastids is not coupled directly to photosynthesis. In addition, various isoforms of both starch synthase and starch branching enzyme are, on the one hand, involved in the synthesis of a distinct starch granule, and on the other hand, needed to synthesize assimilatory and reserve starch granules.

In the last years, significant progress has been made in the understanding of physiology and biochemistry of both glycogen and starch. Dr. Martin Steup and I conceived of the idea to edit a volume that combines knowledge from all areas of polyglucan-related research. Our idea was driven by the many parallels and differences of glycogen and starch metabolism. Dr. Martin Steup is a renowned expert in primary plant metabolism whose research, among other things, led to the discovery of two glucan phosphorylating enzymes in plants. His work set the ground for our current understanding of the role of phosphate in starch metabolism. My own research endeavors started out in his laboratory. However, Dr. Martin Steup's interest was not solely focussed on starch as he was always looking at polyglucans from a wider angle. While I was still his student, he allowed me to shift my research focus from starch toward glycogen metabolism, the field I am still passionate about today. I was honored to be asked to contribute to this volume as co-editor. Dr. Martin Steup and I contacted many experts in both fields, presenting our idea. It is surely because of his name and reputation that many senior researchers contributed excellent chapters to this volume, which now presents broad and current views on both carbohydrate stores, glycogen and starch, in lower and higher organisms.

This volume comprises eleven chapters, beginning with three that focus mainly on select methods and techniques that are relevant to the analysis of polyglucans. The following three chapters give a comprehensive overview on biochemistry and physiology of glycogen in lower and higher organisms including glycogen-related diseases. A chapter on polyglucan phosphorylation and the recently identified starch and glycogen phosphatases, SEX4 and laforin, respectively, shifts the focal point of this volume from glycogen back to starch, the structural complexity of which is comprehensively discussed in the subsequent chapter. Finally, three chapters follow that present the current understanding of enzymology and regulation of assimilatory and of reserve starch metabolism.

I would like to express my deep and sincere gratitude to Dr. Martin Steup for giving me the opportunity to initially co-edit this book, and eventually suggesting that I finalize it after his resignation as editor in the late stages of the editing process. His efforts are invaluable to me and this volume.

Also, I would like to thank all authors for committing their time to contribute manuscripts to this work. It is their individual expertise that lends this book a high quality and makes it a significant source of information. I also thank Dr. Berge Minassian for his support throughout the whole process of completing this book.

It is my hope that this volume will be a rich resource for new and established researchers in all polyglucan-related fields, and that it contributes to the collaboration of researchers that are working on structural features and metabolism of glycogen or starch.

"Glycogen and Starch: So Similar, Yet So Different" (a heading from Dr. Christophe Colleoni's Chapter) summarizes elegantly a fascination with both carbohydrates, and it likewise emphasizes the potential for new discoveries in both fields through fruitful collaboration.

Felix Nitschke
Dallas, USA, July 2020

Contents

Morphological and Structural Aspects of α-Glucan Particles from Electron Microscopy Observations

Jean-Luc Putaux

1.1 Introduction

Although similar in terms of chemical composition (they are both homopolymers of α-D-glucose), native glycogen and starch granules lie at opposite ends of the spectrum regarding morphology and structure. Glycogen occurs in the form of water-soluble hyperbranched amorphous nanoparticles, 30–60 nm in diameter (Manners 1991), whereas native starch granules are several micrometer-large semicrystalline objects, water-insoluble at room temperature, with a fascinating hierarchical and multiscale ultrastructure that has been described in several review articles (French 1984, Zobel 1988, Buléon et al. 1998, Tang et al. 2006, Pérez and Bertoft 2010, Lourdin et al. 2015). The variety of shape and size of α(1,4)-linked, α(1,6)-branched polyglucans gets wider if one includes particles in mutants that lack an enzyme involved in the glycogen/starch metabolism (Woodford and Tso 1980, Leel-Ôssy 2001, Gentry et al. 2009, Brewer et al. 2020) and those prepared *in vitro*, either by enzymatic biosynthesis (Potocki-Véronese et al. 2005, Kajiura et al. 2010, Ciric et al. 2013), enzymatic modification of native particles (Putaux et al. 2006), acid hydrolysis of starch granules (Putaux et al. 2003), or *in vitro* recrystallization of α-glucan solutions (Helbert et al. 1993, Hejazi et al. 2009, Montesanti et al. 2010).

Specific diffraction and imaging techniques are used to characterize each type of glucan particle at different length scales. While X-ray diffraction data provide average information on the crystalline fraction of the specimens (Frost et al. 2009), high-resolution microscopy techniques such as atomic force (Zhu 2017) or electron microscopy must be used to describe their fine local ultrastructure. In this chapter, we describe practical aspects of sample preparation, contrast enhancing protocols and specific imaging and diffraction techniques to resolve the morphology and ultrastructure of a variety of these glucan particles, using scanning and transmission electron microscopy (SEM and TEM, respectively), in imaging and diffraction modes.

Univ. Grenoble Alpes, CNRS, CERMAV, F-38000 Grenoble, France, Email: jean-luc.putaux@cermav.cnrs.fr

1.2 Scanning Electron Microscopy (SEM)

1.2.1 Principle

A thin focused beam of electrons accelerated at a typical voltage < 30 kV is raster-scanned on the surface of a specimen. The incident electrons make elastic and inelastic collisions with the atoms of the material and different types of electrons are reemitted from the surface and collected with specific detectors. The scanning of the specimen is synchronized with that of a computer screen and the intensity of the signal collected from one point of the surface is used to modulate the intensity of a corresponding pixel on the screen (Stokes 2008). The scanning speed must be high enough so that the operator only sees one global image due to retinal persistence. SEM micrographs are thus composite images. Their aspect is rather close to that of optical micrographs and their interpretation is often (but not always) intuitive due to the high depth of field and resulting three-dimensional impression. Although only the surface region of the specimen is imaged, a large variety of samples can be observed, from macroscopic objects and fragments of bulk materials to powders of micro- or nanoparticles.

The gun that emits the electrons and the microscope column are designed to focus the incident beam on the surface of the specimen. The recent generation of SEMs is equipped with so-called field emission (FE) guns that produce a very thin and bright beam of highly coherent electrons, which allows reaching a high resolution of about 1 nm. Two main types of re-emitted electrons can be considered. Backscattered electrons are re-emitted upward by the specimen after elastic collisions with the atoms of the material without any significant energy loss. The number of backscattered electrons depends on the average atomic number at the impact point, which thus generates a chemical contrast. Secondary electrons are electrons from the specimen that are ejected after interactions with incident or backscattered electrons. They have a low energy (< 50 eV) and are emitted from a surface layer about 50 nm-thick. Their number mainly depends on the surface topography but also on the average atomic number of the sample. Images produced from secondary electrons show the surface details with a higher resolution (Stokes 2008). The primary beam can penetrate the specimen to a depth that depends on the accelerating voltage and chemical nature of the material. The volume of interaction is the region of the sample that can produce signals due to interactions with the incident electrons. The effect of the imaging conditions on the "quality" of the resulting SEM micrographs of large α-glucan particles will be discussed in a further section.

Conductive materials can be directly observed by SEM under high vacuum. For non-conductive materials, like many polymers and biological samples, the number of incident electrons can rapidly exceed the number of electrons escaping from the surface, which results in the building of a local charge that may generate artifacts such as image distortions and flash discharges. The first method to cancel such a charging effect is to sputter-coat a thin layer of metal under vacuum on the specimen (typically about 2 nm of gold/palladium alloy) in order to make the surface conductive. In this case, the heavier metal atoms in the surface layer generate more secondary electrons, resulting in a stronger signal and images with a higher contrast. However, since sputter-coating is performed under high vacuum, hydrated specimens are likely to suffer from the treatment.

The recent generation of SEM instruments offers different possibilities to observe non-conductive samples without resorting to metal-coating. First, for each type of material, there is an optimum voltage at which the number of incident electrons is balanced by the number of electrons escaping the sample. This accelerating voltage is low, typically of the order of 1–2 kV, and sensitive detectors have been developed to collect the weak signals that are emitted at such low voltages. In addition, systems have been specifically designed to keep the specimen under a low pressure of gas (a few tens of Pa) during its observation. These "variable pressure" microscopes can be operated under high or low vacuum conditions since the specimen chamber is separated from the upper high-vacuum column by a differential pumping system. The benefit is significant in the case of non-conductive materials since the positive ions resulting from the ionization of the residual gas by the emitted electrons somewhat compensate the charging of the specimen (Stokes 2008).

In so-called "environmental" SEMs (or ESEMs), the sample can be observed in a humid atmosphere maintained via a precise control of the temperature and pressure. In order to keep a significant degree of moisture under a low pressure, the temperature must be kept between 3 and 5 °C and the pressure between a few tens and a few hundred Pa. It is even possible to keep liquid water during the observation. Consequently, with an ESEM, plant material (Stabentheiner et al. 2010) or fragments of freshly cut specimens can be observed without resorting first to dehydration by solvent exchange or critical point drying and subsequent metal-coating (James 2009).

Although this chapter is mainly dedicated to the imaging capabilities of electron microscopes, one can briefly mention the possibility to perform chemical analyses in a SEM using so-called energy-dispersive X-ray spectroscopy (EDX). A specific detector collects the X-rays emitted from the specimen surface upon scanning the electron beam and a multichannel analyzer produces corresponding spectra that allow determining the elemental composition (Watt 1997). However, the technique has only been used in a very limited number of cases on starch granules. In a context of water purification, the ability of starch granules to absorb chromium ions was studied by mapping and quantifying the distribution of chromium atoms at the surface of and inside the granules (Szczygieł et al. 2014), whereas chlorine and sulfur atoms were detected in chemically-functionalized starch granules (Chauchan et al. 2015). SEM-EDX may thus be potentially useful to investigate phosphorylation aspects in starch granules.

1.2.2 SEM Observation of Starch Granules and Large Glucan Particles

Since the average size of native starch granules typically ranges between 1 and 200 μm, SEM is a particularly well-adapted technique to characterize their variety of shapes (Fig. 1.1) (Jane et al. 1994, Pérez and Bertoft 2010, Putaux et al. 2010). Specimens can easily be prepared from dry powders or aqueous dilute dispersions deposited onto metallic stubs, cleaved mica or carbon tape, with or without metal coating, as previously explained. Considering the organic nature of starch, backscattered electrons are rarely used to visualize native granules. Most images found in the literature are secondary electron micrographs that allow comparing granules of different botanical origins, from standard to more "exotic" sources (Ao and Jane 2007, Wang et al. 2009, Kong et al. 2009, Pérez and Bertoft 2010), highlighting the peculiarities of some granules like the rod-like or dumbbell shape of *Euphorbiaceae*

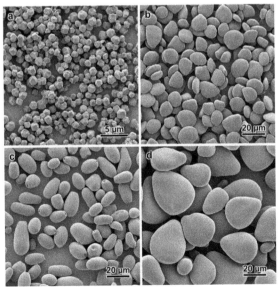

Figure 1.1. Variety of sizes and shapes of native starch granules of various botanical origins: (a) quinoa seeds, (b) ginger root, (c) avocado nut, (d) tulip bulb. Secondary electron FE-SEM images of Au/Pd-coated specimens recorded under high vacuum.

granules (Mahlberg 1973, Seshagiri Rao et al. 1988), or the lateral secondary outgrowth of granules from the *Phajus grandifolius* orchid (Chanzy et al. 2006). Significantly smaller granules purified from leaves, plant chloroplasts, algae or cyanobacteria (typically 1–2 µm) can be observed as well, allowing the identification of their three-dimensional shape that otherwise would be more ambiguous when visualized from thin sections of resin-embedded material (see Section 1.3.2.3) (Izumo et al. 2011, Lin et al. 2013, Vandromme et al. 2019).

While SEM is mostly used to observe the surface of starch granules (Baldwin et al. 2009), the inner structure can be visualized as well, provided that the particles have been fragmented, in particular if they have been submitted to acid hydrolysis or enzymatic digestion that partially etched away some of the material. Such treatments can reveal, for instance, the size and organization of growth rings, or the degradation pattern of acids or amylases (Chanzy et al. 2006, Blazek and Copeland 2010, Blazek and Gilbert 2010, Hasjim et al. 2010, Soares et al. 2011, Huang et al. 2014, Dhital et al. 2015). Apart from native granules, different types of large starchy particles prepared *in vitro* have been observed as well, such as spherocrystals from heat-treated starch (Fanta et al. 2002, Singh et al. 2010), debranched starch (Kiatponglarp et al. 2016), recrystallized amylose (Helbert et al. 1993), microspheres of retrograded starch (Li et al. 2018), B-type axialites of amylose biosynthesized *in vitro* by amylosucrase (Potocki-Véronèse et al. 2005) and A-type single crystals from short-chain amylose (Montesanti et al. 2010).

A survey of the SEM micrographs found in the literature shows that the specimens are not necessarily observed under optimal imaging conditions, which may result in missed information or misinterpreted artifacts. In particular, the choice of accelerating voltage is critical. At high voltage (typically 10–30 kV), the incident electrons can penetrate deeper in the organic material, thus generating cascades of collisions in a larger interaction volume that, in turn, generates secondary electrons from a region of the surface laterally much wider than that irradiated by the primary beam (Stokes 2008). Although the number of emitted electrons is higher and the particles appear brighter, there is an overlap of information in the signals collected from neighboring points, resulting in a loss of resolution. At a lower acceleration, the volume of the primary beam interaction in the specimen is much smaller and the secondary electrons are emitted from a narrower region. The surface topography is thus revealed in more details.

This influence of accelerating voltage is illustrated by Fig. 1.2 that shows FE-SEM micrographs of uncoated waxy maize starch granules observed under high vacuum or low air pressure. Although the contrast of the granules increases with increasing voltage, the high-resolution details on the surface are lost. This "smoothing" effect can be misinterpreted as radiation damage. For this uncoated specimen, the optimum conditions would be close to those corresponding to Fig. 1.2b where the charging is minimal while the contrast is satisfactory and details can still be seen at the granule surface. The effect is even more significant for small granules and particularly flat ones. At high voltage, the electrons can penetrate deeper than the thickness of the granules and partly interact with the supporting material. In Fig. 1.3d, the metal-coated *Arabidopsis* leaf starch granules imaged at 10 kV are very bright but they also look very smooth, while at 1 or 2.5 kV (Figs. 1.3a and 1.3b, respectively), the contrast is lower but the high-resolution texture of the granule surface is visible.

As mentioned earlier, ESEM is an attractive technique to directly observe hydrated plant material and, more particularly, starch granules within their tissue environment. As an example, Fig. 1.4 shows secondary electron images of a freshly cut fragment of potato tuber that was directly introduced into the microscope chamber without fixative treatment. The fragment was first observed under fully hydrated conditions (Fig. 1.4a), then after controlled partial drying (Fig. 1.4b). In both conditions, the starch granules are seen inside cellulosic cells with a very good contrast, even without metal coating. Figure 1.4c to 1.4e show one starch granule that was alternatively imaged in wet, hydrated and dry conditions by slightly changing the specimen temperature and gas pressure in the chamber. Liquid water droplets can thus be reversibly stabilized during observation or dried, which allows studying the dynamic effects of water sorption or dehydration of starch granules and large glucan particles. However, care must be taken to record images at a relatively low magnification since,

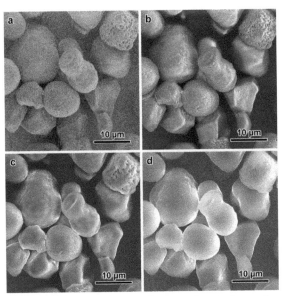

Figure 1.2. Secondary electron FE-SEM images of uncoated waxy maize starch granules recorded at a magnification of 3000×. The same region of the preparation has been imaged at different accelerating voltages of the electrons and air pressures: (a) 2 kV, low pressure of air (40 Pa); (b) 4 kV, 40 Pa; (c) 10 kV, 70 Pa; (d) 20 kV, 70 Pa.

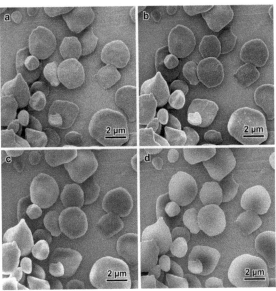

Figure 1.3. Secondary electron FE-SEM images of Au/Pd-coated flat starch granules from *A. thaliana* leaves (*Wassilewskija* ecotype) recorded under high vacuum at a magnification of 10000×. The same region of the preparation has been imaged at different accelerating voltages: (a) 1 kV; (b) 2.5 kV; (c) 5 kV; (d) 10 kV. Sample courtesy of F. Wattebled (UGSF, University of Lille).

under humid conditions, the specimens can be more rapidly damaged under electron illumination due to the enhanced diffusion of the radiation defects. So far, only a few ESEM images of starchy specimens can be found in the literature. In particular, the technique was used to characterize the pores on the surface of starch granules (Fannon et al. 1992), the effect of milling on rice grains (Dang and Copeland 2004) and wheat endosperm (Edwards et al. 2008, Barrera et al. 2013), as well as the effect of relative humidity on the volume change in rice powders (Tang et al. 2007) and starch surfaces (Fechner et al. 2005).

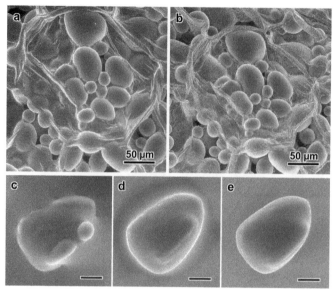

Figure 1.4. (a,b) Secondary electron ESEM images of an uncoated freshly cut fragment of a potato tuber: (a) the specimen temperature was 2 °C and the pressure of humid air 700 Pa, maintaining a 99% humidity around the specimen. The accelerating voltage was 10 kV. (b) The pressure was decreased to 200 Pa and the humidity was 30%. The accelerating voltage was 5 kV. Note how the cell walls get more wrinkled and the granules get closer to each other upon drying. (c–e) Secondary electron ESEM images of a native starch granule for *Phajus grandifolius* recorded at about 3 °C. From "c" to "e", the pressure of humid air was decreased, resulting in the drying of the granule: (c) the granule is soaked and droplets of liquid water can be seen; (d) the granule is hydrated with visible traces of humidity; (e) the granule is under high vacuum. Scale bar in images "c" to "e": 20 μm.

1.3 Transmission Electron Microscopy (TEM)

1.3.1 Principle

A beam of electrons accelerated at a typical voltage of 100–300 kV is transmitted through a thin specimen (< 500 nm for organic materials) to form a magnified image of its contours and projected volume with a potential resolution of a few ångstroms. Scattering of the electrons by the atoms of the material and the use of selective apertures in the microscope column generate contrasts in the images (Watt 1997). Another major advantage of TEM is the possibility to record local electron diffraction patterns of semicrystalline specimens and therefore study the structure and orientation of the crystalline domains. However, organic samples are extremely sensitive to radiation damage from the electron beam. They are rapidly degraded at room temperature and, therefore, must be observed under low illumination conditions. Organic specimens can also be preserved for a longer time if they are observed at higher voltage or kept at low temperature during the observation. Hydrated specimens must be quench-frozen before introduction in the microscope to prevent vacuum drying, and observed at low temperature.

1.3.2 Sample Preparation and Contrast Enhancement Techniques

One of the main constraints for TEM observation is that the specimens must be very thin. Glycogen particles, with a typical diameter of a few tens of nanometers, can be readily imaged from dilute aqueous suspensions (typically 0.001–0.01% w/v, without any buffer salts that would crystallize upon drying). Much larger objects like starch granules, recrystallized spherulites or plant tissues containing glucan particles cannot be observed as such and thin sections must be prepared using ultramicrotomy.

1.3.2.1 Particle Suspensions

1.3.2.1.1 Grid Pretreatment and Negative Staining

Achieving a satisfying distribution of particles from suspensions on the carbon film with limited aggregation or overlapping can be challenging. This is particularly critical when one wants to determine the size distribution of a population of glycogen particles. The supporting carbon film being initially hydrophobic, the direct deposition of aqueous suspensions and subsequent air-drying generally result in locally accumulated material. The so-called "glow discharge" procedure is efficient to treat carbon films prior to sample deposition (Aebi and Pollard 1987, Harris 1997). The carbon-coated TEM grids are placed in a chamber in which a low pressure of air is ionized. The grids are submitted to this mild plasma cleaning during a few seconds, which results in a temporarily hydrophilic carbon surface onto which aqueous suspensions easily spread. As charges are generated on carbon during the treatment, the particles also tend to adsorb on the surface. The excess of liquid is thus gently blotted away with filter paper and the objects should remain homogenously distributed on the supporting film.

1.3.2.1.2 TEM Observation of Negatively Stained Polyglucan Particles

Glycogen has been one of the first biopolymers to be observed in the early age of TEM development. However, the images produced by Husemann and Ruska did not reveal any fine details of the particles (Husemann and Ruska 1940). The reasons of this poor contrast are that (i) drying induces a shrinking of the macromolecules, (ii) the constituting light atoms do not generate any significant scattering contrast, and (iii) the particles are prone to rapid degradation under the electron beam. The most widely used technique to enhance the contrast of individual nano-objects is negative staining. Briefly, a drop of an aqueous solution of heavy-atom salt is deposited on the specimen. Upon drying, a thin layer of heavy atoms concentrates around the nanoparticles, creating an electron-dense cast (Harris 1997). The particles thus appear as clear objects on a darker background. The preparations can be observed at a higher magnification since, although the glucan particles are indeed damaged by the electron beam, the heavy atom cast is resistant and thus reveals the contours and fine details of the surface topography. Up to now, the most commonly used stains have been aqueous uranyl acetate (2% w/v), and phototungstic acid, to a lesser extent. However, due to recent regulations regarding the handling of radioactive compounds, uranyl acetate will soon be replaced by new staining solutions that do not contain uranium (Hosogi et al. 2015, Kuipers and Giepmans 2020). Practically, a homogenous negative staining of particles can be achieved under two conditions: (i) the supporting carbon film has to be glow-discharged before depositing the nanoparticles and the stain, and (ii) the stain solution has to be applied on the specimen before drying. After a few minutes, the stain in excess can be blotted away with a filter paper and the residual thin film of stain allowed to dry.

Examples of TEM images of negatively stained preparations of oyster glycogen and maize phytoglycogen are shown in Figs. 1.5a and 1.5c, respectively. Many other images of negatively stained glycogen and phytoglycogen from various sources can be found in the literature: rat and mouse liver (Drochmans 1962, Sullivan et al. 2010, 2011, Hu et al. 2018), rat muscle (Wanson and Drochmans 1968), pig liver (Sullivan et al. 2012, Powell et al. 2015), oyster (Hata et al. 1984, Matsuda and Hata 1985), slipper limpet (Sullivan et al. 2010), *sugary*-1 maize mutant (Putaux et al. 1999, Huang and Yao 2011, Powell et al. 2014, 2015). On such images, the smaller spheroidal β-particles can clearly be distinguished from the larger multilobular α-particles that have been described as supramolecular clusters of β-particles (Drochmans 1962, Sullivan et al. 2010, 2012). Size distribution histograms of each population can be determined from the images, even though the larger particles may be deformed due to the flattening induced by drying. These morphological data can be compared to the various mathematical models that describe the hyperbranched architecture of the macromolecule and address the question of a possible limitation in size of native glycogen due to peripheral chain crowding (Melendez et al. 1993, 1999, Matsui et al. 1996, Bezborodkina et al. 2018, Zhang et al. 2018, 2020).

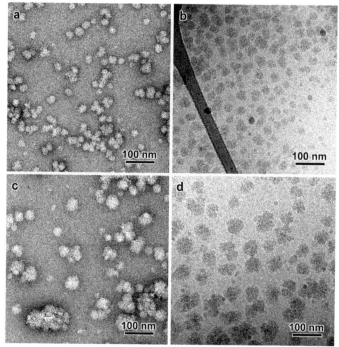

Figure 1.5. TEM images of glycogen particles from oyster (a,b) and phytoglycogen from the *sugary*-1 maize mutant (c,d). In images "a" and "c", the preparations have been negatively stained with uranyl acetate while in "b" and "d", the particles were unstained, embedded in a thin film of vitreous ice and observed at low temperature by cryo-TEM.

Glycogen-like hyperbranched particles have been synthesized *in vitro* using enzymes, and observed after negative staining. Examples of one-pot biosynthesis consisted in incubating short-chain amylose with the branching enzyme (BE) from the hyperthermophilic bacteria *Aquifex aeolicus* (Kajiura et al. 2010, 2011) or that from *Rhodothermus obamensis* (Roussel et al. 2013). Branched polyglucans were also synthesized by the tandem reaction of an elongation enzyme, to yield amylose-like chains, and a BE. Successful combinations contained the amylosucrase from *Neisseria polysaccharea* and the BE from *R. obamensis* in the presence of sucrose (Grimaud et al. 2013), or phosphorylase *b* from rabbit muscle and the BE from *Deinococcus geothermalis* in the presence of glucose-1-phosphate (Ciric et al. 2013).

Starch granules purified from algae, cyanobacteria, parasites or leave chloroplasts can be observed by TEM after deposition on a carbon film, provided that their thickness does not exceed 500 nm. In many cases, the smaller granules are lenticular with a diameter < 1 μm and a thickness not exceeding 200 nm and details of the surface structure can be revealed by negative staining. A short and mild acid pre-treatment that etches away some of the granule surface material can be helpful (Deschamps et al. 2008). In this line, a lamellar organization has been clearly revealed in the semi-amylopectin granules of different species of Porphyridiales (Shimonaga et al. 2008) and in amylopectin granules of *Cryptosporidium parvum* oocysts (Harris et al. 2004). Thicker granules can be directly visualized by SEM (Section 1.2.2) or embedded in a hardening resin and their ultrathin sections observed by TEM (Section 1.3.2.3).

When the starch granules are significantly bigger, a longer treatment in sulfuric acid (Naegeli's method) or hydrochloric acid (so-called "lintnerization"—Robin et al. 1974) results in the progressive disruption of the granule architecture. This has been illustrated in the case of waxy maize starch granules hydrolyzed with 2.2 N HCl at 36 °C (Putaux et al. 2003). After 2 weeks, negatively stained granule fragments exhibited a lamellar organization (Fig. 1.6a) while, after 6 weeks, parallepipedal platelets were individualized, lying flat on the carbon film (Fig. 1.6b). A similar behavior was observed

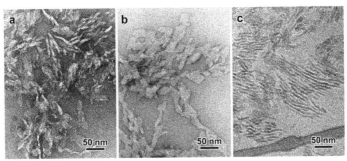

Figure 1.6. (a,b) TEM images of negatively stained nanocrystals prepared by acid hydrolysis of waxy maize starch granules after 2 (a) and 6 (b) weeks of hydrolysis in 2.2 N HCl, at 36 °C. In "a", the clear elongated units are 7–9-nm-thick lamellae seen edge-on, while the platelets in "b" would correspond to individual lamellae lying flat on the carbon film. (c) Cryo-TEM image of unstained fragments after 2 weeks of hydrolysis. Here, the stacked lamella seen edge-on are dark, partly due to diffraction contrast. Image "a" courtesy of H. Angellier-Coussy (CERMAV, LGP2, Grenoble), printed with permission.

for waxy maize starch granules hydrolyzed with 3.2 M H_2SO_4 at 30 °C during six weeks (Yamaguchi et al. 1979) or at 40 °C under constant stirring during one week (Angellier-Coussy et al. 2009). These platelet nanoparticles would correspond to the lamellae in the cluster model of amylopectin where the short linear branches form crystalline arrays of parallel double helices (Robin et al. 1974, Pérez and Bertoft 2010). It was assumed that the acid preferentially severed the more accessible α(1,6) branching point in the interlamellar regions, thus allowing the crystalline platelets to separate (Putaux et al. 2003). However, the chromatography analysis of H_2SO_4-hydrolyzed waxy maize starch granules revealed that the nanocrystals also contained a significant number of branch points (Angellier et al. 2009). Lintners were prepared from potato (Wikman et al. 2014) and barley (Goldstein et al. 2016) starch granules as well, but the shape of the resulting platelets was less defined.

1.3.2.2 Cryo − Transmission Electron Microscopy (Cryo − TEM) of Particle Suspensions

Cryo-TEM can be used to observe nanoparticles in aqueous suspension when drying or staining is thought to induce artifacts such as deformation, aggregation or crystallization of buffer salts. This technique is particularly helpful for soft or water-swollen colloidal nanoparticles whose morphology or structure would be affected by air-drying or staining. As described in more details elsewhere (Dubochet et al. 1988, Harris 1997), droplets of suspensions are deposited on lacey carbon films supported by TEM grids, or on perforated support foils with calibrated holes. Using dedicated automatized workstations, the liquid in excess is blotted with filter paper and the remaining thin film standing over the support holes is quench-frozen in liquid ethane. The frozen specimen is then mounted in a pre-cooled cryo-specimen holder, introduced into the microscope, and observed at low temperature (around –180 °C).

Examples of cryo-TEM images of oyster glycogen and maize phytoglycogen are shown in Figs. 1.5b and 1.5d, respectively. In this case, the concentration of the suspension must be about 100 times higher than for the preparation of dry specimens (typically 0.1 to 1.0 % w/v). The well-dispersed unstained particles are seen embedded in a thin film of transparent vitreous ice. Cryo-TEM is not yet a routine technique compared to negative staining. Although it is perfectly adapted to visualize glucan particles frozen in their water-swollen state, only a few images can be found in the literature. Putaux et al. compared the size distributions of *sugary-1* maize phytoglycogen particles determined from TEM images of negatively stained, metal-coated and quench-frozen preparations (Putaux et al. 1999). Martinez-Garcia et al. compared the morphology of glycogen particles from various bacterial sources (Martinez-Garcia et al. 2016). Cryo-TEM has also been used to visualize core-shell particles prepared by enzymatic extension of oyster glycogen surface chains by the amylosucrase of *N. polysaccharea* in the presence of sucrose (Putaux et al. 2006), as well as glycogen-like hyperbranched particles synthesized *in vitro* by phosphorylase *b* and a glycogen branching enzyme (Ciric et al. 2013). As aforementioned, small fragments of starch granules that partially retain the lamellar organization of amylopectin can be prepared by mild acid hydrolysis of the native granules. Figure 1.6c shows such fragments imaged by cryo-TEM after quench-freezing. In this case, the contrast does not result from any staining but is partly due to the diffraction by the crystalline lamellae.

1.3.2.3 Ultramicrotomy of Bulk Samples and Large Particles

Soft and hydrated materials such as leaf fragments, microalgae or cyanobacteria cannot be directly sectioned. They have to be chemically fixed in paraformaldehyde/glutaraldehyde, post-fixed with osmium tetroxide (OsO_4), dehydrated by exchange with ethanol and embedded in hardening resins (Glauert 1975). Purified starch granules can be directly soaked in the liquid resin before hardening. Ultrathin (50–150 nm) sections of the embedded specimens are then cut at room temperature with a diamond knife, in an ultramicrotome (Reid 1975). The sections float on water and are collected on bare or carbon-coated TEM grids. When the sample is soft at room temperature, ultrasectioning must be performed under cryogenic conditions in a dedicated unit.

Different techniques can be used to enhance the contrast of the organelles or particles in the sample. On the one hand, before the dehydration step, the sample can be treated with uranyl acetate so that it is already stained when it is soaked in the resin. On the other hand, the thin sections can be post-stained with uranyl acetate/lead citrate (Reid 1975). In both cases, the glucan particles generally remain unstained and appear as clear objects on a darker background, as illustrated by the image of *Arabidopsis* chloroplasts in Fig. 1.7a and by many of those found in the literature (Ramazanov et al. 1994, Roldán et al. 2007, Eicke et al. 2017, Seung et al. 2018). Specific staining techniques have been proposed to selectively stain glycogen particles inside animal tissues (Revel et al. 1960, Ryman 1974, Riemersma et al. 1984) like, for instance, the addition of potassium hexacyanoferrate ($K_3Fe(CN)_6$) to the classical OsO_4 post-fixative (De Bruijn 1973). Another technique of interest to reveal the presence of glucan particles in thin sections has been developed by Gallant and collaborators (Gallant et al. 1969). The sections are treated with periodic acid thiosemicarbazide silver proteinate (PATAg) that positively stains molecules that contain sugar units (Thiéry 1967). Condensed glucan particles thus appear as dark objects: glycogen as small dots and starch granules as more extended regions (Gérard et al. 2001, Cenci et al. 2013, Kadouche et al. 2016, Boyer et al. 2016). Note that glycoproteins and glycolipids can be stained as well, thus outlining organelles and cell walls. An example of PATAg-treated Arabidopsis leaf plastids is shown in Fig. 1.7b.

One artifact sometimes observed on ultrathin sections of resin-embedded starch granules is the presence of dark ripples across the granule sections. This effect, which does not depend on the size

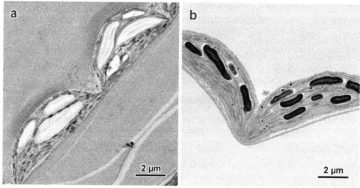

Figure 1.7. TEM images of ultrathin sections of fragments of resin-embedded wild-type *A. thaliana* leaves. (a) Chloroplasts from the Columbia ecotype. The sample has been stained "in block" with uranyl acetate before sectioning. The starch granules are not stained and appear as clear elongated regions. (b) Chloroplasts from the *Wassilewskija* ecotype. The thin section has been treated with the PATAg method and the positively stained starch granules are dark. Samples courtesy of L. Boyer and F. Wattebled (UGSF, University of Lille).

of the granules and may occur on 1–2 µm-large plastidial starch as well as 50–100 µm-large tuber granules, has been studied in detail by Gallant and Guilbot. The ripples would be due to an insufficient permeation of the granules by the inclusion agent and also due to the partial swelling of the granule sections while they float on water, resulting in local folding of the section (Gallant and Guilbot 1971). However, images of thin sections without any folding effect, even from large granules, can be found in the literature (Helbert and Chanzy 1996), which suggests that the effect may also depend on the nature of the embedding resin and operating conditions during ultramicrotomy (angle of the diamond knife, sectioning speed, etc.).

It must be noted that the ultramicrotome can also be used to prepare mirror-like surfaces from blocks of resin-embedded material. This surface can then be observed by SEM using backscattered electrons, sensitive to atomic number, to reveal the regions that were specifically stained by OsO_4 and other contrasting agents. However, since most embedding resins are non-conductive, the blocks must be observed under low-vacuum conditions in order to suppress the charging effects (see Section 1.2.2). Even though the resolution of such block face images is lower than those of TEM micrographs of corresponding ultrathin sections, their contrast is high and the field of view is not limited by the bars of the supporting TEM copper grid. This method is thus particularly useful to check the quality of the embedding and staining before actual preparation of ultrathin sections from the same block. Figure 1.8 shows a comparison of the block face SEM image and corresponding TEM micrograph of an ultrathin section of the same region. Block surfaces of resin-embedded starch granules have also been used to record topography images by AFM, reveal details of the granule inner structure (Peroni-Okita et al. 2015) and study the deformation of artifacts induced by the diamond knife during cutting (Tsukamoto et al. 2012).

1.3.2.4 Electron Diffraction of Crystalline Specimens

Thin sections of starch granules can be observed without any staining, provided that the semicrystalline internal structure has not been affected by a fixative treatment and that the embedding resin does not impregnate the granules (Yamaguchi et al. 1979). Chanzy et al. (1990) and Helbert and Chanzy (1996) used a water-soluble melamine resin to embed native and acid-hydrolyzed corn starch granules. Since the thin sections prepared by ultramicrotomy did not initially exhibit any contrast, they were briefly submitted to a mild HCl hydrolysis, resulting in the partial etching of the resin and revealing the location of the granules (Helbert and Chanzy 1996). The growth rings of previously partially lintnerized granules were visible under low-dose illumination (Fig. 1.9a). Electron diffraction patterns

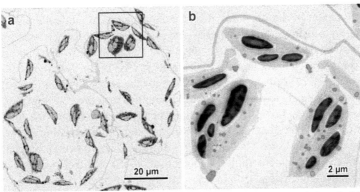

Figure 1.8. (a) Backscattered electron FE-SEM image of chloroplasts in a fragment of *A. thaliana* leaf (Columbia ecotype) embedded in resin. The block has been surfaced with a diamond knife. The contrast of the image has been reversed for clarity. The regions enriched in OsO_4 are dark and the starch granules are unstained. (b) TEM image of an ultrathin section of the same block stained by the PATAg method. The starch granules are dark. The image corresponds to the region framed on the SEM image in "a". The starch granule content appears to be different because the section was cut at a slightly different height in the block. Sample courtesy of M. Facon and F. Wattebled (UGSF, University of Lille).

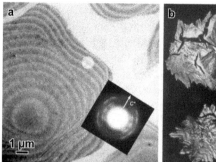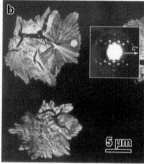

Figure 1.9. (a) Bright-field TEM image of an ultrathin section of partially HCl-hydrolyzed native corn starch granules embedded in melamine resin. No additional staining was used to reveal the growth rings. Inset: electron diffraction fiber pattern recorded from the region of the granule indicated by the clear disk. The pattern shows that the fiber axis c is oriented perpendicular to the growth rings. In order to record this pattern, the hydrated section has been quench-frozen in liquid nitrogen prior to introduction in the microscope and observed at low temperature. (b) Dark-field TEM image of a frozen-hydrated ultrathin section of A-type amylose spherocrystals embedded in melamine resin. Inset: electron diffraction pattern of the region indicated by the clear disc, showing that the c-axis of the crystallites is radially oriented. Images and diffraction patterns courtesy of W. Helbert (CERMAV, Grenoble), printed with permission.

were recorded at low temperature under frozen-hydrated conditions, which corresponded to the A-type and confirmed, at the micron scale, the radial orientation of the double helices in the crystalline domains with respect to the growth rings (Fig. 1.9a) (Helbert and Chanzy 1996). Similar electron fiber diffraction patterns had previously been recorded from frozen-hydrated fragments of B-type potato starch granules (Oostergetel and van Bruggen 1993). Highly crystalline spherocrystals prepared by crystallization of short-chain amylose from solution could also be sectioned after embedding in melamine resin subsequently cured at room temperature to prevent the partial dissolution of the particles. Well-resolved A-type electron diffraction patterns were recorded at low temperature from quench-frozen hydrated sections, revealing the radial orientation of the amylose double helices (Fig. 1.9b) (Helbert et al. 1993).

1.4 Perspectives

As explained earlier, TEM images are 2D projections of the volume of objects along the beam direction. Information along this direction is thus lost. One of the most exciting developments in TEM is the possibility to reconstruct the volume of specimens using series of 2D images recorded at consecutive tilt angles of the specimen. This can be achieved in modern microscopes by taking advantage of highly sensitive digital cameras and software that precisely control the specimen orientation and the image acquisition. After alignment with respect to one another, the collected 2D images are back-projected to calculate a 3D reconstruction of the specimen volume (Nudelman et al. 2011, Fridman et al. 2012). Now widely used to study the morphology and structure of biological systems, this electron tomography approach could be applied to address several questions related to $\alpha(1,4)\alpha(1,6)$ glucan particles. For instance, the ultrastructure of glycogen α-particles could be studied in their water-swollen state by determining the 3D organization of the constituting β-units from series of cryo-TEM images. Visualizing the spatial organization of amylopectin lamellae in very small starch granules like those from chloroplasts (Engel et al. 2015) would also be particularly useful to shed more light on the much-debated architecture of larger starch granules.

Acknowledgments

Except when explicitly indicated, the micrographs displayed in this article have specifically been recorded for this chapter. The SEM images in Figs. 1.1–1.4 and 1.8a have been recorded using a FEI

Quanta 250 microscope equipped with a field-emission gun, while the TEM micrographs in Figs. 1.5–1.7 and 1.8b have been recorded with a FEI-Philips CM200 'Cryo' microscope operating at 80 or 200 kV. The author would like to thank the NanoBio-ICMG Platform (FR 2607, Grenoble) for granting access to the Electron Microscopy facility, Christine Lancelon-Pin (CERMAV, Grenoble) for recording the FE-SEM images, Fabrice Wattebled (UGSF, Lille) for giving a sample of *Arabidopsis* starch granules, and Henri Chanzy (CERMAV) for his critical reading of the manuscript.

References

Aebi, U. and T.D. Pollard. 1987. A glow discharge unit to render electron microscope grids and other surfaces hydrophilic. J. Electron. Microsc. Tech. 7: 29–33.

Angellier-Coussy, H., J.-L. Putaux, S. Molina-Boisseau, A. Dufresne, E. Bertoft and S. Pérez. 2009. The molecular structure of waxy maize starch nanocrystals. 344: 1558–1566.

Ao, Z. and J. Jane. 2007. Characterization and modeling of the A- and B-granule starches of wheat, triticale, and barley. Carbohydr. Polym. 67: 46–55.

Baldwin, P.M., D.J. Gallant and S. Pérez. 2009. Structural features of starch granules I. pp. 149–192. *In*: BeMiller, J. and R. Whistler (eds.). Starch: Chemistry and Technology (Third Edition). Chapter 5.

Barrera, G.N., G. Calderón-Domínguez, J. Chanona-Pérez, G.F. Gutiérrez-López, A.E. León and P.D. Ribotta. 2013. Evaluation of the mechanical damage on wheat starch granules by SEM, ESEM, AFM and texture image analysis. Carbohydr. Polym. 98: 1449–1457.

Bezborodkina, N.N., A.Y. Chestnova, M.L. Vorobev and B.N. Kudryavtsev. 2018. Spatial structure of glycogen molecules in cells. Biochemistry (Moscow) 83: 467–482.

Blazek, J. and L. Copeland. 2010. Amylolysis of wheat starches. II. Degradation patterns of native starch granules with varying functional properties. J. Cereal Sci. 52: 295–302.

Blazek, J. and E.P. Gilbert. 2010. Effect of enzymatic hydrolysis on native starch granule structure. Biomacromolecules 11: 3275–3289.

Boyer, L., X. Roussel, A. Courseaux, O. Mvundza Ndjindji, C. Lancelon-Pin, J.-L. Putaux et al. 2016. Expression of *Escherichia coli* glycogen branching enzyme in an Arabidopsis mutant devoid of endogenous starch branching enzymes induces the synthesis of starch-like polyglucans. Plant, Cell Environ. 39: 1432–1447.

Brewer, M.K., J.-L. Putaux, M. Sullivan, A. Rondon, A. Uittenbogaard and M.S. Gentry. 2020. Polyglucosan body structure in Lafora disease. Carbohydr. Polym. 240: 116260.

Buléon, A., P. Colonna, V. Planchot and S. Ball. 1998. Starch granules: structure and biosynthesis. Int. J. Biol. Macromol. 23: 85–112.

Cai, L. and Y.-C. Shi. 2014. Preparation, structure, and digestibility of crystalline A- and B-type aggregates from debranched waxy starches. Carbohydr. Polym. 105: 341–350.

Cenci, H., M. Chabi, M. Ducatez, C. Tirtiaux, J. Nirmal-Raj, Y. Utsumi et al. 2013. Convergent evolution of polysaccharide debranching defines a common mechanism for starch accumulation in cyanobacteria and plants. Plant Cell 25: 3961–3975.

Chanzy, H., R. Vuong and J.-C. Jésior. 1990. An electron diffraction study on whole granules of lintnerized potato starch. Starch/Staerke 42: 377–379.

Chanzy, H., J.-L. Putaux, D. Dupeyre, R. Davies, M. Burghammer, S. Montanari et al. 2006. Morphological and structural aspects of the giant starch granules from *Phajus grandifolius*. J. Struct. Biol. 154: 100–110.

Chauhan, K., V. Priya, P. Singh, G.S. Chauhan, S. Kumari and R.K. Singhal. 2015. A green and highly efficient sulfur functionalization of starch. RSC Adv. 5: 51762–51772.

Ciric, J., A.J.J. Woortman, P. Gordiichuk, M.C.A. Stuart and K. Loos. 2013. Physical properties and structure of enzymatically synthesized amylopectin analogs. Starch/Stärke 65: 1061–1068.

Dang, J.M.C. and L. Copeland. 2004. Studies of the fracture surface of rice grains using environmental scanning electron microscopy. J. Sci. Food Agric. 84: 707–713.

De Bruijn, W.C. 1973. Glycogen, its chemistry and morphologic appearance in the electron microscope. I. A modified OsO4 fixative which selectively contrasts glycogen. J. Ultrastruct. Res. 42: 29–50.

Deschamps, P., C. Colleoni, Y. Nakamura, E. Suzuki, J.-L. Putaux, A. Buléon et al. 2008. Metabolic symbiosis and the birth of the plant kingdom. Mol. Biol. Evol. 25: 536–548.

Dhital, S., V.M. Butardo, Jr., S.A. Jobling and M.J. Gidley. 2015. Rice starch granule amylolysis—Differentiating effects of particle size, morphology, thermal properties and crystalline polymorph. Carbohydr. Polym. 115: 305–316.

Drochmans, P. 1962. Morphologie du glycogène: Etude au microscope électronique de colorations négatives du glycogène particulaire. J. Ultrastruct. Res. 6: 141–163.

Dubochet, J., M. Adrian, J.J. Chang, J.-C. Homo, J. Lepault, A.W. McDowall and P. Schultz. 1988. Cryo-electron microscopy of vitrified specimens. Q. Rev. Biophys. 21: 129–228.

Edwards, M.A., B.G. Osborne and R.J. Henri. 2008. Effect of endosperm starch granule size distribution on milling yield in hard wheat. J. Cereal Sci. 48: 180–192.

Eicke, S., D. Seung, B. Egli, E.A. Devers and S. Streb. 2017. Increasing the carbohydrate storage capacity of plants by engineering a glycogen-like polymer pool in the cytosol. Metabolic. Eng. 40: 23–32.

Engel, B.D., M. Schaffer, L.K. Cuellar, E. Villa, J.M. Plitzko and W. Baumeister. 2015. Native architecture of the *Chlamydomonas* chloroplast revealed by in situ cryo-electron tomography. eLife 4: e04889.

Fannon, J.E., R.J. Hauber and J.N. BeMiller. 1992. Surface pores of starch granules. Cereal Chem. 69: 284–288.

Fanta, G.F., F.C. Felker and R.L. Shogren. 2002. Formation of crystalline aggregates in slowly-cooled starch solutions prepared by steam jet cooking. Carbohydr. Polym. 48: 161–170.

Fechner, P.M., S. Wartewig, A. Kiesow, A. Heilmann, P. Kleinebudde and R.H.H. Neubert. 2005. Influence of water on molecular and morphological structure of various starches and starch derivatives. Starch/Stärke 57: 605–615.

French, D. 1984. Chapter VII - Organization of starch granules. pp. 183–247. *In*: R.L. Whistler, J.N. BeMiller and E.F. Parschall (eds.). Starch, Chemistry and Technology. Academic Press, NewYork.

Fridman, K., A. Mader, M. Zwerger, N. Elia and O. Medalia. 2012. Advances in tomography: probing the molecular architecture of cells. Nat. Rev. Mol. Cell. Biol. 13: 736–742.

Frost, K., D. Kaminski, G. Kirwan, E. Lascaris and R. Shanks. 2009. Crystallinity and structure of starch using wide angle X-ray scattering. Carbohydr. Polym. 78: 543–548.

Gallant, D. and A. Guilbot. 1969. Etude de l'ultrastructure du grain d'amidon à l'aide de nouvelles méthodes de préparation en microscopie électronique. Die Stärke 6: 156–163.

Gallant, D.J. and A. Guilbot. 1971. Artefacts au cours de la préparation de coupes de grains d'amidon. Etude par microscopie photonique et électronique. Die Stärke 23: 244–250.

Gallant, D.J., B. Bouchet and P.M. Baldwin. 1997. Microscopy of starch: evidence of a new level of granule organization. Carbohydr. Polym. 32: 177–191.

Gentry, M.S., J.E. Dixon and C.A. Worby. 2009. Lafora disease: insights into neurodegeneration from plant metabolism. Trends Biochem. Sci. 34: 628–639.

Gérard, C., P. Colonna, B. Bouchet, D.J. Gallant and V. Planchot. 2001. A multi-stages biosynthetic pathway in starch granules revealed by the ultrastructure of maize mutant starches. J. Cereal Sci. 34: 61–71.

Glauert, A.M. 1975. Fixation, Dehydration and Embedding of Biological Specimens. North-Holland/American, Elsevier.

Goldstein, A., G. Annor, J.-L. Putaux, K.H. Hebelstrup, A. Blennow and E. Bertoft. 2016. Impact of full range of amylose contents on the architecture of starch granules. Int. J. Biol. Macromol. 89: 305–318.

Grimaud, F., X. Roussel, C. Lancelon-Pin, S. Laguerre, A. Viksø-Nielsen, A. Rolland-Sabaté et al. 2013. *In vitro* synthesis of hyperbranched α-glucans using a biomimetic enzymatic toolbox. Biomacromolecules 14: 438–447.

Harris, J.R. 1997. Negative Staining and Cryoelectron Microscopy: The Thin Film Techniques. RMS Microscopy Handbook, Oxford, BIOS Scientific Publishers.

Harris, J.R., M. Adrian and F. Petry. 2004. Amylopectin: a major component of the residual body in *Cryptosporidium parvum* oocysts. Parasitology 128: 269–282.

Hasjim, J., G. Cesbron Lavau, M.J. Gidley and R.G. Gilbert. 2010. *In vivo* and *in vitro* starch digestion: are current in vitro techniques adequate? Biomacromolecules 11: 3600–3608.

Hata, K., M. Hata, M. Hata and K. Matsuda. 1984. A proposed model for glycogen particles. J. Jpn. Soc. Starch Sci. 31: 146–155.

Helbert, W., H. Chanzy, V. Planchot, A. Buléon and P. Colonna. 1993. Morphological and structural features of amylose spherocrystals of A-type. Int. J. Biol. Macromol. 15: 183–187.

Helbert, W. and H. Chanzy. 1996. The ultrastructure of starch from ultrathin sectioning in melamine resin. Starch/Stärke 48: 185–188.

Hejazi, M., J. Fettke, O. Paris and M. Steup. 2009. The two plastidial starch-related dikinases sequentially phosphorylate glucosyl residues at the surface of both the A- and B-type allomorphs of crystallized maltodextrins but the mode of action differs. Plant Physiol. 150: 962–976.

Hosogi, N., H. Nishioka and M. Nakakoshi. 2015. Evaluation of lanthanide salts as alternative stains to uranyl acetate. Microscopy 64: 429–425.

Hu, Z., B. Deng, X. Tan, H. Gane, C. Li, S.S. Nada et al. 2018. Diurnal changes of glycogen molecular structure in healthy and diabetic mice. Carbohydr. Polym. 185: 145–152.

Huang, J., N. Wei, H. Li, S. Liu and D. Yang. 2014. Outer shell, inner blocklets, and granule architecture of potato starch. Carbohydr. Polym. 103: 355–358.

Huang, L. and Y. Yao. 2011. Particulate structure of phytoglycogen nanoparticles probed using amyloglucosidase. Carbohydr. Polym. 83: 1665–1671.

Husemann, E. and H. Ruska. 1940. Versuche zur Sichtbarmachung voxi Glykogenmolekülen. J. prakt. Chemie 156: 1–10.

Izumo, A., S. Fujiwara, T. Sakurai, S.G. Ball, Y. Ishii, H. Ono et al. 2011. Effects of granule-bound starch synthase I-defective mutation on the morphology and structure of pyrenoidal starch in *Chlamydomonas*. Plant Sci. 180: 238–245.

James, B. 2009. Advances in "wet" electron microscopy techniques and their application to the study of food structure. Trends Food Sci. Technol. 20: 114–124.

Jane, J., Y. Kasemsuwan, S. Leas, H. Zobel and J.F. Robyt. 1994. Anthology of starch granule morphology by scanning electron microscopy. Starch/Stärke 46: 121–129.

Kadouche, D., M. Ducatez, U. Cenci, C. Tirtiaux, E. Suzuki, Y. Nakamura et al. 2016. Characterization of function of the GlgA2 glycogen/starch synthase in *Cyanobacterium* sp. Clg1 highlights convergent evolution of glycogen metabolism into starch granule aggregation. Plant Physiol. 171: 1879–1892.

Kajiura, H., H. Takata, T. Kuriki and S. Kitamura. 2010. Structure and solution properties of enzymatically synthesized glycogen. Carbohydr. Res. 345: 817–824.

Kajiura, H., H. Takata, T. Akiyama, R. Kakutani, T. Furuyashiki, I. Kojima et al. 2011. *In vitro* synthesis of glycogen: the structure, properties, and physiological function of enzymatically-synthesized glycogen. Biologia 66: 387–394.

Kiatponglarp, W., S. Rugmai, A. Rolland-Sabaté, A. Buléon and S. Tongta. 2016. Spherulitic self-assembly of debranched starch from aqueous solution and its effect on enzyme digestibility. Food Hydrocolloids. 55: 235–243.

Kong, X., J. Bao and H. Corke. 2009. Physical properties of *Amaranthus* starch. Food Chem. 113: 371–376.

Kuipers, J. and B.N.M. Giepmans. 2020. Neodymium as an alternative contrast for uranium in electron microscopy. Histochem. Cell Biol. 153: 271–277.

Leel-Ôssy, L. 2001. New data on the ultrastructure of the corpus amylaceum (polyglucosan body). Pathol. Oncol. Res. 7: 145–150.

Li, B.-Z., X.Q. Xian, Y. Wang, B. Adhikari and D. Chen. 2018. Production of recrystallized starch microspheres using water-in-water emulsion and multiple recycling of polyethylene glycol solution. LWT - Food Sci. Technol. 97: 76–82.

Lin, Q., M. Facon, J.-L. Putaux, J.R. Dinges, F. Wattebled, C. D'Hulst et al. 2013. Function of isoamylase-type starch debranching enzymes ISA1 and ISA2 in *Zea mays* leaf. New Phytol. 200: 1009–1021.

Lourdin, D., J.-L. Putaux, G. Potocki-Véronèse, C. Chevigny, A. Roland-Sabaté and A. Buléon. 2015. Crystalline structure in starch. pp. 61–90. *In*: Nakamura, Y. (ed.). Starch—Metabolism and Structure. Springer Japan.

Mahlberg, P.G. 1973. Scanning electron microscopy of starch grains from latex of *Euphorbia terracina* and *E. tirucalli*. Planta 110: 77–80.

Manners, D.J. 1991. Recent developments in our understanding of glycogen structure. Carbohydr. Polym. 16: 37–82.

Martinez-Garcia, M., M.C.A. Stuart and M.J.E.C. van der Maarel. 2016. Characterization of the highly branched glycogen from the thermoacidophilic red microalga *Galdieria sulphuraria* and comparison with other glycogens. Int. J. Biol. Macromol. 89: 12–18.

Matsuda, K. and Hata K. 1985. The structure of glycogen particles. J. Jpn. Soc. Starch Sci. 32: 118–127.

Matsui, M., M. Kakut and A. Misaki. 1996. Fine structural features of oyster glycogen: mode of multiple branching. Carbohydr. Polym. 31: 227–235.

Melendez-Hevia, E., T.G. Waddell and E.D. Shelton. 1993. Optimization of molecular design in the evolution of metabolism: the glycogen molecule. Biochem. J. 295: 477–483.

Melendez, R., E. Melendez-Hevia and E.I. Canela. 1999. The fractal structure of glycogen: A clever solution to optimize cell metabolism. Biophys. J. 77: 1327–1332.

Montesanti, N., G. Véronèse, A. Buléon, P.C. Escalier, S. Kitamura and J.-L. Putaux. 2010. A-type crystals from dilute solutions of short amylose chains. Biomacromolecules 11: 3049–3058.

Nickels, J.D., J. Atkinson, E. Papp-Szabo, C. Stanley, S.O. Diallo, S. Perticaroli et al. 2016. Structure and hydration of highly-branched, monodisperse phytoglycogen nanoparticles. Biomacromolecules 17: 735–743.

Nudelman, F., G. de With and N.A.J.M. Sommerdijk. 2011. Cryo-electron tomography: 3-dimensional imaging of soft matter. Soft Matter 7: 17–24.

Oostergetel, G.T. and E.F.J. van Bruggen. 1993. The crystalline domains in potato starch granules are arranged in a helical fashion. Carbohydr. Polym. 21: 7–12.

Pérez, S. and E. Bertoft. 2010. The molecular structures of starch components and their contribution to the architecture of starch granules: A comprehensive review. Starch/Stärke 62: 389–420.

Potocki-Véronèse, G., J.-L. Putaux, D. Dupeyre, C. Albenne, M. Remaud-Simeon, P. Monsan et al. 2005. Amylose synthesized *in vitro* by amylosucrase: morphology, structure and properties. Biomacromolecules 6: 1000–1011.

Powell, P.O., M.A. Sullivan, M.C. Sweedman, D.I. Stapleton, J. Hasjim and R.G. Gilbert. 2014. Extraction, isolation and characterisation of phytoglycogen from su-1 maize leaves and grain. Carbohydr. Polym. 101: 423–431.

Powell, P.O., M.A. Sullivan, J.J. Sheehy, B.L. Schulz, F.J. Warren and R.G. Gilbert. 2015. Acid hydrolysis and molecular density of phytoglycogen and liver glycogen helps understand the bonding in glycogen α (composite) particles. PLoS ONE 10(3): e0121337.

Putaux, J.-L., A. Buléon, R. Borsali and H. Chanzy. 1999. Ultrastructural aspects of phytoglycogen from cryo-TEM and quasi-elastic light scattering data. Int. J. Biol. Macromol. 26: 145–150.

Putaux, J.-L., S. Molina-Boisseau, T. Momaur and A. Dufresne. 2003. Platelet nanocrystals resulting from the acid hydrolysis of waxy maize starch granules. Biomacromolecules 4: 1198–1202.

Putaux, J.-L., G. Potocki-Véronèse, M. Remaud-Simeon and A. Buléon. 2006. Alpha-D-glucan-based dendritic nanoparticles prepared by *in vitro* enzymatic chain extension of glycogen. Biomacromolecules 7: 1720–1728.

Putaux, J.-L., D. Dupeyre, B. Pontoire, J. Davy, A. Buléon, C. d'Hulst et al. 2010. Amidothèque: an online database on the morphology, structure and composition of native starch granules. https://amidotheque.cermav.cnrs.fr.

Ramanazov, Z., M. Rawat, M.C. Henk, CB. Mason, S.W. Matthews and J.V. Moroney. 1994. The induction of the CO_2-concentrating mechanism is correlated with the formation of the starch sheath around the pyrenoid of *Chlamydomonas reinhardtii*. Planta 195: 210–216.

Reid, N. 1975. Practical Methods in Electron Microscopy: Ultramicrotomy", North-Holland/American: Elsevier.

Revel, J.-P., L. Napolitano and D.W. Fawcett. 1960. Identification of glycogen in electron micrographs of thin tissue sections. J. Biophys. Biochem. Cytol. 8: 575–859.

Riemersma, J.C., E.J.J. Alsbach and W.C. De Bruijn. 1984. Chemical aspects of glycogen contrast-staining by potassium osmate. Histochem. J. 16: 123–136.

Robin, J.-P., C. Mercier, R. Charbonnière and A. Guilbot. 1974. Lintnerized starches. Gel filtration and enzymatic studies of insoluble residues from prolonged acid treatment of potato starch. Cereal Chem. 51: 389–406.

Roldán, I., F. Wattebled, M.M. Lucas, D. Delvallé, V. Planchot, S. Jiménez et al. 2007. The phenotype of soluble starch synthase IV defective mutants of *Arabidopsis thaliana* suggests a novel function of elongation enzymes in the control of starch granule formation. Plant J. 49: 492–504.

Roussel, X., C. Lancelon-Pin, A. Viksø-Nielsen, A. Rolland-Sabaté, F. Grimaud, G. Véronèse et al. 2013. Characterization of substrate and product specificity of the purified recombinant glycogen branching enzyme of *Rhodothermus obamensis*. Biochim. Biophys. Acta - Gen. Subj. 1830: 2167–2177.

Ryman, B.E. 1974. The glycogen storage diseases. J. Clin. Pathol. Suppl. (R. Coll. Pathol.) 8: 106–121.

Ryoyama, K., Y. Kidachi, H. Yamaguchi, H. Kajiura and H. Takata. 2004. Anti-tumor activity of an enzymatically synthesized alpha-1,6 branched alpha-1,4-glucan, glycogen. Biosci. Biotechnol. Biochem. 68: 2332–2340.

Seshagiri Rao, K. and M.N.V. Prasad. 1988. Typology of latex starch grains of certain *Euphorbiaceae* and their possible significance in systematics. Plant Syst. Evol. 160: 189–193.

Seung, D., T.B. Schreier, L. Bürgy, S. Eicke and S.C. Zeeman. 2018. Two plastidial coiled-coil proteins are essential for normal starch granule initiation in Arabidopsis. Plant Cell 30: 1523–1542.

Shimonaga, T., M. Konishi, Y. Oyama, S. Fujiwara, A. Satoh, N. Fujita et al. 2008. Variation in storage α-polyglucans of the *Porphyridiales (Rhodophyta)*. Plant Cell Physiol. 49: 103–116.

Singh, J., C. Lelane, R.B. Stewart and H. Singh. 2010. Formation of starch spherulites: Role of amylose content and thermal events. Food Chem. 121: 980–989.

Soares, C.A., F.H.G. Peroni-Okita, M.B. Cardoso, R. Shitakubo, F.M. Lajolo and B.R. Cordenunsi. 2011. Plantain and banana starches: granule structural characteristics explain the differences in their starch degradation patterns. J. Agric. Food Chem. 59: 6672–6681.

Stabentheiner, E., A. Zankel and P. Pölt. 2010. Environmental scanning electron microscopy (ESEM)—a versatile tool in studying plants. Protoplasma 246: 89–99.

Stokes, D.J. 2008. Principles and Practice of Variable Pressure/Environmental Scanning Electron Microscopy (VP-ESEM). John Wiley & Sons LtD, UK.

Sullivan, M.A., F. Vilaplana, R.A. Cave, D. Stapleton, A.A. Gray-Weale and R.G. Gilbert. 2010. Nature of α and β particles in glycogen using molecular size distributions. Biomacromolecules 11: 1094–1100.

Sullivan, M.A., J. Li, C. Li, F. Vilaplana, D. Stapleton, A.A. Gray-Weale et al. 2011. Molecular structural differences between type-2-diabetic and healthy glycogen. Biomacromolecules 12: 1983–1986.

Sullivan, M.A., M.J. O'Connor, F. Umana, E. Roura, K. Jack, D.I. Stapleton et al. 2012. Molecular insights into glycogen α-particle formation. Biomacromolecules 13: 3805–3813.

Szczygieł, J., K. Dyrek, K. Kruczała, E. Bidzińska, Z. Brożek-Mucha, E. Wenda et al. 2014. Interactions of chromium ions with starch granules in an aqueous environment. J. Phys. Chem. B 118: 7100–7107.

Tang, H.J., T.H. Mitsunaga and Y. Kawamura. 2006. Molecular arrangement in blocklets and starch granule architecture. Carbohydr. Polym. 63: 555–560.

Tang, X., M. De Rooij and L. De Jong. 2007. Volume change measurements of rice by environmental scanning electron microscopy and stereoscopy. Scanning 29: 197–205.

Thiéry, J.-P. 1967. Mise en évidence des polysaccharides sur coupes fines en microscopie électronique. J. Microscopie 6: 987–1018.

Tsukamoto, K., T. Ohtani and S. Sugiyama. 2012. Effect of sectioning and water on resin-embedded sections of corn starch granules to analyze inner structure. Carbohydr. Polym. 89: 1138–1149.

Vandromme, C., C. Spriet, D. Dauvillée, A. Courseaux, J.-L. Putaux, A. Wychowski et al. 2019. PII1: a protein involved in starch initiation that determines granule number and size in Arabidopsis chloroplast. New Phytol. 221: 356–370.

Wang, S., J. Yu, Q. Zhu, J. Yu and F. Jin. 2009. Granular structure and allomorph position in C-type Chinese yam starch granule revealed by SEM, ^{13}C CP/MAS NMR and XRD. Food Hydrocoll. 23: 426–433.

Wanson, J.-C. and P. Drochmans. 1968. Rabbit skeletal muscle glycogen. A morphological and biochemical study of glycogen β-particles isolated by the precipitation-centrifugation method. J. Cell Biol. 38: 130–150.

Watt, I.M. 1997. The Principles and Practice of Electron Microscopy. Cambridge University Press.

Wikman, J., A. Blennow, A. Buléon, J.-L. Putaux, S. Pérez, K. Seetharaman et al. 2014. Influence of amylopectin structure and degree of phosphorylation on the molecular composition of potato starch lintners. Biopolymers 101: 257–271.

Woodford, B and M.O.M. Tso. 1980. An ultrastructural study of the corpora amylacea of the optic nerve head and retina. Am. J. Ophthalmol. 90: 492–502.

Yamaguchi, M., K. Kainuma and D. French. 1979. Electron microscopic observations of waxy maize starch. J. Ultrastruct. Res. 69: 249–261.

Zhang, P., S. Nada, X. Tan, B. Deng, M.A. Sullivan and R.G. Gilbert. 2018. Exploring glycogen biosynthesis through Monte Carlo simulation. Int. J. Biol. Macromol. 116: 264–271. 2020. Erratum to "Exploring glycogen biosynthesis through Monte Carlo simulation". Int. J. Biol. Macromol. 144: 1043–1044.

Zhu, F. 2017. Atomic force microscopy of starch systems. Crit. Rev. Food Sci. Nutr. 57: 3127–3144.

Zobel, H.F. 1988. Molecules to granules: a comprehensive starch review. Starch/Staerke 40: 44–55.

CHAPTER 2

Polarimetric Nonlinear Microscopy of Starch Granules
Visualization of the Structural Order of α-Glucan Chains within a Native Starch Particle

Danielle Tokarz,[1,*] *Richard Cisek*[1] and *Virginijus Barzda*[2]

2.1 Introduction

Polarimetric nonlinear optical microscopy is a novel technique allowing a fast quantitative structural analysis of the organization of alpha-glucan chains within a single starch granule and for a larger number of particles. For food or technological applications, an increase in yield or a modification of starch properties, or both, are commonly attempted by genetic engineering of starch metabolizing pathways. In most cases, this is performed by changing the expression of one or several starch related genes or by introducing novel gene functions (Cakir et al. 2015, Fujita 2015, Seung et al. 2015, Pfister and Zeeman 2016). Due to the complexity of starch metabolism, however, genetic engineering does not always result in the desired phenotype. This is especially true when novel target genes are used, or a new plant species is transformed. The microscopic technique described in this chapter represents a highly efficient and sensitive analytical tool for characterizing the resulting structural modifications of α-glucan particles, such as starch.

In general, the catalytic functions of starch-related enzymes are well understood, but cellular and/or plastidial responses to modified or altered gene expression are largely unknown and, therefore, difficult to predict. For example, when expressing a non-plant phosphatase, laforin, in white potato tubers the content of phosphate esters in starch has been reported to increase rather than the expected decrease, and therefore, a compensatory response has been postulated (Xu et al. 2017). In particular, the close integration between starch metabolism and the biology of the entire cell sometimes results in unpredictable phenotypical changes in transgenic plants (Malinova et al. 2014, Feike et al. 2016, Liu et al. 2016). Furthermore, severe reductions in starch yield and/or plant growth often occur when the expression of starch-related genes is modified.

Further uncertainties with genetically induced bioengineering strategies are directly related to native starch properties such as hydro-insolubility of particles. Current research suggests that most

[1] Saint Mary's University, Department of Chemistry, Halifax, NS, B3H 3C3, Canada, Email: richard.cisek@smu.ca.
[2] University of Toronto, Department of Physics, Toronto, ON, M5S 1A7, Canada; University of Toronto Mississauga, Department of Chemical and Physical Sciences, Mississauga, ON, L5L 1C6, Canada, Email: virgis.barzda@utoronto.ca.
* Corresponding author: danielle.tokarz@smu.ca

of the starch-related reactions occur at or near the starch granule surface. Starch biology is likely to include processes by which starch-related proteins are targeted to these surfaces (Feike et al. 2016, Pfister and Zeeman 2016). In leaves of *Arabidopsos thaliana*, three proteins (AtPTST1-AtPTSI3) have been identified that all contain a carbohydrate binding module (CBM) family 48 and a coiled coil domain which is likely to be involved in protein-protein interactions. None of these proteins has any detectable catalytic activity, but one of them appears to be involved in targeting granule-bound starch synthase to the granule. These targeting processes and their capabilities are still poorly understood. Other members of this small family seem to be essential for granule initiation (Seung et al. 2015, 2017, 2018). Furthermore, another starch-binding protein, EARLY STARVATION1, alters the phosphorylation pattern of starch granules. This protein decreases the action of the glucan water dikinase, which esterifies glucosyl residues at carbon 6, but stimulates the phosphoglucan water dikinase, phosphorylating at carbon 3. Thus, our knowledge of protein actions during starch metabolism continues to evolve.

In some cases, the limited substrate specificity of some starch-related enzymes (see also Chapters 8, 10, and 11 in this volume) allows plants to partially compensate for altered gene expression. For example, the various soluble starch synthase isozymes possess partly overlapping specificities for chain lengths of glucosyl acceptors and, therefore to some extent, can minimize the effect of altered gene expression of distinct synthases or other transferases. However, overlapping of distinct enzymes is often functional only in a certain temperature range (Satoh et al. 2008, Fettke et al. 2012, Nakamura 2015a). Finally, the interconvertibility of distinct states of some starch-related enzymes adds to the uncertainty inherent to predictions of the phenotype. Several enzymes associated with assimilatory or reserve starch metabolism have been reported to exist in multiple and convertible states (Bustos et al. 2004, Geigenberger et al. 2005, Hennen-Bierwagen et al. 2008, Tetlow et al. 2008, 2015, Kubo et al. 2010, Utsumi et al. 2011, Sundberg et al. 2013, Crofts et al. 2017). In these cases, the intended alterations of single gene expression may affect the various states in an unexpected way.

In unicellular algae, the accumulation of storage products, such as starch (Garz et al. 2012) and triacylglycerol (Di Caprio et al. 2018) has been reported to exhibit a significant cell-to-cell heterogeneity. These data imply that the accumulation of both storage products is controlled by mechanisms that cannot be appropriately analyzed by data derived from bulk starch particle analysis. Given the high sensitivity of optical methods, the polarimetric nonlinear optical microscopy described in this chapter permits a fast non-invasive analysis of the structure of single starch granules.

2.2 Second Harmonic Generation (SHG) Microscopy

2.2.1 Introduction to SHG

The technique of polarization resolved Second Harmonic Generation (SHG) microscopy enables quantification of the structural ordering of macromolecules in biological materials [for review see (Cisek et al. 2017)]. SHG is a nonlinear optical process restricted to materials with non-central symmetry and is particularly suitable for ordered biological structures in the micro-meter size range. When biological structures of interest possess the required symmetry and sufficiently high nonlinearity constants, such as native starch, SHG microscopy does not require any labeling procedure nor sample modifications. Thus, the technique is versatile and adaptable to various imaging conditions. Furthermore, SHG microscopy allows the investigation of microscopic volumes up to several hundred micrometers deep within samples and provides optical sectioning. The volumes of interest are quickly probed by raster scans revealing the semi-crystalline ordering and thereby three-dimensional maps of ordered structures are obtained, which reveal structural information about the samples.

The molecular hyperpolarizability, β, and macromolecular second-order nonlinear optical susceptibility, $\chi^{(2)}$, are the two nonlinearity tensors that characterize the ability of constituting molecules and the material, respectively, to produce efficient SHG signals in the microscope. Non-central symmetry, also known as non-inversion symmetry, is inherent to both requirements. Biological materials that emit SHG signals have been identified in plants, animals, and bacteria. These materials include microcrystalline structures of starch granules (Mizutani et al. 2000), carotenoid aggregates (Tokarz et al. 2014a), cellulose (Brown et al. 2003), microtubules (Campagnola et al. 2002), light-harvesting pigment-protein complexes from plants and photosynthetic bacteria (Cisek et al. 2009), collagen fibers (Roth and Freund 1981), and myosin filaments in muscle (Campagnola et al. 2002). The molecular hyperpolarizability, β, represents the propensity of molecules to produce SHG, analogous to the fluorescence quantum yield, Φ, in the linear regime that governs the brightness of a fluorophore. SHG from macrostructures requires constituent molecules with a high value of β to be arranged in a non-centrally symmetric structure. The macromolecular susceptibility, $\chi^{(2)}$, is proportional to β, the concentration of molecules in the material and is dependent on their arrangement. For instance, a solution containing molecules with a high β value will typically not emit significant SHG signal because solutions induce an isotropic arrangement, which is centrally symmetric. Therefore, $\chi^{(2)}$ values of solutions are usually very low.

The non-central symmetry requirement has been previously demonstrated (Moreaux et al. 2000). In this study, the authors investigated SHG emitted by chiral molecules that were embedded in giant unilamellar vesicles. When located in the membranes of isolated vesicles, these molecules emitted SHG signals because they were oriented in a parallel fashion, satisfying non-central symmetry. However, in the regions where two vesicles were in close contact, the SHG vanished since molecules in the two opposing membranes were arranged in antiparallel orientations, rendering a centrosymmetric arrangement. Therefore, the sensitivity of SHG to symmetry renders an extremely insightful tool for investigations of the orientation of microcrystalline structures in starch granules.

2.2.2 SHG: Basic Theory

An induced polarization (*P*) occurs in materials interacting with an intense laser light. This process can be expressed as a power series:

$$P \propto \chi^{(1)}E + \chi^{(2)}E^2 + \chi^{(3)}E^3 + \cdots \tag{1}$$

where *E* is the electric field of the incident laser light, and $\chi^{(n)}$ are the susceptibility tensors of order *n*, describing the material response. For example, $\chi^{(1)}$ is the first-order linear optical susceptibility tensor representing commonly observed optical processes such as absorption of photons, and the refraction of light. Nonlinear optical processes are represented by $n \geq 2$. Second-order nonlinear optical processes such as SHG are represented by the $\chi^{(2)}$ tensor and are proportional to the square of the electric field. The $\chi^{(3)}$ tensor accordingly describes third-order nonlinear optical processes such as third harmonic generation (THG), coherent Raman scattering and two-photon absorption.

Essentially, SHG is a quantum process consisting of the interaction of two photons of wavelength λ with matter, resulting in the production of one photon of wavelength $\lambda/2$. Since each SHG photon requires two laser photons for its production, a quadratic power scaling between laser power and SHG intensity is observed. For instance, if the laser intensity is reduced by 50%, the emitted SHG is diminished to 25%. A similar effect also occurs during two-photon excitation fluorescence in multiphoton microscopes. The power scaling of the SHG intensity is particularly important for quantifying results at different sample depths, for example, starch inside plant leaves. As the laser traverses leaf tissue, the laser power is reduced due to scattering. Thus depending on the location of starch inside a leaf, the laser may lose more or less power before reaching the particles. Due to the aforementioned quadratic SHG power scaling, laser intensity variations are magnified (squared), hence they have significant effects on the SHG intensity. Therefore, to properly quantify the SHG intensity of starch particles at different locations in a leaf, such as close to the epidermis versus deep in the mesophyll, the loss in laser intensity should be accounted for. This problem can also occur when comparing SHG intensities from differently sized starch particles.

The process of SHG is energy conserving, or parametric, and hence it differs significantly from non-parametric processes such as absorption and subsequent fluorescence, where the emitted photon has lower energy than absorbed energy in the multiphoton process. In non-parametric processes, some energy is deposited into the sample and, therefore, thermally induced alterations can occur. Furthermore, long lived excited fluorescence states can create oxygen radicals, which can also lead to significant sample damage and photobleaching. Inversely, parameteric processes conserve energy, hence during SHG emission no energy is deposited to the sample. Therefore, SHG is often referred to as a 'photobleach-free' technique. In practice, however, several optical processes can be simultaneously induced in biological structures when high laser powers are used [see Equation (1)]. During experimental design, the biomolecules that may absorb laser light should be identified, and laser power and wavelength should be chosen to ensure that non-parametric processes are minimized. Therefore, for nonlinear optical microscopy, the laser wavelength is typically chosen in the near infrared range, where most biomolecules do not absorb significantly.

Coherence influences the direction of SHG emission via phase matching. Due to the fact that preferential angles exist in crystals, SHG obtains a directional emission profile, meaning that SHG emission has different intensities in different directions. This effect is in contrast to fluorescence which is emitted isotropically. In practice, the SHG mostly occurs in the forward direction along the laser propagation orientation, and therefore, some SHG microscopes utilize a forward detection geometry for optimal sensitivity. In a few cases, SHG signal can also be preferentially emitted in the backward direction, for example, from small cellulose fibers (Nadiarnykh et al. 2007).

2.2.3 Nonlinear Polarimetry

SHG polarimetry is an optical technique that can be used for structural characterization of a material via quantifying its $\chi^{(2)}$ tensor symmetry properties and magnitudes of the tensor elements. Up to 18 unique $\chi^{(2)}$ tensor elements describe the SHG in a material. Measured element values can be analyzed to reveal sample properties such as crystallographic symmetry. For example, in one polarization SHG microscopy study, ZnSe nanowires were differentiated based on hexagonal versus cubic crystallinities (Cisek et al. 2014a). Polarization SHG microscopy is also particularly sensitive to the amount of order within a crystalline material (Simpson and Rowlen 1999). Although the technique is new, it is already proving to be a powerful characterization method in several fields including starch (Cisek et al. 2014b, 2015, 2017, Psilodimitrakopoulos et al. 2016), collagen (Tokarz et al. 2015, Golaraei et al. 2016) and myosin (Rou et al. 2007, Kontenis et al. 2016). While there are several variations of SHG polarimetry, they can be represented by Nonlinear Stokes Mueller Polarimetry (NSMP) theory (Samim et al. 2015, Kontenis et al. 2016). According to NSMP, the SHG induced by a laser can be described by the equation:

$$\tilde{s} = MS \tag{2}$$

where $\tilde{s}$ is a conventional Stokes vector which depicts the outgoing SHG polarization. M is the SHG Mueller matrix, which describes the nonlinear optical properties of the material, and S is a double Stokes vector, which contains 9 elements and expresses the polarization of the incoming laser beam. Equation (2) demonstrates that the nonlinear optical properties of the material (M) can be deduced by measuring the polarization of the SHG ($\tilde{s}$) at specific polarizations of the laser beam (S). To fully characterize the nonlinear optical properties of the material, a set of nine polarization states of the incoming laser beam is required. This includes four linear polarization states at 0°, 45°, 90°, and 135°, left- and right-handed circular polarization states (LCP and RCP), a linear polarization state at –22.5°, as well as left- and right-handed elliptical polarization states (LEP and REP) (Samim et al. 2015).

NSMP allows the maximum amount of information to be obtained from the sample because it utilizes a complete basis for laser polarization, and it measures the complete outgoing polarization profile. Furthermore, since it uses the minimum required polarization states, the technique is quite fast. NSMP is built on Mueller calculus and as a result, scattering of the laser and the SHG can be accounted for in the sample as well as in different optical components. However, since the theory is new, relatively little work has thus far been performed with starch. This chapter focuses on the reduced NSMP theory where circular polarization components as well as scattering are neglected. However, several advancements using circularly polarized light will be reviewed.

In the reduced NSMP theory, the $\chi^{(2)}$ tensor component ratios are determined from polarization-resolved SHG microscopy measurements of linear polarization states. Circular polarization states can be neglected if the $\chi^{(2)}$ tensor components can be assumed to have negligible complex values. One technique for measuring polarization-resolved SHG of only linearly polarized incoming and outgoing light is referred to as polarization-in, polarization-out (PIPO) SHG microscopy. This technique presupposes cylindrical sample symmetry, and several versions of this technique were utilized during the last decade for starch investigations (Cisek et al. 2014b, 2015, 2017).

Although starch granules exhibit circularly polarized SHG when illuminated with linearly polarized light, the amplitude is low and to a first approximation, it can be neglected. In the microscopy setup for PIPO SHG, the polarization state generator (PSG) consists of a polarizer to ensure that the incoming light from the laser is linearly polarized and a half-wave plate to rotate the linear polarization. The polarization state analyzer (PSA) consists of a linear polarizer, which is often referred to as the analyzer. The analyzer probes the linearly polarized SHG signal.

For PIPO SHG microscopy measurements, a Cartesian laboratory coordinate system is used, XYZ, and it is defined with respect to the principal propagation direction of the incoming laser, which occurs along Y while the sample plane is defined as XZ. Lower case labeled coordinates are used for the molecular coordinate system with the cylindrical molecular z-axis. If one dominant nonlinear dipole exists in the z direction along the axis of the cylindrical structure, which lies in the XZ plane, then two unique non-zero molecular tensor components can be assumed, $\chi^{(2)}_{zzz}$ and $\chi^{(2)}_{zxx}$. With this assumption, the general SHG intensity (I_{SHG}) equation for PIPO measurements is as follows:

$$I_{SHG} = A \left| \begin{array}{c} sin^2(\theta - \delta)\, sin(\phi - \delta) + sin^2(\theta - \delta)cos(\phi - \delta) \\[2mm] + \dfrac{\chi^{(2)}_{zzz}}{\chi^{(2)}_{zxx}}\, cos^2(\theta - \delta)cos(\phi - \delta) \end{array} \right|^2 + B \tag{3}$$

where θ and ϕ are the angles of the laser beam polarization and the SHG analyzer orientation, respectively, measured from the laboratory Z-axis. The angle, δ, represents the orientation angle within the image plane, between the molecular cylindrical symmetry axis and the laboratory Z axis. The coefficient, A, is used for fitting the absolute amplitude, while the coefficient B, includes both noise and the depolarized contribution of the signal. If the cylindrical molecular axis is no longer assumed to lie in the XZ (sample) plane but rather, it is tilted out of the XZ plane by an angle, α, which occurs if the optical section is chosen higher or lower than the equatorial plane of the starch granule, then Equation (3) describes the laboratory frame second-order nonlinear optical susceptibility tensor component ratio, $\chi^{(2)}_{ZZZ}/\chi^{(2)}_{ZXX}$, which is related to $\chi^{(2)}_{zzz}/\chi^{(2)}_{zxx}$ by the following equation:

$$\frac{\chi^{(2)}_{ZZZ}}{\chi^{(2)}_{ZXX}} = \left(\frac{\chi^{(2)}_{zzz}}{\chi^{(2)}_{zxx}} - 3 \right) cos^2 \alpha + 3 \tag{4}$$

Equation (4) demonstrates that the angle, α, influences the measured second-order nonlinear optical susceptibility tensor component ratio. Changes in $\chi^{(2)}_{ZZZ}/\chi^{(2)}_{ZXX}$ with angle α were experimentally observed in wheat starch granules by imaging at different heights above and below the equatorial plane (Psilodimitrakopoulos et al. 2012). Many variations in $\chi^{(2)}_{ZZZ}/\chi^{(2)}_{ZXX}$ values have been experimentally observed in starch granules. In addition to the tilt of the cylindrical axis with respect to the image plane, the variations could also originate from differences in the intrinsic $\chi^{(2)}_{ZZZ}/\chi^{(2)}_{ZXX}$ values. The variations in values for $\chi^{(2)}_{ZZZ}/\chi^{(2)}_{ZXX}$ indicate an alteration in the internal structure of the starch due to various factors, including changes in hydration of the crystalline material or coexistence of several structures with different individual $\chi^{(2)}_{ZZZ}/\chi^{(2)}_{ZXX}$ values within the same voxel.

2.2.4 SHG Microscopy Instrumentation

To produce SHG signal in biological structures, specialized ultrafast lasers are utilized. To satisfy the requirement for efficient SHG emission, where two photons must combine at the same point in space and time with sufficient probability, lasers with high peak powers (in the kW range) are required. However, to prevent damage to biological samples, the average laser power must be sufficiently low. Ultrafast pulsed lasers can satisfy these requirements. Typically, lasers are selected that have pulse durations ranging from 50–500 fs, and pulse repetition rates in the 1–100 MHz range, delivering pulses with several nJ of energy, and hence kW of power, at the sample.

A typical SHG microscope consists of a pulsed laser coupled to a laser scanning microscope. Interestingly, optical sectioning is achieved in SHG without the usual requirement of confocal pinholes. This is because the near-IR wavelength of lasers precludes the out of focus interactions with the sample, since wavelengths are chosen outside the sample linear absorption regions. As a result, only the focal spot, where photon density is sufficiently high, induces nonlinear interactions leading to SHG and enabling optical sectioning. This also eliminates out of focus photobleaching that often occurs in confocal microscopes using continuous wave lasers.

A typical SHG microscope for starch analysis is shown in Fig. 2.1. A femtosecond Yb:KGd(WO$_4$)$_2$ laser provides ~ 450 fs duration pulses at a wavelength of 1028 nm with a pulse repetition rate of 14.3 MHz (Major et al. 2006). The laser is coupled into a custom laser scanning microscope capable of detecting SHG signal, changing the incoming laser light polarization, and determining the polarization of emitted SHG signals. Galvanometric mirrors (Cambridge Technology Inc.) are used for raster scanning the laser over the sample, while a high numerical aperture (NA) air objective lens (20 × 0.75 NA, Carl Zeiss Canada Ltd.) is used for sample excitation with the incoming laser beam.

Specialized polarization optics are used for manipulation of the polarization of the laser as well as for determining the polarization of the SHG signal. A linear polarizer (IR 1100 BC4, Laser Components Inc.) is used to ensure the linear polarization of the laser beam. A half-wave plate (custom flat, Comar Optics Ltd.) before the excitation objective lens rotates the linear polarization orientation of the laser. For optimum sensitivity, the SHG signal is collected in the forward direction of the sample through a custom collection objective lens. Subsequently, a polarizer (10LP-VIS-B, Newport Corp.), also known as the analyzer, is placed behind the collection objective lens to determine the SHG polarization. The SHG signal is then separated from the laser by two optical filters, a band pass interference filter (F10-514.5, CVI Laser, LLC.) and a color glass filter (BG-39, CVI Laser, LLC.). For imaging starch inside leaves or photosynthetic bacteria, the interference filter is needed to separate the SHG from two-photon fluorescence, derived from molecules such as chlorophylls. Studies performed with isolated starch granules do not require this filter. The SHG intensity is measured with a photon-counting detector (H7421-40, Hamamatsu Photonics K.K.).

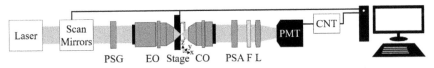

Figure 2.1. A schematic of a polarimetric second harmonic generation microscope used for starch imaging. Scan mirrors raster scan the laser beam in two lateral directions. A polarization state generator (PSG) is placed before the excitation objective lens (EO) and the sample stage. A polarization state analyzer (PSA) is placed after the collection objective lens (CO). The PSG and PSA consist of a fixed polarizer followed by a motorized half-wave plate and optionally, a motorized quarter-wave plate for Nonlinear Stokes Mueller Polarimetry. A lens (L) and filter (F) placed after the PSA ensure detection of second harmonic generation signal by the photomultiplier tube (PMT), which is connected to a counting card (CNT).

The accurate collection of polarization resolved SHG data is critical for the structural analysis of the samples. Polarization resolved SHG data acquisition using the reduced NSMP with PIPO involves imaging at different combinations of laser linear polarization angles and analyzer orientation angles. This is achieved by collecting an SHG image at different orientation combinations of the half-wave plate [θ in Equation (3)], which rotates the laser polarization, and of the analyzer [ϕ in Equation (3)], which measures the outgoing SHG polarization. In typical experiments, images at 81 combinations of θ and ϕ angles are collected, along with 9 additional reference images that possess the same polarization combination. These reference images allow the monitoring of the possible bleaching and deterioration of the signal. For each pixel of the SHG images, the variation in SHG intensity as a function of θ and ϕ was fitted using Equation (3) (MATLAB). Calibration of the initial orientation of the polarizer and the analyzer with respect to the scanning frame in the microscope was performed using collagen from a rat tail tendon.

2.2.5 Sample Preparation

Preparation of starch granule samples for nonlinear optical microscopy imaging requires immobilization of the granules. For imaging of isolated dry starch granules, the starch powder is placed between two microscope coverglasses with no further preparation. To obtain fully hydrated starch particles, the dry granules are suspended in a 20 mM Tris-HCl buffer pH 7.5 with 0.02% (w/v) NaN$_3$ for at least 12 hr at room temperature. Then, the hydrated granules are concentrated by centrifugation (10 min at 10,000 × g, 4°C) forming a pellet which was stored at –20°C.

Crystalline maltodextrins (B-type allomorph) were prepared essentially as described elsewhere (Hejazi et al. 2009). In brief, a solution of 30% (w/v) commercial maltodextrins (Aldrich 419672) was heated for 10 min in a boiling water bath, and subsequently kept overnight at 4°C. The pellet consisting of crystallized maltodextrins was then centrifuged at 4°C (as above), washed twice with water and stored (as above).

For imaging of hydrated granules including maltodextrins, the frozen suspension was thawed. The starch particles were then immobilized in a polyacrylamide gel containing 1.1 M acrylamide (Sigma Aldrich, 99%), 13.8 mM N,N'-methylenebisacrylamide (Sigma Aldrich, 99%), 32.3 mM ammonium persulfate (Sigma Aldrich, 98%), and 33.2 mM tetramethylethylenediamine (Sigma Aldrich, 99%), which was placed between coverglasses (Cisek et al. 2009). Inside leaves (Mizutani et al. 2000) and in green algal cells (Cisek et al. 2009, Tokarz et al. 2014b), starch granules can also be imaged *in situ* by placing them between two coverglasses. Under these conditions, the polyacrylamide gel is omitted.

2.3 Application of SHG Microscopy for Determining Starch Structure

2.3.1 Molecular Origin of SHG in Starch

Initially, the source of SHG signal from starch granules was hypothesized to be amylopectin, the main and semicrystalline constituent of starch granules (Chu et al. 2002). A polarization-resolved SHG study investigating wheat starch granules proposed that the measured $\chi_{zzz}^{(2)}/\chi_{zxx}^{(2)}$ values are related to the helical tilt angle of amylopectin glucan double helices (Psilodimitrakopoulos et al. 2010). According to the calculations presented, the measured $\chi_{zzz}^{(2)}/\chi_{zxx}^{(2)}$ values matched the helical tilt angle in amylopectin, but not in crystallized amylose. This conclusion was supported by a second investigation (Zhuo et al. 2010), which compared the SHG signal intensities of wild-type (wt) and waxy (wx) rice starch granules. A higher SHG intensity was observed for wx rice starch, which has a higher amylopectin, but a significantly lower amylose content, supporting the view that in native starch amylopectin produces SHG. Another study investigated the SHG signal from transgenic barley starch granules that mainly contain a highly modified amylopectin, consisting of less branches and longer inter-branched regions, in addition to normal amylopectin and amylose (Cisek et al. 2015). This modification of barley starch granules was performed by a partial reduction of all branching isozyme levels, and resulted in a modified amylopectin which shares several properties with amylose. SHG signal intensity from this transgenic barely was found to be lower than the wild-type control. While these results are in line with the assumption that amylopectin is the main source of SHG signal, the data also indicated that a further classification was needed for analyzing the relation between SHG signal and the structure of starch components.

To address the origin of SHG signals from starch granules, theoretical modeling was performed using published crystal structures of starch allomorph types A and B. Time-dependent Hartree-Fock calculations were employed using the software, GAMESSUS (Schmidt et al. 1993), to calculate the susceptibility tensor, $\chi^{(2)}$. The calculations revealed that in the two starch allomorph types, the carbon-oxygen and carbon-carbon bonds contribute little to the $\chi^{(2)}$, whereas hydroxide and hydrogen bonds involved in forming intra- and interhelical linkages largely influence the $\chi^{(2)}$ (Cisek et al. 2014b).

To empirically demonstrate the effect of alterations of the hydroxide and hydrogen bonding network on the $\chi^{(2)}$, completely hydrated and dry starch granules were compared using SHG microscopy (Cisek et al. 2014b, 2015). Hydrated starch granules gave rise to higher SHG intensities than dry starch granules. Dry starch granules, however, exhibited a higher second-order nonlinear optical susceptibility ratio with a broader distribution. Furthermore, hydrated starch granules using H_2O were compared to starch granules treated with D_2O (Cisek et al. 2014b). Deuterated starch granules had lower SHG intensities, which is in line with a disruption of the normal hydrogen bonding network as the length and angle between the oxygen-deuteron bonds and the oxygen-hydrogen bonds differ. Finally, lower SHG signals were observed from H_2O treated starch granules that were heated and subsequently cooled in several studies (Slepkov et al. 2010, Cisek et al. 2014b), further suggesting that disturbing the hydrogen bond network affects SHG emission.

Thus, evidence from several studies indicates that in starch granules, the SHG signals largely originate from organized hydroxide/hydrogen bonds, formed by aligned water molecules that co-crystallize with α-glucan chains of amylopectin (Cisek et al. 2014b). The variations in SHG intensity and $\chi_{zzz}^{(2)}/\chi_{zxx}^{(2)}$ values, as affected by starch structure, are discussed in the following sections. However, the basics are as follows: SHG intensity is mainly a function of both concentration and crystalline organization of hydrogen/hydroxide bonds, while the $\chi_{zzz}^{(2)}/\chi_{zxx}^{(2)}$ parameter is independent of the concentration, and is solely a function of their organization.

2.3.2 Polarization Resolved Starch Studies using SHG

Several studies have used polarization SHG microscopy to identify structural differences in various types of starch granules. Figure 2.2 shows typical results of the equatorial optical slice of starch granules isolated from non-green tissues of potato (a-d), maize (e-h) and barley (i-l). The fourth row of Fig. 2.2 presents the first SHG analysis of equatorial optical sections of crystalline maltodextrin particles (B-allomorph, m-p). In the left column, the SHG intensity images are given visualizing those particle regions that emit SHG. Due to inherent optical sectioning of SHG, the optical slices are limited to ~ 2 μm thickness by the numerical aperture of the objective lens, whilst in the regions above and below the optical section, SHG is not induced. SHG intensity is very sensitive to the concentration of SHG emitters, and therefore, it is a useful parameter for denoting the amount of crystallinity. However, care should be taken for its interpretation, since disorder reduces the SHG intensity, and therefore, optimal analysis is performed when also the polarization-dependent parameter, $\chi_{zzz}^{(2)}/\chi_{zxx}^{(2)}$, is considered.

The intensity independent $\chi_{zzz}^{(2)}/\chi_{zxx}^{(2)}$ measurements are a useful tool for analyzing the starch structure. Images in the second column of Fig. 2.2 contain grey scale representations of the fitted $\chi_{zzz}^{(2)}/\chi_{zxx}^{(2)}$ values from polarization resolved SHG imaging and represent the ordering measure of SHG emitters at that pixel. Near the peripheries, starch granules have higher $\chi_{zzz}^{(2)}/\chi_{zxx}^{(2)}$ values as compared to the supposed area of starch hila. When comparing starches from various plant species, the highest values are found for potato close to the periphery. The values for maize and barley starch are lower. The third column

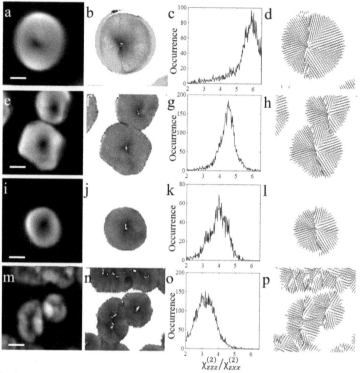

Figure 2.2. Polarization second harmonic generation analysis of hydrated starch granules isolated from various higher plants, and of crystallized maltodextrins (B-type allomorph). Typical starch granules from potato tubers **(a-d)**, maize **(e-h)** and barley **(i-l)** seeds are compared to *in vitro* crystallized maltodextrins (m-p). SHG intensity images (via image sum over polarizations) are in the left column (column 1: a, e, i, m). Fitted $\chi^{(2)}_{zzz}/\chi^{(2)}_{zxx}$ values are indicated for each pixel via grey shade ranging from light-grey ($\chi^{(2)}_{zzz}/\chi^{(2)}_{zxx} = 6.5$) to dark ($\chi^{(2)}_{zzz}/\chi^{(2)}_{zxx} = 2$) in column 2 (b, f, j, n), while occurrence histograms of $\chi^{(2)}_{zzz}/\chi^{(2)}_{zxx}$ values of the corresponding images are given in column 3 (c, g, k, o). The right column (column 4: d, h, l ,p) shows average crystalline orientations (δ) for each pixel obtained by fitting.

(counted from the left side) gives an occurrence histogram of the values from the images shown in column 2, which is useful for finding an average value of $\chi^{(2)}_{zzz}/\chi^{(2)}_{zxx}$ for a starch granule. Another interesting parameter revealed from the polarization resolved data is the orientation of the cylindrical axis for each pixel (Fig. 2.2, column 4). For clarity, only every fourth vector is shown in the images although the fit gives the orientation of each individual pixel. These vector images indicate the radial arrangement of the SHG emitters within starch granules. Additionally, the presence of a hilum in a starch granule may be inferred by observing radial points of the origin of the vectors.

In starch granules from all plant species and conditions observed thus far, an area near the inner core of the starch particle always possesses low $\chi^{(2)}_{zzz}/\chi^{(2)}_{zxx}$ values near 3 (see Fig. 2.2, column 2). SHG polarization imaging of this area, assumed to be the starch hilum, induces a SHG specific imaging effect related to the lower order starch regions (Cisek et al. 2017). During imaging near the supposed starch hilum, the laser focal spot, which has ~ 1 μm^3 volume, encompasses double-helices that have a large spread of orientations, which is in line with the radial architecture of starch granules. As the orientation distribution of SHG emitters broadens within a laser focal spot, a well-known SHG phenomenon occurs, where the measured $\chi^{(2)}_{zzz}/\chi^{(2)}_{zxx}$ converges to a value of 3 with increased average orientation distribution (Simpson and Rowlen 1999). Towards the granule periphery, the radial starch architecture constrains the double helices to increasingly parallel configurations within the focal volume; hence, in these regions, the measured $\chi^{(2)}_{zzz}/\chi^{(2)}_{zxx}$ acquires higher values of 4–6 for different starches. Therefore, the pixels with $\chi^{(2)}_{zzz}/\chi^{(2)}_{zxx}$ values near 3 visualize starch regions with higher spread or disorder of orientations of nonlinear dipoles, similar to the ones observed near the putative hilum.

The distribution of $\chi^{(2)}_{zzz}/\chi^{(2)}_{zxx}$ close to 3 for the *in vitro* crystallized maltodextrins (Fig. 2.2 m-p) indicates that ordering of these nonlinear dipoles is lower than for starch granules.

2.3.2.1 Maize and Potato Starches

The differences in average susceptibility ratio $\chi^{(2)}_{zzz}/\chi^{(2)}_{zxx}$ of starch granules from potato tubers and maize seeds have been reported (Cisek et al. 2014b). The average $\chi^{(2)}_{zzz}/\chi^{(2)}_{zxx}$ of hydrated potato starch granules was found to be 5.6 ± 0.2. This value is significantly higher than the average $\chi^{(2)}_{zzz}/\chi^{(2)}_{zxx}$ of hydrated maize starch (Fig. 2.2b and f). These data support the theoretical modeling results indicating that the starch allomorph type B can have higher $\chi^{(2)}_{zzz}/\chi^{(2)}_{zxx}$ values than starches representing the allomorph type A (as in maize). Barley seed starch (Fig. 2.2j), which has the allomorph type A, shows lower $\chi^{(2)}_{zzz}/\chi^{(2)}_{zxx}$ than potato, which again is in agreement with the results obtained by theoretical modeling.

The SHG intensity of hydrated maize starch (type A allomorph) was previously compared to that of potato (type B allomorph), revealing a 25% higher intensity for maize. Under hydrated conditions, the B type allomorph contains about 3 times as much water as the A allomorph (Imberty and Perez 1988, Popov et al. 2009). The hydroxide and hydrogen bonds of co-crystallized water molecules in the crystal allomorphs are aligned, and if these were the only SHG emitters in starches, the SHG intensity of type B allomorphs could be ~ 15 times greater than type A. Since, however, the SHG intensities are quite similar, a large amount of water is expected to be not aligned within potato starch granules. Overall, the increased average $\chi^{(2)}_{zzz}/\chi^{(2)}_{zxx}$ values and the similar SHG intensity of potato and maize starch strongly suggest that potato starch granules (B-type allomorph) contain a larger portion of non-aligned water than maize granules (A-type allomorph). In addition, the co-crystallized portion of water molecules has higher alignment in potato starch granules (B-type allomorph).

2.3.2.2 Starches from Various Maize Lines

Structural differences between native starches isolated from various maize genotypes can be ascertained by polarization SHG. In particular, several maize genotypes were previously investigated by polarization SHG microscopy including wild type (wt), three mutants [wx, sugary 2 (su2), and amylose extender (ae)], one double mutant (wx ae), as well as the commercial maize Hylon starches that have all been generated as branching enzyme reduced mutants (without further information being available). In this chapter, we report on starch granules of three maize mutant lines (HV, HVII, and HVIII). For further information on some of the mutants, the reader is referred to the following references (BeMiller and Whistler 2009, Nakamura 2015b).

Briefly, the starches from the various maize mutant lines differ in their apparent amylose content. Starch from wx largely consists of amylopectin and has a very low amylose content. This line has a strongly diminished expression of granule-bound starch synthase, which synthesizes mainly amylose. In the other lines, the apparent amylose content is increased. To some extent, the apparent amylose is actually a strongly modified amylopectin and mimics some of the features of amylose, such as elongated inter-branching chains (Vilaplana et al. 2012). In these cases, the actual amylose content of starch is often overestimated by conventional tests (e.g., the iodine binding test).

The su2 line contains a high sucrose content and low amount of starch. It has been reported to have a 10–15% higher content of apparent amylose as determined by the iodine binding test. Internal fractures have also been observed in the su2 starch line (BeMiller and Whistler 2009).

The amylose-extender line expresses much less starch branching isozyme BEIIb and, therefore, reserve starch contains 30% more apparent amylose as measured by the iodine binding test, which is actually largely a highly modified amylopectin. Starch granules formed by the ae line are heterogeneous including spherical birefringent particles and irregular granules with reduced or no birefringence. The apparent amylose content is decreased to varying degrees (BeMiller and Whistler 2009). It seems that this phenotype is related to an unequal distribution of the low amount of branching isozyme BEIIb.

Starch granules from the double mutant wx ae vary with respect to birefringence. Some particles exhibit a more normal birefringence, but irregular shaped wx ae granules only exhibit birefringence in the outer periphery. The commercial Hylon maize starch lines, HV, HVII, and HVIII, produce reserve starch granules containing an apparent amylose content of 50, 70, and 80%, respectively.

Based on the data obtained by polarization SHG microscopy, average $\chi^{(2)}_{zzz}/\chi^{(2)}_{zxx}$ values were determined for hydrated starches from various maize genotypes. Higher $\chi^{(2)}_{zzz}/\chi^{(2)}_{zxx}$ values are consistent with a more ordered crystalline arrangement, while lower values of $\chi^{(2)}_{zzz}/\chi^{(2)}_{zxx}$ indicate a higher degree of disorder. For the wt maize line, an average $\chi^{(2)}_{zzz}/\chi^{(2)}_{zxx}$ value of 4.6 ± 0.2 was found (Fig. 2.2g), but for the wx maize line, a lower average $\chi^{(2)}_{zzz}/\chi^{(2)}_{zxx}$ value (4.1 ± 0.1) was observed. The SHG intensity (*I*) of granules from the wx line was, however, higher than that of the wt by a factor of 2.0 ± 0.8. Presumably, the higher SHG intensity of the wx starch line suggests an increased concentration of crystalline regions with co-crystallized water as compared to wt. On the other hand, the lower average $\chi^{(2)}_{zzz}/\chi^{(2)}_{zxx}$ value of the wx starch line indicates that overall, the water present in the starch granules is more mobile and less structurally ordered. Thus, although the starch of the wx mutant forms more crystalline regions, the packing of water molecules in those regions is suboptimal compared to the wt. Of all the maize genotypes studied, the su2 starch line has both the lowest SHG intensity with respect to wt (*I* = 0.16 ± 0.04) and the lowest $\chi^{(2)}_{zzz}/\chi^{(2)}_{zxx}$ value (3.3 ± 0.2), followed by HVIII (*I* = 0.3 ± 0.1 and $\chi^{(2)}_{zzz}/\chi^{(2)}_{zxx}$ = 3.5 ± 0.1), HV (*I* = 0.3 ± 0.1 and $\chi^{(2)}_{zzz}/\chi^{(2)}_{zxx}$ = 3.6 ± 0.1), ae (*I* = 0.32 ± 0.05 and $\chi^{(2)}_{zzz}/\chi^{(2)}_{zxx}$ = 3.8 ± 0.1), and HVII (*I* = 0.3 ± 0.1 and $\chi^{(2)}_{zzz}/\chi^{(2)}_{zxx}$ = 3.9 ± 0.1). The lower SHG intensities and $\chi^{(2)}_{zzz}/\chi^{(2)}_{zxx}$ values indicate that these starch lines contain crystallites with a reduced capacity for axially aligned hydroxide and hydrogen bonds, and increased disorder, respectively. Starch of the wx ae double mutant has a high SHG intensity with respect to wt (1.3 ± 0.2), but a lower average $\chi^{(2)}_{zzz}/\chi^{(2)}_{zxx}$ value (3.8 ± 0.1) suggesting that the wx ae starch has a higher density of crystallites, but in total, a lower degree of order.

Table 2.1. The $\chi^{(2)}_{zzz}/\chi^{(2)}_{zxx}$ values and the SHG intensity ratio with respect to the SHG intensity of hydrated wt maize for a number of hydrated maize genotypes.

Starch Type	SHG Intensity Ratio (*I*)	$\chi^{(2)}_{zzz}/\chi^{(2)}_{zxx}$
wt	1	4.6 ± 0.2
wx	2.0 ± 0.8	4.1 ± 0.1
su2	0.16 ± 0.04	3.3 ± 0.2
ae	0.32 ± 0.05	3.8 ± 0.1
wx ae	1.3 ± 0.2	3.8 ± 0.1
HV	0.3 ± 0.1	3.6 ± 0.1
HVII	0.3 ± 0.1	3.9 ± 0.1
HVIII	0.3 ± 0.1	3.5 ± 0.1

The $\chi^{(2)}_{zzz}/\chi^{(2)}_{zxx}$ values and the SHG intensity ratio values stated are granule by granule averages ± standard error.

2.3.2.3 Barley Starches with Varying Apparent Amylose Contents

The average $\chi^{(2)}_{zzz}/\chi^{(2)}_{zxx}$ values have been measured for starches from a number of barley genotypes including wt (cv. Golden Promise), wx (cv. Cinnamon), and so called 'amylose only' (ao, cv. Golden Promise). The apparent amylose content of wt barley starch is ~ 30% of the reserve starch dry weight. Similar to wx maize, wx barley starch lacks granule-bound starch synthase, which mainly forms amylose, resulting in starch consisting essentially of amylopectin. Starch granules from the ao barley line are generated by an incomplete suppression of all genes encoding starch branching isozymes (Carciofi et al. 2012). Therefore, starch of this line largely consists of apparent amylose and amylopectin which is modified to generate a loosely branched structure. The branching point frequency, thermal properties and lower apparent molecular weight of amylopectin in the ao line are similar to amylose.

Hydrated wt barley line starch was found to have an average $\chi^{(2)}_{zzz}/\chi^{(2)}_{zxx}$ value of 4.0 ± 0.1 (Fig. 2.2j). Similar to the hydrated starch granules from the wx maize line, the hydrated granules from the wx barley line have a lower average $\chi^{(2)}_{zzz}/\chi^{(2)}_{zxx}$ value (3.6 ± 0.1) and a higher SHG intensity with respect to the SHG intensity of wt granules (1.74 ± 0.08). The average $\chi^{(2)}_{zzz}/\chi^{(2)}_{zxx}$ value of the ao barley line granules is lower than both wt and wx starches (3.1 ± 0.1). Additionally, the SHG intensity ratio of the ao starch with respect to wt was observed to be extremely low (0.09 ± 0.01). Further studies compared the effect of hydration on the average $\chi^{(2)}_{zzz}/\chi^{(2)}_{zxx}$ values and the SHG intensity of wt, wx and ao barley lines by investigating air-dried and ultra-dried starch granules. For ultra-dry conditions, a glove box was used. Air-dried granules were placed in a vacuum chamber, subjected to four air evacuation and nitrogen refill cycles, and subsequently kept in the glove box under a nitrogen atmosphere with 4.6 ppm water and 12.7 ppm oxygen for 48 hours. For imaging, the ultra-dried granules were mounted between microscope coverglasses without polyacrylamide gel, sealed with silicone in the glove box, and were immediately imaged.

Table 2.2. The $\chi^{(2)}_{zzz}/\chi^{(2)}_{zxx}$ values and the SHG intensity ratio with respect to the SHG intensity of hydrated wt for a number of hydrated, air-dried, and ultra-dried barley genotypes.

Starch Type	Hydrated		Air-dried		Ultra-dried	
	SHG Intensity Ratio (I)	$\chi^{(2)}_{zzz}/\chi^{(2)}_{zxx}$	SHG Intensity Ratio (I)	$\chi^{(2)}_{zzz}/\chi^{(2)}_{zxx}$	SHG Intensity Ratio (I)	$\chi^{(2)}_{zzz}/\chi^{(2)}_{zxx}$
wt	1	4.0 ± 0.1	0.48 ± 0.03	4.1 ± 0.1	0.05 ± 0.01	3.8 ± 0.1
wx	1.74 ± 0.08	3.6 ± 0.1	0.63 ± 0.03	4.4 ± 0.2	0.26 ± 0.02	3.5 ± 0.1
ao	0.09 ± 0.01	3.1 ± 0.1	0.029 ± 0.009	3.5 ± 0.1	0.011 ± 0.003	3.1 ± 0.1

The $\chi^{(2)}_{zzz}/\chi^{(2)}_{zxx}$ values are derived from individual granules. Averages ± standard error are given where $n \geq 5$.

2.3.2.4 Hydration of Barley Starch

Overall, it was observed that within each genotype, the SHG intensity of the starch granule itself increases with increasing hydration indicating that water is integral to the generation of SHG signal. The SHG intensities of starch granules dried in air and ultra-dried demonstrated the same trends to observations made with hydrated starch granules. The SHG intensity of air-dried wx barley starch was 0.63 ± 0.1 times reduced compared to hydrated wt barley starch, while the SHG intensity of air-dried ao barley starch was reduced by a factor of 0.029 ± 0.009 times as compared with hydrated wt. Further, the SHG intensity of ultra-dried wx starch was reduced by 0.26 ± 0.02 times compared to hydrated wt barley starch, while the SHG intensity of ultra-dried ao barley starch was much more reduced than that of the hydrated wt starch. The SHG intensity of starch granules from different lines indicates that the wx line starch provides a matrix for a highly ordered crystalline structure, while the ao line starch supports a much lower crystallinity.

The same trend in average $\chi^{(2)}_{zzz}/\chi^{(2)}_{zxx}$ values of hydrated starches from wt, wx, and ao barley lines were observed with ultra-dried conditions where the average $\chi^{(2)}_{zzz}/\chi^{(2)}_{zxx}$ values were found to be 3.8 ± 0.1, 3.5 ± 0.1, and 3.1 ± 0.1, respectively. Similarly, the air-dried starch from the ao line had the lowest average $\chi^{(2)}_{zzz}/\chi^{(2)}_{zxx}$ value (3.5 ± 0.1) indicating that these starches have the lowest degree of order. However, the average $\chi^{(2)}_{zzz}/\chi^{(2)}_{zxx}$ value of the starch from the wx line (4.4 ± 0.2) was slightly higher than the average $\chi^{(2)}_{zzz}/\chi^{(2)}_{zxx}$ value of starch from wt (4.1 ± 0.1). During hydration of the air-dried starches, both the structural and co-crystallized water in the starches result in lower $\chi^{(2)}_{zzz}/\chi^{(2)}_{zxx}$ values than air-dried conditions in the starches of the three barley genotypes. Therefore, the removal of structural water at moderate moisture conditions (air-dried) causes the remaining co-crystallized water to be oriented closer to the axis of various glucan double helices, while ultra-dried conditions decrease structural order (Cisek et al. 2015).

2.3.2.5 Crystalline Maltodextrins

Maltodextrins were crystallized and analyzed using polarization SHG microscopy to determine their optical properties. Figure 2.2 m–p shows the polarization SHG results of imaging the crystalline maltodextrin particles (B-type allomorph). The SHG intensity of hydrated particles was on average 98% lower as compared with hydrated wt maize, about four times less than hydrated ao barley. This observation is in line with: (1) the assertion that starch semi-crystalline amylopectin contains a high number of ordered water molecules, and (2) that most of the SHG intensities from starch are generated by the ordered water molecules. Since significant SHG is emitted by crystalline maltodextrins, this may indicate that in starch, amylose also influences the SHG properties. However, maltodextrins are not amylose, and further it is not known how exactly amylose chains are integrated and positioned in starch granules. Therefore, the ordered structures in maltodextrin crystals may not be the same as those of long amylose chains in starch. If amylose induces SHG, it would seemingly also concur with an earlier study of ao barley starch (Cisek et al. 2015); however, since the ao barley starch contains modified amylopectin rather than real amylose embedded in amylopectin, it remains uncertain whether amylose in starch actually contributes to the SHG signal.

The susceptibility ratios of crystalline maltodextrin particles had an average $\chi_{zzz}^{(2)}/\chi_{zxx}^{(2)}$ value of 3.0, indicating a very disordered SHG emitting structure, which is similar to the ao barley. This observation suggests that the small crystalline clusters within the maltodextrin particles are likely to be mostly disordered within the laser focal spot. Figure 2.2p shows that the maltodextrin particles have a radially oriented structure. Therefore, although the disorder is high within each laser spot, the average orientation within each laser focal spot adheres to the large scale radial arrangement of starch granules, suggesting that self-assembly might be the driving force behind the radial structure of starch granules. A closer look at the summary graph (Fig. 2.2o) reveals the surprising observation that many regions contain $\chi_{zzz}^{(2)}/\chi_{zxx}^{(2)}$ values above four, which is similar to typical wt barley starch. A closer look at different crystalline maltodextrin polarization images (Fig. 2.2n) further reveals that the regions with $\chi_{zzz}^{(2)}/\chi_{zxx}^{(2)}$ values higher than four are always located at or very near the granule periphery, but do not have significantly more SHG intensity. A very similar observation has been made in ao barley starch granules, but in that type of starch, more regions of high $\chi_{zzz}^{(2)}/\chi_{zxx}^{(2)}$ are present. Together, these observations suggest that while crystalline maltodextrins contain a much smaller concentration of crystalline material as compared to wt starch, self-assembly occasionally achieves small regions containing high levels of ordering. It is tempting to speculate that in the ao barley starch, the internal structure is mostly driven by self-assembly. By further optimizing the preparation conditions of crystalline maltodextrins, processing may result in structures similar to the internal alignment of starch granules.

2.4 Conclusion and Outlook

An overview of starch characterization using polarization resolved nonlinear optical microscopy is presented as a noninvasive biocompatible technique. Due to its sensitivity, small structural variations within individual granules and between different granules can be detected and quantified. While the main details of the advanced instrumentation, theory, and experiments are presented, many further details could not be described in this chapter due to the limited scope, and therefore, the reader is directed to specialized literature. While full polarimetric microscopy offers additional parameters to investigate, the established reduced polarimetry techniques, such as PIPO, such as PIPO, can differentiate structural variations in starch. These parameters include SHG intensity and the cylindrical susceptibility ratio $\chi_{zzz}^{(2)}/\chi_{zxx}^{(2)}$, which can be quickly obtained from submicron resolution regions within individual starch granules as well as from different granule populations, without extensive sample preparation, no addition of dyes, and no chemical treatments. Further, the SHG parameters have been proved to have high sensitivity to starch structural changes due to hydration conditions, and especially starch genotypes, making this technique promising for monitoring starch during bioprocessing for bioengineering.

Currently, there is much interest in bioengineering of starch for various applications including, for example, novel biodegradable materials. For such applications, polarization SHG microscopy has the potential to be a very useful tool for industry. SHG could be particularly applicable during large scale starch processing where heterogeneities in the product adversely affect its properties. In these cases, a polarization SHG setup could be made adjacent to a chemical reactor, and connected by microfluidics in order to continuously sample the chamber contents. Since the analysis is purely quantitative, obtaining data and subsequent fitting can be automated, and it can be used as a hardware feedback to trigger changes in chemical additives in order to obtain particular homogeneous desired starch products.

Acknowledgements

The authors thank Dr. Martin Steup (University of Potsdam, Brandenburg, Germany) for edits to the book chapter and for donating crystalline maltodextrins. The authors also thank Dr. Ian J. Tetlow (University of Guelph, Ontario, Canada) for donating starch samples. Authors DT, RC and VB acknowledge funding from the Natural Sciences and Engineering Research Council of Canada.

References

BeMiller, J. and R. Whistler. 2009. Starch Chemistry and Technology. 3rd Edition (J BeMiller and R Whistler, Eds.). Academic Press. Burlington, MA.

Brown, R.M., A.C. Millard and P.J. Campagnola. 2003. Macromolecular structure of cellulose studied by second-harmonic generation imaging microscopy. Opt. Lett. 28: 2207–2209.

Bustos, R., B. Fahy, C.M. Hylton, R. Seale, N.M. Nebane, A. Edwards et al. 2004. Starch granule initiation is controlled by a heteromultimeric isoamylase in potato tubers. Proc. Natl. Acad. Sci. 101: 2215–2220.

Cakir, B., A. Tuncel, S.-K. Hwang and T.W. Okita. 2015. Increase of grain yields by manipulating starch biosynthesis. pp. 371–395. *In*: Nakamura, Y. (ed.). Starch Metabolism and Structure. Springer.

Campagnola, P.J., A.C. Millard, M. Terasaki, P.E. Hoppe, C.J. Malone and W.A. Mohler. 2002. Three-dimensional high-resolution second-harmonic generation imaging of endogenous structural proteins in biological tissues. Biophys. J. 82: 493–508.

Carciofi, M., A. Blennow, S.L. Jensen, S.S. Shaik, A. Henriksen, A. Buléon et al. 2012. Concerted suppression of all starch branching enzyme genes in barley produces amylose-only starch granules. BMC Plant Biol. 12: 223.

Chu, S.W., I.H. Chen, T.M. Liu, C.K. Sun, S.P. Lee, B.L. Lin et al. 2002. Nonlinear bio-photonic crystal effects revealed with multimodal nonlinear microscopy. J. Microsc. 208: 190–200.

Cisek, R., L. Spencer, N. Prent, D. Zigmantas, G.S. Espie and V. Barzda. 2009. Optical microscopy in photosynthesis. Photosynth. Res. 102: 111–141.

Cisek, R., D. Tokarz, N. Hirmiz, A. Saxena, A. Shik, H.E. Ruda et al. 2014a. Crystal lattice determination of ZnSe nanowires with polarization-dependent second harmonic generation microscopy. Nanotechnology 25: 505703.

Cisek, R., D. Tokarz, S. Krouglov, M. Steup, M.J. Emes, I.J. Tetlow et al. 2014b. Second harmonic generation mediated by aligned water in starch granules. J. Phys. Chem. B 118: 14785–14794.

Cisek, R., D. Tokarz, M. Steup, I.J. Tetlow, M.J. Emes, K.H. Hebelstrup et al. 2015. Second harmonic generation microscopy investigation of the crystalline ultrastructure of three barley starch lines affected by hydration. Biomed. Opt. Express 6: 3694.

Cisek, R., D. Tokarz, L. Kontenis, V. Barzda and M. Steup. 2017. Polarimetric second harmonic generation microscopy: An analytical tool for starch bioengineering. Starch/Staerke 70: 1700031.

Crofts, N., Y. Nakamura and N. Fujita. 2017. Critical and speculative review of the roles of multi-protein complexes in starch biosynthesis in cereals. Plant Sci. 262: 1–8.

Di Caprio, F., F. Pagnanelli, R.H. Wijffels and D. Van der Veen. 2018. Quantification of Tetradesmus obliquus (Chlorophyceae) cell size and lipid content heterogeneity at single-cell level. J. Phycol. 54: 187–197.

Feike, D., D. Seung, A. Graf, S. Bischof, T. Ellick, M. Coiro et al. 2016. The starch granule-associated protein EARLY STARVATION1 (ESV1) is required for the control of starch degradation in Arabidopsis thaliana leaves. Plant Cell 28: 1472–1489.

Fettke, J., A.R. Fernie and M. Steup. 2012. Transitory starch and its degradation in higher plant cells. pp. 309–372. *In*: Tetlow, I.J. (ed.). Starch: Origins, Structure and Metabolism. Society for Experimental Biology.

Fujita, N. 2015. Manipulation of rice starch properties for application. pp. 335–369. *In*: Nakamura, Y. (ed.). Starch Metabolism and Structure. Springer.

Garz, A., M. Sandmann, M. Rading, S. Ramm, R. Menzel and M. Steup. 2012. Cell-to-cell diversity in a synchronized chlamydomonas culture as revealed by single-cell analyses. Biophys. J. 103: 1078–1086.

Geigenberger, P., A. Kolbe and A. Tiessen. 2005. Redox regulation of carbon storage and partitioning in response to light and sugars. J. Exp. Bot. 56: 1469–1479.

Golaraei, A., L. Kontenis, R. Cisek, D. Tokarz, S.J. Done, B.C. Wilson et al. 2016. Changes of collagen ultrastructure in breast cancer tissue determined by second-harmonic generation double Stokes-Mueller polarimetric microscopy. Biomed. Opt. Express. 7: 4054.

Hejazi, M., J. Fettke, O. Paris and M. Steup. 2009. The two plastidial starch-related dikinases sequentially phosphorylate glucosyl residues at the surface of both the A- and B-Type allomorphs of crystallized maltodextrins but the mode of action differs. PLANT Physiol. 150: 962–976.

Hennen-Bierwagen, T.A., F. Liu, R.S. Marsh, S. Kim, Q. Gan, I.J. Tetlow et al. 2008. Starch biosynthetic enzymes from developing maize endosperm associate in multisubunit complexes. Plant Physiol. 146: 1892–1908.

Imberty, A. and S. Perez. 1988. A revisit to the 3-dimensional structure of B-type starch. Biopolymers 27: 1205–1221.

Kontenis, L., M. Samim, A. Karunendiran, S. Krouglov, B. Stewart and V. Barzda. 2016. Second harmonic generation double stokes Mueller polarimetric microscopy of myofilaments. Biomed. Opt. Express 7: 559–569.

Kubo, A., C. Colleoni, J.R. Dinges, Q. Lin, R.R. Lappe, J.G. Rivenbark et al. 2010. Functions of heteromeric and homomeric isoamylase-type starch-debranching enzymes in developing maize endosperm. Plant Physiol. 153: 956–969.

Liu, F., Q. Zhao, N. Mano, Z. Ahmed, F. Nitschke, Y. Cai et al. 2016. Modification of starch metabolism in transgenic Arabidopsis thaliana increases plant biomass and triples oilseed production. Plant Biotechnol. J. 14: 976–985.

Major, A., R. Cisek and V. Barzda. 2006. Femtosecond Yb : KGd(WO4)(2) laser oscillator pumped by a high power fiber-coupled diode laser module. Opt. Express 14: 12163–12168.

Malinova, I., S. Mahlow, S. Alseekh, T. Orawetz, A.R. Fernie, O. Baumann et al. 2014. Double knockout mutants of Arabidopsis grown under normal conditions reveal that the plastidial phosphorylase isozyme participates in transitory starch metabolism. Plant Physiol. 164: 907–921.

Mizutani, G., Y. Sonoda, H. Sano, M. Sakamoto, T. Takahashi and S. Ushioda. 2000. Detection of starch granules in a living plant by optical second harmonic microscopy. J. Lumin. 87–9: 824–826.

Moreaux, L., O. Sandre, M. Blanchard-Desce and J. Mertz. 2000. Membrane imaging by simultaneous second-harmonic generation and two-photon microscopy. Opt. Lett. 25: 320–322.

Nadiarnykh, O., R. LaComb, P.J. Campagnola and W.A. Mohler. 2007. Coherent and incoherent SHG in fibrillar cellulose matrices. Opt. Express 15: 3348–3360.

Nakamura, Y. 2015a. Biosynthesis of reserve starch. pp. 161–209. *In*: Nakamura, Y. (ed.). Starch Metabolism and Structure. Springer.

Nakamura, Y. 2015b. Starch: Metabolism and structure. Springer.Tokyo.

Pfister, B. and S.C. Zeeman. 2016. Formation of starch in plant cells. Cell Mol. Life Sci. 73: 2781–2807.

Popov, D., A. Buleon, M. Burghammer, H. Chanzy, N. Montesanti, J.L. Putaux et al. 2009. Crystal structure of A-amylose: a revisit from synchrotron microdiffraction analysis of single crystals. Macromolecules 42: 1167–1174.

Psilodimitrakopoulos, S., I. Amat-Roldan, P. Loza-Alvarez and D. Artigas. 2010. Estimating the helical pitch angle of amylopectin in starch using polarization second harmonic generation microscopy. J. Opt. 12: 1–6.

Psilodimitrakopoulos, S., I. Amat-Roldan, P. Loza-Alvarez and D. Artigas. 2012. Effect of molecular organization on the image histograms of polarization SHG microscopy. Biomed. Opt. Express 3: 2681–2693.

Psilodimitrakopoulos, S., E. Gavgiotaki, K. Melessanaki, V. Tsafas and G. Filippidis. 2016. Polarization second harmonic generation discriminates between fresh and aged starch-based adhesives used in cultural heritage. Microsc. Microanal: 1–12.

Roth, S. and I. Freund. 1981. Optical 2nd-harmonic scattering in rat-tail tendon. Biopolymers 20: 1271–1290.

Rou, D., F. Tiaho, G. Recher and D. Rouede. 2007. Estimation of helical angles of myosin and collagen by second harmonic generation imaging microscopy. Opt. Express 15: 12286–12295.

Samim, M., S. Krouglov and V. Barzda. 2015. Double Stokes Mueller polarimetry of second-harmonic generation in ordered molecular structures. J. Opt. Soc. Am. B. 32: 451.

Satoh, H., K. Shibahara, T. Tokunaga, A. Nishi, M. Tasaki, S.K. Hwang et al. 2008. Mutation of the plastidial alpha-glucan phosphorylase gene in rice affects the synthesis and structure of starch in the endosperm. Plant Cell 20: 1833–1849.

Schmidt, M.W., K.K. Baldridge, J.A. Boatz, S.T. Elbert, M.S. Gordon, J.H. Jensen et al. 1993. General atomic and molecular electronic-structure system. J. Comput. Chem. 14: 1347–1363.

Seung, D., S. Soyk, M. Coiro, B.A. Maier, S. Eicke and S.C. Zeeman. 2015. Protein targeting to starch is required for localising granule-bound starch synthase to starch granules and for normal amylose synthesis in Arabidopsis. PLoS Biol. 13: 1–29.

Seung, D., J. Boudet, J. Monroe, T.B. Schreier, L.C. David, M. Abt et al. 2017. Homologs of protein targeting to starch control starch granule initiation in arabidopsis leaves. Plant Cell 29: 1657–1677.

Seung, D., T.B. Schreier, L. Bürgy, S. Eicke and S.C. Zeeman. 2018. Two coiled-coil proteins are essential for normal starch granule initiation in Arabidopsis. Plant Cell in press.

Simpson, G.J. and K.L. Rowlen. 1999. An SHG magic angle: Dependence of second harmonic generation orientation measurements on the width of the orientation distribution. J. Am. Chem. Soc. 121: 2635–2636.

Slepkov, A.D., A. Ridsdale, A.F. Pegoraro, D.J. Moffatt and A. Stolow. 2010. Multimodal CARS microscopy of structured carbohydrate biopolymers. Biomed. Opt. Express 1: 1347–1357.

Sundberg, M., B. Pfister, D. Fulton, S. Bischof, T. Delatte, S. Eicke et al. 2013. The heteromultimeric debranching enzyme involved in starch synthesis in arabidopsis requires both isoamylase1 and isoamylase2 subunits for complex stability and activity. PLoS One in press.

Tetlow, I.J., K.G. Beisel, S. Cameron, A. Makhmoudova, F. Liu, N.S. Bresolin et al. 2008. Analysis of protein complexes in wheat amyloplasts reveals functional interactions among starch biosynthetic enzymes. Plant Physiol. 146: 1878–1891.

Tetlow, I.J., F., Liu and M.J. Emes. 2015. Protein-protein interactions during starch biosynthesis. pp. 291–313. *In*: Nakamura Y. (eds.). Starch. Springer, Tokyo.

Tokarz, D., R. Cisek, S. Krouglov, L. Kontenis, U. Fekl and V. Barzda. 2014a. Molecular organization of crystalline β-carotene in carrots determined with polarization-dependent second and third harmonic generation microscopy. J. Phys. Chem. B 118: 3814–3822.

Tokarz, D., R. Cisek, O. El-Ansari, G.S. Espie, U. Fekl and V. Barzda. 2014b. Organization of astaxanthin within oil bodies of Haematococcus pluvialis studied with polarization-dependent harmonic generation microscopy. PLoS One 9: 1–8.

Tokarz, D., R. Cisek, A. Golaraei, S.L. Asa, V. Barzda and B.C. Wilson. 2015. Ultrastructural features of collagen in thyroid carcinoma tissue observed by polarization second harmonic generation microscopy. Biomed. Opt. Express 6: 3475.

Utsumi, Y., C. Utsumi, T. Sawada, N. Fujita and Y. Nakamura. 2011. Functional diversity of isoamylase oligomers: the ISA1 homo-oligomer is essential for amylopectin biosynthesis in rice endosperm. Plant Physiol. 156: 61–77.

Vilaplana, F., J. Hasjim and R.G. Gilbert. 2012. Amylose content in starches: Toward optimal definition and validating experimental methods. Carbohydr. Polym. 88: 103–111.

Xu, X., X.-F. Huang, R.G.F. Visser and L.M. Trindade. 2017. Engineering potato starch with a higher phosphate content. PLoS One 12: 1–21.

Zhuo, Z.Y., C.S. Liao, C.H. Huang, J.Y. Yu, Y.Y. Tzeng, W. Lo et al. 2010. Second harmonic generation imaging—A new method for unraveling molecular information of starch. J. Struct. Biol. 171: 88–94.

CHAPTER 3

Analyses of Covalent Modifications in α-Glucans

Felix Nitschke[1,*] and *Peter Schmieder*[2]

3.1 Introduction

Covalent modifications are commonly understood to be chemical groups of variable size that are added to pre-existing macromolecular structures. The largest variety of naturally occurring covalent modifications is certainly found in proteins, also known as post- or co-translational modifications which are mostly incorporated by specific enzymes. Some of the most abundant of the many types of covalent modifications of proteins comprise phosphorylation, acetylation, glycosylation, and amidation (Khoury et al. 2011). In general, covalent modifications alter the function and/or the location of the target protein and play a major role in the regulation of cell metabolism (Bah and Forman-Kay 2016). Another prominent example of covalent modification of macromolecules is DNA methylation. It is accomplished by few DNA methyltransferases and plays a critical role in transcriptional repression (Edwards et al. 2017). Lastly, covalent modifications can be found in the two major types of storage carbohydrates: starch and glycogen.

Both types are polyglucans, i.e., they consist (almost exclusively) of branched chains of glucosyl residues, which are linked via α-1,4 glucosidic bonds within chains and via α-1,6 glucosidic bonds at branching points connecting two chains. In both types of polyglucans, covalent modifications can be found in the form of phosphate esters. In addition to phosphate esters, amino groups have been found in rat liver glycogen which replace the hydroxyl group at carbon two of ten to thirty in a million glucosyl residues (Kirkman and Whelan 1986, Romero et al. 1980). However, these rare events have their origin in the incorporation of glucosamine residues from UDP-glucosamine rather than that of glucosyl residues from UDP-glucose during glycogen biosynthesis. Incorporation is thought to be due to the substrate specificity of one of the few glycogen biosynthesizing enzymes, glycogen synthase. Therefore, amination, as opposed to phosphorylation, is not performed *after* the construction of the glycogen molecule glycogen (Tarentino and Maley 1976).

In vitro various covalent modifications are chemically introduced into starch to adjust physico-chemical properties for technological purposes. These modifications include acetylation and carboxymethylation (Tharanathan 2005).

This chapter focuses on phosphate esters as naturally occurring covalent modifications in both types of polyglucans. After a brief summary of the biological background, we will describe various methods to analyze covalently bound phosphate in polyglucans.

[1] University of Texas Southwestern, Medical Center, Departments of Pediatrics and Biochemistry, Dallas, TX, 75390, USA.
[2] Leibniz-Institut für Molekulare Pharmakologie, 13125 Berlin, Germany, Email: schmieder@fmp-berlin.de
* Corresponding author: felix.nitschke@utsouthwestern.edu

3.1.1 Phosphorylation of Starch

The majority of covalently bound phosphate in starch occurs as monophosphate esters bound to the glucosyl carbons C6 and C3, with a C6:C3 ratio of 0.72 – 0.89 (Blennow et al. 2000a, Haebel et al. 2008, Ritte et al. 2006). Earlier reports give evidence of the existence of infrequent phosphate esters at carbon C2 (Tabata and Hizukuri 1971), which could not be confirmed in more recent studies (Nitschke et al. 2013, Schmieder et al. 2013). Phosphate esters in starch seem to be confined to the amylopectin fraction of starch (Blennow et al. 2000b, Schoch 1942), their frequency varying greatly depending on the origin of the starch. An overview of C6 phosphate in different starches is presented in Table 3.1.

So far, two plastidial enzymes have been identified that incorporate phosphate esters into starch. Glucan water dikinase (GWD) exclusively phosphorylates glucosyl carbons C6, while phosphoglucan water dikinase (PWD) specifically incorporates phosphate at carbons C3 (Ritte et al. 2006). In *Arabidopsis thaliana* chloroplasts, PWD seems to act downstream of GWD as *in vitro* it is largely unable to phosphorylate native starch unless the starch was pre-phosphorylated by GWD (Kötting et al. 2005). Nevertheless, *in vitro* PWD can phosphorylate non-phosphorylated maltodextrins and starch isolated from transgenic potato and rice plants (Hejazi et al. 2009). The ability of PWD to phosphorylate glucan chains seems to be dependent on the physico-chemical properties of the substrate surface, which is altered by GWD-mediated C6 phosphorylation rendering it more suitable for PWD-mediated phosphorylation.

Both crystalline maltodextrins and amylopectin contain regions of highly ordered arrangements of glucan chains, which form double helices that are poorly hydrated, pack together and, thus, form water insoluble particles (Gidley and Bulpin 1987, Imberty et al. 1991). According to molecular modeling studies, C6 phosphate interferes with helix-helix interactions, while C3 phosphate is capable of breaking double helices (Blennow and Engelsen 2010, Hansen et al. 2009). When crystalline maltodextrins are phosphorylated by GWD and/or PWD, glucan chains are released from the crystalline substrate. However, only few phosphate esters seem to be required to disturb the tight packing of chains, which then leads to a massive solubilisation of maltodextrins that are largely non-phosphorylated (Hejazi et al. 2008, 2009). *In vitro* degradation of native leaf starch granules is facilitated by simultaneous phosphorylation (Edner et al. 2007). This indicates that phosphate-

Table 3.1. C6 Phosphate Esters in Starches of Different Origins.

Organ	Species	C6 phosphate esters [mmol/mol glc]	References
Roots and tubers	*Curcuma*	9.5[1]	Blennow et al. 2000a,b
	Potato	2.8 – 4.6[2]	
	Cassava	0.41	
Leaves	Potato	0.43 – 0.73	Blennow et al. 2000b, Ritte et al. 2004
	Arabidopsis thaliana	0.65 – 1.4	Ritte et al. 2006, Haebel et al. 2008, Kötting et al. 2009, Santelia et al. 2011
Seeds	Mung bean	0.57	Blennow et al. 2000b
	Sorghum	0.15	Blennow et al. 2000a,b
	Rice	< 0.02[3]	Blennow et al. 2000b, Tabata et al. 1975
	Maize	< 0.01[3]	
	Wheat	< 0.01	Tabata et al. 1975
	Barley	n.d.	Blennow et al. 2000b
Algae	*Chlamydomonas reinhardtii*	0.45 – 0.55	Nitschke 2013

[1]   Value given for *Curcuma zedoaria*
[2]   Value depending on the variety of *Solanum tuberosum*
[3]   Only Tabata et al. (1975) were able to detect very small amounts of C6 phosphorylation
n.d. Below limit of detection

mediated alterations at the starch granule render its surface more suitable for starch-hydrolysing enzymes, which seem to preferably act on soluble glucans (Hejazi et al. 2008). Accordingly, it could be shown that phosphorylation at the periphery of potato leaf starch granules increases during starch degradation *in vivo* (Ritte et al. 2004). *Arabidopsis thaliana* mutant, deficient in either GWD or PWD, exhibit a starch-excess phenotype and are compromised in growth (Baunsgaard et al. 2005, Caspar et al. 1991, Kötting et al. 2005, Yu et al. 2001). The leaf starch formed by transgenic *Arabidopsis* plants lacking GWD significantly deviates from that of the wild type plants (Mahlow et al. 2014). Though phosphate incorporation into starch granule clearly has a role in starch degradation, it also largely occurs during starch biosynthesis (Nielsen et al. 1994, Ritte et al. 2004). Recent data indicate that phosphorylation during biosynthesis facilitates efficient mobilization of the starch during subsequent periods of starch degradation (Hejazi et al. 2014).

Dephosphorylation of starch granules in higher plants is accomplished by two glucan phosphatases, SEX4 and Like-Sex-Four 2 (LSF2) (Kötting et al. 2009, Santelia et al. 2011). While SEX4 is capable of hydrolyzing phosphate esters at carbons C3 and C6 (Hejazi et al. 2010), LSF2 specifically removes phosphate from carbon C3 (Santelia et al. 2011). Another protein that exists with high sequence similarity to SEX4 and LSF2 is Like-Sex-Four 1 (LSF1). As revealed by a leaf starch excess phenotype, this protein is also involved in starch metabolism, but instead of acting as a phosphatase it seems to function as a scaffold that promotes association of starch-degrading enzymes to the starch particle (Comparot-Moss et al. 2010, Schreier et al. 2019).

Phosphate incorporated by GWD and PWD facilitates access of starch degrading enzymes to the otherwise rigid starch surface, but it also presents an obstacle for the exo-acting β-amylases largely responsible for starch degradation. This is in agreement with the starch-excess and growth retardation phenotype observed in SEX4-deficient transgenic plants from *Arabidopsis thaliana* (Kötting et al. 2009), which is even enhanced in the double knockout of SEX4 and LSF2 (Santelia et al. 2011). SEX4-deficient plants accumulate phosphorylated oligodextrins, which are evidently released by endo-acting starch degrading enzymes, such as α-amylases and debranching enzymes, and cannot be fully metabolised. *In vitro* experiments demonstrated that starch degradation by β-amylase and debranching enzyme in the presence of GWD-mediated phosphorylation is greatly enhanced when SEX4 is additionally used to simultaneously dephosphorylate the starch granules (Kötting et al. 2009). Thus, it appears that during starch degradation, cycles of both phosphorylation and dephosphorylation are required for efficient and complete utilization of starch-derived glucosyl residues (Meekins et al. 2016).

3.1.2 Phosphorylation of Glycogen

The presence of phosphate esters in glycogen was first demonstrated by Fontana (1980). About two decades later, the progressive childhood-onset myoclonic epilepsy, Lafora disease, was associated with mutations in EPM2A and EPM2B, the former gene encoding for the phosphatase laforin (Minassian et al. 1998, Serratosa et al. 1999), the latter for malin, a ubiquitin E3 ligase (Chan et al. 2003). Later it was shown that laforin is a functional equivalent to the plant SEX4 glucan phosphatase (Gentry et al. 2007) and is capable of dephosphorylating amylopectin and glycogen (Tagliabracci et al. 2007, Worby et al. 2006). Glycogen of laforin-deficient mice is characterised by elevated levels of phosphate esters (Tagliabracci et al. 2007). Furthermore, the glycogen is structurally altered, and the mice accumulate insoluble glycogen in the form of so-called Lafora bodies. Together, these findings put glycogen phosphate into the center of investigations into the pathogenesis of Lafora disease.

An early report concluded the existence of phosphate di-esters in glycogen, where phosphate forms two ester bonds with the glucosyl carbon C1 and C6, each on different glucosyl units of glycogen (Lomako et al. 1993). However, these findings could not be confirmed in later studies, which show that phosphate in glycogen exists as mono-esters bound to the glucosyl carbons C3, C6, and C2 (Nitschke et al. 2013, Tagliabracci et al. 2011). Using NMR, the relative amounts of C2, C3, and C6 phosphate were shown to be similar in rabbit muscle glycogen. C6 phosphate, relative to total glycogen phosphate (irrespective of the carbon position), was found in the same range in murine Lafora disease models and WT mice (DePaoli-Roach et al. 2015). As it is currently not possible to determine C3 or C2 phosphate separately (except by NMR), levels of glycogen phosphorylation are largely expressed as total phosphate or as C6 phosphate content of glycogen. They differ depending on the origin of the glycogen (Table 3.2).

Attempts to identify glycogen phosphorylating enzyme(s) were first undertaken in the laboratory of Whelan (Lomako et al. 1993). Upon incubating glycogen with rabbit muscle lysates in the presence of UDP-glucose radiolabeled in the β-phosphate position (adjacent to the glucosyl residue), they found evidence for phosphate di-esters in the glycogen but eventually were unable to identify the underlying enzyme activity. Many years later, the Roach laboratory published work that held evidence that glycogen synthase itself is capable of incorporating phosphate esters from UDP-glucose (Tagliabracci et al. 2011). They postulated that within the active site of the enzyme, UDP-glucose forms an intracyclic glucose phosphate, where the former β-phosphate bridges from carbon C1 to carbon C2 or C3 of the glucosyl residue. Subsequently, cleaving the ester linkage to carbon C1 and transferring the phospho-glucosyl residue to a glycogen chain, a glucosyl residue would be incorporated into glycogen that carries a phosphate mono-ester at carbon C2 or C3. While further work in the same lab provided more evidence for this hypothesis (Chikwana et al. 2013, Contreras et al. 2016), it should be noted that the proposed mechanism does not explain the existence of C6 phosphate, as an intracyclic glucose with phosphate bridging from carbon C1 to C6 has been reported impossible (Khorana et al. 1957). While glycogen synthase-mediated phosphate incorporation is portrayed as a stochastic event, a consequence of an inadvertent side reaction, the presence and distribution of C6 phosphate in glycogen points to the existence of at least one other glycogen phosphorylating enzyme which is yet to be discovered (Nitschke et al. 2013). Furthermore, esterification of glycogen by the proposed mechanism is likely to be restricted to biosynthesis but should be impossible during glycogen degradation.

The role of glycogen phosphate has been investigated largely in relation to studies in Lafora disease mouse models. The fact that laforin acts as glucan phosphatase, and that its absence leads to increased glycogen phosphate, triggered the advancement of the hypothesis that the non-removal of erratically introduced phosphate into glycogen via an unknown mechanism alters the structure of glycogen and promotes its precipitation (Roach 2015, Tagliabracci et al. 2008). This hypothesis was tested by overexpression of catalytically inactive laforin in mice that were deficient in endogenous laforin. In these mice, no Lafora bodies formed and the behavioral phenotype was normalized. The glycogen had a normal branching structure but was, however, still highly phosphorylated due to the catalytic inactivity of the expressed transgene (Gayarre et al. 2014, Nitschke et al. 2017). Interestingly, in mouse models of Lafora disease that are deficient in the E3 ubiquitin ligase malin, phosphate levels are only increased in the insoluble glycogen portion, while those in the soluble glycogen contain WT-like amounts of phosphate (Sullivan et al. 2019). These results indicate that the disease-related glycogen hyperphosphorylation is not the cause of glycogen precipitation and accumulation. On the contrary, regarding what is known from starch metabolism, phosphorylation of glycogen should actually counteract glycogen insolubility. The role of phosphate in starch metabolism is widely

Table 3.2. Total Phosphate Content in Glycogens of Different Origins.

Organ	Species	Total phosphate [mmol/mol glc]	References
Liver	Rat	1.9 – 3.2[1]	Fontana 1980
	Rabbit	0.17	Manners 1991
	Mouse	0.4 – 0.5	Tagliabracci et al. 2007, Turnbull et al. 2010
Skeletal muscle	Rabbit	1.2 – 1.7	Lomako et al. 1993, Manners 1991
	Mouse	0.3 – 1.0	Tagliabracci et al. 2007, 2008, Turnbull et al. 2010, Tiberia et al. 2012, DePaoli-Roach et al. 2015, Nitschke et al. 2017

[1] Calculated from 0.1 – 0.17% by weight (Fontana 1980)

accepted to be in transition of glucan chains from a highly ordered but less hydrated state into a less ordered but more hydrated form, facilitating access of starch degrading enzymes. As both glycogen and amylopectin have a common chemical basis, both being composed of branched α-glucan chains, local physicochemical effects of similar covalent modifications (such as phosphorylation) are expected to have, in principle, similar consequences (Sullivan et al. 2017).

3.2 Biochemical Assays to Analyse Polyglucan Phosphorylation

The aim of this section is to provide a wide overview of currently available biochemical methods to analyse phosphate mono-esters and related enzymatic reactions.

3.2.1 Quantification of Total Phosphate

Phosphate covalently bound to organic matter as mono-esters can be determined as orthophosphate after complete hydrolysis of phosphate ester bonds. Both the hydrolysis of ester bonds and the detection of orthophosphate are achieved by different methods.

3.2.1.1 Hydrolysis of Phosphate Esters

Several methods for chemical hydrolysis of phosphate esters are available. These are similar with respect to the use of hot sulfuric acid with additional oxidants. However, the concentration of sulfuric acid varies between the methods as well as the type of oxidant used. Small amounts of either nitric acid (Fiske and Subbarow 1925), hydrogen peroxide (Morrison 1964), or perchloric acid (Hess and Derr 1975) can be used.

Alternatively, solubilized polyglucans can be subjected to enzymatic treatment using α-amylase and Antarctic phosphatase (Nitschke et al. 2017). Both enzymes work in a similar pH range, the phosphatase hydrolyzing the ester bonds, while α-amylase, by simultaneous degradation of the polyglucan, ensures access to the phosphate esters in the interior of the macromolecular polyglucan.

3.2.1.2 Quantification of Released Orthophosphate

The released orthophosphate can be quantified in several different ways. For instance, the phosphate-dependent formation of phosphomolybdic acid at low pH has been employed in many studies. Some of the molybdenum atoms within this heteropolymolybdate complex are reduced by reducing agents, such as aminonaphtholsulfonic acid (Fiske and Subbarow 1925) or ascorbic acid (Morrison 1964), to yield phosphomolybdenum blue which can be detected photometrically at 822 nm.

By the use of malachite green, the sensitivity of the molybdate-mediated quantification of phosphate can be improved several fold, as at low pH the basic dye malachite green strongly changes its color from yellow to green (absorbance 620 to 650 nm) at relatively small concentrations of phosphomolybdic acid (Itaya and Ui 1966). The malachite green method enables the detection of 0.1 to 5 nmol phosphate per assay (Hess and Derr 1975). Meanwhile, a commercial assay kit became available that uses the same method: PiColorLock (TM) Phosphate Detection Reagent from Novus Biologicals. It greatly improved the stability of the green complex, increasing technical reproducibility while being affordable and having a long shelf-life.

Alternatively, orthophosphate can be determined enzymatically using a two-step procedure (Nelson and Kaufman 1987, Nitschke et al. 2017). In the first step, rabbit muscle phosphorylase and phosphoglucomutase convert orthophosphate to glucose 6-phosphate (G6P) via glucose 1-phosphate in the presence of an excess of glycogen. In the second step, G6P is used by G6P dehydrogenase in the presence of NADP to generate NADPH. The latter can be quantified either by absorbance at 340 nm or by fluorescence at 470 nm following excitation at 340 nm. The sensitivity of this method is comparable to the commercial PiColorLock assay (see above).

3.2.1.3 Some Remarks on the Quantification of Total Glycogen Phosphate

Depending on the origin of the sample, it is possible that besides phosphate mono-esters, varying amounts of contaminating (unbound) orthophosphate are already present before hydrolysing the phosphate esters. To distinguish between bound phosphate and unbound orthophosphate, it is essential to perform orthophosphate assays with and without hydrolyzing the phosphate esters before. This is certainly facilitated when using the phosphatase-assisted hydrolysis method as the same reaction mixture can be generated except for the presence/absence of the phosphatase. Also, it is important that the conditions under which the unbound orthophosphate is determined do not lead to significant release of bound phosphate. The malachite green-based assay is conducted at low pH, which could lead to the release of highly acid labile phosphate esters. Unexpectedly high amounts of apparently unbound orthophosphate as well as continuously increasing dye formation in the range of hours can be indicative of a slow hydrolysis of phosphate esters during the malachite green assay.

Having determined the amount of phosphatase-sensitive phosphate (orthophosphate in the sample that could only be released by phosphatase) in preparations of glycogen or starch does not yet qualify the conclusion that this phosphate was bound to the polyglucan. Small contaminations with highly phosphorylated organic substances, such as nucleic acids, would lead to overestimated levels of polyglucan phosphate. Contamination with nucleic acid is of special relevance for total phosphate determination in glycogen. Native glycogen is hydrosoluble and its purification mainly relies on precipitation by ethanol that is repeated several times to remove contaminations. During the isolation of small amounts of DNA or RNA, glycogen is routinely used as co-precipitant. This clearly points to an interaction between nucleic acids and glycogen, which is unfavorable during the analysis of glycogen phosphate. The determination of total organic phosphate is used as standard method to quantify the glycogen phosphate in glycogen preparations. The purity of the samples regarding for instance nucleic acids is, however, rarely elaborated (Tagliabracci et al. 2007).

Phosphate that derives from nucleic acids can be estimated in concentrated solutions of purified glycogen by absorbance spectrometry (Fig. 3.1). This estimation is based on the absorbance coefficient of nucleic acids of 0.025 µl ng^{-1} cm^{-1} (Gallagher and Desjardins 2001) and an average molar weight of anhydrous nucleotides of 309 g/mol, assuming an average GC content of 42% (Karro et al. 2008).

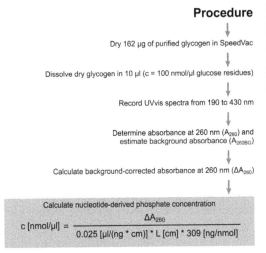
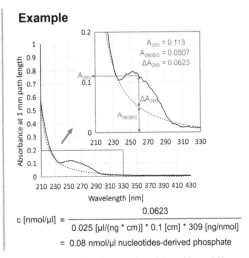

Figure 3.1. Estimation of phosphate derived from nucleic acids. Based on the absorbance of nucleic acids at 260 nm, phosphate derived from DNA or RNA can be estimated in solutions of high glycogen concentrations. After concentrating glycogen to reach 100 nmol glucosyl units per microliter, small volumes, such as 1–1.5 microliter, are sufficient to record UVvis spectra using for instance a NanoDrop. Background absorbance from the glycogen sample has to be estimated and used for correction. As each nucleotide contains one phosphate, a nucleotide-derived concentration of phosphate can be calculated using the average anhydrous molar weight of nucleic acids of 309 g/mol (see text for details).

As each nucleotide contains one phosphate, the estimated concentration of nucleotides equals that of the nucleic acid derived phosphate. If a considerable concentration of nucleotides is detected in a glycogen solution of 100 nmol glucosyl residues per µl, further attempts of purification have to be undertaken. Repeated precipitations of glycogen at 67% ethanol in the presence of 15 mM lithium chloride are effective to reduce nucleic acid contaminations. The use of the phosphatase-mediated hydrolysis (see above) is recommended as this method would only release terminal phosphate esters from nucleic acids.

The specific composition of each glycogen sample may have an impact on the reactions employed during phosphate ester hydrolysis and orthophosphate determination. As such it is recommended to always measure standards and blanks side-by-side under equal conditions and to include several controls for each sample: (1) sample alone, (2) double sample amount, (3) sample with internal standard, such as G6P, 4) double sample amount with internal standard. The sample-based phosphate amount being identical between (1) and (2) indicates completion of both hydrolysis and orthophosphate determination. The recovery of the internal standard being the same in the presence of different amounts of sample indicates that the responsiveness of the orthophosphate determination is unaffected.

3.2.2 Quantification of C6 Phosphate

In starch and glycogen, the amount of glucosyl residues that are phosphorylated at carbon C6 can be determined specifically using an enzymatic assay. The basis of glucosyl C6 phosphate determination in carbohydrates is the specific quantification of glucose 6-phosphate (G6P) in hydrolysates of the respective carbohydrate. An adequate hydrolysis is, therefore, required to cleave glucosidic bonds without hydrolyzing the phosphoester at carbon C6. Enzymatic hydrolyses using hydrolases do not fulfill this aim, since enzymes, such as α-amylase or amyloglucosidase, do not cleave glucosidic linkages in close vicinity of phosphorylated glucosyl residues (Takeda and Hizukuri 1982, Wikman et al. 2011). However, several chemical (acid) hydrolysis methods are available that quantitatively cleave glucosidic bonds irrespective of their relative vicinity to phosphorylated glucosyl residues and preserve glucosyl 6-phosphomonoesters (Haebel et al. 2008, Hizukuri et al. 1970).

The enzymatic quantification of released G6P relies on its conversion to 6-phosphogluconate by G6P dehydrogenase (G6PDH) in the presence of excess $NAD(P)^+$ (Lowry and Passonneau 1972). The reaction yields equimolar amounts of NAD(P)H, which, as opposed to $NAD(P)^+$, absorbs light at 340 nm with a molar absorptivity of 6,300 M^1 cm^1. At completion of the enzymatic reaction, the amounts of NAD(P)H equal that of the initial G6P. Thus, from the difference of absorbance values measured before and after enzymatic conversion of G6P, the concentration of G6P in the sample can be deduced using the Lambert-Beer law. At commonly used photometric path lengths (0.5–1 cm), concentrations of G6P higher than 3 µM are required for reliable measurements using direct NAD(P)H detection (Nisselbaum and Green 1969). Even using modern plate photometers, the molar absorptivity of NAD(P)H limits the sensitivity of G6P detection to a minimum of 500 pmol per assay (Fig. 3.2A).

The abundance of phosphate mono-esters in both starch and glycogen is generally low. C6 phosphate is only a fraction of it (see 3.1.1 and 3.1.2 this monograph). If direct detection of NAD(P)H is used for determination of G6P in hydrolysates of polysaccharides, the consumption of starting material (carbohydrate) is very high. If carbohydrates are analyzed having a very low glucose-based C6 phosphate content and/or if the amount of carbohydrate is limited, more sensitive methods are required.

3.2.2.1 Signal Enhancement by Redox-Coupled G6P Assays

The sensitivity of an enzymatic G6P assay can be enhanced if the resulting NAD(P)H is undergoing subsequent redox reactions that lead to the formation of a product that is detected with greater sensitivity.

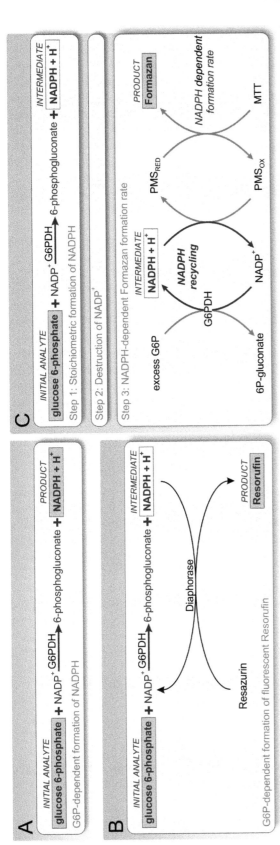

Figure 3.2. Enzymatic assay formats to quantify glucose 6-phosphate. (**A**) After complete conversion of G6P to 6-phosphogluconate, the equimolarly formed NADPH is directly measured by absorbance G6P-dependent absorbance changes at 340 nm. (**B**) Based on the reaction in (A), NADPH in the presence of diaphorase reduces resazurin to form the highly fluorescent dye resorufin, which is detected with excitation and emission at 530 and 590 nm, respectively. (**C**) Based on the reaction in (A) and after excess NADP+ has been destroyed, equimolarly formed NADPH is subjected to a cycling reaction that leads to continuous formation of the blue dye formazan (absorbance 570 nm), while NADPH is maintained essentially constant. The rate of formazan formation is in linear correlation with the NADPH amount introduced into the cycling reaction and hence the initial G6P amount.

An example is the coupling to the redox reaction of resazurin to resorufin. The latter is a highly fluorescent reagent that, by reduction of resazurin, is formed in the presence of NADPH and the enzyme diaphorase (Guilbault and Kramer 1965). This assay concept is quite convenient. In one step, G6P is converted to 6-phosphogluconate, yielding NADPH, which is directly being used to reduce resazurin (Fig. 3.2B). The fluorescence intensity of resorufin (excitation at 530 nm, emission at 590 nm) is measured before and after the addition of G6PDH, the fluorescence difference being in linear correlation with the initial G6P concentration. Using a resazurin-diaphorase-coupled assay, low concentrations of G6P had been previously determined in cell lysates (Zhu et al. 2009) and in acid hydrolysates of glycogen (DePaoli-Roach et al. 2015). When highly pure resazurin is used, the limit of G6P detection can be as low as 3 pmol per assay (DePaoli-Roach et al. 2015).

Even higher sensitivity can be achieved when the formed NADPH is undergoing redox reactions, yielding a sensitively detected product, but is itself continuously being rebuilt in a so-called cycling reaction. This cycling of NADPH leads to a continuous formation of the detected product, the formation rate of which correlates linearly with the initial G6P concentration. The most sensitive NADPH cycling assay involves reduction of phenazine methosulfate (PMS) by NADPH, which subsequently reduces (3-(4,5-dimethylthiazol-2-yl)-2,5-diphenyltetrazolium bromide (MTT), forming a blue formazan dye that is detected photometrically at 570 nm (Nisselbaum and Green 1969). The redox system had been used to measure G6P in plant material (Gibon et al. 2002) and later in hydrolysates of glycogen (Nitschke et al. 2017, 2013). The assay comprises three steps: (1) first G6P is converted to 6-phosphogluconate, yielding equimolar amounts of NADPH; (2) the excess $NADP^+$ is destroyed by alkaline and heat treatment; (3) after neutralisation, PMS and MTT are added alongside G6PDH, which mediates the cycling of $NADP^+$ to NADPH in the presence of an excess of G6P. The G6PDH-mediated cycling reaction maintains a maximal NADPH concentration in the third step and ensures formazan formation rates that are only dependent on the amount of G6P initially introduced in the first step (Fig. 3.2C). Using this PMS-MTT-coupled cycling assay, G6P amounts as low as 0.5 pmol can be reliably determined (Nitschke et al. 2013).

3.2.2.2 Impact of High Glucose Concentrations in Hydrolysates of Polyglucans

Hydrolysates of starch and glycogen, both being homoglucans, comprise almost exclusively glucose, besides small amounts of the target compound, G6P. Any effect of the highly abundant glucose on the G6P determination can lead to severe analytical errors. For instance, an intrinsic glucose dehydrogenase activity has been reported in preparations of G6PDH (Horne et al. 1970). This would lead to glucose-dependent NADPH formation when hydrolysates of glycogen or starch are analysed using an enzymatic assay that depends on G6PDH.

A later study confirmed that indeed the choice of the commercial G6PDH preparation is decisive for the unbiased and sensitive G6P determination in the presence of high concentrations of glucose (Nitschke et al. 2013). The use of several commercial G6PDH preparations led to significant glucose concentration-dependent overestimation of G6P amounts. Only the preparation from *Leuconostoc mesenteroides* (Roche, Cat# 10165875001) showed no impact of glucose on the results of G6P determination even if the glucose concentration exceeded that of G6P by 3 to 5 orders of magnitude. These findings were also confirmed in a separate study (DePaoli-Roach et al. 2015).

3.2.3 Determination of Phosphoglucosyl Residues by Chromatography/MS and Fluorescence-Assisted CE

Depending on the carbon position of the phosphate ester, glucose monophosphates exhibit differential physico-chemical properties, which permit their separation by liquid or gas chromatography (LC or GC, respectively) or capillary electrophoresis (CE).

Under appropriate conditions, glucose monophosphates differ in their mobility on anion exchange chromatography columns (Ritte et al. 2006), or during CE after derivatization with a charged fluorophore (e.g., APTS) (Verbeke et al. 2016). The specific phosphate position can be assigned by comparing retention times with those of authentic standards. When the analytes are detected by pulsed amperometry, quantification is possible after calibrating the detector response for each molecule species. Derivatization with a fluorophore enables detection by laser-induced fluorescence, where the detector response is supposedly in direct proportion to the molar amount of analyte passing the detector, but does not depend on the molecule species. However, inadvertent co-migration of more than one molecule species, which is possible during the analysis of carbohydrate hydrolysates, can undermine the quantification of glucose monophosphates (Ritte et al. 2006).

Unequivocal identification of the separated analyte can be achieved by suitable techniques of mass spectrometry (MS). All glucose monophosphates are structural isomers and have identical molecule masses. Using MS without fragmentation of the analytes is therefore not appropriate to quantify specific glucose monophosphates. However, fragmentation of glucose monophosphate precursor ions, as obtained by collision-induced dissociation (CID) after electrospray ionization and precursor ion selection using tandem MS with ion trap, revealed that fragmentation patterns are characteristic of different glucose monophosphates. Techniques of tandem MS were described for the determination of G6P and glucose 3-phosphate (G3P) in hydrolysates of starch. They are used either without prior separation for determination of ratios between G6P and G3P (Haebel et al. 2008) or, if they are coupled downstream of a suitable liquid chromatography (LC) technique, for separate quantification of both glucose and glucose monophosphates (Carpenter et al. 2012). Chromatography with online quantitative MS combines quantitative and specific qualitative information about components eluting from chromatography columns. High performance anion exchange chromatography, coupled to pulsed amperometric detection (HPAEC-PAD), is not suitable for MS coupling because of the usually high salt concentrations in the eluents used. However, by hydrophilic interaction chromatography (HILIC) glucose, G6P and G3P can be sufficiently separated using eluents that are MS compatible (Antonio et al. 2008, Carpenter et al. 2012). When HILIC is coupled to tandem MS, the HILIC derived eluate is constantly ionized by an electrospray source (Carpenter et al. 2012). Precursor ions of interest (glucose and glucose monophosphate) are selected with an ion trap before they are fragmented by CID. G3P and G6P are determined by alignment of fragmentation patterns with those of authentic standards. Quantification of characteristic fragment ions allows the calculation of initial amounts of glucose and the respective glucose monophosphates on the basis of calibration with known amounts of authentic standards.

The described chromatography- and MS-based methods are believed to be suitable for quantification of glucose phosphates provided they endure chemical hydrolysis of the carbohydrate. As opposed to starch, where the major glucosyl phosphates at carbon C6 and C3 are stable under appropriate hydrolysis conditions (Haebel et al. 2008), a major proportion of glycogen phosphate has been reported to be acid labile and later interpreted as evidence for phosphate di-esters (Fontana 1980, Lomako et al. 1993). Glucosyl phosphates that are largely acid labile, thus, cannot be quantified nor qualitatively analyzed by methods that rely on acid hydrolysis of the polyglucan. Nevertheless, recently a method was published that claims to quantify glucose, and all glucose phosphates in glycogen after hydrolysis in 1 M hydrochloric acid followed by GC/MS (Young et al. 2020). The authors do not comment on the differential stability of the different phosphate esters and determined unusually low glycogen contents with unusually high phosphate per glucosyl unit.

Milder conditions of polyglucan hydrolysis, such as enzymatic (hydrolase-mediated) degradation, however, result in various species of phosphorylated oligoglucans (Tagliabracci et al. 2011, Nitschke et al. 2013). Chromatographic, electrophoretic and MS-based analysis of such degradation products is difficult since all these methods rely on alignment with authentic standards which are practically unavailable in this case.

3.2.4 *Analyses of Glucan Phosphorylation and Dephosphorylation*

Several enzymes in starch and glycogen metabolism have been reported to phosphorylate or dephosphorylate polyglucans. In plants, these include the dikinases GWD and PWD as well as the glucan phosphatases SEX4 and LSF2. In animals, glycogen synthase has been reported to incorporate phosphate into glycogen, while laforin can remove it (see 3.1.1. and 3.1.2. this monograph). Several assays have been described to determine the activities of these enzymes.

3.2.4.1 *Non-Radioactive Assays*

In principle, it is possible to measure enzyme-dependent changes in the phosphate content of a suitable polyglucan substrate. For instance, phosphate incorporation into modified bovine liver glycogen by recombinant GWD results in a marked increase of C6 phosphate in glycogen (Ritte et al. 2002). In a similar fashion, it has been shown that recombinant laforin removes phosphate esters from glycogen glucosyl carbon C6 in a time-dependent manner (Nitschke et al. 2013). The level of glycogen C6 phosphate can be determined after acid hydrolysis of the glycogen using an enzymatic assay for G6P (see 3.2.2. this monograph) (Nitschke et al. 2013, Ritte et al. 2002). In order to detect the enzyme-mediated changes in glycogen C6 phosphate, amounts of glycogen used are high and incubation times long. While the enzymatic activity can be qualitatively proven, the long incubation times make these types of assays inappropriate for activity quantification.

Glucan phosphatase activity can also be measured by quantification of the orthophosphate that is released from a suitable phosphorylated glucan substrate. For instance, Laforin has been shown to release phosphate from both amylopectin and rabbit muscle glycogen (Tagliabracci et al. 2007, Worby et al. 2006). In both cases, the released orthophosphate is quantified using a variant of the malachite green assay (see 3.2.1.2. this monograph).

3.2.4.2 *Assays Utilizing Radiolabelled ATP*

The existence of the radioisotopes ^{33}P and ^{32}P enables measurements of phosphorylation and dephosphorylation reactions with much greater sensitivity and with shorter incubation times. Being dikinases, GWD and PWD incorporate the β-phosphate of ATP into suitable glucan substrates. For instance, phosphorylation of leaf starch from *Arabidopsis thaliana* can be detected after incubation of the starch with recombinant GWD and β-labelled ATP. If the starch was prephosphorylated by GWD but using unlabelled ATP, PWD-mediated incorporation of label into the starch can be detected in the same manner (Kötting et al. 2005). Dikinase activity was even measurable in crude protein extracts from plants that were incubated with commercially available soluble potato starch (Ritte et al. 2003). This type of labelling experiment greatly relies on the removal of labelled compounds (that have not been incorporated into the starch), such as unused ATP. If granular starch is used, this is mostly accomplished by extensive washing of the starch granules before measuring label using a scintillation counter. In the case of a water-soluble polyglucan, ultra-filtration and extensive washing of the filter removes the excess of radiolabel.

Radiolabelling experiments can also give insight into the glucosyl carbon position of the incorporated phosphate. Hydrolysates of GWD- or PWD-labeled starch were subjected to anion exchange chromatography which separates G6P and G3P. GWD-incorporated label was exclusively found in G6P, and PWD-incorporated label in G3P. From these results, it could be concluded that GWD and PWD specifically phosphorylate glucosyl carbon C6 and C3, respectively (Ritte et al. 2006).

The specificity of both known plant dikinases has been utilized to investigate the specificity of the plant glucan phosphatases SEX4 and LSF2. By subsequent incorporation of phosphate into starch granules, first by GWD and subsequently by PWD, starch is generated that contains radiolabelled phosphate esters either in glucosyl carbon C6 or in C3, depending on whether β-labelled ATP is used during GWD- or PWD-mediated phosphorylation. Then the glucan phosphatase is incubated with either of the two types of prelabelled starch, and radiolabel released to the supernatant is quantified with a scintillation counter. SEX4 is able to hydrolyse labelled phosphate esters from both glucosyl carbons but has a preference for C6 phosphate (Hejazi et al. 2010). By targeted mutagenesis, variants of SEX4 were generated that have a preference for dephosphorylating at C3 (Meekins et al. 2014). Using similar experiments, LSF2 was shown to specifically dephosphorylate the glucosyl carbon C3 (Santelia et al. 2011).

The action of GWD and PWD was also studied using a less complex model substrate, i.e., crystalline maltodextrins, which are water-insoluble particles that largely consist of unbranched α-1,4-glucan chains having various length (Gidley and Bulpin 1987). Interestingly, they are phosphorylated at a much higher rate than all of the previously used native starches. Similar to labelling experiments with starch granules, a suspension of crystalline maltodextrins is incubated with the recombinant dikinase in the presence of β-labelled ATP. Radiolabel is incorporated into the crystalline particles and maltodextrin chains are partly solubilized upon phosphorylation. Label in the suspension or after centrifugation in the supernatant and pellet can be quantified to assess the total incorporated radiolabel, or that present in the still crystalline as well as the solubilized maltodextrin fraction. The suspension, supernatant or pellet are subsequently heat treated to stop the enzymatic reaction and, if applicable, for complete solubilisation of the formerly crystalline maltodextrins. Then an aliquot is separated by thin layer chromatography on polyethyllene imine (PEI)-cellulose plate, prior to label quantification by phosphor imaging. In PEI-TLC, negatively charged compounds interact with immobilized ionic groups, and therefore analytes are separated mainly according to their mass/charge ratio. This allows separate assessment of phosphorylated maltodextrins, unused ATP, and released or contaminating orthophosphate and ADP. The mobility of phosphorylated compounds decreases in the following order: phosphomaltodextrins, G6P, orthophosphate, ADP, ATP (Hejazi et al. 2008). Using this kind of experimental setup, it was shown that PWD-mediated phosphorylation is dependent on the degree of prephosphorylation by GWD, but that PWD is nevertheless capable of phosphorylating unphosphorylated chain. This indicates that it is rather the physical arrangement of chains which is greatly impacted by few phosphate esters than the phosphate itself that determines whether or not a glucan can act as a substrate for either of the dikinases (Hejazi et al. 2009).

3.2.5 Enrichment of Phosphoglucans

The generally low abundance of phosphate esters in the polyglucans starch and glycogen can be problematic if insufficiently sensitive methods for phosphate detection are used. An example is the use of nuclear magnetic resonance (NMR) spectroscopy that detects slight changes of a magnetic field due to the presence of phosphorus in a sample. Analyzing the glycogen and starch phosphate by certain techniques of NMR requires the enrichment of phosphoglucan chains from either of the two polyglucans.

Prior to the enrichment, the polyglucan has to be degraded using amylolytic enzymes such as α-amylase and amyloglucosidase. The simultaneous use of both enzymes is advantageous. The exo-acting amyloglucosidase is very efficient to degrade both linkages within chains and branch points. However, it can only remove one glucosyl residue at a time from the non-reducing end of a chain and is not capable of removing phosphorylated residues. The action of the endo-acting α-amylase is not interfered with by the presence of phosphate esters or branch points. Regardless of whether starch or glycogen was used, the result of a successful amylolytic treatment with both enzymes is a mixture of mostly glucose and a few phosphorylated oligoglucans. However, when granular starch is being enzymatically hydrolyzed, as opposed to when glycogen is degraded, combinations of enzymatic and heat treatment have proven useful to facilitate the dismantling of the rigid particles (Nitschke et al. 2013, Ritte et al. 2006).

The phosphorylated and hence negatively charged oligoglucans can be separated from the majority of neutral glucose by anion exchange. Most efficient and reproducible enrichment is achieved when a chromatography is performed using columns with a suitable resin such as DEAE or QFF Sepharose. The bed volume of the resin has to be adjusted to the amount of phosphoglucan applied. Higher bed volumes ensure a high binding capacity of the column and a low loss of phosphorylated analyte. However, it also determines the extent of analyte dilution and the subsequent necessary efforts for concentrating the analyte. For phosphoglucan enrichment, around 200 mg of glycogen bed volumes of 2 mL DEAE and 8 mL QFF Sepharose have been used (DePaoli-Roach et al. 2015, Nitschke et al. 2013). It is important that the anion concentration in the sample is small. Therefore, polyglucans have been degraded at low concentrations of sodium acetate or ammonium acetate (DePaoli-Roach et al. 2015, Nitschke et al. 2013).

Prior to the application of the enzymatic polyglucan hydrolysate, the anion exchange column has to be well equilibrated with a buffer containing low amounts of anions. This allows binding of the phosphoglucans to the column when the sample is added. Subsequent washing of the column with low-anion buffer removes the excess of glucose. Elution of phosphorylated compounds from the column is achieved by step-wise raising of the buffer concentration. Through this process, the phosphoglucans will be substantially diluted in the elution buffer and have to be concentrated before further analyses. Using volatile buffers for elution prevents high concentrations of the elution buffer salt during concentrating of the samples. Buffers of 0–1 M ammonium bicarbonate and acetate have been used for phosphoglucan enrichment on DEAE and QFF Sepharose, respectively (DePaoli-Roach et al. 2015, Nitschke et al. 2013).

Prior to phosphate analyses, the eluted fractions are concentrated by lyophilisation (freeze-drying). Repeated cycles of dilution in water and drying may have to be performed to completely remove the salt derived from the elution buffer. To avoid the enrichment of contaminating orthophosphate, water and buffer salts used should be essentially free of orthophosphate.

3.3 Identification of Phosphorylated Carbons by Nuclear Magnetic Resonance Spectroscopy (NMR)

Using nuclear magnetic resonance spectroscopy (NMR), glucosyl phosphates can be analyzed without acid hydrolysis of the polyglucans. Thus, by NMR, positions of phosphate esters on glucosyl carbons in polyglucans can be determined irrespective of their stability in acidic media. However, NMR is a rather insensitive method which requires relatively high amounts of analyte. This largely excludes NMR from being applied when only very small amounts of analyte are available.

The following sections will provide an introduction into the basics of NMR and aim to give the reader the means to understand and interpret the various types of spectra that are obtained when carbon positions of phosphate esters in polyglucans are analyzed by NMR.

3.3.1 General Principles of NMR Spectroscopy

NMR-spectroscopy is based on the interaction of electromagnetic radiation with the nuclear spin that can only be observed when the nuclear spin is placed in a magnetic field. The majority of atomic nuclei possess such a spin and are thus detectable by NMR. Unfortunately, the biologically important nuclei ^{12}C and ^{16}O cannot be detected by NMR as they possess an even number of protons and neutrons and, hence, no spin. Of particular importance for high resolution NMR-spectroscopy are nuclei with a spin quantum number of I = 1/2, among which are ^{1}H, ^{13}C, ^{15}N and ^{31}P, which are used in chemical or biochemical NMR investigations. If these spins are placed in a magnetic field, they can assume two orientations: either along the direction of the external magnetic field (α-state), or against it (β-state). In addition, the spins perform a precession around the axis of the external magnetic field with the so-called Larmor frequency. This frequency depends on the external magnetic field strength and the gyromagnetic ratio which is an isotope specific constant (Equation 1). The energy difference ($\Delta E_{\alpha,\beta}$) between the two possible spin states (α or β) also depends on the external magnetic field strength and the gyromagnetic ratio (Equation 2). Since this energy difference is generally small compared to the average kinetic energy of samples at room temperature, the α- and β-states are almost equally populated with a small excess of spins oriented along the external magnetic field direction. The slight excess then adds up to a macroscopic magnetic moment that is oriented in the direction of the external magnetic field.

$$\nu_L = \frac{\gamma}{2\pi} \cdot B_0 \tag{1}$$

$$\Delta E_{\alpha,\beta} = \frac{\gamma}{2\pi} \cdot B_0 \cdot h = h \cdot \nu_L \tag{2}$$

$\Delta E_{\alpha,\beta}$ Energy difference between α- and β-state of magnetic moment vector orientation

ν_L Larmor frequency

γ Gyromagnetic ratio

B_0 External magnetic field strength

h   Planck's constant

If energy with a frequency matching the Larmor frequency is introduced at a right angle to the external field, the nuclear spins will absorb energy and flip from α- to β-state, resulting in equal population of both energetic states (nuclear magnetic resonance). A measurement of the energy absorption in a frequency dependent manner then leads to the NMR spectrum.

Due to their different gyromagnetic ratios in a given external magnetic field, different types of nuclei, such as ^1H or ^{31}P, exhibit largely different Larmor frequencies, which allow separate detection of nucleus species by adequate irradiation. Furthermore, within one species of nuclei, Larmor frequencies (and hence excitation energy) are slightly different because the effect of the external magnetic field on a nucleus depends on magnetic shielding of the nucleus by its surrounding electrons. Consequently, the electronic constitution around nuclei of the same type influences the effective magnetic field strength and thus—according to equation (1) and (2)—the Larmor frequency as well as the excitation energy differ. Therefore, even nuclei of the same type can be distinguished by their nuclear magnetic resonance frequency, which then allows conclusions about the chemical environment of the nucleus (Friebolin 1992, Günther 1992). The frequency differences are recorded relative to an arbitrarily defined standard and are then called chemical shift.

Similar to electronic shielding, the magnetic moment of one nucleus can affect the resonance frequency of another nucleus by altering the effective magnetic field strength the other nucleus is exposed to (spin-spin coupling). Generally, the effect of a neighbouring nucleus decreases with the number of chemical bonds between them. For example, in a solution of glucose 3-phosphate (G3P, phosphate ester at carbon C3), the effective external magnetic field at each ^{31}P nucleus is locally altered by the ^1H nucleus that is connected via 3 chemical bonds (vicinal coupling). The spins of the vicinal ^1H can either be in α- or in β-state. In both states, it slightly alters the magnetic field at the neighboring ^{31}P with the same intensity (since both states are almost equally populated) but with different direction (increase or decrease). The same nucleus thus exhibits slightly different frequencies, which result in a splitting in the spectrum. The effect is called J-coupling and is usually much smaller then differences in chemical shift. It therefore leads to a fine structure of the peaks of an NMR spectrum. J-couplings are called either homo- or heteronuclear depending on whether the two interacting nuclei are of the same or a different type. Since J-couplings result from an interaction through chemical bonds, they contain information about bond angles. More importantly, they can be used and analysed using multidimensional NMR spectroscopy.

3.3.2 One-Dimensional NMR Spectroscopy

In a one-dimensional (1D) NMR experiment, the resonance frequency of nuclei of one type, for example ^{31}P, is detected in a sample. The frequencies recorded in a spectrum are equal for all chemically equivalent nuclei but different for those with varying chemical surroundings. For example, in a sample containing glucose 3-phosphate (G3P) and glucose 6-phosphate (G6P), ^{31}P exists as phosphate on carbon C3 or C6 of α- or β-glucopyranose (in this case, 6 α and β refer to the glucosyl anomers). Thus, four groups of ^{31}P nuclei (C3α, C3β, C6α, C6β) exist, each group with a distinct Larmor frequency and thus a different position in the ^{31}P NMR spectrum.

In modern NMR experiments, the excitation of the nuclei is not performed using one frequency at a time but rather using an electromagnetic pulse applied at a right angle to the magnetic field. This leads to simultaneous excitation of all nuclei of one sort (in this example ^{31}P) irrespective of their specific chemical surroundings. This pulse has two main effects: in each set of ^{31}P nuclei, it reduces the difference between the populations of spins in α- and β-state and it forces the spins away from performing a random precession around the axis of the external magnetic field into a coherent movement, which results in an overall magnetization in a right angle relative to the external magnetic field. If a pulse of the right strength and length is applied, the magnetization in the direction of the external magnetic field vanishes and only a magnetic moment perpendicular to the external field is left. This so called 90°-pulses are instrumental for multidimensional NMR spectroscopy.

After the pulse, the spins return to their equilibrium distribution while performing a coherent precession around the magnetic field. This leads to a spiral movement of the overall magnetization that is recorded as an oscillating and decaying current (FID, free induction decay) using a receiver coil at right angle of the external magnetic field. Since the four sets of ^{31}P nuclear magnetic moments in the G3P/G6P sample (see above) rotate with their specific Larmor frequency, the FID is composed of induced currents from four overlapping precessions. The four specific Larmor frequencies can be determined by Fourier transformation (FT), resulting in a 1D ^{31}P NMR spectrum with theoretically four signals representing the four groups of ^{31}P nuclei. The chemical shifts thus recorded are small compared to the external field and are therefore given in parts per million (ΔHz per MHz).

In 1D ^{1}H NMR, the chemical shift of ^{1}H nuclei is affected by the position of the hydrogen on the glucosyl residue (carbon C1 to C6). Each position theoretically gives two signals, one for each anomer state of the glucosyl residue. The attachment of phosphate groups influences the chemical shifts of ^{1}H nuclei depending on the distance of phosphate and the respective hydrogen. Hydrogens in vicinal position to phosphorus (connected via 3 chemical bonds to the same carbon) likely exhibit the most significant change in chemical shift. In addition to the different chemical shifts, the ^{1}H peaks will also exhibit a fine structure due to J-coupling to other ^{1}H and ^{31}P nuclei.

It should be noted that due to the low sensitivity of the technique, it is usually necessary to repeat the recording several times to improve the signal-to-noise ratio of the spectrum. One dimensional experiments are recorded in two stages. In the first preparation period, detectable magnetization is created after waiting for a suitable time for the system to return to equilibrium. In the second detection period, the FID is recorded. These two stages are repeated as often as considered necessary to obtain a sufficient signal-to-noise ratio.

3.3.3 Two-Dimensional (2D) NMR Spectroscopy

3.3.3.1 Principle of 2D NMR Spectroscopy

Essential information about the connectivity of specific nuclei via chemical bonds can be derived from the interaction of nuclei via homo- or heteronuclear couplings. This can be determined using suitable two-dimensional (2D) NMR techniques. For that matter, FIDs are recorded after the application of a sequence of electromagnetic (mostly 90°) pulses. These sequences contain two more stages compared to one-dimensional experiments and thus consist of four stages: preparation period, evolution period t_1, mixing period, and detection period t_2. Pulses applied during the preparation period result in sample magnetization transversal to the external magnetic field in the same manner as in a 1D experiment. During the evolution time, the spins perform a precession with frequencies determined by chemical shift and J-coupling, which leads to a different situation for each group of nuclei at the end of the evolution time. During the mixing period, another series of pulses is applied upon the resulting magnetic moments, which lead to a transfer of magnetization from one nucleus to another via J-coupling. The intensity of the transfer depends on the situation at the end of the evolution time. During the detection period (t_2), an FID is recorded as in a one-dimensional experiment. The pulse sequence is repeated several times with incremented changes of the evolution time (t_1) so that the magnetization at the beginning of the mixing period differs for different values of the t_1-time. As opposed to 1D NMR where only identical FIDs are recorded, added and subjected to Fourier transformation, in 2D NMR several FIDs are now obtained as a function of t_1 which allows to indirectly record chemical shifts and J-couplings. After the Fourier transformation of the FIDs, a series of 1D spectra is obtained in which the intensity of the individual peaks depends on the frequency of the nucleus that was active during the evolution time. A second Fourier transformation results in a two-dimensional spectrum with two frequency axes that derive from either of the two time variables t_1 or t_2 and represent a chemical shift each. Two types of signals are possible in such a spectrum: if no transfer of magnetization has taken place during the mixing time, then the frequencies in both dimensions are identical (diagonal peaks). If, however, a transfer has taken place the two frequencies are different and a cross peak is obtained. Note that such a peak can only occur if an interaction (J-coupling) between the spins was present and that it is thus a direct indication of the presence of a chemical bond. A plot of a 2D NMR spectrum usually shows intensity values for a pair of frequency variables (F_1, F_2), the intensity being represented in a third dimension using contour lines (Friebolin 1992, Martin and Zektzer 1988).

3.3.3.2 *Variants of 2D NMR Spectra in the Analyses of Polyglucan Phosphate Esters*

Two types of 2D NMR experiments are possible: if the chemical shifts of the same type of nucleus correlated, we obtain a homonuclear spectrum; for different types of nuclei, we obtain a heteronuclear spectrum. An example for the first kind is the ^{1}H,^{1}H-COSY (COrrelation SpectroscopY) spectrum that exhibits chemical shifts of ^{1}H nuclei (protons) on both axes (F_1 and F_2). Some signals occur along the diagonal (diagonal peaks) and correlate with identical chemical shifts in F_1 and F_2. The signal pattern on this diagonal is equivalent to that of the 1D ^{1}H spectrum. In these type of spectra, cross peaks are correlations between protons with different chemical shifts that can only be detected when both nuclei are connected via three chemical bonds (vicinal coupling) or, provided they are chemically non-equivalent, via two chemical bonds (geminal coupling). An example for the second kind is the in ^{1}H,^{31}P-HMBC spectra (Heteronuclear Multiple Bond Correlation), where ^{1}H and ^{31}P chemical shifts are on the axes F_2 and F_1, respectively. In these types of spectra, no diagonal peaks are possible and the cross peaks can only occur if there is a J-coupling between a ^{1}H and a ^{31}P nucleus.

In glucosyl residues, hydrogen is bound either to carbon or to oxygen in hydroxyl groups. The chemical shifts of hydroxyl protons can be found over a wide range and is strongly dependent on the acidity of a specific hydroxyl group. To simplify, ^{1}H spectra measurements of carbohydrates are usually performed in D_2O. This way, ^{1}H at hydroxyl groups are quantitatively exchanged by ^{2}H (deuterium) which is not excited when ^{1}H-pulses are applied. Thus, hydroxyl protons are not detected in ^{1}H,^{1}H-COSY plots of carbohydrate compounds. For the analysis of glucosyl residues, this type of experiment serves to assign signals in complex 1D ^{1}H NMR spectra to positions on the carbon backbone. The protons (H1) at glucosyl C1 exhibit the strongest chemical shift because they are de-shielded by two electronegative oxygen nuclei that are bound to the same carbon (Pomin 2012). H1 couples with a vicinal proton the signal of which can consequently be assigned to H2. H2 reveals H3, etc., until all proton signals are assigned to their respective carbon (Friebolin 1992, Martin and Zektzer 1988). Regarding phosphorylated glucosyl residues, each ^{31}P nucleus is vicinal connected to hydrogen. Correlations can be found in an ^{1}H,^{31}P-HMBC and signals indicate the presence of phosphate at the carbon to which the hydrogen is connected.

When the signal pattern in ^{1}H spectra is rather complex and various signals are very close to each other or even overlapping, another type of heteronuclear 2D NMR spectrum is useful that correlates hydrogen and carbon to disentangle the complexity of the ^{1}H dimension. ^{1}H,^{13}C-HSQC

(Heteronuclear Single Quantum Coherence) and ^{1}H,^{13}C-HMQC (Heteronuclear Multiple Quantum Coherence) spectra detect hydrogen-carbon correlations via one chemical bond. Thus, the proton signals in F_2 are separated along the ^{13}C chemical shift in F_1 and direct connectivity of hydrogens and carbons is indicated. Again, no diagonal peaks are possible. Using multiplicity editing, signals indicating methylene (CH_2) can be distinguished from those that point to methine (CH) or methyl groups (CH_3). Since the latter do not occur in polyglucans, the multiplicity editing allows assignment of carbon and hydrogen signals to the position C6 (Friebolin 1992, Martin and Zektzer 1988). Note that in contrast to ^{1}H and ^{31}P which are present with a natural abundance of 100% (or close to it), ^{12}C cannot be used for NMR spectroscopy and that the suitable nucleus, ^{13}C, is only present with 1.1% natural abundance. Therefore, the recording of heteronuclear ^{1}H,^{13}C spectra is more time consuming.

3.3.4 Analyses of Authentic Glucose Monophosphates

In aqueous solution, glucose exists as a mixture of α- and β-glucose. Using an ^{1}H,^{1}H-COSY spectrum, all proton signals can be assigned to their carbon position (Fig. 3.3A). For both anomers, the ^{1}H,^{13}C-HMQC reveals several signals all of which represent H,C correlations via one chemical bond (Fig. 3.3C). The anomeric state influences the chemical shift of the H,C correlations in the ^{1}H,^{13}C-HMQC. For each of the glucosyl carbons 1, 2, 3, and 5, two H,C correlations are observed. By contrast, H4,C4 correlation signals of α- and β-glucose are overlapping and occur as one signal in the spectrum. The two hydrogen atoms at carbon C6 (H6) are diastereotopic, i.e., substitution of either of the two would result in diastereoisomers. Therefore, for each anomer of glucose two H6,C6 signals emerge. In the case of α-glucose, the two H6,C6 signals exhibit similar ^{1}H-shifts. Thus, they appear as one broadened signal, while the respective correlations in β-glucose are clearly separated (Fig. 3.3C).

In the glucose dimers maltose and isomaltose, the number of H,C correlations detected in a ^{1}H,^{13}C-HMQC increases as compared to glucose (Fig. 3.3C–E). The chemical shifts of H,C correlations depend on the carbon position in the glucosyl residue (C1 through C6), on whether the glucosyl carbon belongs to the glucosyl ring with the reducing end, and on the anomeric state of that glucosyl ring. Furthermore, shifts of H,C correlations are significantly influenced at carbon positions at or in close vicinity of a glucosidic bond. Examinations of longer oligodextrins such as maltotriose and maltotetraose revealed that shifts of H,C correlations at specific glucosyl carbons on glucosyl residues with larger distance from the reducing end only slightly differ from those of non-reducing glucosyl ring in maltose (Schmieder et al. 2013).

The influence of phosphate on the shifts of H,C correlations can be examined in ^{1}H,^{13}C-HMQC spectra of glucose X-phosphate standards (GXP; X = 1, 2, 3, 4, 6). H,C correlations are shifted towards higher ppm in both dimensions when phosphate is bound in close vicinity. In G1P, the correlation H1,C1 exhibits the highest change in chemical shift. The same is true for H2,C2 and H3,C3 in G2P and G3P, respectively, as well as for H4,C4 in G4P and H6,C6 in G6P. Except for G1P, this phosphate induced change in chemical shift similarly affects H,C correlations of both anomers. In G1P, the glucosyl ring is fixed in its α-anomeric form due to esterification at C1. Hence, no H,C correlations are detectable for the β-anomer (Fig. 3.4).

Using ^{1}H,^{31}P-HMBC NMR, the coupling of hydrogen and phosphorus is detected and can be aligned with ^{1}H,^{13}C-HMQC spectra of the monoglucosyl phosphates. The H,P correlation signals clearly correspond to those H,C signals that were shifted most in comparison to the respective H,C signal in glucose (see above, Fig. 3.4). The unambiguous assignment of H,P correlations to H,C correlations can be complicated by overlapping hydrogen resonances, which is for instance observed in the ^{1}H,^{13}C-HMQC of G4P, where the signals of αH5, αH3, and α/βH4 exhibit similar ^{1}H chemical shifts, while only the latter is attached to a phosphorylated carbon (Schmieder et al. 2013).

Two types of ^{1}H,^{13}C experiments can be used to find out which positions in a molecule have phosphorus attached. In a conventional ^{1}H,^{13}C-HSQC that has been recorded with high resolution (Fig. 3.5A), those H,C pairs in which an additional coupling to phosphorus is present exhibit an E.COSY-type pattern. This pattern results from the fact that ^{31}P is not changing its spin state during

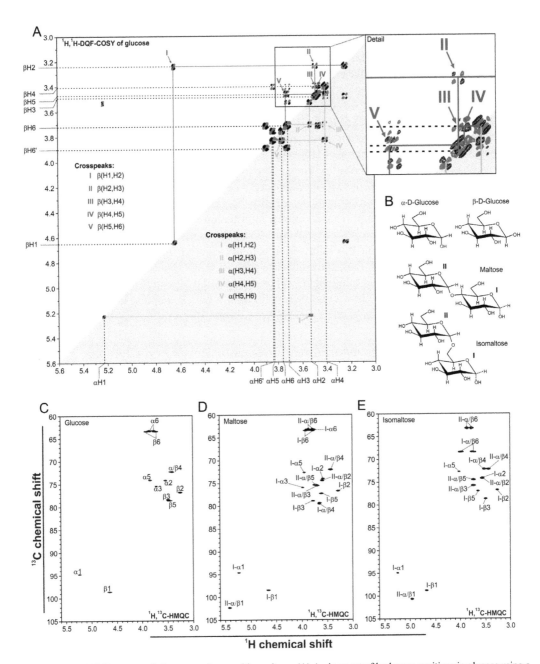

Figure 3.3. 2D NMR spectra of glucose, maltose and isomaltose. **(A)** Assignment of hydrogen positions in glucose using a ¹H,¹H-DQF-COSY spectrum. On both axes, hydrogen chemical shift is given. The spectrum is divided by a diagonal upon which signals occur that reflect the signals of a 1D ¹H-NMR spectrum. So called cross-peaks occur apart from the diagonal, appear mirrored along the diagonal, and reflect coupling of hydrogens via three or less chemical bonds. A cross-peak is always aligned with two signals on the diagonal, derived from the coupling hydrogens. For better visualization, the cross-peaks of the α-anomer are marked in the grey shadowed half of the spectrum, while those of the β-anomer are marked in the white half. Alignment of cross-peaks is indicated by green (α-anomer) and blue (β-anomer) lines. Furthermore, cross-peaks are marked with arrows and Roman numbers (assignment of coupling hydrogens is given in the spectrum). The H1 signals of both anomers are mostly shifted to higher ppm on the diagonal and are, thus, a good starting point for the assignment of other hydrogens. H1 is aligned with a cross-peak (I) that reveals the position of H2, which again is aligned with another cross-peak (II) which reveals H3, etc. This way all hydrogen atoms of both glucose anomers can be assigned to values of ¹H chemical

Figure 3.3 contd. ...

the experiment and only one of the two possible peaks occurs in each dimension. Cross-peak signals are thus shifted in the ^{13}C dimension by half the additional C,P coupling and in the ^{1}H dimension by half the additional H,P coupling (Fig. 3.5F and G). In general, the E.COSY-type pattern is more pronounced the closer ^{31}P is connected to ^{1}H and ^{13}C. E.COSY-type pattern of cross-peaks is observed at C,P and H,P coupling via two and three chemical bonds. For instance, in glucose 3-phosphate C,P coupling occurs for carbon C2, C3, and C4 (Fig. 3.5D–H). However, H,P coupling can only be detected for carbon C3 (Fig. 3.5F and G). Similar effects have been observed for other glucose monophosphates. Thus, those carbon positions that directly carry the phosphate group exhibit two tilted peaks (C,P coupling via two and H,P coupling via three chemical bonds) and can generally be distinguished from those carbons that couple to phosphorus via more than two chemical bonds, where only the C,P coupling leads to a detectable cross-peak splitting.

Since it can be difficult to unambiguously identify these patterns in more crowded ^{1}H,^{13}C-HSQC spectra, a second spectrum can provide additional information. In this triple-resonance experiment (^{31}P-edited ^{1}H,^{13}C-HSQC), the carbon-phosphorus coupling is used to identify carbons carrying a phosphate group and to combine this information with the superior signal dispersion of the ^{1}H,^{13}C-HSQC. The resulting spectrum, thus, corresponds to a ^{1}H,^{13}C-HSQC but contains only resonances of carbons that exhibit a detectable coupling to ^{31}P. For example, the ^{31}P-edited ^{1}H,^{13}C-HSQC spectrum of G3P contains all peaks that show cross-peak splitting due to C,P coupling in the high resolution ^{1}H,^{13}C-HSQC (Fig. 3.5B).

3.3.5 *Identification of Phosphorylated Carbons in Starch and Glycogen*

The determination of phosphate positions in starch and glycogen requires the preparation of phosphoglucans to enrich the lowly abundant phosphate (see 3.2.5 this monograph). The derived phosphoglucan sample is complex in several aspects: (1) varying degree of polymerisation of the glucan chain, (2) phosphate esters can be attached at different glucosyl carbons within one glucosyl residue and/or, (3) at different glucosyl residues within a phosphoglucan chain. Due to the complexity of the sample, complete assignment of all resonances to specific carbon positions in the oligosaccharides is impossible and the low dispersion of the ^{1}H and ^{31}P resonances (signal overlap) prevents an unambiguous identification of phosphorylation sites in one- or two-dimensional approaches that include only ^{1}H and/or ^{31}P resonances. Therefore, two-dimensional high-resolution ^{1}H,^{13}C correlation spectra with superior signal dispersion have to be obtained.

Assignment can initially be based on ^{1}H,^{1}H-DQF-COSY spectra to obtain information about the glucosyl carbon positions 1, 2, and 3, and on multiplicity editing in ^{1}H,^{13}C correlation spectra (see above) to identify the carbons in position 6 (Fig. 3.6A-B).

The attachment of phosphate to carbon atoms is initially detected by ^{1}H,^{31}P-HMBC spectra and confirmed by ^{31}P-edited ^{1}H,^{13}C-HSQC spectra. While only the protons directly attached to the carbon carrying the phosphate group are visible in the ^{1}H,^{31}P-HMBC (Fig. 3.6C), correlations

---

...Figure 3.3 contd.

shift. To compensate for the low signal dispersion between 3.35 and 3.55 ppm, a range has been cut out and shown in greater detail (as indicated). **(B)** Representation of glucose (both anomers), maltose, and isomaltose in chair conformations. For the latter two, the glucosyl residues were indexed, and glucosyl ring I contains the reducing end, which can occur as α- or β-anomer indicated by the dashed bonds at the anomeric carbon. **(C–E)** ^{1}H,^{13}C-HMQC spectra were obtained for glucose, maltose, and isomaltose. All chemical shifts are given in ppm. Signals reflect connectivity of carbon with hydrogen via one chemical bond. H,C correlations are labeled according to their specific glucosyl carbon positions. While in the spectrum of glucose only the two anomers α and β can be distinguished, signals in the spectra from maltose and isomaltose additionally derive from two glucosyl residues (I and II denoted as in (B)), one of which (I) can occur as α- or β-anomer. Thus, the label of signals includes 'I' or 'II' to mark the glucosyl ring the carbon is attached to. Often signals of H,C correlations in glucosyl ring II show the same position irrespective of the anomeric state of glucosyl ring I in the same molecule. As an example: II-α/β2 denotes the position of two overlapping C,H correlation signals, one representing C2H2 on ring II while ring I occurs as α-anomer, the second representing C2H2 on ring II while ring I occurs as β-anomer. The assignment of hydrogen and carbon positions has been achieved by additional spectra, such as ^{1}H,^{1}H-COSY (as shown in (A) for glucose) and ^{1}H,^{13}C-HMQC-COSY.

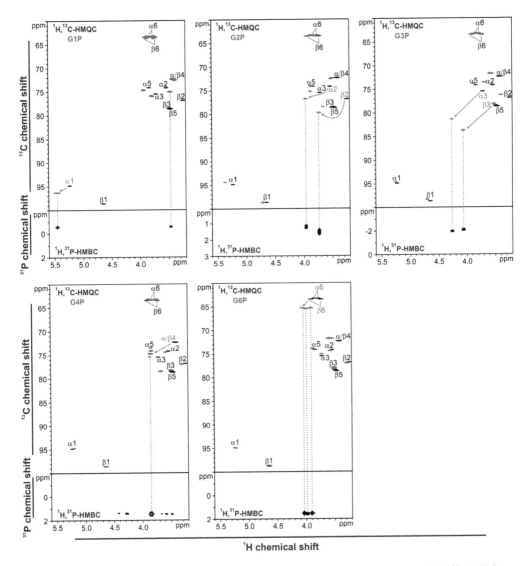

Figure 3.4. 2D NMR spectra of glucosyl monophosphates. ^{1}H,^{13}C-HMQC spectra are aligned with ^{1}H,^{31}P-HMBC spectra of G1P, G2P, G3P, G4P, and G6P (as indicated). Gray signals in the ^{1}H,^{13}C-HMQC spectra represent H,C correlations of the respective glucosyl monophosphate. Black signals derive from ^{1}H,^{13}C-HMQC of glucose (assignment of signals as in Fig. 3.3C). H,C signals of monoglucosyl phosphates were assigned utilizing additional spectra, such as ^{1}H,^{1}H-COSY (as shown in Fig. 3.3C for glucose) and ^{1}H,^{13}C-HMQC-COSY. H,C signals exhibit offset to higher ppm (in both dimensions) from the respective signal of glucose when phosphate is attached (offset is indicated by gray arrows). Signals with most offset in ^{1}H,^{13}C-HMQC spectra align with H,P signals in the ^{1}H,^{31}P-HMBC, verifying direct phosphate connection. Usually, the ^{1}H,^{31}P-HMBC only shows signals when hydrogen and phosphate are bound to the same carbon ($^3J_{H,P}$ coupling) because signal intensity depends on the coupling constant, and coupling via 4 chemical bonds ($^4J_{H,P}$) is usually too small to cause signals. However, in the case of G1P the coupling of ^{31}P at C1 with H2 ($^4J_{H,P}$) is relatively strong, causing an additional signal.

from these carbons as well as potentially their neighbors are visible in the ^{31}P-edited-^{1}H,^{13}C-HSQC (Fig. 3.6D). Combining the assignment information of these spectra leads to the identification of the phosphorylated positions. The ^{1}H,^{31}P-HMBC spectra (Fig. 3.6C) reveal that in phosphoglucan preparations of glycogen, one phosphate couples with a hydrogen that has been assigned to carbon position 3 using the information of the DQF-COSY spectra (Fig. 3.6A). They also occur in the ^{31}P-edited ^{1}H,^{13}C-HSQC spectra of the preparation verifying their proximity to phosphate. Likewise, both the ^{31}P-edited ^{1}H,^{13}C-HSQC and the ^{1}H,^{31}P-HMBC indicate that phosphate is bound to some of

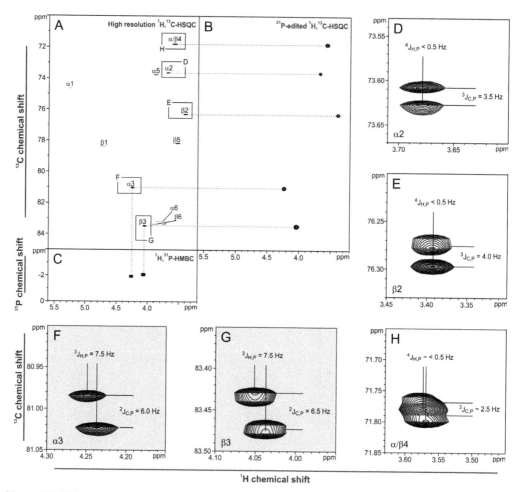

Figure 3.5. High resolution ^{1}H,^{13}C-HSQC spectrum **(A)**, ^{31}P-edited ^{1}H,^{13}C-HSQC spectrum **(B)**, and ^{1}H,^{31}P-HMBC spectrum **(C)** of authentic G3P. H,C signals at glucosyl carbon position C3 in (A) align with H,P signals in (C), indicating the attachment of phosphate at carbon C3. Due to high resolution data sampling, signals of H1,C1 and H6,C6 correlations (grey) were outside the sampling range and are folded into the spectrum with opposite sign. Signals in (B) indicate H,C correlations where additional C,P coupling is detected. The signals in (A) that show C,P coupling in (B) are marked with the letters D through H, processed with high resolution, and shown in greater detail in the panels **(D)–(H)**. The H,C signals exhibit an E.COSY-type pattern, which results from the fact that ^{31}P does not change its spin state during the experiment and thus only one of the two possible signals occurs: cross-peak signals are thus shifted in the ^{13}C dimension by half the C,P coupling and in the ^{1}H dimension by half the H,P coupling. Coupling constants J are given in Hz for C,P coupling via two ($^{2}J_{C,P}$) or three ($^{3}J_{C,P}$) chemical bonds and for H,P coupling via three ($^{3}J_{H,P}$) and four ($^{4}J_{H,P}$) chemical bonds. $^{4}J_{H,P}$ as in the case of H3,C3 signals of G3P (F and G), are always significantly smaller than $^{3}J_{H,P}$. Thus, connection of phosphate and hydrogen to the same carbon can be proven by the inspection of the H,P coupling induced tilt of E.COSY-type C,H signals.

the methylene groups whose signal appear offset from those H6,C6 correlations without phosphate association. However, another H,P correlation exclusively appears in the glycogen sample which has been assigned to carbon position 2 using the information of the DQF-COSY spectra (Fig. 3.6A).

Signals appearing in the ^{31}P-edited ^{1}H,^{13}C-HSQC indicate C,H correlations where the carbon either directly carries a phosphate group or is the neighbor of an adjacent phosphorylated carbon. Combining the assignment information of ^{1}H,^{31}P-HMBC and ^{31}P-edited ^{1}H,^{13}C-HSQC spectra already indicates which signals arise from direct connectivity of phosphate and carbon. This can be verified by inspecting the pattern of peaks from a high-resolution ^{1}H,^{13}C-HSQC spectrum (Fig. 3.6E-J). Those carbon positions that directly carry the phosphate group exhibit tilted cross-peaks, while the correlations of carbons adjacent to phosphorylation sites show two peaks almost aligned in the

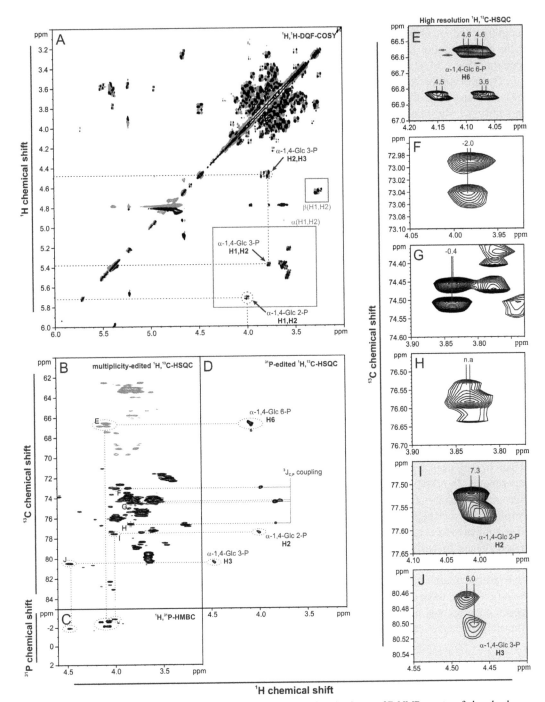

Figure 3.6. NMR analyses of phosphate esters in glycogen-derived phosphoglucans. 2D NMR spectra of phosphoglucans derived from mouse liver glycogen with higher levels of glycogen phosphate due to laforin-deficiency. **(A)** ¹H,¹H-DQF-COSY with cross-peaks indicating correlations of vicinal hydrogens on glucosyl residues existing either as α- or β-anomer. Between 5.00 and 5.80 ppm (y-axis), αH1 couple with αH2 occurring between 3.50 and 4.10 ppm (x-axis). βH1 at 4.65 ppm (y-axis) couple with βH2 at 3.28 ppm (x-axis). Circles highlight the hydrogen correlations that are strongly shifted toward higher ppm as compared to their counterparts in glucose, maltose, and isomaltose (see Fig. 3.3). The offset resonance of αH2 residing at 4.01 ppm is exactly that of αH2 in αG2P (see Fig. 3.4), indicating that the offset H1,H2 signal is caused by α-1,4-glucosyl 2-phosphate residues. That of αH3 occurs at 4.47 ppm, resembling the position of αH3 in α-G3P

Figure 3.6 contd. ...

[1]H direction, as the H,P coupling via four chemical bonds is usually quite small and only the C,P coupling via three chemical bonds leads to a detectable splitting. Observation of tilted cross-peaks in the high resolution ^{1}H,^{13}C-HSQC is consistent with the occurrence of peaks in the ^{1}H,^{31}P-HMBC spectrum (Fig. 3.6C).

The presence of C6 phosphorylation in glycogen and starch can be conclusively demonstrated by the occurrence of phosphorylated methylene groups. *Curcuma* starch, as well as glycogen, contains an additional phosphate at position C3. For glycogen, a third phosphorylation site exists at C2. However, monophosphate at C2 is below the limit of detection in phosphoglucans from *Curcuma* starch (Nitschke et al. 2013). The previously published NMR data reveals no evidence for the occurrence of phosphodiesters in phosphoglucan preparation of starch and glycogen. No phosphate signals were observed that correlate with two hydrogens at different glucosyl carbon positions (DePaoli-Roach et al. 2015, Nitschke et al. 2013, Tagliabracci et al. 2011).

3.3.6 Limitations of NMR Spectroscopy

NMR relies on the detection of minimal changes in the magnetic field as induced by the excitation of specific isotopes in a sample. By irradiation, small population differences of magnetic moments in α- and in β-state are equalized and brought to coherence, which results in detected changes of the overall sample magnetization. These changes in sample magnetization depend on the absolute number of the specific isotope in the sample and on the extent of population differences between α- and β-state. The latter is dependent on the isotope species and the external magnetic field strength, which was improved in the last decades by the utilization of stronger magnets. Furthermore, sensitivity was increased by the use of cryogenic NMR probes, where the transmit/receive coils as well as the tuning and matching circuits are maintained at a very low temperature in order to reduce the noise contributions resulting from the random thermal motion of electrons in the conductors.

^{31}P is an NMR detectable isotope with 100% natural abundance. However, the abundance of phosphate in polyglucans is very low. Therefore, comparably large amounts of polyglucans are required to enable the detection of the ^{31}P nuclei. When Lim and Seib (1993) analyzed phosphoesters in starch, they used 1 g of starch. Ritte et al. (2006) consumed 225 mg of starch for similar NMR experiments. In NMR studies using glycogen, 150 to 350 mg was used depending on the level of glycogen phosphorylation (DePaoli-Roach et al. 2015, Nitschke et al. 2013, Tagliabracci et al. 2011). Despite using such high amounts of source material, spectra of many repetitive measurements had to be accumulated to obtain acceptable signal-to-noise ratios. Thus, analysis of the rare phosphoesters in polyglucans by NMR is highly time and material consuming.

Another problem is the quantification of the amount of specific phosphate esters present. While the relative amount of a species can be obtained from one-dimensional spectra by integration of the peaks (given that there is no overlap with other peaks in the spectrum), the intensity in two-dimensional spectra does not only depend on the relative amount of species present in the sample but also on the efficiency of magnetization transfer during the mixing time, which in turn depends on the size of the

---

...Figure 3.6 contd.

(see Fig. 3.4), indicating that the offset H2,H3 signal is caused by α-1,4-glucosyl 3-phosphate residues. **(B)** multiplicity-edited ^{1}H,^{13}C-HSQC: grey signals indicate methylene groups, black signals point to methine groups. **(C)** ^{1}H,^{31}P-HMBC: signals indicate vicinal correlation of phosphate and hydrogen. **(D)** ^{31}P-edited ^{1}H,^{13}C-HSQC: occurring signals indicate either $^2J_{C,P}$ or $^3J_{C,P}$ coupling. The former proves connection of phosphate and hydrogen to the same carbon (dotted circles), while the latter occurs when hydrogen and phosphate are bound to adjacent carbons. The assignment to either of the two possibilities was obtained by alignment with signals occurring in the respective ^{1}H,^{31}P-HMBC (C) and by high-resolution HSQC (E-J). The assignment of H2 and H3 is based on ^{1}H,^{1}H-DQF-COSY NMR (A). **(E–J)** Labeled signals that occur in the ^{31}P-edited ^{13}C,^{1}H-HSQC are shown in a high-resolution ^{1}H,^{13}C-HSQC spectrum and shown in greater detail. The phosphorus-induced tilt in the hydrogen dimension of each signal is given in Hz. Larger values for $^3J_{H,P}$ coupling indicate that phosphate and hydrogen are connected to the same carbon. The assignment of H6 is done with respect to the multiplicity-edited ^{13}C,^{1}H-HSQC (B) showing a methylene group at that position. H2 and H3 are assigned by ^{1}H,^{1}H-DQF-COSY experiments (A).

coupling constants and the width of the lines which are all difficult to obtain. Thus, the amount of phosphate esters cannot be quantified using the methods described here.

Acknowledgement

The authors thank Dr. Martin Steup for helpful discussion and suggestions that improved this chapter.

References

Antonio, C., T. Larson, A. Gilday, I. Graham, E. Bergström and J. Thomas-Oates. 2008. Hydrophilic interaction chromatography/electrospray mass spectrometry analysis of carbohydrate-related metabolites from Arabidopsis thaliana leaf tissue. Rapid Commun. Mass Spectrom. 22: 1399–1407.

Bah, A. and J.D. Forman-Kay. 2016. Modulation of intrinsically disordered protein function by post-translational modifications. J. Biol. Chem. 291: 6696–6705.

Baunsgaard, L., H. Lutken, R. Mikkelsen, M.A. Glaring, T.T. Pham and A. Blennow. 2005. A novel isoform of glucan, water dikinase phosphorylates pre-phosphorylated alpha-glucans and is involved in starch degradation in Arabidopsis. Plant J. 41: 595–605.

Blennow, A., A.M. Bay-Smidt, C.E. Olsen and B.L. Moller. 2000a. The distribution of covalently bound phosphate in the starch granule in relation to starch crystallinity. Int. J. Biol. Macromol. 27: 211–218.

Blennow, A., S.B. Engelsen, L. Munck and B.L. Moller. 2000b. Starch molecular structure and phosphorylation investigated by a combined chromatographic and chemometric approach. Carbohydr. Polym. 41: 163–174.

Blennow, A. and S.B. Engelsen. 2010. Helix-breaking news: fighting crystalline starch energy deposits in the cell. Trends Plant Sci. 15: 236–240.

Carpenter, M., N. Joyce, R. Butler, R. Genet and G. Timmerman-Vaughan. 2012. A mass spectrometric method for quantifying C3 and C6 phosphorylation of starch. Anal. Biochem. 431: 115–119.

Caspar, T., T.P. Lin, G. Kakefuda, L. Benbow, J. Preiss and C. Somerville. 1991. Mutants of Arabidopsis with altered regulation of starch degradation. Plant Physiol. 95: 1181–1188.

Chan, E.M., E.J. Young, L. Ianzano, I. Munteanu, X. Zhao, C.C. Christopoulos et al. 2003. Mutations in NHLRC1 cause progressive myoclonus epilepsy. Nat. Genet. 35: 125–127.

Chikwana, V.M., M. Khanna, S. Baskaran, V.S. Tagliabracci, C.J. Contreras, A. DePaoli-Roach et al. 2013. Structural basis for 2'-phosphate incorporation into glycogen by glycogen synthase. Proc. Natl. Acad. Sci. U.S.A. 110: 20976–20981.

Comparot-Moss, S., O. Kotting, M. Stettler, C. Edner, A. Graf, S.E. Weise et al. 2010. A putative phosphatase, LSF1, is required for normal starch turnover in Arabidopsis leaves. Plant Physiol. 152: 685–697.

Contreras, C.J., D.M. Segvich, K. Mahalingan, V.M. Chikwana, T.L. Kirley, T.D. Hurley et al. 2016. Incorporation of phosphate into glycogen by glycogen synthase. Arch. Biochem. Biophys. 597: 21–29.

DePaoli-Roach, A.A., C.J. Contreras, D.M. Segvich, C. Heiss, M. Ishihara, P. Azadi et al. 2015. Glycogen phosphomonoester distribution in mouse models of the progressive myoclonic epilepsy, Lafora disease. J. Biol. Chem. 290: 841–850.

Edner, C., J. Li, T. Albrecht, S. Mahlow, M. Hejazi, H. Hussain et al. 2007. Glucan, water dikinase activity stimulates breakdown of starch granules by plastidial beta-amylases. Plant Physiol. 145: 17–28.

Edwards, J.R., O. Yarychkivska, M. Boulard and T.H. Bestor. 2017. DNA methylation and DNA methyltransferases. Epigenetics Chromatin 10: 23.

Fiske, C.H. and Y. Subbarow. 1925. The colorimetric determination of phosphorus. J. Biol. Chem. 66: 375–400.

Fontana, J.D. 1980. The presence of phosphate in glycogen. FEBS Lett. 109: 85–92.

Friebolin, H. 1992. Ein- und zweidimensionale NMR-Spektroskopie. VCH Verlagsgesellschaft mbH, Weinheim, Basel, Cambridge, New York.

Gallagher, S.R. and P.R. Desjardins. 2001. Quantitation of DNA and RNA with absorption and fluorescence spectroscopy. Curr. Protoc. Mol. Biol. 76: A.3D.1-A.3D.21.

Gayarre, J., L. Duran-Trio, O. Criado Garcia, L. Aguado, L. Juana-Lopez, I. Crespo et al. 2014. The phosphatase activity of laforin is dispensable to rescue Epm2a-/- mice from Lafora disease. Brain 137: 806–818.

Gentry, M.S., R.H. Dowen, C.A. Worby, S. Mattoo, J.R. Ecker and J.E. Dixon. 2007. The phosphatase laforin crosses evolutionary boundaries and links carbohydrate metabolism to neuronal disease. J. Cell Biol. 178: 477–488.

Gibon, Y., H. Vigeolas, A. Tiessen, P. Geigenberger and M. Stitt. 2002. Sensitive and high throughput metabolite assays for inorganic pyrophosphate, ADPGlc, nucleotide phosphates, and glycolytic intermediates based on a novel enzymic cycling system. Plant J. 30: 221–235.

Gidley, M.J. and P.V. Bulpin. 1987. Crystallisation of malto-oligosaccharides as models of the crystalline forms of starch: minimum chain-length requirement for the formation of double helices. Carbohydr. Res. 161: 291–300.

Guilbault, G.G. and D.N. Kramer. 1965. Fluorometric procedure for measuring the activity of dehydrogenases. Anal. Chem. 37: 1219–1221.

Günther, H. 1992. NMR-Spektroskopie. Grundlagen, Konzepte und Anwendungen der Protonen und Kohlenstoff-13 Kernresonanz-Spektroskopie in der Chemie. Georg Thieme Verlag, Stuttgart, New York.

Haebel, S., M. Hejazi, C. Frohberg, M. Heydenreich and G. Ritte. 2008. Mass spectrometric quantification of the relative amounts of C6 and C3 position phosphorylated glucosyl residues in starch. Anal. Biochem. 379: 73–79.

Hansen, P.I., M. Spraul, P. Dvortsak, F.H. Larsen, A. Blennow, M.S. Motawia et al. 2009. Starch phosphorylation—maltosidic restrains upon 3'- and 6'-phosphorylation investigated by chemical synthesis, molecular dynamics and NMR spectroscopy. Biopolymers 91: 179–193.

Hejazi, M., J. Fettke, S. Haebel, C. Edner, O. Paris, C. Frohberg et al. 2008. Glucan, water dikinase phosphorylates crystalline maltodextrins and thereby initiates solubilization. Plant J. 55: 323–334.

Hejazi, M., J. Fettke, O. Paris and M. Steup. 2009. The two plastidial starch-related dikinases sequentially phosphorylate glucosyl residues at the surface of both the A- and B-allomorph of crystallized maltodextrins but the mode of action differs. Plant Physiol. 150: 962–976.

Hejazi, M., J. Fettke, O. Kotting, S.C. Zeeman and M. Steup. 2010. The laforin-like dual-specificity phosphatase SEX4 from Arabidopsis hydrolyzes both C6- and C3-phosphate esters introduced by starch-related dikinases and thereby affects phase transition of {alpha}-glucans. Plant Physiol. 152: 711–722.

Hejazi, M., S. Mahlow and J. Fettke. 2014. The glucan phosphorylation mediated by alpha-glucan, water dikinase (GWD) is also essential in the light phase for a functional transitory starch turn-over. Plant Signal. Behav. 9.

Hess, H.H. and J.E. Derr. 1975. Assay of inorganic and organic phosphorus in the 0.1–5 nanomole range. Anal. Biochem. 63: 607–613.

Hizukuri, S., S. Tabata, Kagoshima and Z. Nikuni. 1970. Studies on starch phosphate part 1. Estimation of glucose-6-phosphate residues in starch and the presence of other bound phosphate(s). Starch/Staerke 22: 338–343.

Horne, R.N., W.B. Anderson and R.C. Nordlie. 1970. Glucose dehydrogenase activity of yeast glucose 6-phosphate dehydrogenase. Inhibition by adenosine 5'-triphosphate and other nucleoside 5'-triphosphates and diphosphates. Biochemistry 9: 610–616.

Imberty, A., A. Buléon, V. Tran and S. Péerez. 1991. Recent advances in knowledge of starch structure. Starch/Staerke 43: 375–384.

Itaya, K. and M. Ui. 1966. A new micromethod for the colorimetric determination of inorganic phosphate. Clin. Chim. Acta 14: 361–366.

Karro, J.E., M. Peifer, R.C. Hardison, M. Kollmann and H.H. von Grünberg. 2008. Exponential decay of GC content detected by strand-symmetric substitution rates influences the evolution of isochore structure. Mol. Biol. Evol. 25: 362–374.

Khorana, H.G., G.M. Tener, R.S. Wright and J.G. Moffatt. 1957. Cyclic phosphates. III. Some general observations on the formation and properties of five-, six- and seven-membered cyclic phosphate esters. J. Am. Chem. Soc. 79: 430–436.

Khoury, G.A., R.C. Baliban and C.A. Floudas. 2011. Proteome-wide post-translational modification statistics: frequency analysis and curation of the swiss-prot database. Sci. Rep. 1: 90.

Kirkman, B.R. and W.J. Whelan. 1986. Glucosamine is a normal component of liver glycogen. FEBS Lett. 194: 6–11.

Kötting, O., K. Pusch, A. Tiessen, P. Geigenberger, M. Steup and G. Ritte. 2005. Identification of a novel enzyme required for starch metabolism in Arabidopsis leaves. The phosphoglucan, water dikinase. Plant Physiol. 137: 242–252.

Kötting, O., D. Santelia, C. Edner, S. Eicke, T. Marthaler, M.S. Gentry et al. 2009. Starch-Excess4 is a laforin-like Phosphoglucan phosphatase required for starch degradation in Arabidopsis thaliana. Plant Cell 21: 334–346.

Lim, S. and P.A. Seib. 1993. Location of phosphate esters in a wheat starch phosphate by ^{31}P-nuclear magnetic resonance spectroscopy. Cereal Chem. 70: 145–152.

Lomako, J., W.M. Lomako, W.J. Whelan and R.B. Marchase. 1993. Glycogen contains phosphodiester groups that can be introduced by UDPglucose: glycogen glucose 1-phosphotransferase. FEBS Lett. 329: 263–267.

Lowry, O.H. and J.V. Passonneau (eds.). 1972. A flexible system of enzymatic analysis. Academic Press, New York.

Mahlow, S., M. Hejazi, F. Kuhnert, A. Garz, H. Brust, O. Baumann et al. 2014. Phosphorylation of transitory starch by alpha-glucan, water dikinase during starch turnover affects the surface properties and morphology of starch granules. New Phytologist 203: 495–507.

Manners, D.J. 1991. Recent developments in our understanding of glycogen structure. Carbohydr. Polym. 16: 37–82.

Martin, G.E. and A.S. Zektzer. 1988. Two-dimensional NMR methods for establishing molecular connectivity. A chemist's guide to experiment selection, performance, and interpretation. VCH Publishers, Inc., New York.

Meekins, D.A., M. Raththagala, S. Husodo, C.J. White, H.-F. Guo, O. Koetting et al. 2014. Phosphoglucan-bound structure of starch phosphatase Starch Excess4 reveals the mechanism for C6 specificity. Proc. Natl. Acad. Sci. U. S. A. 111: 7272–7277.

Meekins, D.A., C.W. Vander Kooi and M.S. Gentry. 2016. Structural mechanisms of plant glucan phosphatases in starch metabolism. FEBS J. 283: 2427–2447.

Minassian, B.A., J.R. Lee, J.A. Herbrick, J. Huizenga, S. Soder, A.J. Mungall et al. 1998. Mutations in a gene encoding a novel protein tyrosine phosphatase cause progressive myoclonus epilepsy. Nat. Genet. 20: 171–174.

Morrison, W.R. 1964. A fast, simple and reliable method for the microdetermination of phosphorus in biological materials. Anal. Biochem. 7: 218–224.

Nelson, T.J. and S. Kaufman. 1987. Two enzymatic methods for determination of the phosphate content of phosphoproteins. Anal. Biochem. 161: 352–357.

Nielsen, T.H., B. Wischmann, K. Enevoldsen and B.L. Moller. 1994. Starch phosphorylation in potato tubers proceeds concurrently with de novo biosynthesis of starch. Plant Physiol. 105: 111–117.

Nisselbaum, J.S. and S. Green. 1969. A simple ultramicro method for determination of pyridine nucleotides in tissues. Anal. Biochem. 27: 212–217.

Nitschke, F. 2013. Phosphorylation of polyglucans, especially glycogen and starch. Ph. D. Thesis, University of Potsdam, Potsdam, Germany.

Nitschke, F., P. Wang, P. Schmieder, J.-M. Girard, Donald E. Awrey, T. Wang et al. 2013. Hyperphosphorylation of glucosyl C6 carbons and altered structure of glycogen in the neurodegenerative epilepsy Lafora disease. Cell. Metab. 17: 756–767.

Nitschke, F., M.A. Sullivan, P. Wang, X. Zhao, E.E. Chown, A.M. Perri et al. 2017. Abnormal glycogen chain length pattern, not hyperphosphorylation, is critical in Lafora disease. EMBO Mol. Med. 9: 906–917.

Pomin, V.H. 2012. Unravelling glycobiology by NMR spectroscopy. *In*: Petrescu S. (ed.). Glycosylation. InTech, Available from: http://www. intechopen. com/books/glycosylation/unravelling-glycobiology-by-nmr-spectroscopy.

Ritte, G., J.R. Lloyd, N. Eckermann, A. Rottmann, J. Kossmann and M. Steup. 2002. The starch-related R1 protein is an alpha -glucan, water dikinase. Proc. Natl. Acad. Sci. U. S. A. 99: 7166–7171.

Ritte, G., M. Steup, J. Kossmann and J.R. Lloyd. 2003. Determination of the starch-phosphorylating enzyme activity in plant extracts. Planta 216: 798–801.

Ritte, G., A. Scharf, N. Eckermann, S. Haebel and M. Steup. 2004. Phosphorylation of transitory starch is increased during degradation. Plant Physiol. 135: 2068–2077.

Ritte, G., M. Heydenreich, S. Mahlow, S. Haebel, O. Kotting and M. Steup. 2006. Phosphorylation of C6- and C3-positions of glucosyl residues in starch is catalysed by distinct dikinases. FEBS Lett. 580: 4872–4876.

Roach, P.J. 2015. Glycogen phosphorylation and Lafora disease. Mol. Aspects Med. 46: 78–84.

Romero, P.A., E.E. Smith and W.J. Whelan. 1980. Glucosamine as a substitute for glucose in glycogen metabolism. Biochem. Int. 1: 1–9.

Santelia, D., O. Kötting, D. Seung, M. Schubert, M. Thalmann, S. Bischof et al. 2011. The phosphoglucan phosphatase like sex Four2 dephosphorylates starch at the C3-position in Arabidopsis. Plant Cell 23: 4096–4111.

Schmieder, P., F. Nitschke, M. Steup, K. Mallow and E. Specker. 2013. Determination of glucan phosphorylation using heteronuclear 1H,13C double and 1H,13C,31P triple-resonance NMR spectra. Magn. Reson. Chem. 51: 655–661.

Schoch, T.J. 1942. Fractionation of starch by selective precipitation with butanol. J. Am. Chem. Soc. 64: 2957–2961.

Schreier, T.B., M. Umhang, S.-K. Lee, W.-L. Lue, Z. Shen, D. Silver et al. 2019. LIKE SEX4 1 acts as a β-amylase-binding scaffold at the starch granule during starch degradation. The Plant Cell. In press.

Serratosa, J.M., P. Gomez-Garre, M.E. Gallardo, B. Anta, D.B. de Bernabe, D. Lindhout et al. 1999. A novel protein tyrosine phosphatase gene is mutated in progressive myoclonus epilepsy of the Lafora type (EPM2). Hum. Mol. Genet. 8: 345–352.

Sullivan, M., S. Nitschke, M. Steup, B. Minassian and F. Nitschke. 2017. Pathogenesis of Lafora disease: transition of soluble glycogen to insoluble polyglucosan. Int. J. Mol. Sci. 18: 1743.

Sullivan, M.A., S. Nitschke, E.P. Skwara, P. Wang, X. Zhao, X.S. Pan et al. 2019. Skeletal muscle glycogen chain length continuum correlates with insolubility in polyglucosan body associated neurodegenerative disease. Cell Rep. (in press).

Tabata, S. and S. Hizukuri. 1971. Studies on starch phosphate. Part 2. Isolation of glucose 3-phosphate and maltose phosphate by acid hydrolysis of potato starch. Starch/Staerke 23: 267–272.

Tabata, S., K. Nagata and S. Hizukuri. 1975. Studies on starch phosphates. Part 3. On the esterified phosphates in some cereal starches. Starch/Staerke 27: 333–335.

Tagliabracci, V.S., J. Turnbull, W. Wang, J.M. Girard, X. Zhao, A.V. Skurat et al. 2007. Laforin is a glycogen phosphatase, deficiency of which leads to elevated phosphorylation of glycogen *in vivo*. Proc. Natl. Acad. Sci. U. S. A. 104: 19262–19266.

Tagliabracci, V.S., J.M. Girard, D. Segvich, C. Meyer, J. Turnbull, X. Zhao et al. 2008. Abnormal metabolism of glycogen phosphate as a cause for Lafora disease. J. Biol. Chem. 283: 33816–33825.

Tagliabracci, V.S., C. Heiss, C. Karthik, C.J. Contreras, J. Glushka, M. Ishihara et al. 2011. Phosphate incorporation during glycogen synthesis and Lafora disease. Cell. Metab. 13: 274–282.

Takeda, Y. and S. Hizukuri. 1982. Location of phosphate groups in potato amylopectin. Carbohydr. Res. 102: 321–327.

Tarentino, A.L. and F. Maley. 1976. Direct evidence that D-galactosamine incorporation into glycogen occurs via UDP-glucosamine. FEBS Lett. 69: 175–178.

Tharanathan, R.N. 2005. Starch—Value addition by modification. Crit. Rev. Food Sci. Nutr. 45: 371–384.

Tiberia, E., J. Turnbull, T. Wang, A. Ruggieri, X.C. Zhao, N. Pencea et al. 2012. Increased laforin and laforin binding to glycogen underlie lafora body formation in malin-deficient Lafora disease. J. Biol. Chem. 287: 25650–25659.

Turnbull, J., P. Wang, J.M. Girard, A. Ruggieri, T.J. Wang, A.G. Draginov et al. 2010. Glycogen hyperphosphorylation underlies lafora body formation. Ann. Neurol. 68: 925–933.

Verbeke, J., C. Penverne, C. D'Hulst, C. Rolando and N. Szydlowski. 2016. Rapid and sensitive quantification of C3-and C6-phosphoesters in starch by fluorescence-assisted capillary electrophoresis. Carbohydr. Polym. 152: 784–791.

Wikman, J., F.H. Larsen, M.S. Motawia, A. Blennow and E. Bertoft. 2011. Phosphate esters in amylopectin clusters of potato tuber starch. Int. J. Biol. Macromol. 48: 639–649.

Worby, C.A., M.S. Gentry and J.E. Dixon. 2006. Laforin, a dual specificity phosphatase that dephosphorylates complex carbohydrates. J. Biol. Chem. 281: 30412–30418.

Young, L.E.A., C.O. Brizzee, J.K.A. Macedo, R.D. Murphy, C.J. Contreras, A.A. DePaoli-Roach et al. 2020. Accurate and sensitive quantitation of glucose and glucose phosphates derived from storage carbohydrates by mass spectrometry. Carbohydrate Polymers 230: 115651.

Yu, T.S., H. Kofler, R.E. Hausler, D. Hille, U.I. Flugge, S.C. Zeeman et al. 2001. The Arabidopsis sex1 mutant is defective in the R1 protein, a general regulator of starch degradation in plants, and not in the chloroplast hexose transporter. Plant Cell 13: 1907–1918.

Zhu, A., R. Romero and H.R. Petty. 2009. An enzymatic fluorimetric assay for glucose-6-phosphate: application in an *in vitro* Warburg-like effect. Anal. Biochem. 388: 97–101.

CHAPTER 4

Storage Polysaccharide Metabolism in Micro-Organisms

Christophe Colleoni

4.1 Introduction

The metabolic regulation of energy-storage compounds is a critical issue for any living organism. The balance between catabolism and anabolism pathways enables both prokaryotic and eukaryotic cells to sustain their energy homeostasis. A good indicator of energy state within the cell consists of the measurement of adenylate energy charge (AEC)[1]. This AEC concept, proposed by Atkinson, reflects the physiological state of cells by inferring the cellular amounts of ATP, ADP and AMP (Atkinson 1968). Thus, at the exponential phase of the culture, the AEC is approximately 0.8 and reaches slowly 0.6 upon nearing the stationary phase. When, however, the AEC falls to 0.5, cells die rapidly (Chapman et al. 1971).

Homopolymers of D-glucose units, lipids, poly-β-hydroxybutyrate (PHB)/polyhydroxyalakanoate (PHA) and polyphosphate represent the major energy-carbon or energy storage compounds found in prokaryotes. They fulfill the criteria for energy-carbon function because (i) they are polymers with a high molecular weight, reducing the internal osmotic pressure within cells, (ii) they are synthesized when there is an excess of energy, (iii) they are catabolized for the maintenance of the cells in times of starvation. For this reason, the accumulation of disaccharides in prokaryotes (e.g., trehalose, sucrose) may not be regarded as energy-carbon storage compounds but rather as transitory carbohydrate compounds formed in response to thermal or salinity stress.

In this book chapter, we will restrict our attention to polymers of D-glucose units linked through two types of α-glycosidic linkages, α-1,4 and α-1,6, which belong to α-glucan as do glycogen, starch and granulose.[2] We will exclude other polysugars (such as dextrans, fructans), PHB/PHA and polyphosphate. Nevertheless, energy-storage functions of PHB/PHA and polyphosphate have been discussed in recent reviews (Jendrossek and Pfeiffer 2014, Albi and Serrano 2016).

α-Glucan polysaccharides, such as glycogen, starch, constitute the most widely distributed form of carbon storage across the living world. They are composed of α-1,4 residues and branched in α-1,6. In contrast to glycogen, starch granules are confined to some photosynthetic eukaryotes and small group of unicellular cyanobacteria capable of performing nitrogen fixation in aerobic conditions. Within eukaryotic cells, the appearance of starch granules coincides with the acquisition of the

University of Lille, CNRS, UMR8576-UGSF-Unité de Glycobiologie Structurale et Fonctionnelle, F-59000 Lille, France.
  Email: Christophe.colleoni@univ-lille.fr
[1] Adenylate Energy Charge is defined as (ATP + ½ ADP)/(ATP+ADP+AMP) and varies from 0 to 1.
[2] Storage polysaccharide produced by Clostridium species (see 4.5.1).

photosynthetic apparatus about 1.5 billion years ago. This unique event called primary endosymbiosis occurred when a phagotrophic heterotroph eukaryote engulfed a cyanobacterium (Yoon et al. 2004, Eme et al. 2014, Ponce-Toledo et al. 2017) and established a symbiosis based on photosynthetic carbon export from the cyanobacterial symbiont to the host cytosol. Over time, cyanobacterium has evolved into a new organelle, plastid (e.g., chloroplast), that enables the eukaryotic cell to conduct oxygenic photosynthesis. It is widely accepted today that plants and green algae, red algae and Glaucophyta (freshwater algae) derive from this common ancestor. In the early 90s, several laboratories suggested that starch crystallization was allowed by the presence of an isoamylase type debranching enzyme removing α-1,4 chains that otherwise prevent double helices formation (James et al. 1995, Mouille et al. 1996, Kubo et al. 1999). More recently, the characterization of mutants in starch accumulating *Cyanobacterium* sp. CLg1 and phylogeny analyses outline convergent evolution of starch crystallization mechanism. At first glance, it was tempting to postulate that direct[3] debranching enzymes in the plant had been inherited from the closest ancestor of starch-accumulating *Cyanobacterium* sp. CLg1 during primary endosymbiosis. Nevertheless, the phylogeny analysis reveals that this critical activity was in fact inherited through lateral gene transfer (LGT) from obligatory intracellular bacteria, chlamydiales (Cenci et al. 2013). This finding contributes to "ménage à trois" hypothesis, which seeks to clarify the role of chlamydiales in plastid endosymbiosis. In addition, it should be noted that in some non-photosynthetic eukaryotes derived from a secondary endosymbiosis event consisting of the engulfment of red-alga by another phagotrophic eukaryotic lineage followed by photosynthesis loss, carbon storage often occurs as starch granules. Unlike Archaeplastida lineage, the starch metabolism pathways involved in these cases are a different network of starch metabolizing enzymes, reflecting their complex evolutionary origin. For details on the evolution of starch metabolism in Archaeplastida and non-photosynthetic derived red-alga organisms, we invite the reader to a recent review (Ball et al. 2015).

At variance with starch, glycogen particles are ubiquitous in the three domains of life. They form hydrosoluble particles in the cytosols of Archaebacteria, Prokaryota and Eukaryota phyla. Until recently, our understanding of glycogen metabolism pathways essentially relied on the studies in *Escherichia coli* and yeast *Saccharomyces cerevisiae*, which are the usual representatives of prokaryotes and heterotroph eukaryotes, respectively (for review see Wilson et al. 2010). At present, the high-throughput genome sequencing provides a unique opportunity to compare genomes and biochemical pathways. Thus, the plethora of bacterial genomes outlines the diversity of glycogen pathways as well as the fact that some bacterial species have lost the ability to synthesize carbohydrate storage polysaccharides. The aim of this chapter is to give an overview of alternative glycogen metabolism pathways in prokaryotes, and how storage polysaccharide affects cell physiology.

4.2 Glycogen and Starch: So Similar, yet so Different

Both glycogen and starch are made up of glucan chains consisting of glucose residues linked in α-1,4 position and those glucan chains are hooked together by α-1,6 linkages. Although they share the same primary structure, their physicochemical properties are very distinct. Glycogen is localized into the cytosol of prokaryotes and eukaryotic organisms as tiny hydrosoluble particles of 40–60 nm. Interestingly, mathematical modeling predicts a size of 42 nm in diameter and justifies this self-limitation due to homogenous distribution of branching points. This particular organization leads quickly to an exponential increase in the number of non-reducing ends at the surface of polysaccharide (Fig. 4.1A) and then will limit the access to the catalytic sites of biosynthetic enzymes and *per se* the size of glycogen particles (Meléndez-Hevia et al. 1993, Meléndez et al. 1999). Interestingly, further

---

[3] The term "direct" refers to the ability of debranching enzymes of plants or prokaryotes to cleave directly the α-1,6 linkages of α-glucan polysaccharides. On the other hand, the term "indirect" refers to debranching enzymes of yeast or animal cells which involve the coordinated action between an α-glucanotransferase and amylo-1,6 glucosidase domains to catabolize the α-1,6 linkages

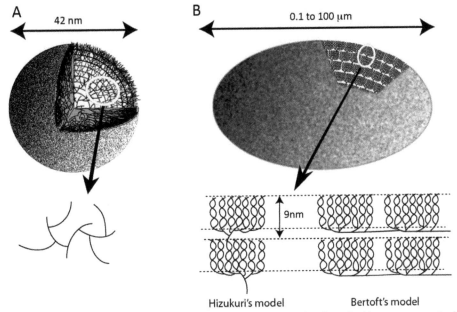

Figure 4.1. Structural comparison between glycogen and amylopectin. Both polysaccharides are composed of glucan chains made of α-1,4 glucose residues and branched in α-1,6 position. **(A)** The uniform distribution of branches leads to an exponential increase that self-limit glycogen particle size to 42 nm. Mathematical modeling suggests that glycogen particles are fractal objects made of repeating patterns of glucan chains (black lines) harboring two glucan chains (intersection lines). **(B)** The size of semi-crystalline starch granule is variable depending on the source. Clusters of amylopectin are generated through the asymmetric distribution of branches, localized in the amorphous lamellae, while the intertwined glucan chains define crystalline lamellae. The sum of one amorphous and crystalline lamella is constant to 9 nm, independent of amylopectin clusters examined. In the model proposed by Hizukuri (1986), a long glucan chain interconnects two clusters, whereas in the alternate model of Bertoft (2010), the clusters are anchored to a backbone consisting of a long glucan chain.

mathematical modeling infers that glycogen particles can be considered as fractal objects, made by iteration of a single motif: one glucan chain supporting two branches. Thanks to this particular glucan chain organization, a single catabolic enzyme can release 19 000 glucose molecules in 20 seconds (Meléndez et al. 1997). Hence, glycogen particles are dynamic polysaccharides, which can meet the immediate need of eukaryotic and prokaryotic cells and can be used as an intermediate buffer or osmotically inert sink for carbon.

On the contrary, starch granules harbor a much more complex organization (Fig. 4.1B). This semi-crystalline non-aqueous particle is usually composed of two types of polysaccharides, amylopectin and amylose. The former, the major fraction, is a branched polysaccharide containing approximately 5% of the total interglucose linkages as branching points while amylose, a dispensable fraction of starch, is a complex mixture of strictly linear and slightly branched polyglucans. First proposed by Hizukuri (1986) and then updated by Bertoft (2004; see also chapter 8 of this volume), these models propose a specific localization of branching points enabling the formation of double helices of glucan. Thus, this particular "cluster" organization allows storing glucose residues more efficiently without the size limitations imposed on glycogen particles. However, the downside is that synthesis and degradation of starch are more complex and less dynamic.

4.2.1 Glycogen/Starch Biosynthesis Pathway in Prokaryotes

In the last decades, our understanding of glycogen biosynthetic pathway in prokaryotes was largely focused on functional approaches using *Escherichia coli* as a model. Currently, most of the activities involving carbohydrate metabolism pathways and their regulation are well understood (for reviews see Wilson et al. 2010, Preiss 2014). In most prokaryotes, classical glycogen biosynthetic pathway

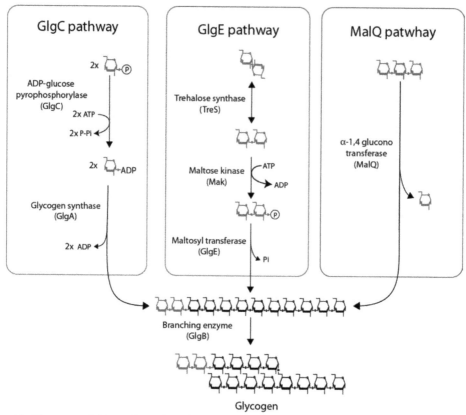

Figure 4.2. Glycogen anabolism pathways in prokaryotes. The most predominant glycogen biosynthesis pathway, GlgC-pathway, involves an ADP-glucose pyrophosphorylase (GlgC) and glycogen synthase (GlgA) for synthesizing the glucan chains. The former synthesizes the nucleotide-sugar, ADP-glucose, from ATP and G-1-P. Glycogen synthase then transfers the glucose moiety of ADP-glucose onto the non-reducing end of growing glucan chain. In the GlgE pathway, non-reducing disaccharide, trehalose, is converted reversibly into maltose by trehalose synthase (TreS) followed by a phosphorylation onto the reducing-end of maltose. The latter reaction processed by maltose kinase (Mak) is found in some bacterial species fused with TreS activity. Maltose-1-phosphate (M1P) is incorporated onto the non-reducing end of glucan chain, thanks to maltosyl transferase (GlgE). MalQ pathway, described in few species, relies on pre-existing short glucan chains. α-1,4 glucanotransferase (MalQ) activity disproportionates short glucan chains. Then, in both GlgC- GlgE- pathways as well as malQ-pathway, an a-1,6 linkage is produced when the neo-glucan chain fits the catalytic site of branching enzyme (GlgB). The incorporation of two glucose moieties (in gray) requires 2, 1 and 0 ATP for GlgC-, GlgE and MalQ pathways, respectively.

(GlgC) is based on the use of nucleotide-sugars as activated substrates for the polymerization. Nevertheless, both high-throughput genomes sequencing and comparative genome analyses have evidenced two alternative glycogen biosynthetic pathways in bacteria known as the GlgE-pathway and the MalQ-pathway, respectively (Fig. 4.2).

4.2.1.1 Classical Glycogen Biosynthetic Pathway in Bacteria: : The GlgC-Path

In the classical path, the biosynthesis of glycogen contains a set of three enzymatic reactions: (i) the nucleotide-sugar synthesis mediated by ADP-glucose pyrophosphorylase (GlgC), (ii) the formation of α-1,4 linkage catalyzed by glycogen synthase (GlgA), and (iii) the formation of α-1,6 linkages mediated by glycogen branching enzyme (GlgB). Glycogen metabolizing genes are commonly organized in operon such as the *glgBXCAP* well conserved in Enterobacteriales and the sister lineage Pasteurales (Almagro et al. 2015). Both *glgX* and *glgP* genes encode for glycogen catabolic enzymes (see below Catabolic Pathway in Prokaryotes).

4.2.1.2 *Nucleotide-Sugar Synthesis: UDP-/ADP-Glucose Pyrophosphorylase (EC 2.7.7.27)*

In 1957, Leloir and Cardini were the first to establish the importance of nucleotide-sugar in the biosynthesis of storage polysaccharides rather than glucose-1-phosphate, which was previously thought to define the polymerization substrate (Leloir and Cardini 1957). In prokaryotes, genetic evidence of ADP-glucose pyrophosphorylase (GlgC) as a pivotal enzyme for glycogen biosynthesis in *E. coli* was published in 1968 (Damotte et al. 1968). Following this study, correlation between GlgC activity and glycogen biosynthesis has been confirmed in many other prokaryotes as well as in cyanobacteria (Iglesias et al. 1991). With regard to its role in the glycogen pathway, ADP-glucose pyrophosphorylase is an allosteric enzyme that is either negatively or positively regulated by fructose-1,6-biphosphate or AMP, respectively, as well as in cyanobacteria subjected to redox regulation (Ballicora et al. 2003, Díaz-Troya et al. 2014). It should be noted that some bacteria species such as *Prevotella bryantii* B14, a common bacterium found in the rumen, apparently rely on UDP-glucose pyrophosphorylase activity for glycogen synthesis instead of ADP-glucose pyrophosphorylase. Accordingly, glycogen anabolism in this species depends on UDP-glucose dependent glycogen synthase (Lou et al. 1997). Another possible but yet to be confirmed exception to the classical pathways outlined above was revealed by the study of sucrose metabolism in the filamentous nitrogen fixing cyanobacterium *Anabaena* ATCC29213. Sucrose synthesis is induced in response to osmotic stress in cyanobacteria and then catabolized either by invertase or by sucrose synthase (SuSy) activities. In plants, several studies suggest that SuSy activity is involved in the cell wall synthesis by catalyzing the release of UDP-glucose and fructose from UDP and sucrose. Unexpectedly, the phenotype of mutants of sucrose synthase (Susy) in Anabaena appears more complex. Indeed, Susy null mutants harbor a decrease of glycogen content, despite two-fold increases of ADP-glucose pyrophosphorylase activity and are impaired in the nitrogen fixation process (Porchia et al. 1999). Further biochemical characterizations of Susy in both *Anabaena* ATCC29413 and *Thermosynechococcus elongatus* reveal that Susy catalyzes preferentially the synthesis of ADP-glucose and fructose by cleaving sucrose in the presence of ADP (Porchia et al. 1999, Figueroa et al. 2013). These amazing results contrast with the dedicated role of SuSy of plants in the production of UDP-glucose for cell wall synthesis. Subsequent studies based on metabolic network modeling in Anabaena may suggest that ADP-glucose pyrophosphorylase activity alone is insufficient to meet the need in ADP-glucose for glycogen synthesis (Cumino et al. 2007). Unexpectedly, these results suggest that sucrose in Anabaena possesses an additional function besides being an osmotic protectant. Sucrose, among other metabolites, was proposed to be a carrier metabolite linking the vegetative cells and the heterocyst cells. This would be similar to some plants where sucrose is a carrier metabolite that links source organs (leaves) and sink organs (tubers). Indeed, in the filamentous cyanobacterium Anabaena, energy costly nitrogen fixation pathway is catalyzed by an oxygen-sensitive nitrogenase activity localized in a specialized cell called the heterocyst. In turn, both reductive pentose phosphate pathway and photosynthesis II activity (oxygen evolution) are shut down in this dedicated cell to maintain an anoxic environment and high ATP level. In this context, the heterocyst cells may be considered as heterotrophic cells, comparable to sink tissue cells in plants that rely heavily on adjacent photosynthetic cells for ATP and sucrose intakes.

4.2.1.3 *Formation of α-1,4 Linkages: Glycogen/Starch Synthase*

Glycogen/starch synthase activities transfer the glucose moiety of nucleotide-sugar onto the non-reducing end of α-glucan chains. Structural analysis of the enzymes suggests that there are two families named according to Carbohydrate Active Enzyme Classification (CAZy, www.cazy.org): glycosyl transferase 3 (GT3) and glycosyl transferase 5 (GT5) (Coutinho et al. 2003). The former is found exclusively in glycogen metabolism of fungi and animal cells, while the GT5 family is widespread amongst Archaeplastida (plants/green algae, red alga and glaucophytes) and prokaryotes. There is one exception: *Dictyostelium discoideum*, a social-amoeba that possesses both a GT3 and GT5 glycogen synthase (Ball et al. 2015). Due to the general lack of ADP-glucose in glycogen-storing eukaryotes, the majority of glycogen synthases GT3 characterized in eukaryotes are UDP-glucose and oligosaccharide-primer dependent (Ball et al. 2011). In prokaryotes, the specificity of glycogen synthase GT5 is not so stringent. For instance, the characterization of glycogen synthase of *Prevotella bryantii* clearly shows an elongation activity using UDP-glucose (Lou et al. 1997). More recently, the characterization of GlgA of *Chlamydia trachomatis,* an obligate intracellular bacterium that infects eukaryotic cell, has shown that GlgA activity efficiently elongates glucan chain either from ADP-glucose or UDP-glucose as substrate (Gehre et al. 2016). In parallel, this study indicates that GlgA of *C. trachomatis* and other glycogen metabolizing enzymes are secreted in the cytosol of eukaryotic cells through type three-secretion system during the infectious cycle. These results suggest that the specificity of ADP-glucose dependent GlgA of *C. trachomatis* has evolved in order to use the UDP-glucose produced by the host cell within the cytosol.

If most prokaryotes possess a *glgA* gene, cyanobacteria genomes encode two isoforms named *glgA1* and *glgA2* (Ball et al. 2015). Surprisingly, both *Cyanobacterium* sp. CLg1 and *Crocosphaera watsonii* encode a third *glgA* gene, which is phylogenetically related to the Granule Bound Starch Synthase (GBSS) of plants. Mass-spectrometry analyses onto protein attached to the starch granules of *Cyanobacterium* sp. CLg1 as well as incubation experiments of purified starch granules with 14C-ADP-glucose indicate the synthesis of labeled long glucans. Like in plants, the GBSS-like of cyanobacteria appear dedicated to the synthesis of the amylose fraction (Deschamps et al. 2008). The functions of glgA1 and glgA2 activities in glycogen biosynthesis were investigated mainly in Synechocystis PCC6803 (Gründel et al. 2012, Yoo et al. 2014). Single knockout mutants are not affected in growth and glycogen biosynthesis suggesting an overlap in function of these two enzymes. However, a detailed biochemical analysis of glycogen synthases suggests that glgA2 is responsible for long glucan chains synthesis, while the distributive activity of GlgA1 is responsible for short chains glucan (Yoo et al. 2014). More recently, the characterization of a mutant of the starch accumulating *Cyanobacterium* sp. CLg1 strain impaired in GlgA2 activity comforts the idea of the processive nature of GlgA2 (Kadouche et al. 2016). Indeed, the absence of GlgA2 activity results specifically in the disappearance of starch granules without affecting glycogen biosynthesis (Kadouche et al. 2016). The simplest explanation for this phenotype is to propose that GlgA2 activity is mandatory for the synthesis of long glucan chains indispensable for the establishment of the cluster organization of amylopectin. In comparison to the widespread distribution of the GlgA1 isoform amongst prokaryotes, GlgA2 is restricted to a small number of bacterial species including Chlamydiales. More surprisingly, GlgA2 is very similar to the genes of soluble starch synthases III and IV (SSIII and SSIV) found in Archaeplastida (Kadouche et al. 2016). This suggests that the SSIII/SSIV genes of plants have been inherited from a prokaryote and may have been maintained in the Archaeplastida due to its remarkable ability to synthesize long glucan chains.

4.2.1.4 *Formation of α-1,6 Linkages: Glycogen/Starch Branching Enzyme*

Branching enzyme (GlgB)—α-1,4-glucan: α-1,4-D-glucan 6-α glucosyl transferase—catalyzes the transglucosylation reaction by cleaving α-1,4 linkages and by transferring the generated reducing ends to C6 hydroxyls. Branching enzyme plays a critical role in determining the branching pattern of glycogen and starch (Sawada et al. 2014). However, no clear correlation has been found between amino acid sequence of GlgB and the molecular properties of the branched glucan produced. In prokaryotes, branching enzyme activities are divided into two CAZy families: Glycosyl Hydrolase 13 (GH13) and 57 (GH57) families (Suzuki and Suzuki 2016). GH13-branching enzymes of eukaryotic and prokaryotic cells share a common $(\alpha/\beta)_8$ barrel domain with 30 enzymes activities, such as α-amylase, cyclodextinase, debranching enzyme, etc., and are further divided into two sub-families: GH13_8 and GH13_9, respectively (Stam et al. 2006). In 2000, Binderup and his collaborators have shown that the N-terminal end of the *E. coli* branching enzyme influences the chain transfer pattern (Binderup et al. 2000). The deletion of 112 amino acids at the N-terminus leads to a truncated form of *E. coli* branching enzymes, which preferentially transfers longer chains with a degree of polymerization

(DP) of more than 20 DP than the unmodified branching enzyme (DP < 14). Subsequent studies have confirmed the role of the N-terminus of glycogen branching enzymes in *E. coli* as well as for other GlgB proteins of other species (Devillers et al. 2003, Palomo et al. 2009, Jo et al. 2015, Wang et al. 2015). Branching enzymes belonging to the GH57 family, first described in thermophilic bacteria *Dictyoglomus thermophilum* and *Pyrococcus furiosus*, are now described in cyanobacteria and other prokaryote species (Fukusumi et al. 1988, Laderman et al. 1993, Suzuki and Suzuki 2016). Like GH13 family, the GH57 family contains several carbohydrate-active enzyme activities: amylopullulanase, α-galactosidase, α-1,4 glucanotransferase, α-amylase, branching enzyme and uncharacterized activities sharing a catalytic $(\beta/\alpha)_7$ barrel fold and five conserved domains (Zona et al. 2004, Murakami et al. 2006, Santos et al. 2011). In cyanobacteria, the distribution of GH57 family enzymes among species is variable and its exact catalytic functions are yet to be defined (Colleoni and Suzuki 2012). Interestingly, putative GH57 branching enzyme sequences are well conserved among cyanobacteria, including in the reduced genomes of the *Synechococcus/Prochlorococcus* genus as well as in the early diverging cyanobacteria species such as *Gloeobacter violaceus*, *Synechococcus* JA-3-3Ab and *Gloeomargarita lithophora* (Fig. 4.3). So far, the role of this GH57 family in the physiology of cyanobacteria has not yet been explored. Because the GH57 family is mainly composed of thermostable enzymes, it is tempting to suggest that this family of carbohydrate active enzymes plays an important role in thermophilic strains (e.g., *Thermosynechococcus elongatus* BP1 and *Synechococcus* JA3-3Ab) or under thermal stress conditions requiring a set of stable enzymes.

The fact that most of glycogen-accumulating prokaryotes possess at least one GH13-branching enzyme activity or one GH57-branching enzyme (Suzuki and Suzuki 2016) emphasizes the mandatory role of branching enzymes in storage polysaccharide biosynthesis. Cyanobacteria species have both GH13 and GH57 families and additional GH13 branching enzyme isoforms are observed for some cyanobacterial species. It is worth noting that starch-accumulating cyanobacteria harbor in total 3 to 4 candidate branching enzyme activities. Crystallization and biochemical characterization of GH13-branching enzyme activities of *Cyanobacterium* sp. NBRC 102756 and *Cyanothece* ATCC51142 have revealed different branching patterns that may be correlated with the ability of these strains to synthesize starch-like granules (Hayashi et al. 2015, Suzuki et al. 2015, Hayashi et al. 2017).

4.2.2 De novo Biosynthesis of Glucan Chains

Following the discovery of glycogen synthase by Leloir and Cardini in 1957, the main question for many years concerned their ability of priming glucan synthesis from nucleotide sugar. Apparently, glycogen synthase GT3 and GT5 can be distinguished by their ability to prime synthesis *de novo* of a glucan. In eukaryotic cells, the initiation of glycogen synthesis appears to require two activities: (i) a UDP-glucose specific glycosyl transferase of CAzy family GT8 (CAzy classification), which displays self-glycosylation properties (i.e., glycogenin), and are capable to synthesize a short glucan made of 8 up to 36 glucose residues using UDP-glucose as nucleotide-sugar (Albrecht et al. 2004), and (ii) a glycogen synthase GT3, which then elongates the glycogenin dependent primer in order to further elongate a glucan, which thereby becomes accessible for branching enzyme. The study of yeast null mutants affected in both glycogenin isoforms, Glg1p and Glg2p, supports the mandatory nature of glycogenin-controlled glucan priming for synthesis of glycogen in fungi (Torija et al. 2005). However, it should be stressed that glycogen synthesis occurs spontaneously in 2–3% of glycogenin-null mutant of yeast depending on growth conditions (i.e., nitrogen limitation) and it might reach up to 98% if the amount of UDP-glucose is raised up into the cytosol (Torija et al. 2005). More recently, the ability of GT3-glycogen synthase to initiate glycogen synthesis in the absence of glycogenin has been experimented in animals (Testoni et al. 2017). Against all odds, not only knockout mice accumulate glycogen, but an excess of large glycogen particles is also observed in both homozygous and heterozygous strains. Interestingly, these results enlighten, first, that the lack of glycogenin proteins does not prevent glycogen synthesis, and second, that glycogenin may have an

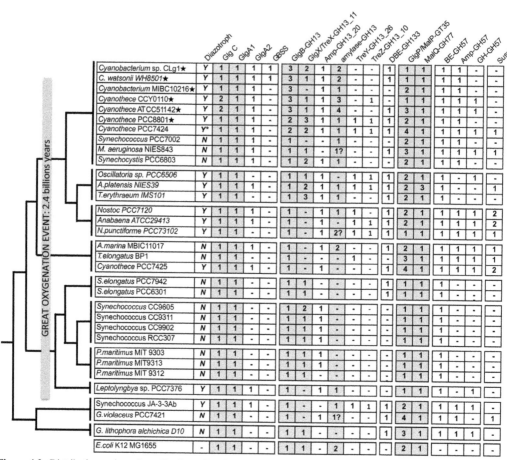

Figure 4.3. Distribution and number of isoforms involved in the storage polysaccharide pathway in cyanobacteria. Black lines depict the evolutionary relationship between cyanobacterial species with respect to the Great Oxygenation Event (GOE). Representatives of each cyanobacteria clade were selected based on the availability of genome sequences. The number of enzyme isoforms was determined using blat searches on NCBI, Cyanobase and CAZy. Cyanobacteria strains labeled with a star synthesized starch-like granules. The ability of nitrogen fixation in aerobic conditions is mentioned by Y (Yes) or N (No) and by Y* (Yes) in anoxic conditions. Key: ADP-glucose pyrophosphorylase (GlgC); glycogen synthase isoforms (GlgA1, GlgA2); Granule bound starch synthase (GBSS); Branching enzyme GH13 family (GlgB-GH13); Debranching enzyme-GH13 subfamily 11(GlgX/TreX-GH13_11); Amylopullulanase-GH13 subfamily 20 (Amp-GH13_20); amylase-GH13; maltooligosyltrehalose synthase GH13 subfamily 26 (TreY-GH13_26); maltooligosyltrehalose hydrolase GH13 subfamily 10 (TreZ-GH13_10); Debranching enzyme or amylo-1,6 glucosidase (DBE-GH133); Glycogen phosphorylase / maltodextrin phosphorylase (GlgP/MalP-GT35); α-1,4 glucanotransferase (MalQ-GH77); putative branching enzyme GH57 family (BE-GH57); putative amylopullulanase GH57 family (Amp-GH57); putative glycosyl hydrolase family GH57 (GH-GH57). *Escherichia coli* is used as reference for the type of isoforms found in cyanobacteria (gray columns).

unexpected function in the glycogen biosynthesis pathway such as a regulator of glycogen synthase activity through protein-protein interactions (Testoni et al. 2017).

In prokaryotes, preliminary studies reported by Krisman and his colleagues suggested that the initiation of glycogen synthesis in *E. coli* is controlled by the formation of glucoprotein such as glycogenin-like protein in the eukaryotic pathway (Barengo et al. 1975, Barengo and Krisman 1978). However, detailed enzymatic analyses suggested that both functions, primer initiation and elongation activity, could probably be under the sole control of *E. coli* glycogen synthase (Kawaguchi et al. 1978, Holmes and Preiss 1979). The discovery in 1997 of a 7.9 kD protein (60 amino acids) named GlgS (S stands for Stimulate) questioned the existence of a glycogenin-like function in *E. coli* (Beglova et al. 1997). Historically, the GlgS mutant of *E. coli* was first identified as glycogen-less phenotype through random transposon insertion mutagenesis and iodine screening of cell patches (Hengge-Aronis and

Fischer 1992). The *glgS* gene does not belong to the *glgBXCAP* operon; its overexpression and its up-regulation by sigmaS during entry into the stationary phase suggested a function in the glycogen metabolism pathway. Therefore, the authors proposed that GlgS could be dedicated to the glycogen initiation. However, in 2003, the expression of recombinant glycogen synthase of *Agrobacterium tumefaciens* in glycogen-less background of *E. coli* confirmed the ability of prokaryote glycogen synthase to synthesize *de novo* a glucan (Ugalde et al. 2003). This result was further confirmed by the restoration of starch synthesis in the double mutant SSIII/SSIV of *A. thaliana* by the transgenic expression of the *A. tumefaciens* glycogen synthase (Crumpton-Taylor et al. 2013).

More recently, both transcriptomic and molecular analyses have definitely invalidated the direct function of GlgS in the glycogen metabolism of *E. coli* (Rahimpour et al. 2013). It has been shown that GlgS renamed ScoR (Surface composition Regulator) negatively regulates the cell wall composition on the surface (flagella, adhesins, exopolysaccharides). Thus, in the absence of ScoR or GlgS, mutant strains harbored an increased number of flagella and a higher synthesis of exopolysaccharides at the expense of glycogen synthesis. This pleiotropic effect disappears when GlgS is overexpressed. The authors proposed to explain glycogen-less phenotype of ScorR/GlgS mutants that both abnormal cell mobility and exopolysaccharides synthesis drained all carbon sources normally stored at the onset of stationary phase in the form of glycogen.

Although the ability of prokaryote glycogen synthase to synthesize *de novo* a glucan is well established, it appears that the recombinant GlgA2 isoform of *Cyanobacterium* sp. CLg1 produced in a glycogen-less background of *E. coli* is not capable of initiating a glucan synthesis by itself. The expression of both GlgA2 and GlgA1 enables the restoration of glycogen biosynthesis in the null glgA mutant only when *E. coli* cells were cultivated in the presence of maltose as a carbon source and not in the presence of glycerol or mannitol. In addition, electrophoretic separation of enzyme followed by *in situ* activity assays (i.e., zymograms) indicate that the priming reaction inside a native-polyacrylamide gel depends on the presence of a glucosylated protein, named X-factor, that can be supplied by *cyanobacterium* CLg1 crude extracts (Kadouche et al. 2016). Further investigation will be necessary to prove the exact nature of this X-Factor.

4.2.3 Other Glycogen Biosynthesis Pathways in Prokaryotes: The GlgE-Path

In contrast to the "classical nucleotide-sugar based glycogen synthesis pathway, this alternative glycogen pathway, named the GlgE-pathway, relies on a maltosyl transferase activity (GH13 subfamily 3 according to CAzy classification) that transfers maltosyl groups from maltose-1-phosphate onto the non-reducing end of glucan chains. The evidence supporting the physiological significance of this alternative pathway in prokaryotes arose from the study of *Mycobacterium tuberculosis*. Mycobacteria are pathogenic bacteria surrounded by a capsule, composed of 80% of glucan made of glucose residue linked in α-1,4 and α-1,6 positions (Lemassu and Daffé 1994). To determine whether the mycobacterial capsule contributes significantly in the virulence and depends on the putative orthologous *glgC*, *glgA* and *glgB* genes that has been identified in the genome of *M. tuberculosis*, a series of knockout mutants were generated either in ADP-glucose pyrophosphorylase activity (GlgC) or glycogen synthase (GlgA) (Sambou et al. 2008). Surprisingly, both inactivation of GlgC and GlgA led only to a reduction but not to the expected wipe-out of both glycogen and extracellular polysaccharides, thus emphasizing the existence of an alternative route for glycogen synthesis. In 2016, a better understanding and clarification of glycogen metabolism in *M. tuberculosis* was achieved when Koliwer-Brandl and collaborators included null *glgE* mutants in their studies in various genetic background (Koliwer-Brandl et al. 2016). Historically, this GlgE activity was first thought to consist of a glucanase activity releasing maltose-1-phosphate from glycogen (Belanger and Hatfull 1999). Unexpectedly, the inactivation of GlgE activity leads to both an accumulation of M-1-P and a decrease of glycogen content, suggesting a function of GlgE in the biosynthesis rather than degradation. In order to determine the source of M1P, further investigations established that GlgA is surprisingly not involved in the elongation process of glucan chain like normal glycogen synthase. On the contrary, GlgA preferentially catalyzes the formation of M1P from ADP-glucose and glucose-1-phosphate. Finally, combination of mutations points out two independent routes: GlgA-GlgC and TreS-Mak (trehalose synthase and maltose kinase) activities that converge on the synthesis of M1P. Despite the name, trehalose synthase activity reversibly converts trehalose (alpha-D-glucose-1,1-alpha-D-glucose) into maltose. The non-reducing disaccharide trehalose functions as an important intracellular osmoprotectant in a wide range of bacterial species, fungi and amoeba. In the GlgE-pathway, maltose produced from trehalose by the TreS activity is then phosphorylated into M1P by maltose kinase activity (Mak or Pep). Interestingly, M1P synthesis is favored by the formation of an heterocomplex between TreS and Mak (Roy et al. 2013) or by the appearance of TreS-Mak fused protein in some bacterial species (Fraga et al. 2015). M1P is then incorporated onto the non-reducing end of glucan chains via GlgE (Kalscheuer et al. 2010). The pivotal role of GlgE in the glycogen synthesis pathway of Mycobacteria and the high cell toxicity of M1P explain the growing interest in the understanding of its regulation (Leiba et al. 2013) and its 3D-structure (Syson et al. 2014) as a potential target for new drugs (Leiba et al. 2013). At the present, comparative genomic analysis suggests that the GlgE pathway is not restricted to mycobacteria species, but is found in 14% of sequenced genomes from diverse bacteria species such as *Pseudomonas*, *Xanthomonas*, and *Burkholderia* (Chandra et al. 2011). In addition, both GlgE-pathway and GlgC-pathway can co-exist in the same bacterium cell such as the gram-positive *Corynebacterium glutamicum*. We have little information about the coordination of these pathways in the glycogen synthesis of such bacteria. Recently, Clermont and colleagues observed that the *C. glutamicum* strain deleted in both *glgC* and *glgA* genes was unable to synthesize glycogen when grown in the presence of glucose. However *glgC glgA* null mutants growing on maltose synthesize a normal amount of glycogen in comparison to wild type strain (Clermont et al. 2015). It is tempting to suggest that *C. glutamicum* synthesizes glycogen via either GlgE-pathway or the GlgC-pathway depending on the carbon source available. However, since maltose is used as a precursor to the GlgE-pathway as well as in the MalQ pathway (see below), further investigations are required to evaluate the contribution of these pathways in the glycogen biosynthesis of *C. glutamicum*.

4.2.4 *Other Glycogen Biosynthesis Pathways in Prokaryotes: the MalQ-Path*

As third glycogen biosynthetic path in prokaryotes, the MalQ pathway relies on the ability of MalQ—an α-1,4 glucanotransferase (GH77) or amylomaltase—to disproportionate short glucans into longer glucans (Monod and Torriani 1950). When long enough to fit the catalytic site of branching enzyme, intra-molecular or extra-molecular transfer generates branched glucan. In the *E. coli* model, the uptake of extracellular maltose induces the transcription of several genes of the maltose metabolism. Among them, both maltodextrin glucanotransferase (MalQ) and maltodextrin phosphorylase (MalP) contribute to the catabolism of maltodextrin. In response to maltose or malto-oligosaccharides (MOS: glucan chains from 3 to 12 glycosyl residues), the *malPQ* operon as well as other genes (such as *malZ* and *malS* encoding α-glucosidase and periplasmic α-amylase, respectively) are transcriptionally activated through the regulator MalT. In contrast to glycogen phosphorylase (GlgP), MalP activity is poorly active on branched polysaccharide, while it releases glucose-1-phosphate from the non-reducing end of MOS with a minimum of four residues of glucose. In the absence of MalP, *E. coli* accumulates long glucan chains responsible for the dark-blue iodine staining of cell patches (Schwartz 1967). Interestingly, the lack of MalP activity leads to abnormal cell morphology ("snake" morphology) in an *E. coli* mutant (Schwartz et al. 1967) as well as in a mutant of *Corynebacterium glutamicum* (Seibold et al. 2009).

MalQ catalyzes the transfer of maltosyl group from maltotriose onto the non-reducing end of maltose or higher complexity malto-oligosaccharide MOS having a higher DP (3 to 7 glucosyl units). Therefore, MalQ generates larger malto-oligosaccharides (DP > 4) suitable for MalP activity. MalQ mutants of *E. coli* do not grow in the presence of maltose but can grow in the presence of MOS with a higher DP. This growth appears dependent on the MalP activity since double mutants D*malPQ* are capable of growing in maltose or MOS as carbon source. By a combination of mutation involving maltose catabolizing enzymes and glycogen synthase activity, Park and his collaborators have shown that normal glycogen synthesis occurs in a *glgA* mutant background (glycogen-less strain) when MalP activity is missing (Park et al. 2011). Interestingly, recent survey of glycogen gene content in sequenced genomes reveals that the artificial situation described by Park and his collaborators might arise in some bacterial species belonging to β-proteobacteria. Thus, glycogen synthesis in both *Ralstonia eutrophora* H16 and *Bordetella parapertussis* 1282 relies on the *glgBXmalQ* operon, while in *Burkolderia* sp. 383 the *TreZ* gene is inserted between *glgX* and *malQ* genes (Almagro et al. 2015). Besides the lack of GlgC and GlgA activities, open questions remain concerning the nature of glycogen metabolism regulation in those species.

4.2.5 Glycogen Catabolic Paths in Prokaryotes

In enterobacteria, the glycogen catabolism pathway is pretty well understood. Glycogen phosphorylase (GlgP) and glycogen debranching enzyme (direct debranching; GlgX) work in synergy with α-1,4 glucanotransferase (MalQ) and maltodextrin phosphorylase (MalP) to essentially convert glycogen to glucose-1-phosphate. Glycogen phosphorylase catabolizes the first catabolic step by releasing G-1-P from the non-reducing ends of glucan chains. This reaction stops around four residues of glucose before a branch point. Then short branched glucans are specifically trimmed by the GlgX activity (Dauvillée et al. 2005, Suzuki et al. 2007). Short glucans released in the cytosol are further metabolized by MalP and MalQ activities. The latter disproportionates short glucans (DP < 4) in longer glucans accessible to the catalytic site of MalP. Thus, in the model organism *E. coli*, two phosphorylase activities are involved in the glycogen catabolism pathway; GlgP and MalP show clear substrate preference for glycogen and for maltooligosaccharides, respectively. Interestingly, both GlgP and GlgX enzymes are produced during glycogen biosynthesis. In order to avoid a futile cycle, ADP-glucose acts as a competitive inhibitor with respect to G1P for the GlgP activity in most prokaryotes during the biosynthesis of glycogen (Chen and Segel 1968, Takata et al. 1998). More recently, subsequent studies on GlgP reveal that the latter forms a tight interaction with the Histidine phosphocarrier protein (Hpr) (Seok et al. 1997). The latter is a component of the phosphoenol-pyruvate: sugar phosphotransferase system (PTS). This PTS system is involved in the transport and the phosphorylation of many sugars (Deutscher et al. 2006). According to the phosphorylation state of Hpr, the affinity of Hpr will change for GlgP. Thus, the phosphorylated form of Hpr (P-Hpr) displays a four-fold higher affinity for GlgP than the unphosphorylated form. However, only the Hpr form allosterically activates GlgP by 2.5 fold. Interestingly, because the concentration of Hpr is much higher than that of GlgP, this indicates that GlgP is always complexed with either P-Hpr or Hpr (Mattoo and Waygood 1983). The phosphorylation state of Hpr varies with the physiological state of the cell. Thus, at the onset of stationary phase, which coincides with glycogen operon activation and glycogen synthesis, most of Hpr is found to be phosphorylated form. Both the predominant form of P-Hpr and the increase in ADP-glucose level will prevent the dimerization and activity of GlgP, respectively, and therefore glycogen degradation (Chen and Segel 1968, Seok et al. 2001).

In contrast to *E. coli*, the set of genes involved in the catabolic pathway varies greatly in cyanobacteria species (Colleoni and Suzuki 2012). So far, the reasons for this multiplicity of degradation pathways are unclear. Recently, the functions of phosphorylase activities were investigated in *Synechocystis* PCC8803 (Fu and Xu 2006). The characterization of knockout mutants in the two phosphorylase genes sll1367 and sll1356 reveal that GlgP (sll1367) is clearly involved in glycogen catabolism, while GlgP (sll1356) is essential for growth at high temperature. Interestingly, despite the increase of GlgP-sll1356 activity in the null GlgP-sll1367 mutant, glycogen content decreased only slightly due to mobilization at night (Fu and Xu 2006). More recently, the functions of glycogen phosphorylase isoforms were investigated in the resuscitation or awakening process, which consists of the ability of non-nitrogen fixing cyanobacterium to shift from the dormant stage to the vegetative stage. In nitrogen limitation condition, light harvesting pigments are degraded and glycogen synthesis is induced until photosynthesis as well as metabolic activities are strongly reduced. Hence, this survival cell response, termed "chlorosis", is maintained until the nitrogen source is supplied. The resuscitation process requires the mobilization of glycogen as carbon-energy source until the photosynthetic apparatus becomes functional. Characterization of single or double mutants of glycogen phosphorylase isoforms in *Synechocystis* PCC6803 reveals a minor role of GlgP-sll1356, while GlgP-sll1367 null mutants were impaired in the resuscitation process and revealed a major contribution of this enzyme to glycogen catabolism during resuscitation. Further investigations are required to understand the function of the GlgP-sll1356 isoform. One reasonable explanation might be that sll1356 gene encodes a maltodextrin phosphorylase, which as described previously is more active on maltodextrin than glycogen. Hence, as with GlgP mutants in *E. coli* showing a defect in glycogen breakdown, the putative-malP (sll1356) isoform cannot compensate for the loss of function of GlgP (Alonso-Casajús et al. 2006).

Surprisingly, several cyanobacterial genomes, if complete, do not seem to encode any classical GlgX-GH13-type debranching enzyme (Colleoni and Suzuki 2012). However, such genomes appear to contain amylopullulanase (GH13-20 and/or GH57) or amylo-1,6 glucosidase (GH133) genes that could exert the corresponding functions. For instance, the early diverging Synechococcus JA-3-3Ab strain contains two candidate debranching enzymes activities: the putative amylopullulanase-GH57 and amylo-1,6-glucosidase-GH133. Little is known, however, about their role in the glycogen catabolism pathway. Recently, an amylopullulanase type of GH13 enzyme has been characterized from the filamentous cyanobacterium *Nostoc ponctiforme* (Choi et al. 2009). In contrast to the GlgX activity, this enzyme displays hydrolysis activity toward both α-1,6 and α-1,4 linkages when incubated with soluble starch or amylopectin. Incubation experiments indicate that amylopullanase-GH13 hydrolyzes first α-1,6 linkages of long branched glucans (4 > DP > 10) and chews up from the reducing-end long glucan chains up to 8 residues of glucose. Because *N.punctiforme* does not contain any GlgX or TreX GH13 activity but harbors both malto-oligosyl trehalose synthase (TreY) and malto-oligosyl trehalose hydrolase (TreZ), it is quite possible that the amylopullulanase-GH13 substitutes for the missing TreX activity, which is required to produce the suitable malto-oligosaccharide for TreY activity. Amylo-1,6-glucosidase (GH133) shows similarity to the amino acid sequence of the C-terminal domain of indirect debranching enzyme activity of animals and fungi. By analogy with this eukaryotic enzyme, the function of this activity may possibly reside in debranching single α-1,6 glucose residues branched on linear glucan chains. So far, no biochemical characterization has ever been carried out on these cyanobacterial amylo-1,6-glucosidases. However, the characterization of the null GlgX mutant of Synechococcus PCC7942 brought some insights on the putative function of the amylo-1,6 glucosidase activity (Suzuki et al. 2007).

As described above, characterizations of glgX mutants in both *E. coli* and *Synechococcus* PCC7942 confirm its function in the catabolism pathway. Nevertheless, characterization of GlgX deficient mutants of the starch-accumulating *Cyanobacterium* sp. CLg1 suggests its involvement in the crystallization process of storage polysaccharide.

4.3 Convergent Evolution of Glucan Trimming Mechanism in the Starch Biosynthetic Pathway of *Cyanobacterium* sp. CLg1

Until the end of the twentieth century, the mechanisms underlying the synthesis of starch rather than glycogen in plants were poorly understood. The characterization of mutants substituting starch by glycogen biosynthesis in green algae and different plant species surprisingly revealed a defect in the isoamylase type of debranching enzyme. In order to explain this unexpected function for hydrolytic enzymes in the amylopectin synthesis, a glucan trimming model was proposed in 1996 (Mouille et al. 1996). This model suggests that an isoamylase type debranching enzyme trims a highly branched precursor named pre-amylopectin and thereby allows the formation of closely packed double helices of glucan that aggregate out of solution. It implies that this isoamylase trims faster those loosely branched glucans that interfere with the formation clustered double helices of glucan. In the absence of this enzyme, a highly branched hydrosoluble polysaccharide accumulates in the deficient mutants. Until 1994, starch accumulation had been described solely in lower photosynthetic eukaryotes. In 1994, time course experiments consisting of the observation of thin sections of nitrogen-fixing cyanobacterium *Cyanothece* ATCC51142 through day-night cycles revealed that aggregates of large polysaccharides accumulate between the cyanobacterial thylakoid membranes during the day and disappear at night (Schneegurt et al. 1994). Those large bodies were partially characterized and described as abnormal glycogen particles and not as starch-like material (Schneegurt et al. 1997). The authors of the study did not realize at the time that they were dealing in fact with starch-like structures. Later on, a survey of storage polysaccharides in different species of cyanobacteria reported the presence of solid granules in others cyanobacterial species (Nakamura et al. 2005). Subsequently, detailed characterization indicated that such carbohydrate granules, including those of *Cyanothece* ATCC51142, are composed of high-molecular weight polysaccharides similar to amylopectin (Suzuki et al. 2013). Around the same time, *Cyanobacterium* sp. CLg1, a new strain isolated in the tropical North Atlantic Ocean phylogenetically related to *Crocosphaera watsonii*, was shown to accumulate starch granules made of both amylopectin and amylose fractions (Falcon et al. 2002, Deschamps et al. 2008). Interestingly, as in plants, the presence of amylose was correlated with the identification of a polypeptide showing a high similarity in amino acid sequence to GBSS (Granule Bound Starch Synthase) of plants (Deschamps et al. 2008). To the present day, less than a dozen of starch-accumulating cyanobacteria have been identified. They are all unicellular, capable of nitrogen fixation and belong to the Chroococcales order (Colleoni and Suzuki 2012). Deciphering the storage polysaccharide metabolism pathways is pretty challenging since those strains are refractory to all available transformation protocols. Nevertheless, in order to tackle this question, a UV mutagenesis was carried out on wild-type *Cyanobacterium* sp. CLg1. Based on the iodine staining of cell patches, more than a hundred mutants were identified and subsequently categorized according to the ratio of soluble to insoluble polysaccharide. Among them, a dozen mutants harboring an increase in the water-soluble glycogen like fraction and the disappearance of starch granules were impaired in one of the two debranching enzyme isoform (s) GlgX2. This striking result suggests that, like in green alga and plants, crystallization process of amylopectin relies on the presence of a debranching enzyme activity. However, detailed phylogeny analysis built with debranching enzyme sequences of Archaeplastida and Bacteria indicates that the eukaryotic sequences do not have a cyanobacterial origin, but are derived from obligate intracellular pathogens and symbionts of the order chlamydiales. Hence, Cenci and his collaborators point out an example of convergent evolution, based on the trimming step of glucose chains required for starch crystallization in two different systems (Cenci et al. 2013). This striking observation opens the door to new research area, which is beyond the scope of this chapter. Nevertheless, it should be stressed out that a model named "ménage à trois" has been proposed to take into account the involvement of Chlamydiales species in the establishment of primary endosymbiosis (Ball et al. 2013 review by Cenci et al. 2017).

4.4 Why do Unicellular Nitrogen Fixing Cyanobacteria Synthesize Starch Granules rather than Glycogen?

Cyanobacteria represent one of the oldest phyla of prokaryotes on earth (Summons et al. 1999). They are unique microorganisms capable of performing oxygenic photosynthesis and some of them reduce molecular nitrogen through nitrogenase activity. Nitrogenase activity loses covalent bonds between two nitrogen (and two carbon or one carbon and one nitrogen) atoms in an ATP-dependent manner and requires anoxic condition. Because of the lack of substrate specificity, it is hypothesized that the function of nitrogenase activity might have evolved over time; first, by reducing the effects of toxic cyanide compounds, then to reduce atmospheric nitrogen in ammonium. Over billion years of evolution, oxygenic-photosynthesis activities of cyanobacteria have changed Earth's atmosphere composition, thereby generating the great oxygen event (GOE) 2.4 billion years ago (Kopp et al. 2005). The transition from a reductive to oxidative environment was a trigger for diversification of cyanobacteria lineages and the appearance of new traits (e.g., size and cell morphology) (Sánchez-Baracaldo et al. 2014). A remarkable adaptation was achieved in nitrogen-fixing cyanobacteria or diazotrophic cyanobacteria. Indeed, the reduction of dinitrogen to ammonium is catalyzed by a protein complex called nitrogenase that displays an extreme sensitivity to oxygen. As oxygen level rose, diazotrophic cyanobacteria developed different strategies to protect the nitrogenase activity: (i) by inhabiting anoxic biotopes, (ii) by confining nitrogenase activity in specialized cells called heterocysts, and (iii) by separating temporally oxygenic photosynthesis and nitrogen fixation. In diazotrophic filamentous cyanobacteria, nitrogenase activity is located in heterocysts harboring a thick cell wall and lacking of photosystem II, which is responsible for photosynthetic oxygen evolution. Neighbor cells performing normal oxygenic photosynthesis activity supply the large amount of energy required to fuel the nitrogenase complex. Hence, physical separation of two exclusive biological processes, i.e., photosynthesis and nitrogen fixation, is possible through the evolution of dedicated cell. Consistent with this view, it was believed that nitrogen fixation could not occur in unicellular cyanobacteria. The discovery of two unicellular cyanobacteria species, *Gloeothece* sp. and *Cyanothece* sp., capable of fixing dinitrogen challenged this paradigm and, for the first time, provided evidence that in microorganisms primary metabolic processes are regulated by the circadian clock regulation (Wyatt and Silvey 1969, Singh 1973). In contrast to filamentous diazotrophic cyanobacteria that perform nitrogen fixation during the day, unicellular diazotrophic cyanobacteria carry out this activity exclusively at night. Unicellular cyanobacteria thus have developed a temporal separation of those two incompatible processes, which take place for some of them exclusively in micro-aerobic and for a small group of cyanobacteria also in aerobic conditions. In the latter case, unicellular nitrogen fixing cyanobacteria exhibit high rates of dark respiration, providing not only the energy required for nitrogen fixation, but also a microaerophilic or anoxic environment (Compaoré and Stal 2010). In order to achieve the required high rates of dark respiration, unicellular diazotrophic cyanobacteria are speculated to have evolved a more efficient storage polysaccharide, allowing an increase in carbon storage in the light phase. Recently, the measurement of nitrogen fixation in six Cyanothece species under various growth and incubations conditions strengthens this assumption. In this report, starch-accumulating *Cyanothece* spp. exhibit a higher rate of nitrogen fixation than glycogen-accumulating Cyanothece species in aerobic growth conditions (Bandyopadhyay et al. 2013). This indirect proof reinforces the idea that the GOE, 2.4 billion years ago, was probably a driving force for the transition from glycogen to semi-crystalline storage polysaccharides. It should be emphasized that distribution of GlgX/isoamylase type debranching enzyme sequences does not necessarily correlate with the accumulation of starch in cyanobacteria. This suggests that the transition from glycogen to starch occurred in theses cyanobacteria by either recruiting another debranching enzyme or calling for a different, yet to be described, mechanism of granule aggregation. Another striking observation from these bioinformatic genome-mining approaches consists of the absence of candidate debranching enzyme sequences of all known types from some starch accumulating organisms. Genome analyses of starch accumulating *Cyanobacterium* MIBC10216 and red algal-derived organisms do not reveal any gene encoding a GH13-type debranching enzyme. In such a case, the crystallization of amylopectin may indeed follow a different mechanism. A preliminary answer to this question has been addressed through the characterization of debranching enzyme and chloroplastic amylase null mutants in Arabidopsis. In the absence of chloroplastic amylase, DBE null mutants accumulate switch from glycogen-like polysaccharide to starch like polysaccharides (Streb et al. 2008). Overall, those studies may suggest that the combination of different cyanobacterial branching enzymes alone possibly offers a specific branching pattern that promotes a cluster organization of amylopectin. In turn, this could explain why usually three to four branching enzymes are usually found in these genomes.

4.5 Is there Another Example of Transition from Soluble to Insoluble Storage Polysaccharide?

Or, in other words, is there another selection pressure, which led to transition from soluble to insoluble storage polysaccharide? Few cases of abnormal glycogen/carbohydrate granules are described in the literature. It should be stressed that in the examples described below, the carbohydrate granules were not characterized structurally. Thus, this absence of information forbids us to distinguish between insoluble glycogen particles (absence of crystallinity) and true starch-like (semi-crystalline) granules. The evidence that they are actually carbohydrate granules is based on glucose assay, electronic microscope observations (size, electron density) and density (sedimentation at low speed). Therefore, further characterization of storage polysaccharides in these microorganisms should definitively improve our understanding of alternative aggregation mechanisms.

4.5.1 Firmicutes Phylum: Clostridium sp.

Clostridium species belong to the Firmicute phylum and define gram-positive anaerobic endospore-forming bacteria mainly used to produce acid and organic solvents (Dash et al. 2016). Back in 1950, Hopson and his collaborators reported in *C. butyricum* the accumulation of amylopectin-like polysaccharides that exert a strong interaction with iodine. The wavelength of the maximum absorbency of the iodine-polysaccharide complex (lmax) was reported to average 545 nm, which is close to the lmax value of amylopectin (550 nm to 560 nm) (Bergère et al. 1975). This material named "granulose" was shown to be distributed in most Clostridium species (Reysenbach et al. 1986). Based on microscopy observations, granulose sizes vary from 210 and 270 nm, which is by far larger than normal glycogen particles (Tracy et al. 2008). Granulose biosynthesis occurs prior to the end of the exponential growth phase when the shift from acid to solvent production takes place, while its degradation coincides with the endospore formation (Johnstone and Holland 1977). Characterization of mutants in *C. pasteurianum* and *C. acetobutylicum* ATCC824 impaired in granulose synthase harbor granulose-less phenotypes and is incapable of initiating sporulation, suggesting that granuloses act as major carbon and energy source (Robson et al. 1974, Ehsaan et al. 2016). A rapid survey of Clostridium genomes emphasizes the presence of one branching isoform belonging either to the GH13 family or the GH57 family (Suzuki and Suzuki 2016) and one glycogen synthase isoform named "granulose synthase". Interestingly, the latter is co-purified with native granulose (Robson et al. 1974). At the present day, the genetic determinants responsible for the synthesis of granulose are still unknown.

4.5.2 Candidatus Methylacidiphilum fumariolicum

The second example concerns *Ca. M. fumariolicum*, which belongs to the Verrucomicrobia phylum (Pol et al. 2007). This phylum forms with Planctomycetes, Chlamydiae and along with the Lentiphaerae and Poribacteria, the PVC superphylum. *Ca. M. fumariolicum* is a methanotrophic bacterium that uses methane as both carbon and energy source. Interestingly, *Ca. M. fumariolicum* is able to fix nitrogen and carbon dioxide through the nitrogenase complex and the Calvin-Benson cycle, respectively (Khadem et al. 2010, 2011). In the absence of the nitrogen source (ammonium or dinitrogen gas), *Ca. M. fumariolicum* accumulates both electron dense bodies of 100–200 nm with an elliptical or circular shape (that have been identified as abnormal glycogen particles) and soluble polysaccharide. The physiological function of the abnormal glycogen particles were further investigated by growth experiments, which consisted of depletion of methane and/or nitrogen source (Khadem et al. 2012). Overall, these experiments suggest that in the absence of the major carbon dioxide and energy source (i.e., methane), the carbohydrate granules are the main source of carbon for maintaining the survival of cells. Although little is known about the fine structure of abnormal glycogen, the glycogen metabolism pathway relies on set of enzymes that is similar to that of *E. coli* (Khadem et al. 2012).

4.6 Interaction between Storage Polysaccharide and Fitness in Prokaryotes

A large number of studies on enterobacteria indicates an interconnection of glycogen metabolism with many cellular processes, including nitrogen, iron and magnesium metabolisms, stress responses, and RNA metabolisms. We invite readers to read the excellent review on these aspects (Wilson et al. 2010). Similar interconnections probably occur in cyanobacteria species. Furthermore, both photosynthetic activity and carbon fixation add a layer of complexity in terms of regulation and cross talks between metabolic networks. At present, there is compelling evidence that glycogen is critical for abiotic stress responses (e.g., salinity, temperature) and for survival under day-night growth conditions (Miao et al. 2003, Suzuki et al. 2010, Gründel et al. 2012). Why do glycogen-less cyanobacteria grow normally in continuous light and not under day-night conditions? The answer starts with the characterization of *rpaA*-null mutant in Synechococcus. RpaA is an important transcriptional factor interconnecting circadian clock and different metabolic pathways. It activates the transcription of key genes belonging to the glycogen catabolism pathway (glycogen phosphorylase: *glgP*), glycolysis (glyceraldehyde-3-phosphate dehydrogenase: *gap*, fructose-1,6-biphosphatase: *fbp*) and the oxidative pentose phosphate pathway (glucose-6-phosphate dehydrogenase: *zwf*, 6-phosphogluconate dehydrogenase: *gnd*). Like glycogen-less mutants, *rpaA* null mutants grow at the same rate as the wild type strain in continuous light but they are not viable under day-night cycle. Furthermore, despite normal GlgC and GlgA activities, the amount of glycogen is strongly reduced in *rpaA* mutants (Puszynska and O'Shea 2017). Two independent research groups investigated the *rpaA* mutant phenotype in Synechococcus. First, Diamond and colleagues posit that abnormal accumulation of reactive oxidative species (ROS) produced at night in *rpaA* mutants might explain day-night lethality. Their hypothesis is sustained by the fact that by modulating light intensity *per se* by modifying ROS levels or/and by reducing NADPH, H+ consuming pathways (knockout of valine, leucine and isoleucine biosynthetic pathways), the *rpaA* mutant is enabled to grow under day-night cycle (Diamond et al. 2017). Thus, they propose that glycogen degradation and the OPP pathway fuel the cell in reducing power (NADPH, H+) at night which prevents the accumulation of ROS. Further investigations conducted by Puszynska and O'Shea suggest that the inability of *rpaA* mutant to maintain the Adenylate Energetic Charge at night might explain the lethality (Puszynska and O'Shea 2017). In order to validate their hypothesis, the glucose transporter (GalP) as well as GlgP, Gap, Zwf activities were expressed in *rpaA* mutant. The authors correlate the phenotype rescue with a restoration of Adenylate Energetic Charge in the *rpaA* mutant. Unfortunately, no information is available concerning the ROS levels in the rescued *rpaA* mutant. Nevertheless, both articles from Diamond et al. (2015) as well as Puszynska and O'Shea (2017) emphasize the critical role of carbon stores at night. Like enterobacteria, the macroelement limitation (i.e., N, P) induces the accumulation of a large amount of glycogen between thylakoid membranes in non-nitrogen fixing cyanobacteria. Nitrogen starvation triggers the degradation of photosynthetic pigments in many photosynthetic organisms. This mechanism, named the "chlorosis response", induces dormant stage, where basal metabolic and photosynthesis activities occur as long as the nitrogen source is not available. Interestingly, only 48 hours are required for a dormant cyanobacterium to be fully metabolically active and perform photosynthesis activity again when nitrogen is supplied. In addition, there is now compelling evidence that both chlorosis response and survival at the dormant stage depend explicitly on glycogen metabolism and not on alternative carbon source such as poly-β-hydroxybutyrate stores (Damrow et al. 2016). More recently, Doello and his collaborators investigated the importance of glycogen pool during this resuscitation or awakening process. This study sheds light on the importance of glycogen catabolism in the few hours of awakening when the photosynthetic apparatus is not active yet to supply the cells in ATP (Doello et al. 2018). Glycogen is degraded when the glycogen phosphorylase isoform, GlgP-slr1367, is activated. The glycogen phosphorylase isoform is synthesized during the biosynthesis of glycogen.

So far, the mechanisms that control glycogen phosphorylase (GlgP-slr1367) activation or inhibition during degradation and synthesis, respectively, are still unknown. In addition, this work outlines that the main glycolytic pathway routes consist of both oxidative pentose phosphate pathway and Entner-Doudoroff (ED) pathways, recently discovered in cyanobacteria and not as expected, the Embden-Meyerhof-Parnas (EMP) pathway, that yields twice more ATP than ED pathway (Chen et al. 2016). Despite the absence of alternative sources of ATP (i.e., photosynthesis activity) during the awakening process, the thermodynamic constraints of the EMP pathway might favor the use of the ED pathway (Flamholz et al. 2013).

Another intriguing aspect concerns the interconnection between glycogen metabolism and aging or senescence process in prokaryotes. The latter is defined by an accumulation of cellular damage within the mother cell, which leads to a decrease in reproductive rates and increase of mortality with age. First described in eukaryotes including yeast, bacteria were considered immortal and free of aging because it was supposed that the cell division gives rise to two identical cells. However, early in 2000, two studies revealed that aging processes occur in bacteria as well (Ackermann et al. 2003, Stewart et al. 2005). Asymmetric partitioning of cellular components in rod-shape bacteria such as *E. coli* and *Caulobacter crescentus* enables to distinguish mother and daughter cells. Thus, by tracking cell divisions of rod-shape bacteria and more precisely by following the polar cells, it appears that bacteria are affected by aging like most cells. More recently, thanks to microfluidic system devices, Boehm and his collaborators have pointed out that the replicative lifespan of *E. coli* is genetically controlled and is linked to the glycogen metabolism pathway (Boehm et al. 2016). In this study, a mutation in the carbon storage regulator gene (*csrA*) drastically reduces the number of cell division from 150 to 5 for a mother's cell. Null *csrA* mutants in *E. coli* result in an increase both in neoglucogenesis and glycogen biosynthesis and an inhibition of glycolysis as well (Sabnis et al. 1995). That this *csrA* phenotype is rescued in glycogen less strain mutants deleted in the glycogen synthase gene (GlgA) further emphasizes the importance of glycogen in the aging process. How can we explain that an excess of glycogen accumulation is responsible for a decrease of replicative lifespan in *E. coli*?

Based on previous reports on effect of partitioning of protein aggregates or inclusion bodies in the aging process of *E. coli* cells (Lindner et al. 2008), the authors localized glycogen particles upon cell division. Interestingly, they observed through fluorescent probe GlgA-GFP in both wild type and *csrA* mutant that most of the glycogen particles occur at the pole of the mother cell, while new released cell is free of glycogen. In the case of *csrA* mutant, the glycogen accumulation gradually increases from the pole of the mother cell more rapidly after each cell division than wild-type cell. Interestingly, GlgA-GFP signal appears across the entire cell usually after the fifth cell division suggesting that *csrA* mutants are devoid of genomes, which would explain why these cells then stop growing and dividing (Boehm et al. 2016).

4.7 Conclusion and Perspectives

The survival of free-living cells depends on their ability to maintain their energy status when faced with fluctuating environmental conditions. This critical issue requires the synthesis of energy-carbon storage compounds such as α-polysaccharides, for fueling the cell in the absence of exogenous energy. Glycogen particles represent the most primitive and abundant form of carbon storage in the living world. Described in prokaryotes, Archaebacteria and eukaryotic cells, it seems clear that glycogen particles are optimized to meet the specific needs of the cell and that evolution has spawned various biochemical pathways to achieve their synthesis. Hence, mathematical modeling and biological evidence outline the tight relationship between structure and function of glycogen particles. In addition, a large body of evidence suggests that any structural alteration of glycogen leads to deleterious effects for glycogen-accumulating cells. Nevertheless, in some cases, glycogen particle was not adapted in certain physiological circumstances. As a consequence, transition states occur from tiny hydrosoluble polysaccharide, glycogen, to water-insoluble starch-like polysaccharide. First described in Archaeplastida lineages (plants/green algae; red algae, glaucophytes), unicellular nitrogen fixing cyanobacteria synthesize amylopectin/starch granules as well. Although there is no direct proof, it is tempting to suggest that both the ATP cost and anoxia required by nitrogen fixation pathway at night was the main driving force to substitute glycogen toward a better storage polysaccharide like starch. We can hypothesize that the same selection pressure led the nitrogen-fixing *Candidatus Methylacidiphylum fumariolicum* to synthesize carbohydrate granules rather than glycogen particles. Interestingly, the energy cost of endospore formation during solvent production in Clostridium might be, in that case, the driving force to substitute glycogen by granulose. Although little is known about the degradation pathway of carbohydrate granules produced in both organisms, those reflect again a compromise between storage capacity and availability of carbon where solid granules, starch and granulose are optimized to store a maximum of glucose moieties in a limited space.

In addition, we have to keep in mind that storage polysaccharides are usually associated with others polymeric materials like polyphosphate granules, lipids, and poly-β-hydroxybutyrate that can be used as energy-carbon store or as energy store exclusively (Achbergerová and Nahálka 2011). Thus, it is worth noting that some organisms are completely devoid of storage polysaccharide. In 2002, Henrissat and his associates pointed out that the loss of carbohydrate active enzymes concerns human or mammalian pathogenic bacteria (Henrissat et al. 2002). This observation can be extended to symbiotic bacteria either with bugs (Degnan et al. 2009, Nikoh et al. 2011) or with eukaryotes (Newton et al. 2007).

Acknowledgement

The authors thank Dr. M. Steup for the invitation to write this chapter. We are grateful to Dr. Steven Ball for helpful comments and suggestions on this chapter.

References

Achbergerová, L. and J. Nahálka. 2011. Polyphosphate—an ancient energy source and active metabolic regulator. Microb. Cell Fact. 10: 1–14.

Ackermann, M., S.C. Stearns and U. Jenal. 2003. Senescence in a bacterium with asymmetric division. Science 300: 1920–1920.

Albi, T. and A. Serrano. 2016. Inorganic polyphosphate in the microbial world. Emerging roles for a multifaceted biopolymer. World J. Microbiol. Biotechnol. 32: 1–12.

Albrecht, T., S. Haebel, A. Koch, U. Krause, N. Eckermann and M. Steup. 2004. Yeast glycogenin (Glg2p) produced in *Escherichia coli* is simultaneously glucosylated at two vicinal tyrosine residues but results in a reduced bacterial glycogen accumulation. Eur. J. Biochem. 271: 3978–3989.

Almagro, G., A.M. Viale, M. Montero, M. Rahimpour, F.J. Muñoz, E. Baroja-Fernández et al. 2015. Comparative genomic and phylogenetic analyses of gammaproteobacterial glg genes traced the origin of the *Escherichia coli* glycogen glgBXCAP operon to the last common ancestor of the sister orders Enterobacteriales and Pasteurellales. PLoS One 10: 1–30.

Alonso-Casajús, N., D. Dauvillée, A.M. Viale, F.J. Muñoz, E. Baroja-Fernández, M.T. Morán-Zorzano et al. 2006. Glycogen phosphorylase, the product of the glgP gene, catalyzes glycogen breakdown by removing glucose units from the nonreducing ends in *Escherichia coli*. J. Bacteriol. 188: 5266–5272.

Atkinson, D.E. 1968. The energy charge of the adenylate pool as a regulatory parameter. Interaction with feedback modifiers. Biochemistry 7: 4030–4034.

Ball, S., C. Colleoni, U. Cenci, J.N. Raj and C. Tirtiaux. 2011. The evolution of glycogen and starch metabolism in eukaryotes gives molecular clues to understand the establishment of plastid endosymbiosis. J. Exp. Bot. doi: 10.1093/jxb/erq411.

Ball, S., C. Colleoni and M.C. Arias. 2015. The transition from glycogen to starch metabolism in cyanobacteria and eukaryotes. Starch Metab. Struct. doi: 10.1007/978-4-431-55495-0_4.

Ballicora, M.A., A.A. Iglesias and J. Preiss. 2003. ADP-glucose pyrophosphorylase, a regulatory enzyme for bacterial glycogen synthesis. Microbiol. Mol. Biol. Rev. 67: 213–225, table of contents.

Bandyopadhyay, A., T. Elvitigala, M. Liberton and H.B. Pakrasi. 2013. Variations in the rhythms of respiration and nitrogen fixation in members of the unicellular diazotrophic cyanobacterial genus cyanothece. Plant Physiol. 161: 1334–1346.

Barengo, R., M. Flawiá and C.R. Krisman. 1975. The initiation of glycogen biosynthesis in *Escherichia coli*. FEBS Lett 53: 274–278.

Barengo, R. and C.R. Krisman. 1978. Initiation of glycogen biosynthesis in *Escherichia coli* studies of the properties of the enzymes involved. BBA - Gen Subj 540: 190–196.

Beglova, N., D. Fischer, R. Hengge-Aronis and K. Gehring. 1997. 1H, 15N and 13C NMR assignments, secondary structure and overall topology of the *Escherichia coli* GlgS protein. Eur. J. Biochem. 246: 301–310.

Belanger, A.E. and G.F. Hatfull. 1999. Exponential-phase glycogen recycling is essential for growth of Mycobacterium smegmatis. J. Bacteriol. 181: 6670–8.

Bergère, J., M. Rousseau and C. Mercier. 1975. Polyoside Intracellulaire Impliqué dans la sporulation de Clostridium butyricum. I. Cytologie, Production et analyse enzymatique préliminaire. Ann Microbiol (Paris). 126: 295–314.

Binderup, K., R. Mikkelsen and J. Preiss. 2000. Limited proteolysis of branching enzyme from *Escherichia coli*. Arch. Biochem. Biophys. 377: 366–371.

Boehm, A., M. Arnoldini, T. Bergmiller, T. Röösli, C. Bigosch and M. Ackermann. 2016. Genetic Manipulation of Glycogen Allocation Affects Replicative Lifespan in *E. coli*. PLoS Genet. 12: 1–17.

Chandra, G., K.F. Chater and S. Bornemann. 2011. Unexpected and widespread connections between bacterial glycogen and trehalose metabolism. Microbiology 157: 1565–1572.

Chapman, A.G., L. Fall and D.E. Atkinson. 1971. Adenylate energy charge in *Escherichia coli* during growth and starvation. J. Bacteriol. 108: 1072–1086.

Chen, G.S. and I.H. Segel. 1968. Purification and properties of glycogen phosphorylase from *Escherichia coli*. Arch. Biochem. Biophys. 127: 175–186.

Chen, X., K. Schreiber, J. Appel, A. Makowka, B. Fähnrich, M. Roettger et al. 2016. The Entner–Doudoroff pathway is an overlooked glycolytic route in cyanobacteria and plants. Proc. Natl. Acad. Sci. 113: 5441–5446.

Choi, J.-H., H. Lee, Y.-W. Kim, J.-T. Park, E.-J. Woo, M.-J. Kim et al. 2009. Characterization of a novel debranching enzyme from Nostoc punctiforme possessing a high specificity for long branched chains. Biochem. Biophys. Res. Commun. 378: 224–229.

Clermont, L., A. Macha, L.M. Müller, S.M. Derya, P. von Zaluskowski, A. Eck et al. 2015. The α-glucan phosphorylase MalP of *Corynebacterium glutamicum* is subject to transcriptional regulation and competitive inhibition by ADP-glucose. J. Bacteriol. 197: 1394–1407.

Colleoni, C. and E. Suzuki. 2012. Storage polysaccharide metabolism in cyanobacteria. pp. 217–254. *In*: Tetlow, Ian J. (ed.). Starch: Origins. Structure and Metabolism. Society for experimental biology.

Compaoré, J. and L.J. Stal. 2010. Oxygen and the light-dark cycle of nitrogenase activity in two unicellular cyanobacteria. Environ. Microbiol. 12: 54–62.

Coutinho, P.M., E. Deleury, G.J. Davies and B. Henrissat. 2003. An evolving hierarchical family classi cation for glycosyltransferases. J. Mol. Biol. 328: 307–317.

Crumpton-Taylor, M., M. Pike, K.J. Lu, C.M. Hylton, R. Feil, S. Eicke et al. 2013. Starch synthase 4 is essential for coordination of starch granule formation with chloroplast division during Arabidopsis leaf expansion. New Phytol. 200: 1064–1075.

Cumino, A.C., C. Marcozzi, R. Barreiro and G.L. Salerno. 2007. Carbon Cycling in *Anabaena* sp. PCC 7120. Sucrose synthesis in the heterocysts and possible role in nitrogen fixation. PLANT Physiol. 143: 1385–1397.

Damotte, M., J. Cattanéo, N. Sigal and J. Puig. 1968. Mutants of *Escherichia coli* K12 altered in their ability to store glycogen. Biochem. Biophys. Res. Commun. 32: 916–920.

Damrow, R., I. Maldener and Y. Zilliges. 2016. The multiple functions of common microbial carbon polymers, glycogen and PHB, during stress responses in the non-diazotrophic cyanobacterium *synechocystis* sp. PCC 6803. Front Microbiol. doi: 10.3389/fmicb.2016.00966.

Dash, S., C.Y. Ng and C.D. Maranas. 2016. Metabolic modeling of clostridia: Current developments and applications. FEMS Microbiol. Lett. doi: 10.1093/femsle/fnw004.

Dauvillée, D., I.S. Kinderf, Z. Li, B. Kosar-Hashemi, M.S. Samuel, L. Rampling et al. 2005. Role of the *Escherichia coli* glgX gene in glycogen metabolism. J. Bacteriol. 187: 1465–1473.

Degnan, P.H., Y. Yu, N. Sisneros, R.A. Wing and N.A. Moran. 2009. *Hamiltonella defensa*, genome evolution of protective bacterial endosymbiont from pathogenic ancestors. Proc. Natl. Acad. Sci. U. S. A. 106: 9063–9068.

Deschamps, P., C. Colleoni, Y. Nakamura, E. Suzuki, J.L. Putaux, A. Buléon et al. 2008. Metabolic symbiosis and the birth of the plant kingdom. Mol. Biol. Evol. 25: 536–548.

Deutscher, J., C. Francke and P.W. Postma. 2006. How phosphotransferase system-related protein phosphorylation regulates carbohydrate metabolism in bacteria. Microbiol. Mol. Biol. Rev. 70: 939–1031.

Devillers, C.H., M.E. Piper, M.A. Ballicora and J. Preiss. 2003. Characterization of the branching patterns of glycogen branching enzyme truncated on the N-terminus. Arch. Biochem. Biophys. 418: 34–38.

Diamond, S., B.E. Rubin, R.K. Shultzaberger, Y. Chen, C.D. Barber and S.S. Golden. 2017. Redox crisis underlies conditional light–dark lethality in cyanobacterial mutants that lack the circadian regulator, RpaA. Proc. Natl. Acad. Sci. 114: E580–E589.

Díaz-Troya, S., L. López-Maury, A.M. Sánchez-Riego, M. Roldán and F.J. Florencio. 2014. Redox regulation of glycogen biosynthesis in the *Cyanobacterium synechocystis* sp. PCC 6803: Analysis of the AGP and glycogen synthases. Mol. Plant 7: 87–100.

Doello, S., A. Klotz, A. Makowka, K. Gutekunst and K. Forchhammer. 2018. A specific glycogen mobilization strategy enables rapid awakening of dormant cyanobacteria from chlorosis. Plant Physiol. 177: 594–603.

Ehsaan, M., W. Kuit, Y. Zhang, S.T. Cartman, J.T. Heap, K. Winzer et al. 2016. Mutant generation by allelic exchange and genome resequencing of the biobutanol organism Clostridium acetobutylicum ATCC 824. Biotechnol. Biofuels. doi: 10.1186/s13068-015-0410-0.

Eme, L., S.C. Sharpe, M.W. Brown and A.J. Roger. 2014. On the age of eukaryotes: evaluating evidence from fossils and molecular clocks. Cold Spring Harb Perspect. Biol. doi: 10.1101/cshperspect.a016139.

Falcon, L.I., F. Cipriano, A.Y. Chistoserdov and E.J. Carpenter. 2002. Diversity of diazotrophic unicellular cyanobacteria in the tropical North Atlantic Ocean. Appl. Environ. Microbiol. 68: 5760–5764.

Figueroa, C.M., M.D. Asención Diez, M.L. Kuhn, S. McEwen, G.L. Salerno, A.A. Iglesias et al. 2013. The unique nucleotide specificity of the sucrose synthase from Thermosynechococcus elongatus. FEBS Lett. 587: 165–169.

Flamholz, A., E. Noor, A. Bar-Even, W. Liebermeister and R. Milo. 2013. Glycolytic strategy as a tradeoff between energy yield and protein cost. Proc. Natl. Acad. Sci. 110: 10039–10044.

Fraga, J., A. Maranha, V. Mendes, P.J.B. Pereira, N. Empadinhas and S. Macedo-Ribeiro. 2015. Structure of mycobacterial maltokinase, the missing link in the essential GlgE-pathway. Sci. Rep. 5: 8026.

Fu, J. and X. Xu. 2006. The functional divergence of two glgP homologues in *Synechocystis* sp. PCC 6803. FEMS Microbiol. Lett. 260: 201–209.

Fukusumi, S., A. Kamizoto, S. Horinouchi and B. Teruhiko. 1988. Cloning and nucleotide sequence of a heat-stable amylase gene from an anaerobic thermophile, Dictyoglomus thermophilum. Eur. J. Biochem. 174: 15–21.

Gehre, L., O. Gorgette, S. Perrinet, M.-C. Prevost, M. Ducatez, A.M. Giebel et al. 2016. Sequestration of host metabolism by an intracellular pathogen. Elife. doi: 10.7554/eLife.12552.

Gründel, M., R. Scheunemann, W. Lockau and Y. Zilliges. 2012. Impaired glycogen synthesis causes metabolic overflow reactions and affects stress responses in the cyanobacterium *Synechocystis* sp. PCC 6803. Microbiol. (United Kingdom) 158: 3032–3043.

Hayashi, M., R. Suzuki, C. Colleoni, S.G. Ball, N. Fujita and E. Suzuki. 2015. Crystallization and crystallographic analysis of branching enzymes from *Cyanothece* sp. ATCC 51142. Acta Crystallogr Sect. Struct. Biol. Commun. doi: 10.1107/S2053230X1501198X.

Hayashi, M., R. Suzuki, C. Colleoni, S.G. Ball, N. Fujita and E. Suzuki. 2017. Bound substrate in the structure of cyanobacterial branching enzyme supports a new mechanistic model. J. Biol. Chem. doi: 10.1074/jbc. M116.755629.

Hengge-Aronis, R. and D. Fischer. 1992. Identification and molecular analysis of glgS, a novel growth-phase-regulated and rpoS-dependent gene involved in glycogen synthesis in *Escherichia coli*. Mol. Microbiol. 6: 1877–1886.

Henrissat, B., E. Deleury and P.M. Coutinho. 2002. Glycogen metabolism loss: A common marker of parasitic behaviour in bacteria? Trends Genet. 18: 437–440.

Holmes, E. and J. Preiss. 1979. Characterization of *Escherichia coli* B glycogen synthase enzymatic reactions and products. Arch. Biochem. Biophys. 196: 436–448.

Iglesias, A., G. Kakefuda and J. Preiss. 1991. Regulatory and structural properties of the cyanobacterial adpglucose pyrophosphorylases. Plant Physiol. 97: 1187–1195.

James, M., D. Robertson and A. Myers. 1995. Characterization of the maize gene sugary1, a determinant of starch composition in kernels. Plant Cell 7: 417–429.

Jendrossek, D. and D. Pfeiffer. 2014. New insights in the formation of polyhydroxyalkanoate granules (carbonosomes) and novel functions of poly(3-hydroxybutyrate). Environ. Microbiol. 16: 2357–2373.

Jo, H.J., S. Park, H.G. Jeong, J.W. Kim and J.T. Park. 2015. Vibrio vulnificus glycogen branching enzyme preferentially transfers very short chains: N1 domain determines the chain length transferred. FEBS Lett. 589: 1089–1094.

Johnstone, K. and K.T. Holland. 1977. Ultrastructural changes during sporulation of *Clostridium bifermentans*. J. Gen. Microbiol. 100: 217–20.

Kadouche, D., M. Ducatez, U. Cenci, C. Tirtiaux, E. Suzuki, Y. Nakamura et al. 2016. Characterization of function of the GlgA2 glycogen/starch synthase in *Cyanobacterium* sp. CLg1 highlights convergent evolution of glycogen metabolism into starch granule aggregation. Plant Physiol. 171: 1879–1892.

Kalscheuer, R., K. Syson, U. Veeraraghavan, B. Weinrick, K.E. Biermann, Z. Liu et al. 2010. Self-poisoning of *Mycobacterium tuberculosis* by targeting GlgE in an α-glucan pathway. Nat. Chem. Biol. 6: 376–384.

Kawaguchi, K., J. Fox, E. Holmes, C. Boyer and J. Preiss. 1978. *De Novo* synthesis of *Escherichia coli* glycogen is due to primer associated with glycogen synthase and activation by branching enzyme. Arch. Biochem. Biophys. 190: 385–397.

Khadem, A.F., A. Pol, M.S.M. Jetten and H.J.M. Op Den Camp. 2010. Nitrogen fixation by the verrucomicrobial methanotroph "*Methylacidiphilum fumariolicum*" SolV. Microbiology 156: 1052–1059.

Khadem, A.F., A. Pol, A. Wieczorek, S.S. Mohammadi, K.-J. Francoijs, H.G. Stunnenberg et al. 2011. Autotrophic Methanotrophy in Verrucomicrobia: *Methylacidiphilum fumariolicum* SolV Uses the Calvin-Benson-Bassham cycle for carbon dioxide fixation. J. Bacteriol. 193: 4438–4446.

Khadem, A.F., M.C.F. van Teeseling, L. van Niftrik, M.S.M. Jetten, H.J.M. Op den Camp and A. Pol. 2012. Genomic and physiological analysis of carbon storage in the verrucomicrobial methanotroph Ca. *Methylacidiphilum fumariolicum* SolV. Front Microbiol. 3: 345.

Koliwer-Brandl, H., K. Syson, R. van de Weerd, G. Chandra, B. Appelmelk, M. Alber et al. 2016. Metabolic Network for the Biosynthesis of Intra- and Extracellular α-Glucans Required for Virulence of *Mycobacterium tuberculosis*. PLoS Pathog 12: 1–26.

Kopp, R.E., J.L. Kirschvink, I.A. Hilburn and C.Z. Nash. 2005. The Paleoproterozoic snowball Earth: A climate disaster triggered by the evolution of oxygenic photosynthesis. Proc. Natl. Acad. Sci. 102: 11131–11136.

Kubo, A., N. Fujita, K. Harada, T. Matsuda, H. Satoh and Y. Nakamura. 1999. The starch-debranching enzymes isoamylase and pullulanase are both involved in amylopectin biosynthesis in rice endosperm. Plant Physiol. 121: 399–410.

Laderman, K.A., K. Asada, T. Uemori, H. Mukai, Y. Taguchi, I. Kato et al. 1993. α-Amylase from the hyperthermophilic archaebacterium Pyrococcus furiosus. Cloning and sequencing of the gene and expression in *Escherichia coli*. J. Biol. Chem. 268: 24402–24407.

Leiba, J., K. Syson, G. Baronian, I. Zanella-Cléon, R. Kalscheuer, L. Kremer et al. 2013. *Mycobacterium tuberculosis* maltosyltransferase GlgE, a genetically validated antituberculosis target, is negatively regulated by Ser/Thr phosphorylation. J. Biol. Chem. 288: 16546–16556.

Leloir, L.F. and C.E. Cardini. 1957. Biosynthesis of glycogen from uridine diphosphate glucose. J. Am. Chem. Soc. 79: 6340–6341.

Lemassu, A. and M. Daffé. 1994. Structural features of the exocellular polysaccharides of *Mycobacterium tuberculosis*. Biochem J 297, Pt 2: 351–357.

Lindner, A.B., R. Madden, A. Demarez, E.J. Stewart and F. Taddei. 2008. Asymmetric segregation of protein aggregates is associated with cellular aging and rejuvenation. Proc. Natl. Acad. Sci. 105: 3076–3081.

Lou, J., K.A. Dawson and H.J. Strobel. 1997. Glycogen biosynthesis via UDP-glucose in the ruminal bacterium Prevotella bryantii B14. Appl. Environ. Microbiol. 63: 4355–4359.

Mattoo, R.L. and E.B. Waygood. 1983. Determination of the levels of HPr and enzyme I of the phosphoenolpyruvate-sugar phosphotransferase system in *Escherichia coli* and *Salmonella typhimurium*. Can. J. Biochem. Cell. Biol. 61: 29–37.

Melendez-Hevia, E., T.G. Waddellt and E.D. Sheltont. 1993. Optimization of molecular design in the evolution of metabolism : the glycogen molecule. J. Bacteriol. 483: 477–483.

Meléndez, R., E. Meléndez-Hevia and M. Cascante. 1997. How did glycogen structure evolve to satisfy the requirement for rapid mobilization of glucose? A problem of physical constraints in structure building. J. Mol. Evol. 45: 446–455.

Meléndez, R., E. Meléndez-Hevia and E.I. Canela. 1999. The fractal structure of glycogen: A clever solution to optimize cell metabolism. Biophys. J. 77: 1327–1332.

Miao, X., Q. Wu, G. Wu and N. Zhao. 2003. Changes in photosynthesis and pigmentation in an agp deletion mutant of the cyanobacterium *Synechocystis* sp. Biotechnol. Lett. 25: 391–396.

Monod, J. and A.-M. Torriani. 1950. De l'amylomaltase *d'Escherichia coli*. Ann. Inst. Pasteur. 78: 65–77.

Mouille, G., M.-L. Maddelein, N. Libessart, P. Talaga, A. Decq, B. Delrue et al. 1996. Preamylopectin processing—A mandatory step for starch biosynthesis in plants. Plant Cell 8: 1353–1366.

Murakami, T., T. Kanai, H. Takata, T. Kuriki and T. Imanaka. 2006. A novel branching enzyme of the GH-57 family in the hyperthermophilic archaeon Thermococcus kodakaraensis KOD1. J. Bacteriol. 188: 5915–5924.

Nakamura, Y., J.I. Takahashi, A. Sakurai, Y. Inaba, E. Suzuki, S. Nihei et al. 2005. Some cyanobacteria synthesize semi-amylopectin type α-polyglucans instead of glycogen. Plant Cell Physiol. 46: 539–545.

Newton, I.L.G., T. Woyke, T.A. Auchtung, G.F. Dilly, R.J. Dutton, M.C. Fisher et al. 2007. The Calyptogena magnifica Chemoautotrophic Symbiont Genome. Science 315: 998–1001.

Nikoh, N., T. Hosokawa, K. Oshima, M. Hattori and T. Fukatsu. 2011. Reductive evolution of bacterial genome in insect gut environment. Genome Biol. Evol. 3: 702–714.

Palomo, M., S. Kralj, M.J.E.C. Van Der Maarel and L. Dijkhuizen. 2009. The unique branching patterns of Deinococcus glycogen branching enzymes are determined by their N-terminal domains. Appl. Environ. Microbiol. 75: 1355–1362.

Park, J.T., J.H. Shim, P.L. Tran, I.H. Hong, H.U. Yong, E.F. Oktavina et al. 2011. Role of maltose enzymes in glycogen synthesis by *Escherichia coli*. J. Bacteriol. 193: 2517–2526.

Pol, A., K. Heijmans, H.R. Harhangi, D. Tedesco, M.S.M. Jetten and H.J.M. Op Den Camp. 2007. Methanotrophy below pH 1 by a new *Verrucomicrobia* species. Nature 450: 874–878.

Ponce-Toledo, R.I., P. Deschamps, P. López-García, Y. Zivanovic, K. Benzerara and D. Moreira. 2017. An early-branching freshwater cyanobacterium at the origin of plastids. Curr. Biol. 27: 386–391.

Porchia, A.C., L. Curatti and G.L. Salerno. 1999. Sucrose metabolism in cyanobacteria: Sucrose synthase from *Anabaena* sp. strain PCC 7119 is remarkably different from the plant enzymes with respect to substrate affinity and amino-terminal sequence. Planta 210: 34–40.

Preiss, J. 2014. Glycogen: Biosynthesis and Regulation. EcoSal Plus. doi: 10.1128/ecosalplus.ESP-0015-2014.

Puszynska, A.M. and E.K. O'Shea. 2017. Switching of metabolic programs in response to light availability is an essential function of the cyanobacterial circadian output pathway. Elife. doi: 10.7554/eLife.23210.

Rahimpour, M., M. Montero, G. Almagro, A.M. Viale, Á. Sevilla, M. Cánovas et al. 2013. GlgS, described previously as a glycogen synthesis control protein, negatively regulates motility and biofilm formation in *Escherichia coli*. Biochem. J. 452: 559–573.

Reysenbach, A.L., N. Ravenschroft and S. Long. 1986. Characterization, biosynthesis, and regulation of granulose in *Clostridium acetobutylicum*. Appl. Environ. Microbiol. 52: 185–190.

Robson, R.L., R.M. Robson and J.G. Morris. 1974. The biosynthesis of granulose by *Clostridium pasteurianum*. Biochem. J. 144: 503–11.

Roy, R., V. Usha, A. Kermani, D.J. Scott, E.I. Hyde, G.S. Besra et al. 2013. Synthesis of α-glucan in mycobacteria involves a hetero-octameric complex of trehalose synthase tres and maltokinase Pep2. ACS Chem. Biol. 8: 2245–2255.

Sabnis, N.A., H. Yang and T. Romeo. 1995. Pleiotropic regulation of central carbohydrate metabolism in *Escherichia coli* via the gene csrA. J. Biol. Chem. 270: 29096–29104.

Sambou, T., P. Dinadayala, G. Stadthagen, N. Barilone, Y. Bordat, P. Constant et al. 2008. Capsular glucan and intracellular glycogen of *Mycobacterium tuberculosis* : biosynthesis and impact on the persistence in mice. Mol. Microbiol. 70: 762–774.

Sánchez-Baracaldo, P., A. Ridgwell and J.A. Raven. 2014. A neoproterozoic transition in the marine nitrogen cycle. Curr. Biol. 24: 652–657.

Santos, C.R., C.C.C. Tonoli, D.M. Trindade, C. Betzel, H. Takata T. Kuriki et al. 2011. Structural basis for branching-enzyme activity of glycoside hydrolase family 57: Structure and stability studies of a novel branching enzyme from the hyperthermophilic archaeon Thermococcus Kodakaraensis KOD1. Proteins Struct. Funct. Bioinforma. 79: 547–557.

Sawada, T., Y. Nakamura, T. Ohdan, A. Saitoh, P.B. Francisco, E. Suzuki et al. 2014. Diversity of reaction characteristics of glucan branching enzymes and the fine structure of α-glucan from various sources. Arch. Biochem. Biophys. doi: 10.1016/j.abb.2014.07.032.

Schneegurt, M.A., D.M. Sherman, S. Nayar and L.A. Sherman. 1994. Oscillating behavior of carbohydrate granule formation and dinitrogen fixation in the cyanobacterium *Cyanothece* sp. strain ATCC51142. J. Bacteriol. 176: 1586–1597.

Schneegurt, M.A., D.M. Sherman and L.A. Sherman. 1997. Composition of the carbohydrate granules of the cyanobacterium, *Cyanothece* sp. strain ATCC 51142. Arch. Microbiol. 167: 89–98.

Schwartz, M. 1967. Expression phenotypique et localisation génétique de mutations affectant le métabolism du maltose chez *Escherichia coli* k12. Ann. Inst. Pasteur. 112: 673.

Seibold, G.M., M. Wurst and B.J. Eikmanns. 2009. Roles of maltodextrin and glycogen phosphorylases in maltose utilization and glycogen metabolism in *Corynebacterium glutamicum*. Microbiology 155: 347–358.

Seok, Y.J., M. Sondej, P. Badawi, M.S. Lewis, M.C. Briggs, H. Jaffe et al. 1997. High affinity binding and allosteric regulation of *Escherichia coli* glycogen phosphorylase by the histidine phosphocarrier protein, HPr. J. Biol. Chem. 272: 26511–26521.

Seok, Y.J., B.M. Koo, M. Sondej and A. Peterkofsky. 2001. Regulation of *E. coli* glycogen phosphorylase activity by HPr. J. Mol. Microbiol. Biotechnol. 3: 385–393.

Singh, P.K. 1973. Nitrogen fixation by the unicellular green-alga Aphanothece. Arch. Mikrobiol. 92: 59–62.

Stam, M.R., E.G.J. Danchin, C. Rancurel, P.M. Coutinho and B. Henrissat. 2006. Dividing the large glycoside hydrolase family 13 into subfamilies: Towards improved functional annotations of α-amylase-related proteins. Protein Eng. Des. Sel. 19: 555–562.

Stewart, E.J., R. Madden, G. Paul and F. Taddei. 2005. Aging and death in an organism that reproduces by morphologically symmetric division. PLoS Biol. 3: 0295–0300.

Streb, S., T. Delatte, M. Umhang, S. Eicke, M. Schorderet, D. Reinhardt et al. 2008. Starch granule biosynthesis in arabidopsis is abolished by removal of all debranching enzymes but restored by the subsequent removal of an endoamylase. Plant Cell 20: 3448–3466.

Suzuki, E., K. Umeda, S. Nihei, K. Moriya, H. Ohkawa, S. Fujiwara et al. 2007. Role of the GlgX protein in glycogen metabolism of the cyanobacterium, *Synechococcus elongatus* PCC 7942. Biochim. Biophys. Acta - Gen. Subj. 1770: 763–773.

Suzuki, E., H. Ohkawa, K. Moriya, T. Matsubara, Y. Nagaike, I. Iwasaki et al. 2010. Carbohydrate metabolism in mutants of the *Cyanobacterium synechococcus* elongatus PCC 7942 defective in glycogen synthesis. Appl. Environ. Microbiol. 76: 3153–3159.

Suzuki, E., M. Onoda, C. Colleoni, S. Ball, N. Fujita and Y. Nakamura. 2013. Physicochemical variation of cyanobacterial starch, the insoluble α-glucans in cyanobacteria. Plant Cell Physiol. doi: 10.1093/pcp/pcs190.

Suzuki, E. and R. Suzuki. 2016. Distribution of glucan-branching enzymes among prokaryotes. Cell Mol. Life Sci. 73: 2643–2660.

Suzuki, R., K. Koide, M. Hayashi, T. Suzuki, T. Sawada, T. Ohdan et al. 2015. Functional characterization of three (GH13) branching enzymes involved in cyanobacterial starch biosynthesis from *Cyanobacterium* sp. NBRC 102756. Biochim. Biophys. Acta - Proteins Proteomics 1854: 476–484.

Syson, K., C.E.M. Stevenson, A.M. Rashid, G. Saalbach, M. Tang, A. Tuukkanen et al. 2014. Structural insight into how Streptomyces coelicolor maltosyl transferase GlgE binds α-maltose 1-phosphate and forms a maltosyl-enzyme intermediate. Biochemistry 53: 2494–2504.

Takata, H., T. Takaha, S. Okada, M. Takagi and T. Imanaka. 1998. Purification and characterization of α-glucan phosphorylase from *Bacillus stearothermophilus*. J. Ferment. Bioeng. 85: 156–161.

Testoni, G., J. Duran, M. García-Rocha, F. Vilaplana, A.L. Serrano, D. Sebastián et al. 2017. Lack of glycogenin causes glycogen accumulation and muscle function impairment. Cell Metab. 26: 256–266.e4.

Torija, M.J., M. Novo, A. Lemassu, W. Wilson, P.J. Roach, J. François et al. 2005. Glycogen synthesis in the absence of glycogenin in the yeast *Saccharomyces cerevisiae*. FEBS Lett. 579: 3999–4004.

Tracy, B.P., S.M. Gaida and E.T. Papoutsakis. 2008. Development and application of flow-cytometric techniques for analyzing and sorting endospore-forming clostridia. Appl. Environ. Microbiol. 74: 7497–7506.

Ugalde, J.E., A.J. Parodi and R.A. Ugalde . 2003. *De novo* synthesis of bacterial glycogen: *Agrobacterium tumefaciens* glycogen synthase is involved in glucan initiation and elongation. Proc. Natl. Acad. Sci. U. S. A. 100: 10659–10663.

Wang, L., A. Regina, V.M. Butardo, B. Kosar-Hashemi, O. Larroque, C.M. Kahler et al. 2015. Influence of *in situ* progressive N-terminal is still controversial truncation of glycogen branching enzyme in *Escherichia coli* DH5α on glycogen structure, accumulation, and bacterial viability. BMC Microbiol. doi: 10.1186/s12866-015-0421-9.

Wilson, W.A., P.J. Roach, M. Montero, E. Baroja-Fernández, F.J. Muñoz, G. Eydallin et al. 2010. Regulation of glycogen metabolism in yeast and bacteria. FEMS Microbiol. Rev. 34: 952–985.

Wyatt, J.T. and J.K.G. Silvey. 1969. Nitrogen fixation by Gloeocapsa. Science 165: 908–909.

Yoo, S.-H., B.-H. Lee, Y. Moon, M.H. Spalding and J. Jane. 2014. Glycogen synthase isoforms in *Synechocystis* sp. PCC6803: Identification of different roles to produce glycogen by targeted mutagenesis. PLoS One 9: e91524.

Yoon, H.S., J.D. Hackett, C. Ciniglia, G. Pinto and D. Bhattacharya. 2004. A molecular timeline for the origin of photosynthetic eukaryotes. Mol. Biol. Evol. 21: 809–818.

Zona, R., F. Chang-Pi-Hin, M.J. O'Donohue and Š. Janeček. 2004. Bioinformatics of the glycoside hydrolase family 57 and identification of catalytic residues in amylopullulanase from Thermococcus hydrothermalis. Eur. J. Biochem. 271: 2863–2872.

CHAPTER 5

Mammalian Glycogen Metabolism
Enzymology, Regulation, and Animal Models of Dysregulated Glycogen Metabolism

Bartholomew A. Pederson

5.1 Introduction

Glycogen synthesis and degradation require several enzymes, with glycogen synthase and glycogen phosphorylase being key enzymes in elongating and shortening, respectively, glucose chains in the glycogen macromolecule (Fig. 5.1). Complex, reciprocal regulation of glycogen synthase and glycogen phosphorylase controls the amount of glycogen. Mammalian glycogen metabolism has been studied most extensively in muscle and liver. In the muscle, glycogen serves as a fuel for contraction and as a storage depot for glucose removed postprandially from the circulation. In the liver, glycogen is synthesized to remove excess postprandial glucose from the bloodstream and subsequently degraded during times of fasting to provide glucose to the body. Glycogen metabolism has also been studied to a lesser degree in other tissues, notably heart (Taegtmeyer 2004) and kidney (Delaval et al. 1983), and especially of late, the brain (DiNuzzo and Schousboe 2019). This chapter focuses primarily on mammalian glycogen metabolism in muscle, liver, and brain. In addition to discussion of the enzymology and regulation of glycogen metabolism, select glycogen storage diseases are discussed, focusing on those where glycogen synthesis has been modulated in animal models of the disease.

Glycogen is stored throughout the body, with the largest reserves found in liver and skeletal muscle. The metabolic pathway for the synthesis of glycogen from glucose (glycogenesis) requires the sequential actions of glucose transport, hexokinase, phosphoglucomutase, UDP-glucose pyrophosphorylase, glycogen synthase, and glycogen branching enzyme. Glycogen synthase catalyzes the transfer of glucose from UDP-glucose to a growing glycogen chain and the glycogen branching enzyme introducing branches in the glycogen molecule. Regulation of synthesis is primarily controlled by glucose transport and modulation of glycogen synthase enzymatic activity, with activity generally being highest when glucose levels are relatively high. In contrast, when glucose levels are relatively low (liver) or when energy demands increase (muscle), degradation of glycogen is favored. There are two pathways for the breakdown of glycogen. One pathway (glycogenolysis) is catalyzed by glycogen phosphorylase and the glycogen debranching enzyme. The former enzyme sequentially removes glucose residues, as glucose-1-P, from a glycogen chain. The latter enzyme both remodels the degrading glycogen molecule and releases the final unit from each chain as free glucose. Glucose-1-P is then acted on by phosphoglucomutase to form glucose-6-P. In liver, and to a lesser extent in

Ball State University/Indiana University School of Medicine-Muncie, Muncie, IN 47303, Email: bapederson@bsu.edu

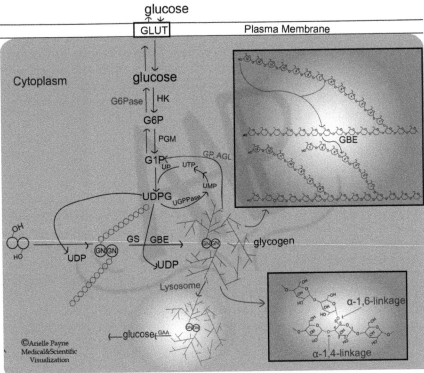

Figure 5.1. Glycogen metabolic pathway. GBE, glycogen branching enzyme; AGL, debranching enzyme; G6P, glucose-6-phosphate; G6Pase, glucose-6-phosphatase; GP, glycogen phosphorylase; GAA, alpha-glucosidase; G1P, glucose-1-phosphate; GLUT, glucose transporter; GN, glycogenin; GS, glycogen synthase; HK, hexokinase; PGM, phosphoglucomutase; UDPG, uridine diphosphate glucose; UGPPase, UDP-glucose pyrophosphatase; UMP, uridine monophosphate; UP, UDP-glucose pyrophosphorylase; UTP, uridine triphosphate. Hexagons depict glucose molecules. Illustration by Arielle Payne.

kidney and small intestine, glucose-6-phosphatase converts this metabolite to free glucose which is released into the circulation to increase blood glucose levels. In muscle, glucose-6-P is metabolized in the glycolytic pathway to generate ATP for the cell. Glycogenolysis is regulated primarily through modulation of glycogen phosphorylase activity. A second pathway (glycophagy) also exists for glycogen degradation. In this pathway, glycogen is taken up into lysosomes and degraded to free glucose by the action of alpha-glucosidase.

5.1.1 Role of Muscle Glycogen

Blood glucose maintenance. Increases in postprandial insulin levels promote glucose uptake into muscle cells and activation of glycogen synthase. In humans, this results in 26 to 35 percent of glucose from a meal being stored as glycogen in skeletal muscle (Taylor et al. 1993).

Exercise. The importance of muscle glycogen during exercise is borne out by both physiological and pathological studies. Impaired performance is observed in many patients with glycogen storage diseases that affect availability of muscle glycogen (Preisler et al. 2015). The benefit of muscle glycogen for exercise is well established and exploited by athletes (reviewed in Hargreaves 2015, Hawley et al. 2018, Jensen and Richter 2012, Nielsen et al. 2011). Saltin and colleagues conducted seminal studies where they found that exercise reduced muscle glycogen levels in contracting muscle (Bergstrom and Hultman 1966, 1967), muscle glycogen levels prior to exercise correlated with endurance exercise capacity (Bergstrom et al. 1967), the utilization of muscle glycogen was related to the intensity and duration of exercise (Hermansen et al. 1967), and the depletion of muscle glycogen was associated with exercise fatigue (Karlsson et al. 1974). Three pools of glycogen have been observed in muscle—intermyofibrillar, intramyofibrillar, and subsarcolemmal—with each being proposed to serve a distinct function (Ortenblad and Nielsen 2015). In addition to serving as a source of ATP to power muscle contraction, glycogen has been proposed to serve in a structural role as well as being a regulator of cell signaling (reviewed in Hawley et al. 2018, Hearris et al. 2018). The breakdown of muscle glycogen during exercise is stimulated by increases in the levels of inorganic phosphate and AMP and possibly increased calcium, all of which promote activation of glycogen phosphorylase (reviewed in Jensen and Richter 2012).

5.1.2 Role of Liver Glycogen

Blood glucose maintenance. The primary role of glycogen metabolism in the liver is in the maintenance of blood glucose levels. During times of excess glucose in the blood stream, such as after a carbohydrate-containing meal, the combination of elevated glucose and insulin levels promotes glucose transport into the liver, glucose phosphorylation, and synthesis of glycogen (Roach et al. 2001). The synthesis of liver glycogen after ingestion of a meal is either from glucose released during digestion, termed the direct pathway, or alternatively glycogen is synthesized from gluconeogenic substrates, termed the indirect pathway (Kurland and Pilkis 1989). In humans, approximately 19 percent of the glycogen synthesized after a meal is stored in the liver (Taylor et al. 1996). When blood glucose levels drop, such as after a period of fasting, insulin levels decrease, and glucagon levels increase. These changes promote liver glycogen degradation and release of glucose into the blood stream to provide fuel to other tissues of the body (Roach et al. 2001).

Exercise. While the role of muscle glycogen in exercise has received much study, the importance of glycogen in the liver for exercise has garnered less attention. Exercise-induced hepatic glucose output from gluconeogenesis and glycogenolysis is regulated by a decrease in hepatocyte energy status in conjunction with changes in levels of circulating hormones including increased glucagon, decreased insulin, and possibly elevated catecholamines (Gonzalez et al. 2016, Trefts et al. 2015). The glucose released from liver glycogen stores can be used by muscle cells to fuel contraction. Significant liver glycogen depletion impairs the body's ability to maintain blood glucose levels and is generally accepted to be a major cause of endurance exercise fatigue (Gonzalez et al. 2016, Hearris et al. 2018).

5.1.3 Role of Brain Glycogen

As compared to muscle and liver, glycogen metabolism in brain is less well studied, due in part to the relatively low concentrations of glycogen found in this organ (Brown 2004). However, brain glycogen stores have been proposed to play a role in diverse processes such as sleep regulation (Kong et al. 2002, Petit et al. 2015), exercise (Matsui et al. 2012), and learning (Bak et al. 2018), as well as in pathological situations such as seizure (Cloix and Hevor 2009, Dalsgaard et al. 2007), ischemia (Fern 2015, Rossi et al. 2007), hypoglycemia (Brown and Ransom 2015, Suh et al. 2007b), and hypoxia (Czech-Damal et al. 2014, Lopez-Ramos et al. 2015).

Learning. Brain glycogen metabolism has been proposed to contribute to normal neural function. Notably, the majority of the glycogen is found in astrocytes with the highest levels in areas of high synaptic density (Koizumi and Shiraishi 1970a,b, Phelps 1972), suggesting that astrocytic glycogen may be involved in neuronal activity. Brain glycogen is mobilized during memory formation in day-old chicks and is reported to be regulated by noradrenaline and serotonin (reviewed in Gibbs 2015). In addition to serving as a source of rapid ATP generation, brain glycogen may act as a preferred substrate for glutamate synthesis (Gibbs et al. 2006, Gibbs et al. 2007). Inhibition of glycogen mobilization in day-old chicks impaired learning ability (Gibbs et al. 2006). Additionally, rats treated with an inhibitor of glycogen phosphorylase had impairments in long-term memory, but not short-term memory in one study (Suzuki et al. 2011), or had impairments of both long and short term memory in a different study (Newman et al. 2011). Most recently, mice with a brain specific disruption of brain glycogen synthase (GYS1) retained the ability to learn, though at a slower pace (Duran et al. 2013). These findings suggest a role for brain glycogen in learning, but our understanding remains incomplete.

Astrocytes play an active role in the memory process, and evidence suggests there is metabolic coupling between astrocytes and neurons. Besides the direct utilization of glucose by neurons and astrocytes, it has been proposed that astrocytes can provide fuel to neurons in the form of lactate produced from the metabolism of either glucose or glycogen (Chih and Roberts 2003). The shuttling of lactate from the astrocyte to the neuron is known as the astrocyte neuron lactate shuttle hypothesis (ANLSH) (Pellerin et al. 1998). In the context of the ANLSH model, glycogen in astrocytes could provide fuel to neurons through degradation to glucose-6-P and subsequent conversion to lactate through the glycolytic pathway. Lactate could be shuttled across the extracellular space to adjacent neurons where pyruvate generated from lactate would enter the TCA cycle to generate ATP for the neuron (Fig. 5.2). Lactate serves as a fuel in the brain, being preferred over glucose when both substrates are present (Wyss et al. 2011), and disruption of brain lactate transporter expression in rats resulted in learning impairments (Suzuki et al. 2011). Additionally, intrahippocampal lactate administration overcame memory impairment elicited by glycogen phosphorylase inhibition (Newman et al. 2011), further supporting the importance of lactate in learning and metabolic coupling between neurons and astrocytes. However, interpretation of these studies is not without its challenges (Dienel and Cruz 2016). The ANLSH model does not accommodate several experimental findings (reviewed in Dienel 2017, 2019) and thus remains quite controversial. An alternative model proposes that when

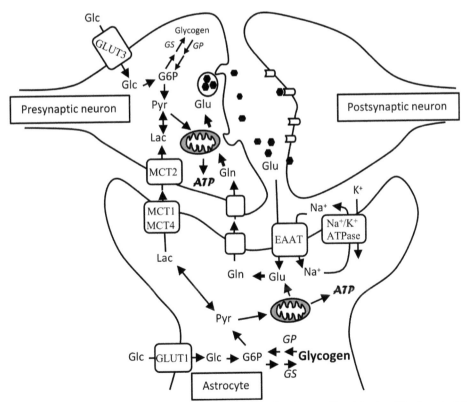

Figure 5.2. Brain glycogen metabolism. G6P, glucose-6-phosphate; Glc, glucose; Gln, glutamine; GS, glycogen synthase; GP, glycogen phosphorylase; Glu, glutamate; Lac, lactate; MCT1, MCT2, MCT4, monocarboxylate transporters; Pyr, pyruvate; EAAT, glutamate transporters (reviewed in Amara and Fontana 2002). There are multiple glutamine transporters (reviewed in Leke and Schousboe 2016).

glycogen is broken down, glucose-6-P concentration increases in astrocytes resulting in greater hexokinase inhibition (DiNuzzo et al. 2011). This model proposes that hexokinase inhibition leads to reduced glucose uptake in astrocytes, increasing availability of glucose for neurons. These models are not mutually exclusive and remain actively under investigation.

Hypoglycemia. The brain is dependent on glucose as its principal fuel (Amaral 2013) and acute hypoglycemia, a common occurrence with intensive insulin treatment of diabetes, can lead to substantial cognitive impairments. These impairments are numerous and include short- and long-term memory impairment (Sommerfield et al. 2003a,b), drowsiness, lack of coordination, confusion, and even coma (Smith and Amiel 2002). In some studies, recurrent hypoglycemia is associated with permanent functional brain abnormalities (reviewed in Warren and Frier 2005) including neuronal cell death (Auer 1986). In contrast, other studies found no cognitive impairment in humans with recurrent hypoglycemia (reviewed in McNay and Cotero 2010). In rodent models, acute hypoglycemia impaired learning (Won et al. 2012b). However, recurrent hypoglycemia either impaired learning (Won et al. 2012c) or enhanced memory during euglycemia but impaired memory during hypoglycemia (McNay and Sherwin 2004, McNay et al. 2006). A protective role for glycogen has been suggested (reviewed in Belanger et al. 2011). Glycogen could mitigate hypoglycemic impairments by providing a source of energy and providing substrates for neurotransmitter synthesis when glucose levels are insufficient (Magistretti and Pellerin 1996). Further studies are needed to determine whether brain glycogen mitigates hypoglycemia-induced learning impairment.

Glycogen in the brain has been hypothesized to be a contributor to hypoglycemia-associated autonomic failure (HAAF) which occurs with recurrent hypoglycemia. In HAAF, the counterregulatory response to hypoglycemia is impaired, increasing the risk of complications (Geddes et al. 2008). In rats, brain glycogen levels decreased with hypoglycemia (Choi et al. 2003). Upon restoration of blood glucose to normal levels, brain glycogen levels were higher than normal, termed supercompensation. This finding led to the hypothesis that these elevated brain glycogen levels would be available for metabolism during a subsequent bout of hypoglycemia, making the brain less dependent on glucose, effectively lowering the glucose threshold at which the brain mounts a counterregulatory response. Brain glycogen supercompensation after hypoglycemia was also observed in mice (Canada et al. 2011) but not observed in another study with rats (Herzog et al. 2008) or in a study involving patients with type 1 diabetes (Oz et al. 2012). Therefore, the importance of brain glycogen during HAAF is incompletely understood.

Severe hypoglycemic events result in substantial neuronal damage. Patients who suffered from either fatal or non-fatal bouts of severe hypoglycemia (blood glucose less than ~ 1.5 mM) showed signs of brain injury, particularly in the frontal lobes and hippocampus (Languren et al. 2013, Warren and Frier 2005). Cortical atrophy was found in 45 percent of type 1 diabetic patients with a history of recurrent severe hypoglycemia (Perros et al. 1997). Acute severe hypoglycemia, 1.7–1.9 mM (Tkacs et al. 2005) or ~ 0.4 mM (Suh et al. 2007a, Won et al. 2012a,b) also elicits neuronal cell death in rats. Several mechanisms have been implicated in the process of hypoglycemia-induced neuronal cell death, but remain incompletely understood (reviewed in Amaral 2013). A negative correlation between hypoglycemia-induced cell death and brain glycogen has been suggested. Neuronal cultures survive better when co-cultured in the presence of glycogen-containing astrocytes (Swanson and Choi 1993). If astrocytes are depleted of glycogen, this benefit is compromised. Treatment of rats with a glucose-dependent inhibitor of glycogen phosphorylase caused an increase in brain glycogen during normoglycemia that correlated with increased neuronal survival upon exposure to hypoglycemia followed by restoration of blood glucose levels (Suh et al. 2007a). During hypoglycemia, low levels of brain glucose (neuroglycopenia) reduce the inhibitor's effectiveness and allow the degradation of the elevated brain glycogen stores. Brain glycogen has the capacity to sustain glycolysis from 10 min (Clark and Sokoloff 1999) to over 100 min (Gruetter 2003). Additionally, brain glycogen mobilization was impaired in Goto-Kakizaki insulin resistant rats (Duarte et al. 2016). Glycogen mobilization may be neuroprotective by (1) providing ATP for Na^+/K^+-ATPase dependent maintenance of ion gradients (Xu et al. 2013) and the uptake of excitotoxic glutamate from the synaptic cleft (Gruetter 2003, Obel et al. 2012), (2) providing lactate as a fuel for neurons (Magistretti and Pellerin 1996) and (3) sparing glucose for use by neurons (DiNuzzo et al. 2011) (Fig. 5.2). These studies suggest that brain glycogen could provide a readily available source of energy for brain function when local glucose supplies are limiting, as occurs in hypoglycemia/neuroglycopenia.

Hypoxia. In post-natal rats, survival in hypoxic conditions was inversely correlated to cerebral energy consumption (Duffy et al. 1975). Initial brain glycogen levels correlated with tolerance to hypoxia and both decreased with the age. Diving seals, which are exposed to repeated episodes of hypoxia, have brain glycogen levels several fold higher than mice (Czech-Damal et al. 2014). When brain slices were subjected to hypoxia, seal neurons survived significantly longer than mouse neurons. Similar elevated levels of brain glycogen were reported in anoxia-tolerant turtles (Lutz et al. 2003). Cultured mouse neurons lacking the ability to synthesize glycogen were more susceptible to hypoxia-induced death (Saez et al. 2014).

5.1.4 Glycogen During Fetal Development

In addition to the postnatal roles of glycogen described above, glycogen metabolism is dynamic during fetal development. The distribution of glycogen in fetal tissues was first described in 1859 (Bernard 1859). Glycogen content in unfertilized and newly fertilized mouse eggs is very low but increases with each cleavage until plateauing at the 8-cell stage (Stern and Biggers 1968). The concentration of glycogen changes during fetal development and/or postnatally in many mammalian tissues (Dawes 1968, Shelley 1961). Notably, the concentration of liver glycogen increases dramatically before birth to levels 2-3 fold higher than that found in the adult (Bhavnani 1983, Dawes and Shelley 1968, Devos and Hers 1974, Tye and Burton 1980). Cardiac glycogen levels during fetal development are as much as forty-fold higher than that found in adult (Bhavnani 1983, Dawes and Shelley 1968, Devos and Hers 1974, Tye and Burton 1980). Skeletal muscle glycogen levels are also elevated in the fetus relative to the adult (Bhavnani 1983, Dawes and Shelley 1968, Devos and Hers 1974, Tye and Burton 1980). The level of fetal lung glycogen has a transient increase (Bhavnani 1983, Dawes and Shelley 1968, Devos and Hers 1974, Tye and Burton 1980) with its decline correlating with increased synthesis of lung surfactant (Bourbon et al. 1982, Brehier and Rooney 1981, Carlson et al. 1987, Compernolle et al. 2002). Brain glycogen levels increased ~ 2 fold from 14.5 to 18.5 days post coitum (dpc) in rats (Gutierrez-Correa et al. 1991) but this increase was not observed in mice (Tye and Burton 1980). Histological detection of glycogen in developing rat brain revealed regional differences and age dependent changes (Bruckner and Biesold 1981). In general, large glycogen stores accumulate late in fetal life and have been postulated to serve as a fuel source during parturition and the transition of the newborn from dependence on placental glucose to independence (Dawes 1968). That glycogen is important for fetal development and newborn survival is indicated by the high newborn lethality observed in mice unable to synthesize glycogen (except in liver) due to the knockout of GYS1 (Pederson et al. 2004a). In contrast, disruption of GYS2, which synthesizes glycogen in the liver, does not reduce newborn survival (Irimia et al. 2010).

5.2 Glycogen Synthesis

The synthesis of glycogen from glucose in muscle, liver, and brain requires the sequential actions of glucose transport, hexokinase, phosphoglucomutase, UDP-glucose pyrophosphorylase, glycogen synthase, and the glycogen branching enzyme. In addition, the initial formation of a glycogen molecule is facilitated by glycogenin. Regulation of synthesis is primarily controlled by glucose transport and modulation of glycogen synthase enzymatic activity.

5.2.1 Glucose Uptake

The synthesis of glycogen from glucose involves movement of glucose, via facilitated diffusion, into cells through the GLUT family of glucose transporters. These transporters are expressed in a tissue specific manner.

Muscle. While GLUT1, GLUT3 and GLUT4 are expressed in developing human skeletal muscle, only GLUT4 is reported to be expressed in adult skeletal muscle (Gaster et al. 2000). GLUT4 is stored in intracellular vesicles which translocate to the plasma membrane upon muscle contraction (Richter and Hargreaves 2013) and insulin stimulation (Jaldin-Fincati et al. 2017). Overexpression of GLUT4 in skeletal muscle increased glycogen levels while decreasing glycogen synthase activity (Bao and Garvey 1997). Overexpression of GLUT1 in mouse skeletal muscle increased glycogen concentration several fold without effects on glycogen synthase activity, suggesting that glucose transport can be limiting in glycogen synthesis, at least in muscle (Ren et al. 1993). Exercise increases GLUT4 expression in muscle, which may contribute to enhanced glycogen storage after exercise (reviewed in Richter and Hargreaves 2013). GLUT4 presence in the plasma membrane is associated with muscle glycogen depletion, though this association does not appear to be direct (Murphy et al. 2018).

Liver. There are a number of GLUT transporters expressed in the liver (Karim et al. 2012), with GLUT2 serving as the primary glucose transporter in hepatocytes (Gould and Holman 1993). This high Vmax, high Km transporter facilitates glucose influx and efflux in hepatocytes based on the intracellular/extracellular glucose concentration gradient. However, the requirement of GLUT2 is challenged by the observation that hepatic glucose output was not decreased in mice disrupted for GLUT2 (Guillam et al. 1998, Hosokawa and Thorens 2002). GLUT2 gene expression is increased with increasing glucose concentrations (Rencurel et al. 1996) and levels of GLUT2 protein in the plasma membrane are decreased with increased insulin levels (Andersen et al. 1994) through an endocytic process (Nathan et al. 2001). Mutations of GLUT2 result in the glycogen storage disease termed Fanconi-Bickel syndrome (Santer et al. 1997). This syndrome is characterized by growth retardation, fasting hypoglycemia, hepatomegaly and increased glycogen levels in liver and other tissues expressing GLUT2 (Manz et al. 1987).

Brain. In the brain, glucose first moves from the blood through the blood brain barrier and then into astrocytes and neurons (reviewed in McEwen and Reagan 2004, Vannucci et al. 1997). GLUT1 and GLUT3 appear to be responsible for the majority of glucose uptake. The 55 kDa isoform of GLUT1, which is expressed in the blood-brain barrier (Farrell and Pardridge 1991), transports glucose across this barrier. The 45 kDA isoform of GLUT1 is expressed in astrocytes (Leino et al. 1997) and facilitates glucose entry into this cell type, while GLUT3 transports glucose into neurons (Leino et al. 1997). Several other glucose transporters are expressed in the brain, but their function is not completely clear (reviewed in Choeiri et al. 2002). The brain has generally been considered to be insulin-independent, but the insulin sensitive glucose transporters GLUT4 and GLUT8 along with the insulin receptor are expressed (reviewed in McEwen and Reagan 2004). Blood glucose levels tend to correlate with brain glycogen levels (Poitry-Yamate et al. 2009). Cultured astrocytes incubated in high concentrations of glucose lead to increased glycogen accumulation (Magistretti et al. 1993). Both glutamate and aspartate stimulated glycogen synthesis in cultured astrocytes by increasing glucose uptake (Hamai et al. 1999). However, glycogen content did not increase in mouse neurons when glucose levels were increased (Saez et al. 2014).

5.2.2 Glucose Phosphorylation

After entering the cell, glucose is phosphorylated by hexokinase to form glucose-6-P. This branchpoint metabolite is an activator of glycogen synthase and can serve as a substrate for multiple pathways including glycolysis, the pentose phosphate pathway, hexose amine biosynthetic pathway, and glycogenesis. Glucose transporters in the muscle, liver, and brain are facilitative, so that glucose uptake is connected with the activity of hexokinase. There are four isoforms of hexokinase (HK1-4), which exhibit different kinetic and regulatory characteristics (Agius 2016, Wilson 2003). Notably, HK1-3 are inhibited by glucose-6-P, while HK4, commonly referred to as glucokinase, is not.

Muscle. HK2 is the primary hexokinase expressed in skeletal muscle (Wilson 2003). Overexpression of HK2 in muscle skeletal muscle resulted in an increase in glycogen accumulation, but only when GLUT1 was also overexpressed (Hansen et al. 2000). Global disruption of HK2 in mice was embryonic lethal (Heikkinen et al. 1999), while heterozygous mice had reduced glycogen levels in the gastrocnemius (Fueger et al. 2003).

Liver. Liver expresses glucokinase (HK4) which, compared to HK1-3, has a higher Km for glucose and is not inhibited by its product, glucose-6-P. This isoform of hexokinase is regulated by a number of mechanisms including gene expression, metabolite levels, and translocation (reviewed in Agius 2016). The glucokinase regulatory protein, which is located in the nucleus, binds glucokinase when glucose levels are low. This results in inactivation of glucokinase. When glucose levels are high, glucokinase is released from the glucokinase regulatory protein and moves to the cytosol where it is active. Glucokinase deficiency in humans leads to reduced liver glycogen synthesis after a meal (Velho et al. 1996). Liver specific disruption of glucokinase in mice resulted in decreased hepatic glycogen synthesis and glycogen accumulation (Postic et al. 1999).

Brain. Glucose taken up into neurons and astrocytes is phosphorylated by the HK1 isoform of hexokinase (Wilson 2003). Hexokinase is inhibited 90 percent by glucose-6-P under basal conditions (DiNuzzo et al. 2015).

Glycogen Shunt. Glucose-6-P produced by the action of hexokinase has multiple fates, including glycolysis and glycogenesis. While the traditional pathway for glycolysis is separate from glycogenesis, Shulman and Rothman (Shulman and Rothman 2001) proposed a hybrid of these pathways termed the glycogen shunt. This shunt was proposed to explain the importance of glycogen in skeletal muscle during exercise. The hypothesis is that rather than glucose-6-P proceeding directly through glycolysis, it is instead used to first synthesize glycogen. Glycogen is then degraded to glucose-1-P and converted to glucose-6-P, which reenters the glycolytic pathway to generate ATP for muscle contraction. Support for this concept is provided by the observed exercise intolerance in patients with McArdle disease, caused by a deficiency in muscle glycogen phosphorylase. In addition, mice disrupted for glycogen synthase specifically in the skeletal muscle of adult mice have impaired exercise endurance capacity (Xirouchaki et al. 2016). In contrast, mice with global disruption of the glycogen synthase isoform expressed in muscle (GYS1) had normal exercise capacity (Pederson et al. 2005b). However, this latter animal model has metabolic changes in the skeletal muscle which enhance oxidative capacity. Thus, these adaptations perhaps overcome a normal reliance on the glycogen shunt. The glycogen shunt has also been reported to play a significant role in astrocytes (Walls et al. 2009).

5.2.3 Phosphoglucomutase

Phosphoglucomutase catalyzes the reversible conversion of glucose-6-P to glucose-1-P. PGM1 is the primary isoform expressed in most tissues and is encoded by a highly polymorphic gene (March et al. 1993). Glucose-1,6-bisphosphate serves as coenzyme and activator for the enzyme (Beitner 1985, Yip et al. 1988). PGM1 is also regulated at the transcriptional level as well as by glycosylation, which affects localization through association of the enzyme with membranes (Dey et al. 1994). Enzyme deficiency results in a glycogen storage disease (GSDXIV). The first patient identified had abnormal accumulation of muscle glycogen and exercise intolerance (Stojkovic et al. 2009). In another study of 19 patients, hypoglycemia was observed while fibroblast glycogen levels were normal (Tegtmeyer et al. 2014). There are several PGM isozymes which complicates the understanding of glycogen metabolism in patients with PGM1 deficiency.

5.2.4 UDP-Glucose Pyrophosphorylase

Glucose-1-P reacts with UTP to produce UDP-glucose in a reaction catalyzed by UDP-glucose pyrophosphorylase. The crystal structure of the human enzyme was determined and shown to form octamers (Yu and Zheng 2012). Disruption of UDP-glucose pyrophosphorylase in *Saccharomyces cerevisiae* indicated that this enzyme is essential for viability (Daran et al. 1995). The enzyme does not appear to be rate limiting for glycogen synthesis, as indicated by the observation that overexpression in mouse skeletal muscle increased levels of UDP-glucose 3-fold, but did not impact muscle glycogen levels (Reynolds et al. 2005). In addition to serving as the substrate for glycogen synthase, UDP-glucose can be converted to glucose-1-P and UMP by UDP-glucose pyrophosphatase. The NUDT14 gene, which encodes this protein, is expressed in several tissues including skeletal muscle, liver, and brain (Heyen et al. 2009). Though no regulation has yet been demonstrated, evidence suggesting post-translational modification was reported. If confirmed, regulation of this enzyme could potentially serve as a regulator of glycogen synthesis.

5.2.5 Glycogenin

The initial formation of a glycogen molecule is facilitated by glycogenin. This enzyme catalyzes the addition of glucose residues to itself, and through subsequent actions of glycogen synthase and branching enzyme, a glycogen molecule is synthesized with a dimer of glycogenin at the core.

GYG1. There are two genes, GYG1 and GYG2, encoding glycogenin in humans. GYG1 is widely expressed, with the highest expression in cardiac and skeletal muscle (Barbetti et al. 1996). A patient with an inactivating GYG1 mutation (Thr83Met) was found to have depleted glycogen stores in skeletal muscle, muscle weakness, and a switch to slow-twitch oxidative muscle fibers (Moslemi et al. 2010), with no cognitive abnormalities observed. This mutation prevents the formation of the glucose-O-tyrosine linkage at Tyr195 (Nilsson et al. 2012). In another study, 7 patients with mutations in GYG1 were found to have either depleted glycogenin protein, or in one case impaired interaction with glycogen synthase, varying levels of normal glycogen, and an accumulation of polyglucosan in skeletal muscle (Malfatti et al. 2014). Genome-wide association studies have identified that GYG1 is associated with myocardial infarction (Lee et al. 2017).

GYG2. GYG2 is expressed in the liver, heart, and pancreas (Mu et al. 1997). Imagawa et al. (Imagawa et al. 2014) reported that GYG2, but not GYG1, is expressed in human brain. Two male siblings with Leigh syndrome, a progressive neurodegenerative disorder, were found to have a hemizygous missense mutation in GYG2 (Imagawa et al. 2014). Both brothers suffered from seizures and neurodegeneration with ketonemia, but normal blood lactate levels. According to the study's authors, it remains to be established whether the GYG2 mutation is causative of Leigh syndrome in these patients.

Non-Primate GYG. In contrast to humans, non-primates appear to have a single gene for glycogenin, GYG. In the mouse, glycogenin was identified as an essential gene. Embryos with GYG disrupted died perinatally and exhibited severe fetal brain and cardiac abnormalities (Dickinson et al. 2016). In a separate report, GYG disruption resulted in perinatal death of approximately 85 percent of pups born, due to cardiopulmonary defects (Testoni et al. 2017). This lethality is reminiscent of what the author observed with the disruption of the GYS1 isoform of glycogen synthase (Pederson et al. 2004a). Fetal tissue glycogen levels were not reported by Testoni et al. (Testoni et al. 2017), but perhaps the lethality of the disruption is due to the inability to synthesize glycogen. Surprisingly, surviving GYG knockout mice had high levels of glycogen in cardiac and skeletal muscle and normal levels of glycogen in liver and brain, despite undetectable expression of GYG in these tissues. These two studies demonstrate the importance of GYG for development and survival, and along with the findings in human GYG1 deficiency (Malfatti et al. 2014), challenge the idea that glycogenin is required for the synthesis of glycogen. Mice lacking GYG were impaired in treadmill exercise performance and had a switch of oxidative muscle fibers toward glycolytic metabolism. This same switch occurred with the elevation of muscle glycogen due to overexpression of glycogen synthase. These findings, coupled with the report by the author and colleagues (Pederson et al. 2005c) that disruption of GYS1 resulted in a switch from glycolytic to oxidative fibers, indicate that the level of muscle glycogen impacts energy metabolism in this tissue.

Structure and Regulation. The crystal structure for rabbit muscle glycogenin indicates that it is a member of the GT8 glycosyltransferase family with a single Rossman-fold domain (Gibbons et al. 2002). UDP-glucose binds in a metal-dependent manner in the central beta-sheet structure transiently transferring the glucose moiety to Asp162 and then to Tyr195. This protein has interesting properties in that it serves in the roles of enzyme, substrate, and product. Initially, glycogenin self-glucosylates, adding a glucose molecule to tyrosine 195. Subsequently, it extends this glucose chain up to 10–20 residues in length (Roach and Skurat 1997, Smythe and Cohen 1991). This protein-glucose complex then serves as a scaffold for glycogen synthase to continue elongation of the glucose chains (Skurat et al. 2006). Glycogen synthase also associates with the C-terminal 33 amino acids, which purify glycogen synthase from tissue extracts. The structure of *Caenorhabditis elegans* glycogen synthase in complex with a minimal targeting region of glycogenin indicated that glycogenin binds to the first of two Rossman fold domains of glycogen synthase in a region not involved in tetramer interactions or binding to glucose-6-P or UDP-glucose (Zeqiraj and Sicheri 2015, Zeqiraj et al. 2014). Mutation of either glycogen synthase or glycogenin to prevent interaction between the two proteins eliminated the synthesis of glycogen in *S. Cerevisiae*, demonstrating the importance of this interaction. The glycogen synthase binding domain and the catalytic domain are separated by a region with variable length and subject to alternative splicing in humans (Zhai et al. 2000). Zeqiraj et al. (Zeqiraj and Sicheri 2015, Zeqiraj et al. 2014) hypothesize that the length of this linker region may determine the size of glycogen particles. Expression of a GFP-glycogenin fusion protein in hepatocytes and muscle cells (C2C12) indicated nuclear and cytosolic distribution and partial co-localization with the actin cytoskeleton (Baque et al. 1997). Evidence for regulation of glycogenin is limited. Glycogenin mRNA levels were increased in cerebral cortex of mice that underwent gentle sleep deprivation (Petit et al. 2010).

5.2.6 Glycogen Synthase

Glycogen synthesis is catalyzed by glycogen synthase (EC2.4.1.11) which uses UDP-glucose as the glucosyl donor to form alpha-1,4 glycosidic linkages (Leloir et al. 1959, Villar-Palasi and Larner 1958). There are two genes encoding this enzyme in mammals. Tissue protein expression was first documented in rat, with GYS1 being expressed in muscle, heart, fat, kidney, and brain (Kaslow and Lesikar 1984), while expression of GYS2 appeared to be limited to liver (Kaslow et al. 1985). Consistent with this pattern, Genevestigator (Hruz et al. 2008) analysis of transcriptome profiles in mice and humans indicates that GYS1 is broadly expressed with the highest expression in skeletal and cardiac muscle, while high GYS2 expression occurs only in the liver.

5.2.6.1 Gene Structure

GYS1. The sequence of rabbit skeletal muscle glycogen synthase was determined (Zhang et al. 1989), encoding a protein of 734 amino acids and a molecular weight of 83,480 daltons. The 5' untranslated and coding regions were 79 and 90 percent identical, respectively, to human muscle glycogen synthase while the 3' untranslated region was less similar. cDNA for human muscle glycogen synthase was also cloned and sequenced (Browner et al. 1989), and found to encode a protein of 737 amino acids with a predicted molecular weight of 83,645 daltons. In humans, the GYS1 gene is located on chromosome

19 (19q13.3) (Groop et al. 1993, Lehto et al. 1993) spanning 32,229 bp and containing 16 exons. The promoter region does not contain TATA or CAAT boxes but does include a number of putative transcription factor binding sites (Fredriksson et al. 2004). These include sites for Sp1, activator protein 2 (Ap-2), octamer binding protein 1 (Oct1), myocyte enhancer factor 2 (MEF2), CAAT/enhancer binding protein (CEBP), and enhancer factor-1. In addition were cAMP responsive elements (CRE), muscle initiator sequences (MINI), sterol responsive elements (SRE), E-boxes, Ets-like motifs and phosphoenolpyruvate carboxykinase (PEPCK)-like motifs. The first 250 bp of the 5' flanking region contain elements characteristic of housekeeping genes and account for basal promoter activity. A hypoxia response element (HRE) was identified at position –314 in GYS1 (Pescador et al. 2010).

GYS2. GYS2 is composed of 16 exons, spans more than 30 kb (Orho et al. 1998), and is located on human chromosome 12 (Nuttall et al. 1994). Two putative PPAR response elements were identified, one in the upstream promoter and another, which is a response element for hepatic nuclear factor 4 alpha, in intron 1 (Mandard et al. 2007). Two E boxes were identified in intron 1 that are important for transcriptional activation by CLOCK (Doi et al. 2010).

5.2.6.2 Protein Structure

GYS1. Human GYS1 contains 737 amino acids. The molecular weight of the GYS1 holoenzyme, as judged by HPLC gel filtration, was 274 kD from muscle, while the molecular weight of the subunit was 87 kDa (Inoue et al. 1987). In addition to catalytic residues, the protein contains several phosphorylation sites and a region that confers sensitivity to glucose-6-P.

GYS2. Glycogen synthase purified from rabbit liver had a molecular weight of 86 or 90 kD depending on the phosphorylation state (Camici et al. 1982, 1984). The human enzyme contains 703 amino acids with a molecular weight of 80–85 kDa and a higher specific activity than the rat enzyme (Westphal and Nuttall 1992). The Km for UDP-glucose was lowered with increasing glucose-6-P concentrations for both human and rat GYS2.

Crystal Structure. The crystal structure for mammalian glycogen synthase has not been determined, but efforts using glycogen synthase from baker's yeast, *S. cerevisiae*, have been fruitful both in this regard as well as in contributing to our understanding of glycogen metabolism (reviewed in Wilson et al. 2010). *S cerevisiae* expresses two isoforms of glycogen synthase, Gsy1 and Gsy2 (Farkas et al. 1991). The Gsy2p isoform of glycogen synthase exists as a tetramer (Baskaran et al. 2010) (Fig. 5.3). Two arginine residues in the C-terminus confer sensitivity to glucose-6-P, which when bound causes translations and rotations among the subunits which permit easier substrate

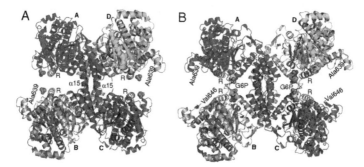

Figure 5.3. Glycogen synthase crystal structure. Structures of the basal and activated states of Gsy2p. **(A)** Ribbon diagram of the basal state conformation in which the individual subunits are labeled (A–D) and colored separately. The regulatory helices (α22) are colored cyan and labeled R, while the intersubunit helices are labeled α15 as is the last ordered residue at the C terminus of subunits B and D. Ordered sulfate ions in this structure are represented using space filling atoms. **(B)** Ribbon diagram of the activated state of Gsy2p. The color scheme, subunit, and regulatory helix labeling is identical to panel A. The bound glucose-6-P molecules at the interface are represented using space filling atoms and labeled G6P. [Produced using Pymol (DeLano 2002) for Windows.] Used with permission from (Baskaran et al. 2010).

access. Mahalingan et al. (Mahalingan et al. 2017) showed that Arg589 and Arg592 are important for in allowing the conversion to and from the T state of the enzyme. Four additional arginine residues in this same helix are involved in the enzyme's response to phosphorylation (Baskaran et al. 2010). Glycogen synthase is associated with glycogen regardless of the activity state of the enzyme (reviewed in Roach et al. 2012). Based on studies with Gsy2p, this affinity between glycogen synthase and glycogen is facilitated by four glycogen association sites distributed across the surface of the protein (Baskaran et al. 2011). One of these sites (site-1) is in the N-terminal domain, two sites (site-2 and site-3) are in the C-terminal domain, and the fourth (site-4) is near the active site of the enzyme. Mutation of site-1 or site-2 reduced glycogen binding and catalytic efficiency. Expression of site-1, site-2 or site-4 mutants in yeast lacking endogenous glycogen synthase resulted in greatly reduced glycogen accumulation. These findings support a mechanism by which site-1 and site-2 facilitate glycogen synthase association with glycogen, while site-4 positions the non-reducing end of the glycogen chain during catalysis.

5.2.6.3 Glycogen Synthase Localization

Muscle. Skeletal muscle is composed of several fiber types with different metabolic characteristics. Glycogen synthase protein expression was similar in human type I, type IIA, and type IIX fibers in vastus lateralis, soleus, and triceps brachii (Daugaard and Richter 2004). In contrast, glycogen synthase expression in type IIA/IID fibers from rabbit tibialis anterior muscle is double that found in type IIB fibers (Azpiazu et al. 2000).

In skeletal muscle fibers from rats, glycogen synthase activity was located in a glycogen-enriched membrane fraction under conditions of relatively high glycogen concentrations and a cytoskeleton fraction when glycogen levels were low (Nielsen et al. 2001). Using confocal microscopy, glycogen synthase localization was found to be influenced by glycogen content. In rabbit skeletal muscle, the pattern of glycogen synthase localization was either diffuse or found in the perinuclear region and myofibrillar striations depending on the phosphorylation state of the enzyme (Prats et al. 2005, 2009).

Liver. GYS2 mRNA was uniformly distributed throughout mouse liver (Ghafoory et al. 2013). In isolated hepatocytes, glycogen synthase was located in cytosol and the cell periphery depending on the glucose concentration (Fernandez-Novell et al. 1997).

Brain. Glycogen synthase mRNA is expressed widely in mouse brain, with the highest levels found in the olfactory bulb, the granular layer of the cerebellum, and the hippocampus and lower levels in the cerebral cortex and striatum (Pellegri et al. 1996). Glycogen synthase protein is widely expressed in rat brain with the highest levels observed in hippocampus, cerebral cortex, caudate-putamen, and cerebellar cortex (Inoue et al. 1988). In humans, glycogen synthase activity in the hippocampus was higher than in either white or gray matter, though these measurements were made in tissue from patients with epilepsy so they may not be reflective of normal activity (Dalsgaard et al. 2007). In both rat and rabbit brain, glycogen synthase activity was highest in the cerebellum (Breckenridge and Crawford 1961) (Knull and Khandelwal 1982). Past studies have shown an agreement between glycogen synthase expression, glycogen synthase activity, and glycogen content (Oe et al. 2016, Sagar et al. 1987). Glycogen synthase mRNA (Pellegri et al. 1996) and protein (Inoue et al. 1988) were present in both neurons and astrocytes in mouse brain. Glycogen synthase mRNA was present in both cytosol and nucleus of cultured astrocytes and neurons from rat hippocampus (Mamczur et al. 2015). Using fluorescence *in situ* hybridization (FISH) analysis on cultured cortical neurons, glycogen synthase mRNA was detected in both the soma and processes (Pfeiffer-Guglielmi et al. 2014). In the latter, the mRNA had a granular appearance. Glycogen synthase activity in rat brain was highest in the cytoplasmic fraction (Knull and Khandelwal 1982). Within neurons, glycogen synthase in mouse embryonic neuronal cultures was present as aggregates in the cytoplasm (Saez et al. 2014). Under normal conditions, the enzyme is relatively inactive in this cell type, due to phosphorylation (Vilchez et al. 2007). Dephosphorylation of glycogen synthase in mouse neurons led to glycogen deposition and apoptosis. However, oxidative stress increased glycogen synthase activity which was neuroprotective (Rai et al. 2018). Thus, the role of glycogen synthase in neurons remains to be established.

5.2.6.4 Catalytic Mechanism

The catalytic mechanism for glycogen synthase is not well understood. Unlike another glycosyltransferase, glycogenin, glycogen synthase is metal-ion independent, instead using hydrogen bonds to amino acids to stabilize UDP (reviewed in Roach et al. 2012). The particular residues in the yeast glycogen synthase, Gsy2p, implicated in catalysis are Arg320, Lys326, residues 513–521, and Glu509. Kinetic characterization of the enzyme was conducted on the enzyme in skeletal muscle (Rosell-Perez et al. 1962), liver (Hizukuri and Larner 1964), and brain (Goldberg and O'Toole 1969). The primary function of glycogen synthase is catalyzing the addition of glucose from UDP-glucose to an elongating glucose chain in a glycogen molecule. In addition, *in vivo* glycogen synthase has been shown to incorporate the beta phosphate of UDP-glucose into glycogen as glucosyl phosphate (Fig. 5.4) at a frequency of ~ 1 phosphate per 10,000 glucose residues (Tagliabracci et al. 2011). Phosphate incorporation into glycogen has been proposed to contribute to the abnormal glycogen accumulation in Lafora disease (Roach 2015, Tagliabracci et al. 2008). Lafora disease is characterized by myoclonus epilepsy in early adolescence (Lafora and Glueck 1911) with declining cognitive function, increased seizure frequency, and death typically occurring within 10 years after diagnosis (Minassian 2001, Serratosa 1999, Turnbull et al. 2016). Patients with this disorder accumulate high levels of poorly branched, hyperphosphorylated glycogen, termed polyglucosan or Lafora bodies, in many tissues (Minassian 2001, Striano et al. 2008). The majority of Lafora disease cases (The Lafora Progressive Myoclonus Epilepsy Mutation and Polymorphism Database, http://projects.tcag.ca/lafora/) are the result of mutations in the EPM2A gene which encodes laforin, a dual specificity phosphatase (Minassian et al. 1998, Serratosa et al. 1999) and the EPM2B/NHLRC1 gene which encodes malin (Chan et al. 2003), an E3 ubiquitin ligase (Gentry et al. 2005). Laforin binds to glycogen through a carbohydrate binding domain (Chan et al. 2004, Wang et al. 2002, Wang and Roach 2004) and dephosphorylates glycogen (Tagliabracci et al. 2007). Disruption of Epm2a in a mouse model resulted

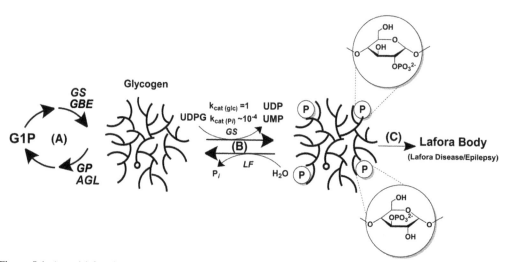

Figure 5.4. A model for glycogen phosphate metabolism and Lafora body formation. **(A)** Glycogen is synthesized by glycogen synthase (GS) and the glycogen branching enzyme (GBE) and degraded by glycogen phosphorylase (PH) and glycogen debranching enzyme (AGL). **(B)** Glycogen synthase infrequently (1 in ~ 10,000) incorporates phosphate residues into glycogen. The kcat values given simply denote the relative rates of glucose versus glucose phosphate incorporation. Excessive incorporation of phosphate as C2 or C3 phosphomonesters, which disrupts glycogen structure, is normally kept in check by the action of the laforin phosphatase (LF). **(C)** When laforin is defective, excessive phosphorylation results in the formation of Lafora bodies and Lafora disease. Used with permission from (Tagliabracci et al. 2011).

in muscle glycogen that contained ~ 5-fold higher levels of phosphate as compared to wild type controls (Tagliabracci et al. 2008, 2007). This phosphate is located at the C2-, C3- and C6-positions (DePaoli-Roach et al. 2015). Incorporation of phosphate at C2- and C3- can be explained by the action of glycogen synthase, while the mechanism for incorporation at the C6-position is unknown (Chikwana et al. 2013, Contreras et al. 2016, Tagliabracci et al. 2011). Phosphorylation of glycogen correlated with decreased branching and solubility through a poorly understood mechanism (Roach 2015). Roach (Roach 2011) proposed that the incorporation of phosphate into glycogen may be a catalytic error and that laforin acts as a repair enzyme. An alternative hypothesis for Lafora body formation is an imbalance of glycogen synthase to branching enzyme activity, leading to synthesis of poorly branched glycogen (Vilchez et al. 2007). Both of these hypotheses have recently been challenged by the reports of normal glycogen synthase activity in both malin and laforin knockout mice (DePaoli-Roach et al. 2010, Tagliabracci et al. 2008) and the prevention of lafora body formation when phosphatase-inactive laforin was overexpressed in laforin knockout mice (Gayarre et al. 2014). In the latter mouse model, the abnormal glycogen chain length pattern characteristic of Lafora bodies is corrected, but glycogen remains hyperphosphorylated (Nitschke et al. 2017). Both soluble and insoluble glycogen isolated from muscle tissue harvested from mice deficient for laforin had increased C6 phosphorylation, while only insoluble glycogen had elevated C6 phosphorylation in malin-deficient mouse muscle (Sullivan et al. 2019). Thus, the mechanism underlying the formation of Lafora bodies remains an active area of investigation. Nonetheless, mouse models demonstrate the importance of glycogen synthesis in the pathology of Lafora disease. Disruption of PTG, an activator of glycogen synthase, in a mouse model resulted in a 70 percent reduction in brain glycogen (Turnbull et al. 2011). Crossing these mice with a laforin knockout mouse largely prevented the formation of Lafora bodies and resolved both myoclonic epilepsy and neurodegeneration. Similarly, disruption of GYS1 in a laforin deficient background prevented Lafora body formation and neurodegeneration and seizure susceptibility was normal (Pederson et al. 2013). In another study, disrupting GYS1 in malin deficient mice prevented the deleterious effects associated with Lafora disease (Duran et al. 2014). Thus, glycogen synthase inhibition may be therapeutic for patients with this fatal disease.

5.2.6.5 Glycogen Synthase Gene Regulation

GYS1—Muscle. Exposing C2C12 skeletal muscle myotubes to hypoxia increases GYS1 expression (Pescador et al. 2010). This response requires the hypoxia inducible factor (HIF) and a hypoxia response element (HRE) in the GYS1 promoter. Increased expression was accompanied by increased glycogen synthase activity and glycogen accumulation. Insulin also increases GYS1 expression. In C2C12 myotubes, GYS1 promoter activity was modestly decreased 24 hr after insulin treatment (Fredriksson et al. 2004). In human muscle biopsies, insulin increased GYS1 mRNA levels ~ 10-fold (Huang et al. 2000).

GYS1—Brain. Glycogen synthase mRNA expression was upregulated by two neurotransmitters, vasoactive intestinal peptide (VIP) and noradrenaline (NA) in cultured cortical astrocytes, but not neurons (Pellegri et al. 1996). Adenosine, which accumulates in the extracellular space following increased neuronal activity (Mitchell et al. 1993), increased glycogen levels in cultured astrocytes by a mechanism that was blocked when transcription was inhibited (Allaman et al. 2003). Adenosine increased mRNA expression of protein targeting glycogen (PTG) as well as C/EBP beta and C/EBP delta. Though not reported, adenosine could possibly work through these mechanisms to affect expression and/or activity of glycogen synthase.

GYS2—Liver. GYS2 mRNA and protein levels exhibit circadian rhythms, consistent with changes in liver glycogen content, and mutation of the CLOCK gene dampens the circadian rhythm (Doi et al. 2010). CLOCK activates transcription of GYS2 through two tandemly located E boxes. The period 2 (PER2) protein plays a role in regulating GYS2 expression during refeeding (Zani et al. 2013). Disruption of PER2 in mouse reduced liver GYS2 mRNA and protein expression to half of that measured in wild type mice. Expression of GYS2 is upregulated by peroxisome proliferator-activated receptors (PPARs) (Mandard et al. 2007). Disruption of PPAR alpha in liver lead to reduced GYS2 expression and glycogen levels.

5.2.6.6 Glycogen Synthase Protein Regulation

Glycogen accumulation in the cell is determined by the balance between synthesis by glycogen synthase and degradation by both glycogen phosphorylase and lysosomal alpha-glucosidase. The two mechanisms best characterized by glycogen synthase regulation are activation by glucose-6-P and inhibition by covalent phosphorylation. Glucose-6-P overcomes phosphorylation-induced inhibition and can restore maximal activity. In early studies of glycogen synthase, the enzyme was classified as either the I form or the D form. The I, independent form, is active without glucose-6-P, while the D, dependent form, is relatively inactive in the absence of glucose-6-P. Because of the effect by glucose-6-P, standard assays to monitor glycogen synthase enzymatic activity are conducted in both the presence and absence of added glucose-6-P, or alternatively, in the presence of low and high concentrations of glucose-6-P (Guinovart et al. 1979). When these assays are conducted in cell or tissue extract, activity is interpreted as providing an index of the *in vivo* phosphorylation state of glycogen synthase (Roach et al. 2012, Roach and Larner 1977). However, it is important to note that the *in vivo* activity of the enzyme is affected not only by the phosphorylation state, but also by other factors, including but not limited to, the concentration of glycogen, glucose-6-P, and UDP-glucose.

The control of the enzyme by glucose-6-P and phosphorylation has been described by a three-state model (Fig. 5.5) (Baskaran et al. 2010, Pederson et al. 2000). When unphosphorylated, glycogen synthase exists in an intermediate state (I state or state II). Upon binding of glucose-6-phosphate, the enzyme is converted to a high activity state (R state or state III), while phosphorylation converts the enzyme to a low activity state (T state or state I). In the low activity state, glycogen synthase is less sensitive to glucose-6-P binding, but saturating levels of this metabolite convert glycogen synthase to the high activity state. Because the structure of the enzyme is tetrameric, hybrid complexes of phosphorylated and unphosphorylated subunits are present, which may lead to more than three activity states. In addition, this model is based on studies with yeast Gsy2p, which has only three phosphorylation sites. Mammalian GYS1 and GYS2 have 9 and 7 phosphorylation sites, respectively,

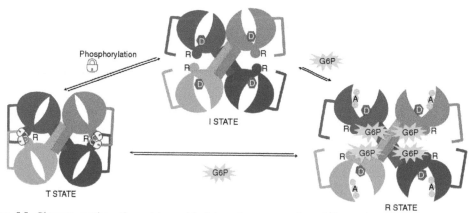

Figure 5.5. Glycogen synthase three state model. Schematic representation of the conformational states underlying regulation of Gsy2p activity. The individual subunits are colored according to the coloring scheme in Fig. 5.3. The regulatory helices containing the arginines are labeled R and the approximate positions of nucleotide-donor sugar (D) and glycogen acceptor (A) are shown for the I- and R-states, respectively. Phosphorylation of Thr668 is shown as "locking" the enzyme in the T-state conformation through intersubunit interactions across the regulatory interface. Glucose-6-phosphate binding frees these constraining interactions to fully activate the enzyme in the R-state conformation. Used with permission from (Baskaran et al. 2010).

which may provide for more complex regulation via phosphorylation/dephosphorylation (Hanashiro and Roach 2002).

Allosteric Activation. Physiologic levels of glucose-6-P allosterically activate glycogen synthase in muscle and liver (reviewed in Villar-Palasi and Guinovart 1997) as well as in brain (Goldberg and O'Toole 1969). Glucose-6-P sensitivity is conferred by arginine residues in glycogen synthase in *S. cerevisiae* (Pederson et al. 2000) and rabbit skeletal muscle (Hanashiro and Roach 2002). Expression in *S. cerevisiae* of Gsy2p, with arginine residues 579,581, and 582 mutated to alanine, resulted in reduced glycogen accumulation (Pederson et al. 2004b). In a knockin mouse expressing constitutively active GSK-3, insulin was unable to promote glycogen synthase activation through desphosphorylation. In spite of losing this mechanism of regulation, normal amounts of glycogen still accumulated, suggesting a major role for glucose-6-P in the regulation of glycogen synthesis (Bouskila et al. 2008). Furthermore, a knockin mouse with Arg 582 of GYS1 modified to alanine expressed glycogen synthase that was insensitive to glucose-6-P, had ~ 80 percent reduction in muscle glycogen synthesis induced by insulin, and a 50 percent reduction in skeletal muscle glycogen levels (Bouskila et al. 2010). When this same mutation was made in GYS2, resulting mice hemizygous for the Arg 582 mutation had impaired glucose-6-P mediated activation of glycogen synthase and impaired glycogen synthesis in liver after refeeding (von Wilamowitz-Moellendorff et al. 2013).

Phosphorylation/Dephosphorylation. One of the first enzymes identified to be phosphorylated at multiple sites was mammalian glycogen synthase (Smith et al. 1971). The much-studied muscle isoenzyme has nine phosphorylation sites (Fig. 5.6). Mutagenesis studies on rabbit GYS1 indicated that sites 2, 2a, 3a, and 3b are the most important for inhibiting enzymatic activity (Skurat and Roach 1995, Skurat et al. 1994). For the liver isoenzyme, GYS2, mutation of sites 2, 3a, 3b, 3c, 4, and 5 to alanine resulted in increased glycogen synthesis in hepatocytes (Kadotani et al. 2007). Subsequent mutagenesis studies indicated that site 2 phosphorylation is the most important for inhibiting enzymatic activity (Ros et al. 2009).

Studies of glycogen synthase phosphorylation found that in certain instances, the addition of one phosphate was the prerequisite for the addition of a second phosphate, a concept termed hierarchal phosphorylation (Roach 1990, 1991). For example, casein kinase 2 (CK2) phosphorylation glycogen synthase is required prior to subsequent phosphorylation by glycogen synthase kinase 3 (GSK3). GSK3 beta appears to be more important than GSK3 alpha for phosphorylation of GYS1 in muscle and is itself regulated by phosphorylation (Patel et al. 2008). Several other kinases phosphorylate glycogen synthase including AMPK, Ca+2/calmodulin-dependent protein kinase II (CAMKII), phosphorylase kinase (PhK), PKA, proteins kinase C (PKC), casein kinase I (CK1), Per/Arnt/Sim domain-containing protein kinase (PASK), dual-specificity tyrosine-phosphorylated and regulated kinase (DYRK), stress-activated protein kinase 2b (p38beta) (reviewed in Roach et al. 2012). AMPK

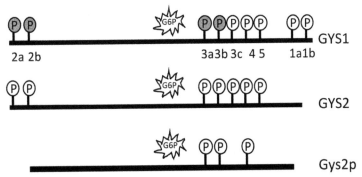

Figure 5.6. Glycogen synthase protein structure. Residue 582 in mouse GYS1 and GYS2 is required for glucose-6-P (G6P) sensitivity. Phosphorylation sites correspond to the following residues in mouse GYS1: site 2, 8; site 2a, 11; site 3a, 641; site 3b, 645; site 3c, 649; site 4, 653; site 5, 657; site 1a, 698; site 1b, 711. Gsy2p is one of the two isoforms of glycogen synthase expressed in *S. cerevisiae*.

has attracted attention due to the presence of a domain which binds glycogen, its well-studied role as a cellular energy sensor, and its proposed role as a glycogen sensor (McBride and Hardie 2009, Roach et al. 2012). A knockout of the alpha 2 subunit of AMPK reduces glycogen synthase phosphorylation at site 2 plus 2a concomitant with activation of glycogen synthase (Jorgensen et al. 2004). AMPK phosphorylation is increased, which results in activation of the kinase, in skeletal muscle from mice that lack glycogen due to disruption of GYS1, consistent with a role of AMPK acting as glycogen sensor (Pederson et al. 2005c). However, this effect appears to be tissue specific; in the brains of mice lacking glycogen due to disruption of GYS1, the activation state of AMPK was not different from mice that contained normal levels of brain glycogen (Duran et al. 2013).

Dephosphorylation of glycogen synthase activates glycogen synthase and thus stimulates glycogenesis. The removal of phosphate is catalyzed by protein phosphatase 1 (PP1c) bound to one of seven glycogen-targeting subunits (PPP1R3A to PPP1R3G, reviewed in Ceulemans and Bollen 2004, Korrodi-Gregorio et al. 2014). PPP1R3A (also termed GM) is primarily expressed in muscle and consistent with its role in regulating glycogen synthesis, disruption in mice reduces glycogen level in skeletal muscle (Delibegovic et al. 2003). PPP1R3B (GL) expression is predominantly in liver. Disruption of GL in liver leads to predominantly phosphorylated and inactive glycogen synthase, along with a reduction in hepatic glycogen levels (Mehta et al. 2017). Consistent with this glycogen targeting subunit playing an important role in liver glycogen metabolism, overexpression of PPP1R3B in liver leads to increased hepatic glycogen levels. PPP1R3C (PPP1R5, PTG) is widely expressed and disruption in mice did not alter insulin-stimulation of glycogen synthase in skeletal muscle (Zhai et al. 2007). However, PTG disruption resulted in reduced glycogen accumulation in skeletal muscle and brain (Turnbull et al. 2011, 2014). PPP1R3D is expressed at highest levels in neurons (Rubio-Villena et al. 2013). PPP1R3F is also highly expressed in brain and was reported to regulate glycogen synthase in astrocytoma cells (Kelsall et al. 2011). When PPP1RR3F was mutated to prevent binding to PP1c, glycogen synthase was hyperphosphorylated on Ser640 and Ser644. PPP1R3G is expressed in liver during the fast to fed transition, consistent with repression by insulin and induction by glucagon (Luo et al. 2011). When disrupted in a mouse model, liver glycogen levels were reduced (Zhang et al. 2017). The expression of GL and PTG in the liver is repressed by fasting and induced by refeeding (reviewed in Agius 2015). Consistent with these observations, GL and PTG expression is induced by both insulin and high glucose levels (Petrie et al. 2013). The circadian rhythmic regulation of glycogen metabolism also appears to involve GL and PTG, as the liver expression of both was reduced in mice lacking functional PER2 (Zani et al. 2013).

Insulin—Muscle. In skeletal muscle, insulin activates glycogen synthase by promoting dephosphorylation of the protein (Lawrence 1992). This reduces the concentration of glucose-6-P required to activate glycogen synthase (Lawrence 1992). Phosphorylation at sites 3 (a+b+c) and site 2 and/or 2a are decreased in response to insulin (reviewed in Lawrence and Roach 1997). Early studies suggested that MAPK was important for insulin-stimulated dephosphorylation of glycogen synthase (reviewed in Lawrence and Roach 1997). However, more recent studies suggest a GSK3-dependent mechanism (reviewed in Roach et al. 2012). *In vitro*, GSK3 phosphorylates sites 3a, 3b, 3c, and 4 in glycogen synthase (Hemmings et al. 1981). Insulin acts through the Akt/PKB signaling pathway to promote phosphorylation of GSK3 resulting in inhibition of the kinase (Woodgett 2005). Evidence for this mechanism was provided by studies in mouse models in which the beta isoform of GSK3 was either overexpressed or disrupted in skeletal muscle. GSK3 overexpression resulted in glycogen synthase that was less active (Pearce et al. 2004), while disruption resulted in enhanced insulin-stimulated activation of glycogen synthase (Patel et al. 2008). Mice expressing constitutively active GSK3-alpha and beta had normal glycogen synthase activity in skeletal muscle, but a blunted insulin-stimulated increase in glycogen synthase activity (McManus et al. 2005). Further, this study indicated that GSK3-beta is the primary isoform though which insulin exerts its effect on glycogen synthase in muscle. While these studies indicate an important role of GSK3 in insulin-mediated regulation of glycogen synthase, other kinases along with PP1c, in conjunction with glycogen targeting subunits, may also be involved (reviewed in Roach et al. 2012).

Insulin—Liver. The role of insulin in liver glycogen metabolism differs from that in muscle. Genetically modified mouse models suggest that insulin does not affect glycogen synthesis through regulation of the phosphorylation state of glycogen synthase. Knockout models of GSK3 had normal levels of liver glycogen after feeding (McManus et al. 2005, Patel et al. 2008, 2011). Dephosphorylation of GYS2 by PP1c in association with GL and glycogen phosphorylase has been proposed to occur with elevated glucose levels. However, genetic manipulation of this interaction did not affect liver glycogen levels after feeding, despite increasing the activity of GYS2 (Kelsall et al. 2009). It appears that activation of GYS2 by glucose-6-P is an important regulatory mechanism for increasing liver glycogen storage after a meal. Knockin mice expressing GYS2 that was unresponsive to glucose-6-P exhibited reduced liver glycogen accumulation after feeding (von Wilamowitz-Moellendorff et al. 2013).

Insulin—Brain. Insulin increased glycogen accumulation in cultured astrocytes (Dringen and Hamprecht 1992). The activation of glycogen synthase by dephosphorylation was stimulated by insulin (Hamai et al. 1999). These effects on phosphorylation were blocked by Wortmannin, a PI-3 kinase inhibitor. In a separate study, the elevated glycogen level was shown to occur without an increase in glucose uptake (Muhic et al. 2015). The phosphorylation of both GSK3 alpha and beta in mouse brain increased with insulin administration, though effects on glycogen synthase were not reported (Clodfelder-Miller et al. 2005).

Other Covalent Modifications. While phosphorylation is the best studied covalent modification of glycogen synthase, acetylation and glucosylation have also been reported. Zhao et al. (Zhao et al. 2010) found that many human liver metabolic enzymes, including glycogen synthase, are acetylated. In addition, O-linked attachment of N-acetylglucosamine was reported to inactivate glycogen synthase (Parker et al. 2003).

5.2.6.7 Translocation

In muscle and liver, glycogen synthase translocates between different subcellular compartments. This provides another mechanism for regulating the enzyme. There is evidence that the localization of glycogen synthase is regulated by the levels of glucose, glycogen, and glucose-6-P, as well as phosphorylation of the enzyme.

GYS1. GYS1 labeled with green fluorescent protein (GFP) was localized to the nucleus of COS-1 and C2C12 cells in the absence of glucose (Ferrer et al. 1997). When glucose was added, the enzyme was localized in the cytosol. Consistent with glycogen content affecting glycogen synthase location, GYS1 translocated from a glycogen enriched membrane fraction to a cytoskeleton fraction with a decrease in glycogen levels (Nielsen et al. 2001). In single muscle fibers, immunofluorescence indicated the presence of larger aggregates of glycogen synthase protein when glycogen levels were high. There are conflicting reports regarding a role for glycogen synthase phosphorylation in affecting the subcellular distribution of the enzyme. GYS1 was found primarily in the perinuclear region and myofibrillar cross-striations if the protein was phosphorylated at site 1a or at sites 3a and 3b (Prats et al. 2005, 2009). However, a diffuse distribution pattern, not associated with the cross-striations or perinuclear region, was observed with phosphorylation of GYS1 at sites 2 and 2a. In contrast, Cid et al. (Cid et al. 2005) reported that phosphorylation of GYS1 does not affect subcellular distribution. When all 9 serine residues were mutated to alanine, localization of glycogen synthase was unaffected. However, mutating the arginine residues in GYS1 that are involved in conferring glucose-6-P sensitivity resulted in the protein quickly moving from the nucleus to the cytoplasm after addition of glucose. Nuclear GYS1 in cultured cells is observed as spherical aggregates co-localizing with p80-coilin (Cid et al. 2005). The latter protein associates with PML bodies which have been implicated in transcriptional regulation and other processes (Ruggero et al. 2000). Cid et al. (Cid et al. 2005) hypothesized that glycogen synthase may function as an energy sensor, with glycogen depletion resulting in translocating to the nucleus where the enzyme performs a moonlighting function. In HeLa cells, depletion of GYS1 affected the translation of a subset of mRNA's, prompting the study's authors to suggest that GYS1 provides a feedback loop between translation and cellular energy status (Fuchs et al. 2011). In cultured neurons and astrocytes, a decrease in the nucleus/cytosol ratio of glycogen synthase was observed when astrocytes and neurons were co-cultured (Mamczur et al. 2015).

GYS2. GYS2 in isolated rat hepatocytes was localized in the cytosol in the absence of added glucose (Fernandez-Novell et al. 1997). When glucose was added, GYS2 translocated to the cell periphery. This effect was compromised when the actin cytoskeleton was disrupted (Garcia-Rocha et al. 2001). The changes in GYS2 cellular localization correlated with intracellular levels of glucose-6-P (Fernandez-Novell et al. 1996). To determine whether phosphorylation of the enzyme affected localization, serine residues at sites 2, 2a, 3a, 3b, 3c, 4, and 5 were individually mutated to alanine (Ros et al. 2009). In the absence or presence of glucose, mutation at site 2 resulted in a localization pattern of GYS2 that was similar to that seen with wild type GYS2 in the presence of added glucose. Mutations at the other phosphorylation sites did not affect translocation.

5.2.6.8 Animal Models with Enhanced Glycogen Accumulation

Animal models that over-accumulate glycogen are beneficial for examining the roles of proteins and metabolites involved in glycogen metabolism.

Muscle. A key question regarding the amount of glycogen synthesized in muscle was the role of glucose availability versus glycogen synthase activity, a push versus pull concept. Overexpression of a constitutively active form of GYS1 in skeletal muscle increased glycogen accumulation up to 5-fold, decreased UDP-glucose levels, and increased glucose-6-P levels (Azpiazu et al. 2000, Manchester et al. 1996). Total glycogen phosphorylase activity was increased 3-fold. GLUT4 levels either decreased or were unchanged as was basal and insulin-stimulated uptake of 2-deoxyglucose (Lawrence et al. 1997). Glycogen accumulation in soleus and quadriceps in these mice is cytosolic and interrupts the normal myofibrillar ultrastructure (Pederson et al. 2005a). These animals with increased muscle glycogen stores were found to perform the same amount of work as wild type mice when subjected to exhaustive treadmill exercise. However, mice overexpressing glycogen synthase used ~ 8x more muscle glycogen but less liver glycogen as compared to wild type controls. This increased use of muscle glycogen was accompanied by an enhanced level of blood lactate. Horses with the neuromuscular disease, polysaccharide storage myopathy (PSSM), have an activating mutation in GYS1 (R309H) (McCue et al. 2008). This mutation increases activity in the absence of glucose-6-P and leads to glycogen accumulation in skeletal muscle. Paradoxically, phosphorylation of sites 2 and 2a in glycogen synthase was increased, but the Km for UDP-glucose was decreased both in the presence and absence of glucose-6-P (Maile et al. 2017). Taken together, these studies demonstrate that regulation of glycogen synthase activity impacts glycogen levels. Increasing glucose uptake can also enhance glycogen accumulation. GLUT1 overexpression in skeletal muscle increased glucose uptake and glycogen accumulation, without affecting glycogen synthase activity (Ren et al. 1993). In contrast, overexpression of GLUT4 in muscle increased insulin-stimulated glucose uptake but did not increase glycogen levels (Brozinick et al. 1996, Hansen et al. 1995).

Liver. Adenoviral expression in rat primary hepatocytes of GYS2 containing serine to alanine substitutions at sites 2, 3a, 3b, 3c, 4 and 5 resulted in a constitutively active enzyme, increased glycogen synthesis, and reduced glucose production (Kadotani et al. 2007). Adenoviral overexpression of wild type GYS2 in rat liver did not lead to increased liver glycogen stores nor affect blood glucose homeostasis in either the fed or fasted states (Ros et al. 2010). When GYS2 made constitutively active by incorporating serine to alanine substitution at sites 2 and 3a was overexpressed, liver glycogen levels were elevated above wild type in both the fed and fasted state. This occurred without a change in GLUT2 expression. These animals also exhibited lower fed blood glucose levels and improved glucose clearance in an oral glucose tolerance test. These findings were replicated in transgenic mice expressing constitutively active GYS2.

Brain. Overexpression of constitutively active glycogen synthase, specifically in neurons in mouse, resulted in dramatic brain glycogen accumulation, of which approximately 40% was insoluble, along with deleterious effects including loss of neurons and impaired locomotion (Duran et al. 2012). It is not clear whether these detrimental consequences are due to the accumulation of insoluble glycogen or the greater than 10-fold increase in total brain glycogen levels. In contrast to other tissues, the skull prevents the brain from increasing in size with the accumulation of glycogen and associated water (associated at a proportion of approximately 1:3 (Fernandez-Elias et al. 2015)). Perhaps this contributes to the pathogenic effect of excess glycogen in the brain.

5.2.6.9 Animal Models with Impaired Glycogen Synthesis

Muscle. A global knockout of GYS1 in mice (muscle glycogen synthase knockout, i.e., MGSKO mice) revealed that glycogen synthase is important for newborn survival (Pederson et al. 2004a). Cardiopulmonary defects led to perinatal death of ~ 90 percent of newborn pups lacking GYS1. MGSKO animals that survived had a 5 to 10 percent reduction in body weight and lacked glycogen in tissues where GYS1 is expressed, including cardiac and skeletal muscle. Heart morphology at 11.5 dpc was normal, but at 14.5 dpc hearts had a thin ventricular wall, due to a decrease in cell proliferation, along with an abnormal ventricular septum, and reduced trabecular structure. Heart morphology, echocardiography, and electrocardiography were similar between adult and wild type controls. However, MGSKO mice 12–16 months of age exhibited significant fibrosis as compared to wild type controls. The loss of glycogen synthase resulted in an increase of active glycogen phosphorylase in both cardiac and skeletal muscle. Overexpressing GYS1 under the control of the muscle creatine kinase promoter in MGSKO mice restored glycogen synthesis in cardiac and skeletal muscle. In skeletal muscle from MGSKO mice, there was an increase in oxidative type I fibers and a trend towards a decrease in glycolytic type II fibers. The levels of AMPK and acetyl-CoA carboxylase phosphorylation were increased, suggesting an increased ability to oxidize fatty acids (Pederson et al. 2005c). This effect on AMPK phosphorylation in the absence of muscle glycogen is notable considering the hypothesis that AMPK serves as glycogen sensor (McBride and Hardie 2009). Fed and fasted blood glucose levels were normal in MGSKO mice, and unexpectedly MGSKO mice cleared glucose better than wild type controls in a glucose tolerance test. Muscle uptake of glucose was decreased in MGSKO mice after a euglycemic-hyperglycemic clamp suggesting peripheral insulin resistance. Numerous other changes in gene expression were found in anterior tibias and medal gastrocnemius muscle from MGSKO animals (Parker et al. 2006). Despite lacking glycogen in muscle, MGSKO animals had normal locomotor activity and performed the same amount of work during exhaustive treadmill exercise compared to wild type controls (Pederson et al. 2005b). Exercise-induced elevation of blood lactate was blunted in MGSKO animals, while liver glycogen usage was similar between genotypes. More recently, a tamoxifen-inducible model was generated to disrupt GYS1 specifically in skeletal muscle of adult mice (Xirouchaki et al. 2016). GYS1 protein and muscle glycogen levels were reduced 85 and 70 percent, respectively. These mice displayed a decrease in insulin-stimulated glucose uptake in muscle, postprandial hyperglycemia, and hyperinsulinemia, indicative of insulin resistance. Liver glycogen levels in the fed state were higher than control animals. Knockout animals also exhibited impaired glucose clearance in a glucose tolerance test. In contrast to MGSKO mice, exercise performance was impaired in the tamoxifen inducible model. Pre- and post-exercise glucose tolerance was also impaired in the latter model.

Liver. A mouse model disrupted for GYS2 in the liver was created by Roach and co-workers (Irimia et al. 2010). These mice exhibited dramatically reduced liver glycogen stores, mild hypoglycemia, impaired glucose tolerance, and shortened time to hypoglycemia after fasting. In addition, the knockout mice had reduced capacity for exhaustive exercise in the fed state but performed normally in the fasted state. Gluconeogenesis was increased and insulin suppression of glucose production was impaired. The decrease in liver glycogen was accompanied by reduced liver insulin signaling and diversion of glucose towards lipid synthesis (Irimia et al. 2017). As mice aged from 4 to 15 months, increased fat accumulation was observed in liver correlating with increased activity of lipogenic enzymes.

Brain. Surviving MGSKO animals lack glycogen in brain and other tissues where GYS1 is expressed (Pederson et al. 2005c). Brain uptake of glucose was decreased in MGSKO mice after a euglycemic-hyperglycemic clamp, suggesting insulin resistance. The ability to synthesize glycogen may contribute to neurological decline with aging (Sinadinos et al. 2014). Global disruption of GYS1 in mouse prevents the formation of polysaccharide-based aggregates, which resemble corpora amylacae normally observed in aged human brain. Reduction of *Drosophila melanogaster* glycogen synthase specifically in neurons resulted in neurological improvement with age and an extended lifespan. A mouse model (GYS1Nestin-KO) with central nervous system-specific disruption of glycogen synthase had undetectable levels of glycogen with normal brain morphology (Duran et al. 2013). The lack of brain GYS1 did not affect protein expression of brain or muscle isoforms of glycogen phosphorylase, but glycogen debranching enzyme was decreased. AMPK was increased but the level of phosphorylated AMPK was normal. Consistent with the findings that brain glycogen is important for learning (Suzuki et al. 2011), GYS1Nestin-KO mice were impaired in the Skinner box associative learning task and exhibited decreased long-term potentiation. When a CNS-specific knockout of GYS1 was generated in the author's laboratory, performance on the inhibitory avoidance test was dependent on the strength of the aversive stimulus (unpublished data). In contrast, MGSKO and MGSKO/GSL30 mice are not impaired in the inhibitory avoidance test (unpublished data). Brain glycogen decreases in rats during exercise and is hypothesized to play a role in central fatigue (Matsui et al. 2011). However, despite lacking glycogen in brain and other tissues, MGSKO animals had normal exercise capacity (Pederson et al. 2005b). Further studies examining adaptations in mice lacking brain glycogen synthase will be useful for understanding the differing phenotypes between the various models.

5.2.6.10 Glycogen Storage Disease Type 0

GYS1 deficiency (GSD0b). The first case of a deficiency of GYS1 was only recently reported (Kollberg et al. 2007). Three siblings had a homozygous stop mutation (R426X). The eldest male child died of cardiac arrest at 10.5 yr of age. He experienced an episode of tonic-clonic seizures six years prior and was subsequently reported to have impaired gross motor performance. A brother exhibited heart abnormalities and muscle fatigue at 11 yr of age and an IQ at the low end of normal. A 2-yr-old sister was asymptomatic. These siblings lacked muscle glycogen, and showed a predominance of oxidative fibers, and mitochondrial proliferation. In another family, an 8-yr-old patient had a homozygous two base pair deletion in exon 2 (c.162-163AG) of GYS1, which is expected to produce a truncated protein (Cameron et al. 2009). He collapsed and died during a bout of exercise. Skeletal muscle lacked glycogen, had a predominance of type 1 fibers, and proliferation of mitochondria. Glycogen synthase protein was absent in fibroblasts. In contrast to the first family, no neurological issues were noted. Other studies in humans and rodents have examined the importance of brain glycogen for both seizure and cognitive function. Patients with temporal lobe epilepsy had two to three fold higher levels of glycogen in the hippocampus than in gray and white matter, leading the study's authors to propose that glycogen is a prerequisite for the sustained neuronal activity that occurs in epileptic seizures (Dalsgaard et al. 2007). Regarding a role for brain glycogen for cognitive function, disruption of glycogen synthesis or inhibition of glycogen utilization in rodents resulted in learning impairment (Duran et al. 2013, Suzuki et al. 2011).

GYS2 Deficiency (GSD0a). Deficiency in GYS2 was first reported in 1963 by Lewis et al. (Lewis et al. 1963) with less than 30 cases reported to date. The disease typically presents in early childhood and results in fasting hyperketotic hypoglycemia and postprandial hyperglycemia (Weinstein et al. 2006). In liver biopsies from these patients, glycogen synthase activity was found to be low while glycogen levels were only moderately lower (Aynsley-Green et al. 1977, Gitzelmann et al. 1996, Lewis et al. 1963). Orho et al. (Orho et al. 1998) characterized the gene structure of GYS2 and identified mutations in the GYS2 gene in five families. Mutations in GYS2 were unique for each family and expression of glycogen synthase with these mutations in COS7 cells resulted in glycogen synthase activity no more than approximately four percent of wild type. Several additional patients with GYS2 deficiency have since been identified (Bachrach et al. 2002, de Kremer et al. 1990, Kasapkara et al. 2017, Laberge et al. 2003, Rutledge et al. 2001, Spiegel et al. 2007, Szymanska et al. 2015). In most other glycogen storage diseases, glycogen overaccumulates. The use of RNAi to silence hepatic GYS2 in a model of GSD III resulted in inhibition of glycogen synthesis and accumulation, hepatomegaly, fibrosis, and nodule development (Pursell et al. 2018). When this approach was used in a model of GSD1a, liver glycogen accumulation and steatosis were decreased. Thus, targeting GYS2 could be therapeutic for other glycogen storage diseases affecting the liver.

5.2.7 Glycogen Branching Enzyme

Amylo-(1,4 to 1,6) transglycosylase, commonly referred to as the glycogen branching enzyme, introduces branch points in the glycogen molecule resulting in glucose chains that range from 1 to more than 30 glucose residues in length (Nitschke et al. 2017). The protein contains both hydrolase and transglycosylase activities. The former cleaves a segment of alpha-1,4 linked glucose residues from the non-reducing end of a glycogen chain, and the latter re-attaches this glucose segment via an alpha-1,6 linkage to the glycogen molecule on the same or neighboring glucose chain (Caudwell and Cohen 1980, Gibson et al. 1971). Branching both increases the solubility of glycogen and provides multiple points of attack for glycogen phosphorylase to degrade the polymer (Melendez-Hevia et al. 1993). The 702 amino acid human protein was recently crystallized and found to be composed of four domains, including a carbohydrate binding module (CBM48) which may facilitate binding to glycogen (Froese et al. 2015). The ratio of branching enzyme to glycogen synthase activity is important for normal glycogen branching. This is supported by studies where overexpressing constitutively active GYS1 in skeletal muscle changed this ratio resulting in the overaccumulation of poorly branched glycogen, i.e., polyglucosan bodies (Pederson et al. 2003). Complete loss of branching enzyme (GBE1) activity is lethal in utero in mice (Lee et al. 2011), mimicking the severe form of glycogen storage disease type IV in humans. In another mouse model, which has only 10 percent of normal branching enzyme activity, polyglucosan bodies accumulate in several tissues including muscle, liver, and brain (Akman et al. 2011b). A knockin of the most common human GBE1 mutation, Y329S, in mice resulted in progressive muscular dysfunction and neuropathy as well as premature death, similar to human adult polyglucosan body disease (Orhan Akman et al. 2015). Reducing the glycogen synthetic capacity in hypomorphic GBE1 mutant mice by crossing with mice globally heterozygous for GYS1 reduced formation of polyglucosan bodies in brain, skeletal muscle, heart and liver, improved mobility, and extended lifespan (Chown 2018).

5.3 Glycogen Degradation

Degradation of glycogen occurs in both the cytosol and in lysosomes. In the cytosol, glycogen phosphorylase works in conjunction with the glycogen debranching enzyme to release glucose-1-P from glycogen. Glucose-1-P is then converted to glucose-6-P. The fate of the glucose-6-P is dependent on the tissue in which it is made; it is either converted to glucose or proceeds through glycolysis. In the lysosomes, alpha-glucosidase degrades glycogen to release free glucose.

5.3.1 Glycogen Phosphorylase

In the cytosol, glycogen phosphorylase cleaves alpha-1,4 glycosidic bonds to release glucose-1-P from the terminal branches of glycogen. There are muscle (mGP), liver (lGP), and brain (bGP) isoforms of glycogen phosphorylase in humans which are encoded by the PYGM, PYGL, and PYGB genes, respectively. At the amino acid level, muscle and brain are ~ 80 percent identical and both share ~ 80 percent sequence identity with the liver isoform. There is ~ 97 percent sequence identity between human and corresponding rodent isoforms (Hudson et al. 1993). All isoforms are allosterically regulated by small molecules and by phosphorylation.

5.3.1.1 Glycogen Phosphorylase Gene Structure

Muscle. The PYGM gene maps to chromosome 11 in humans (Lebo et al. 1984) and 19 in mouse (Glaser et al. 1989). PYGM cloned from human muscle (Gautron et al. 1987) contains 20 exons which encode the 842 amino acid mGP protein (Kubisch et al. 1998). The 5' flanking region of the gene contains a TATA box, GC box, 20 E box motifs, two C rich regions, and sequences similar to CArG and MEF-2 motifs (Froman et al. 1994).

Liver. The PYGL gene maps to chromosome 14 in humans (Newgard et al. 1987) and chromosome 12 in mouse (Glaser et al. 1989). The coding sequence of human liver glycogen phosphorylase was originally reported by Newgard et al. (Newgard et al. 1986) with corrections later reported by Burwinkel et al. (1998a). The encoded lGP protein is comprised of 847 amino acids in humans and 850 amino acids in mouse. The 5' flanking region of the gene contains four consensus binding sites each for SP1 and CTF-NFI, 18 potential sites for API, and a region with alternating purine-pyrimidine (APP) residues (Herrick et al. 1993).

Brain. The PYGB gene maps to chromosome 20 in humans (Newgard et al. 1988) and chromosome 2 in mouse (Glaser et al. 1989). PYGB was cloned from human brain astrocytoma cell line (Newgard et al. 1988) and encodes the bGP protein with a length of 843 aa in both mouse and human.

5.3.1.2 Glycogen Phosphorylase Protein Structure

Glycogen phosphorylase is a glycosyltransferase that transfers glucose from glycogen to inorganic phosphate to form glucose-1-P. Glycogen phosphorylase activity has been described in terms of a four-state model (Fig. 5.7). GPa is used to denote the phosphorylated form and GPb to denote

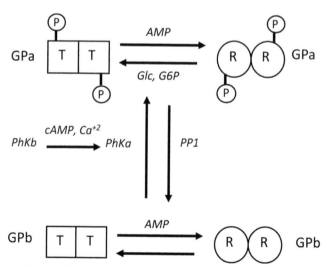

Figure 5.7. Glycogen phosphorylase four state model. Glycogen phosphorylase (GP) transitions between taut (T) and relaxed (R) states and between unphosphorylated (GPb) and phosphorylated (GPa). AMP is an allosteric activator, while glucose (Glc) and glucose-6-phosphate (G6P) are allosteric inhibitors. Glycogen phosphorylase is phosphorylated by phosphorylase kinase (PhK) which activates the enzyme. PhK is regulated by phosphorylation which is stimulated by calcium and cAMP. PhKb; unphosphorylated PhK; PhKa, phosphorylated PhK. Dephosphorylation of GP is catalyzed by protein phosphatase 1 (PP1c) in conjunction with a glycogen targeting subunit.

the unphosphorylated form. Further, each of these forms can exist in either the taut (T)- or relaxed (R)-state. The presence of glucose, glucose-6-P, and uric acid promotes the T-state. AMP promotes the R-state. Phosphorylated glycogen phosphorylase in the presence of AMP (GPa-R-state) is most active, while unphosphorylated phosphorylase in the absence of AMP (GPb-T-state) is least active (Agius 2015). The protein is described as having a regulatory face, containing the AMP binding site and phosphorylation site, and a catalytic face containing the active site (Mathieu et al. 2017).

Muscle. Muscle glycogen phosphorylase (mGP) from rabbit skeletal muscle was first crystallized by Johnson et al. (Johnson et al. 1974). The protein, also called myophosphorylase, has now been crystallized in the various forms described above (reviewed in Johnson and Barford 1993, Nogales-Gadea et al. 2015). mGP is a dimer composed of identical 97 kDa subunits (Fig. 5.8). The N-terminal domain contains the phosphorylation site (ser15) along with AMP, ATP, glucose, glucose-6-P, and glycogen binding sites, and a portion of the catalytic site (Johnson and Barford 1993). The C-terminal portion includes the remainder of the catalytic site, the pyridoxal phosphate binding site and nucleoside inhibitor sites. The catalytic site is in a cleft at the center of each subunit (Johnson and Barford 1993). The co-factor pyridoxal phosphate, which is linked to Lys681 in the active site, is essential for activity (Oikonomakos et al. 1987, Parrish et al. 1977). In the T-state, the catalytic site of the enzyme is blocked by the 280s loop (residues 282–286). This loop is displaced in the transition to the R-state. Phosphorylation at serine 15 causes tertiary and quaternary structural changes including ordering of the N-terminal residues and a change in their contacts from intrasubunit to intersubunit (reviewed in Johnson et al. 1992, Nogales-Gadea et al. 2015). AMP binding increases enzymatic activity to approximately 80 percent of the phosphorylated enzyme but does so without ordering the N-terminus. AMP can bind to both GPa and GPb, but the conformation changes elicited by phosphorylation increase the affinity of the binding site for AMP. The inhibitor glucose-6-P binds to this same effector site but using contacts that are similar but not identical to those used by AMP. The enzyme also binds glycogen, through a binding site consisting of residues 397–437. The nucleotide inhibitor site is located at the entrance to catalytic site and also binds adenosine, caffeine, FMN, NADH, and AMP (Kasvinsky et al. 1978a, Sprang et al. 1982).

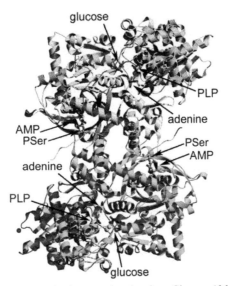

Figure 5.8. Crystal structure of human muscle glycogen phosphorylase. Glucose, AMP, PLP, P-Ser 15, and adenine are shown and labeled. The positions of amino acids that differ between the human and rabbit muscle enzyme are shown in the blue monomer as yellow spheres. Figures were made with MOLSCRIPT (Kraulis 1991) and Raster3D (Merritt and Bacon 1997). Used with permission from (Lukacs et al. 2006).

Liver. The crystal structures of the active and inactive forms of liver glycogen phosphorylase (lGP) were determined (Rath et al. 2000). The enzyme exists as a dimer with each monomer containing two domains. Like mGP, lGP binds AMP, but lGP is not cooperatively activated by this allosteric effector.

Brain. The brain isoform (bGP) was recently crystallized by Mathieu et al. (Mathieu et al. 2016a). The protein is a homodimer with a structure more similar to mGP than to lGP. The region of glycogen phosphorylase containing the regulatory phosphorylation site is observed in crystal structures of the AMP-bound mGP, while no electron density is observed in bGP (Mathieu et al. 2016a, 2017). This difference may explain why bGP has a lower sensitivity to activation by phosphorylation. The binding of AMP to bGP has some characteristics similar to those observed with the muscle isoform, such as the formation of several hydrogen bonds. However, the interaction of helix 2 with AMP in the bGP is more similar to what is observed with the liver isoform, which could explain the non-cooperative binding of AMP observed in bGP.

5.3.1.3 Glycogen Phosphorylase Localization

Muscle. Muscle expresses PYGM at high levels as well as low levels of PYGB (Kato et al. 1989). While glycogen phosphorylase does not contain a specific glycogen binding domain, it does associate with glycogen in a transient manner allowing it to move between glycogen granules (Murphy et al. 2012). In the muscle, glycogen is located primarily between myofibrils close to mitochondria and sarcoplasmic reticulum, with lower levels within myofibrils and beneath the sarcolemma (Nielsen et al. 2011, Prats et al. 2011). Glycogen phosphorylase has been reported to associate with sarcoplasmic reticulum in fast-twitch muscle (Entman et al. 1980). In rabbit muscle, glycogen phosphorylase exhibited a dotted distribution primarily at cross-striations and in the perinuclear region (Prats et al. 2005). Following electrical stimulation, the enzyme translocated to spherical structures composed of β-actin, α-actinin, and smooth muscle tropomyosin. These structures were located primarily in the sub-sarcolemmal region and the I bands of the sarcomeres. Glycogen phosphorylase activity was 2.5 fold higher in type II as compared to type I muscle fibers, consistent with greater exercise-induced glycogen depletion in this fiber type (Harris et al. 1976).

Liver. Liver expresses primarily PYGL (Newgard et al. 1989) as well as low levels of PYGB (Kato et al. 1989). The enzyme is associated with glycogen in the liver (Stapleton et al. 2010). Within the hepatocyte, increasing glucose-6-P levels resulted in the translocation, and inactivation, of phosphorylated glycogen phosphorylase from a soluble to a particulate fraction (Aiston et al. 2004).

Brain. PYGB is the only glycogen phosphorylase isoform that is expressed during development and is later replaced by the muscle and liver isoforms in the respective organs (Newgard et al. 1989). Both PYGB and PYGM are expressed throughout the rat brain (Pfeiffer-Guglielmi et al. 2003). Neurons express PYGB, while astrocytes express both PYGM and PYGB, which co-localize (Pfeiffer-Guglielmi et al. 2003, Saez et al. 2014).

5.3.1.4 Glycogen Phosphorylase Regulation

Glycogen phosphorylase is activated during stress, exercise, hypoxia, and hypoglycemia. All three isoforms are regulated by several allosteric effectors, both activators and inhibitors, as well as by reversible covalent phosphorylation. This phosphorylation, which activates the enzyme, is catalyzed by phosphorylase kinase. Dephosphorylation is catalyzed by PP1c in conjunction with a number of glycogen targeting subunits.

Muscle. Glycogen phosphorylase in the muscle degrades glycogen to fuel muscle contraction. Regulation of mGP is through covalent phosphorylation as well as binding of several metabolites. Phosphorylation, AMP, and glycogen activate the enzyme, while glucose-6-P, glucose, and ATP inhibit mGP. Thus, regulation is determined by integration of all of these factors. AMP binds to mGP in a cooperative manner (Crerar et al. 1995), activating the enzyme (Cori et al. 1938), and is considered to be the most important allosteric effector in muscle (Newgard et al. 1989). AMP activates unphosphorylated mGP (mGPb) to ~ 80 percent of phosphorylated mGP (mGPa) (Morgan and Parmeggiani 1964). The phosphorylated form can be further activated (~ 25 percent) by AMP. Both glucose (Kasvinsky et al. 1978b) and glucose-6-P (Melpidou and Oikonomakos 1983) inhibit the enzyme, with this inhibition counteracted by AMP. Glucose-6-P is an important inhibitor that keeps the enzyme in the inactive state in resting skeletal muscle (Johnson and Barford 1993). Glycogen levels correlate with glycogen degradation in both rat and human muscle, suggesting that glycogen activates mGP (Hespel and Richter 1992, Munger et al. 1993). Inorganic phosphate is a substrate for glycogen phosphorylase and it has been proposed that its availability from ATP and phosphocreatine phosphate turnover may regulate phosphorylase activity (Hargreaves and Richter 1988). Activation of mGP by AMP or phosphorylation promotes subunit association from dimers to less active tetramers, but glycogen promotes disassociation back to dimers (Johnson 1992). Rac1, a member of the Rho family of small GTPases, activates mGP, potentially functioning as an allosteric effector (Arrizabalaga et al. 2012).

Muscle glycogen phosphorylase was the first enzyme shown to be regulated by covalent phosphorylation (Krebs and Fischer 1956). Phosphorylation of ser15 is catalyzed by phosphorylase kinase and increases phosphorylase activity (Brushia and Walsh 1999). Phosphorylase kinase is itself regulated and thus control of mGP phosphorylation is meditated by regulation of phosphorylase kinase. Phosphorylation of mGP in the muscle is promoted during exercise both by elevated local calcium levels and systemic epinephrine levels, via activation of phosphorylase kinase (Jensen and Richter 2012). Phosphorylation of mGP is also influenced by the structural fast twitch muscle protein alpha-actinin-3 which interacts with mGP (Chowrashi et al. 2002). Mice disrupted for the *ACTN3* gene, which encodes the alpha-actinin-3 protein, had a 50 percent reduction in muscle mGP activity and glycogen levels (Quinlan et al. 2010). This reduction in activity appears to be due to alteration of the phosphorylation state of the enzyme.

PYGM gene expression is regulated during myogenesis, though the regulator elements are not clear (Froman et al. 1998). Insulin stimulated GP mRNA expression in cultured L6 myoblasts (Reynet et al. 1996). Both insulin and dibutryl cAMP stimulated expression of a reporter gene under control of the phosphorylase promoter in differentiating myotubes, suggesting that insulin and epinephrine may induce glycogen phosphorylase expression in myogenesis.

A deficiency of PYGM results in McArdle disease. A knock-in mouse containing the most common human mutation (R50X) has been created (Nogales-Gadea et al. 2012). These mice lacked glycogen phosphorylase activity, exhibited massive muscle glycogen accumulation, were exercise intolerant, and myoglobinuria occurred with exercise. While these characteristics align with the human disease, in mice the magnitude of glycogen accumulation was dramatically higher, and the sub cellular location of the glycogen differed.

Liver. The role of glycogen phosphorylase in the liver is to degrade glycogen for the maintenance of blood glucose homeostasis. As with muscle glycogen phosphorylase, lGP is regulated by allosteric regulators and phosphorylation, though this regulation differs between the isoforms. Purified rat and recombinant human glycogen phosphorylase were activated by AMP, and inhibited by ADP, ATP, UDP-glucose, glucose-6-P, and glucose (Ercan-Fang et al. 2002). In intact hepatocytes, glucose-6-P inactivates glycogen phosphorylase (lGP) and inhibits glycogenenolysis (Aiston et al. 2003). In liver, AMP only activates unphosphorylated lGP (lGPb) to ~ 20 percent of phosphorylated lGP (lGPa) (Johnson 1992). The maximal level of lGpa is only 34 percent of the muscle enzyme (Johnson 1992). In contrast to mGP and bGP, which are controlled more by allosteric effectors, lGP appears to be primarily regulated through phosphorylation (from Matheiu 2017). This is supported by the finding that patients with deficiencies of liver phosphorylase kinase exhibit a phenotype including fasting hypoglycemia, liver cirrhosis and growth retardation, which is similar to humans with PYGL deficiency (Burwinkel et al. 1998b). As with mGP and bGP, phosphorylation at ser15 by phosphorylase kinase activates lGP, converting lGPb to lGPa. Hormones that raise cAMP levels (glucagon) or calcium (adrenergic agonists) stimulate rapid conversion of GPb to GPa (Agius 2015). Glucose inhibits lGP by promoting conversion of GPa to GPb (Stalmans et al. 1974). Insulin also stimulates the conversion of GPa to GPb through the action of the PKB/Akt pathway, though the mechanism is unclear (Aiston et al. 2006). In addition to the effects of glucose and insulin on conversion of lGPa to lGPb, neurotransmitters including acetylcholine and serotonin have been proposed to play a role (reviewed in Aiston et al. 2006). Acetylation on lys470 in lGP enhances interaction of lGP with the glycogen targeting protein, GL, promoting dephosphorylation of lGP (Zhang et al. 2012). Acetylation is inhibited by glucagon and stimulated by both inulin and glucose. Liver glycogen levels change during the diurnal cycle due to reciprocal regulation of glycogen synthase and glycogen phosphorylase enzymatic activity and gene expression (Udoh et al. 2015). Cyclic AMP-responsive element binding protein, hepatocyte specific (CREBH) regulates circadian homeostasis of liver glycogen storage (Kim et al. 2017). In mice disrupted for this transcription factor, both mRNA and protein expression of PYGL was reduced. When CREBH was overexpressed, PYGL expression was increased.

Deficiency of PYGL leads to GSDVI (Her's disease), which results in mild to moderate hypoglycemia, enlarged liver, mild ketosis, and growth retardation (Akman et al. 2011a). No information regarding a mouse model of the disorder is available.

Brain. In the brain, degradation of glycogen by glycogen phosphorylase has been proposed to play a role in a number of processes including memory formation, sleep-wake cycle, exercise, seizure, ischemia, hypoxia, and hypoglycemia. The regulation of glycogen phosphorylase in the brain is complex due to the expression of not only bGP in astrocytes and neurons, but also mGP and perhaps lGP in astrocytes (reviewed in Nadeau et al. 2018). Information about the regulation of bGP by allosteric inhibitors is limited, but due to the overlap of the binding sites for AMP and the inhibitors ATP and glucose-6-P, along with the report that AMP binding differs between mGP and bGP, it is reasonable to expect that regulation of these two isoforms may differ with respect to allosteric effectors (Mathieu et al. 2017). bGP is activated by AMP, which binds non-cooperatively and with greater affinity than to mGP (Crerar et al. 1995). Binding of AMP also decreases the enzyme's Km for glycogen (Lowry et al. 1967). bGP is reversibly inhibited by H_2O_2 through the formation of a disulfide bond between cysteines 318 and 326, which interferes with AMP-dependent activation of the enzyme (Mathieu et al. 2016b). Brain glycogen phosphorylase (bGP) is less potently activated by phosphorylation, as compared to mGP and lGP, and also requires AMP binding for full activation (Mathieu et al. 2016a, 2017, Nadeau et al. 2018). Because bGP is more sensitive than mGP to AMP, and mGP is more sensitive than bGP to phosphorylation, bGP has been proposed to regulate brain glycogenolysis in response to the energy status of the cell and mGP in response to neurotransmitters (Mathieu et al. 2016a). Vasointestinal peptide, norephinephrine, and adenosine, via a cAMP-dependent process, increase glycogenolysis in cultured mouse astrocytes (Sorg and Magistretti 1991). Due to its high expression in brain and the presence of a binding site on bGP, Rac1 has recently been proposed as a potential regulator of glycogenolysis in the brain (Nadeau et al. 2018). No human mutations in the PYBG gene have been reported and a knockout mouse has not been generated.

5.3.1.5 Phosphorylation of Glycogen Phosphorylase

Glycogen phosphorylase is phosphorylated and activated by phosphorylase kinase. Phosphorylase kinase is a 1.3 MDa hexadecameric complex composed of 4 copies each of the alpha, beta, gamma, and delta subunits (reviewed in Brushia and Walsh 1999, Pickett-Gies and Walsh 1986). There is one gene for the beta subunit, two genes each for the alpha and gamma subunits, and three genes for the delta subunit (Nadeau et al. 2018). The 45 kDa gamma subunit is the catalytic subunit, while the others are regulatory. Two isoforms of the gamma subunit have been identified, a muscle isoform (PHKG1) and a liver isoform (PHKG2). The regulatory alpha subunit (138 kDa) also exists as two isoforms, PHKA1 and PHKA2. Expression of the former is enriched in muscle and the latter in liver. The 125 kDa beta subunit is encoded by the PHKB gene but three isoforms (muscle, liver, and brain) are made via alternative splicing (Nadeau et al. 2018). The 16.5 kDa δ subunit is calmodulin, which is encoded by three genes (CALM1-3) in humans and is ubiquitously expressed. The alpha and beta subunits inhibit phosphorylase kinase activity but the specific roles of each in this regulation have not been established. Both subunits are farnesylated and contain several phosphorylation sites. PKA-dependent phosphorylation of alpha (ser1018) and beta (ser27, ser701) subunits leads to increased activity of phosphorylase kinase (Meyer et al. 1990). The significance of the other phosphorylation sties is unclear. Hormones that raise cAMP levels activate phosphorylase kinase, through the PKA pathway, resulting in phosphorylation and activation of glycogen phosphorylase (Brushia and Walsh 1999). Calmodulin mediates calcium-dependent regulation of the enzyme, though with several distinctive features as compared to other calmodulin dependent kinases (reviewed in Nadeau et al. 2018). Calcium (Brostrom et al. 1971) and phosphorylation (Krebs and Fischer 1956) are co-dependent activators, with both being required to significantly activate phosphorylase kinase (Cohen 1980, Nadeau et al. 2018). The enzyme is also activated by binding to glycogen (DeLange et al. 1968). Deficiencies of PHKA1 (GSDVIII) and PHKA2 typically result in impaired muscle and liver glycogenolysis, respectively (reviewed in Adeva-Andany et al. 2016). Deficiency of PHKB (GSDIXb) results in impaired glycogenolysis in both muscle and liver. In addition to phosphorylating ser15 of mGP, PhK phosphorylates ser7 of glycogen synthase *in vitro*, but this has not been established *in vivo* (reviewed in Nadeau et al. 2018).

5.3.1.6 Dephosphorylation of Glycogen Phosphorylase

Dephosphorylation of phosphorylase is catalyzed by PP1c in conjunction with glycogen targeting proteins that belong to the PPP1R3 family. This is the same complex described above for the dephosphorylation of glycogen synthase. Thus, this complex works to simultaneously regulate both the synthesis and degradation of glycogen. Overexpression of the various glycogen targeting proteins favors the synthesis of glycogen (reviewed in Agius 2015). GL has a unique feature, relative to the other glycogen targeting proteins, in that it contains an allosteric binding site for GPa (Armstrong et al. 1998). This binding reduces PP1 phosphatase activity towards both GPa and glycogen synthase (Danos et al. 2009). A knock-in of GL with a Y248F mutation, which disrupts the interaction between GL and GPa, resulted in increased glycogen synthase and glycogen phosphorylase activity, along with improved glucose tolerance (Kelsall et al. 2009).

5.3.2 Glycogen Debranching Enzyme

Glycogen phosphorylase removes glucose units from the outer layer of glycogen particles, but stalls when reaching four glucosyl residues from the branch point. The bifunctional debranching enzyme acts on these residues through its transferase and glucosidase activities (Bates et al. 1975, Gordon et al. 1972). The 4-alpha-glucanotransferase activity removes a unit of three glucose residues and adds them to the reducing end of a longer glucose chain through an alpha-1,4 glycosidic bond. The amylo-1,6 glucosidase activity hydrolyzes the remaining glucosyl residue to release free glucose. The human gene (AGL), located on chromosome 1 (Yang-Feng et al. 1992), is composed of 35 exons (Bao et al. 1996). Alternative slicing results in six isoforms, with isoform 1 being widely expressed while expression of informs 2–4 is restricted to muscle. Two promoter regions were identified that confer tissue-specific expression. Nakayama et al. (Nakayama et al. 2001) used site directed mutagenesis to identify residues involved in catalysis. D535, E564, and D670 were required for transferase activity, but not glucosidase activity, while D1086 and D1147 were required for glucosidase activity but not transferase activity. A putative glycogen binding domain was identified in the C-terminal portion of the protein (Yang et al. 1992). The crystal structure of the glycogen debranching enzyme from *Candida glabrata* was recently solved which, when coupled with measurements of enzymatic activity of several point mutants, provided further information about the interaction of the enzyme with glycogen (Zhai et al. 2016). Both the transferase active site and the glucosidase active site exhibited high specificity towards oligosaccharides representative of their physiologic substrates. The enzyme is reported to be activated by cyclodextrins (Watanabe et al. 2006, Yamamoto et al. 2009). Debranching enzyme was found to be cytosolic in HepG2 cells, but glycogen depletion resulted in the protein being present in the nucleus where it interacts with malin and is ubiquitinated (Cheng et al. 2007). The debranching enzyme is reported to bind to the beta-1 subunit of AMPK (Sakoda et al. 2005). AGL was disrupted in mice to model GSDIII (Pagliarani et al. 2014). These animals had increased glycogen stores in liver, muscle, and brain and were exercise intolerant.

5.3.3 Lysosomal Alpha-Glucosidase

In addition to the degradation of glycogen by glycogen phosphorylase in the cytosol, glycogen is taken up into lysosomes where it is degraded to glucose by lysosomal alpha-glucosidase (GAA, EC 3.2.1.20) (Belenky and Rosenfeld 1975, Rosenfeld 1975). This glucose is then released into the cytosol (Jonas et al. 1990, Mancini et al. 1990).

5.3.3.1 GAA Structure

Human GAA was cloned by Hoefsloot (Hoefsloot et al. 1990) and the promoter was found to contain four potential Sp-1 binding sites, and two potential AP-2 binding sites. The crystal structure of the human enzyme was recently solved in both the unbound form as well in complexes with substrate analogs (Roig-Zamboni et al. 2017). Human GAA cleaves both alpha-1,4 and alpha-1,6 linkages but exhibited an ~ 30 fold higher specificity constant for the former. GAA is synthesized as a 110 kDa precursor and processed initially to a 95 kDa intermediate and then to 76 and 70 kDa mature enzymes (Wisselaar et al. 1993). This process begins in the endoplasmic reticulum with the protein being trafficked through the Golgi complex and ending in the lysosome (Lim et al. 2014).

5.3.3.2 GAA Regulation

GAA mRNA expression was relatively low in mouse fetal tissues (9.5–12 dpc) but increased progressively at 14 and 16 dpc (Ponce et al. 1999). In adult tissues, mRNA levels were highest in brain with moderate expression in liver, heart, and skeletal muscle. Intron 1 was found to have a negative regulatory element that contains two tandem E boxes, to which the transcriptional repressor Hes-1 binds, and a potential core Wing Yang 1 binding site (Raben et al. 1996, Yan et al. 2001). Glucagon (Kondomerkos et al. 2004) and cAMP increase GAA enzymatic activity in hepatocytes. Rapamycin, an inhibitor of mTOR, increased the activity of GAA and the degradation of hepatic lysosomal glycogen (Kalamidas et al. 2004).

5.3.3.3 Glycophagy

The process by which glycogen enters the lysosomes for degradation was termed glycophagy by Roach and co-workers (Jiang et al. 2011) and is a form of autophagy selective for glycogen (Kaur and Debnath 2015, Kotoulas et al. 2004, Zhao et al. 2018). This process of cellular self-devouring recycles nutrients to promote survival during times of stress and starvation. The process was once thought to be random, but evidence for substrate selectivity and regulation has accumulated. Glycophagy has been studied primarily in newborn liver, Pompe disease, and in baker's yeast. In newborn hepatocytes, autophagic vacuoles, primarily containing glycogen, surround glycogen stores in the cytosol at a time when large amounts of glycogen are degraded (Kotoulas and Phillips 1971). Accumulation of lysosomal glycogen in a mouse model of Pompe disease was accompanied by an increase in glycogen-containing autophagosomes and late endosomes (Raben et al. 2007). In addition, two genes involved in yeast autophagy, Atg1 and Atg13, restored defective glycogen accumulation in a yeast strain lacking Snf1, the ortholog of mammalian AMPK (Wang et al. 2001). STBD1, also called genethonin, has been implicated in trafficking glycogen to the lysosomes in liver. STBD1, an internal membrane protein, is most highly expressed in liver and muscle (Bouju et al. 1998). The protein contains a carbohydrate binding domain (CBM20) (Janecek 2002) that confers glycogen binding (Jiang et al. 2010, Stapleton et al. 2010). Levels of the protein are decreased in liver and muscle from mice lacking GYS2 and GYS1, respectively (Jiang et al. 2010). When overexpressed in cultured cells, STBD1 co-localized with glycogen, the autophagy protein GABARAPL1, and the lysosomal marker LAMP1 at perinuclear structures. This suggests that STBD1 tethers glycogen to membranes affecting its location and potentially lysosomal trafficking. Disrupting STBD1 in GAA knockout mice reduced fasted liver glycogen levels by 60 percent or more but did not affect glycogen levels in cardiac or skeletal muscle (Sun et al. 2016). Expression of mutant STBD1 in these double knockout mice suggested that the CBM20 and transmembrane regions are critical for transporting glycogen to lysosomes in the liver.

The terminal step in the degradation of this pathway is the transport of glucose from the lysosome to the cytosol. While this transport has been kinetically characterized, the responsible protein has not been identified, though GLUT8 is implicated in the brain and testis (Chou et al. 2010a, Mancini et al. 1990).

Regulation and Role of Glycophagy. Glycogen accumulates in several organs prior to birth and is utilized during the transition from living in the womb to the autonomy that occurs with parturition (Dawes and Shelley 1968). This pathway for glycogen utilization is important in liver during the hours after birth (Phillips et al. 1967). During this time, autophagic vacuoles containing glycogen increase in both number and size along with an increase in GAA activity (reviewed in Kotoulas et al. 2004). During this postnatal period, hypoglycemia develops, triggering an increase in plasma glucagon levels. The hormones glucagon and adrenalin promote autophagy in liver and heart (Kondomerkos et al. 2005, Kotoulas and Phillips 1971). In contrast, the administration of glucose reverses both hypoglycemia and the induction of autophagy (Kotoulas et al. 1971, 2004). Glucagon increases cAMP levels which activate cAMP-dependent protein kinase. This activation promotes calcium entry into lysosomes, which stimulates autophagy and activity of GAA (reviewed in Kotoulas et al. 2004, Zhao et al. 2018). Fewer studies have examined the importance of glycophagy in the adult. Transgenic mice expressing green fluorescent protein (GFP) fused to LC3 were created to monitor autophagosomes (Mizushima et al. 2004). When these mice were subjected to starvation, they had increased production of autophagosomes in glycogen-rich skeletal muscle as compared to oxidative muscle. Starvation also increased the production of autophagosomes in liver, heart and pancreas. Disrupting the autophagy gene ATG7 in liver resulted in death during the neonatal starvation period (Komatsu et al. 2005), while a whole-body conditional knock out in adult mice resulted in mice that experienced hypoglycemia-induced fatality upon fasting (Karsli-Uzunbas et al. 2014). In fed mice, exercise was found to increase autophagy in both skeletal and cardiac muscle (He et al. 2012). In human Pompe patients that underwent exercise training, peak oxidative capacity was 50 percent of healthy subjects, but oxidation of fat and carbohydrate in muscle was normal, suggesting that lysosomal degradation of muscle glycogen is not a significant source of fuel for exercise (Preisler et al. 2012). While the relative contribution of glycogen breakdown by the cytosolic and lysosomal pathways is not clear, these studies indicate the importance of the latter pathway, at least in mice.

Pompe Disease. A deficiency of GAA results in Pompe disease (GSDII). Individuals with the disease accumulate lysosomal glycogen mainly in cardiac and skeletal muscle which leads to hypertrophic cardiomyopathy and respiratory failure (Hirschhorn and Reuser 2001, Kohler et al. 2018). Disruption of GAA in a mouse model closely resembled human Pompe disease, though clinical symptoms developed late in the lifespan of the mice (Raben et al. 1998). These mice accumulated glycogen in skeletal muscle lysosomes and exhibited reduced mobility and strength as young as 3.5 wk of age. At an age of eight to nine months, animals developed muscle wasting. Overexpressing GYS1 or GLUT1 in skeletal muscle of the GAA-disrupted mice further increased glycogen levels and resulted in the development of muscle wasting at an earlier age (Raben et al. 2001). Glycogenesis has been modulated in an attempt to reduce glycogen accumulation and clinical symptoms. Administration of the mTORC1 inhibitor rapamycin to GAA knockout mice increased glycogen synthase phosphorylation and decreased glycogen content in skeletal muscle, but not liver (Ashe et al. 2010). shRNA-mediated reduction of cytoplasmic and lysosomal glycogen accumulation was accomplished by targeting glycogenin or GYS1 in C2C12 cells and primary myoblasts from GAA mice (Douillard-Guilloux et al. 2008). In addition, intramuscular injection of adenovirus vector expressing shRNA into GAA mice reduced glycogen accumulation. Disrupting GYS1 in GAA knockout mice resulted in reduction in cardiac and skeletal lysosomal glycogen and lysosomal swelling, along with correction of cardiomegaly, less pronounced muscle atrophy, and improved exercise capacity (Douillard-Guilloux et al. 2010). These findings are being exploited for substrate reduction therapy; systemic administration of a phosphorodiamide morpholine oligonucleotide targeted to GYS1 led to a reduction in GYS1 transcript and reduction of lysosomal glycogen accumulation in quadriceps and heart in a mouse model of Pompe disease (Clayton et al. 2014).

5.3.4 Glucose-6-Phosphatase

Glucose-1-P produced by glycogen phosphorylase is converted to glucose-6-phosphate by phosphoglucomutase, the same enzyme that catalyzes the reverse reaction in the synthesis of glycogen. Primarily in the liver, but also in the kidney and intestine, glucose-6-P can be hydrolyzed to release glucose into the circulation. The terminal step in the production of glucose from glycogenolysis, as well as from gluconeogenesis, is catalyzed by glucose-6-phosphatase. This multi-subunit enzyme complex hydrolyzes glucose-6-phosphate releasing glucose and inorganic phosphate (Foster et al. 1997, van Schaftingen and Gerin 2002). The catalytic unit is associated with the endoplasmic reticulum and two models have been proposed to explain the structure and function of the enzyme complex (reviewed in Foster et al. 1997). In one model, the active site of the catalytic unit is located in the ER lumen, while in the other, the active site of the catalytic unit resides within the ER membrane. In both cases, transport proteins were proposed to transport substrate (glucose-6-phosphate) from the cytosol to the catalytic unit and products (glucose and inorganic phosphate) from the catalytic unit to the cytosol. The former model has become favored (Fig. 5.9). While as many as five protein

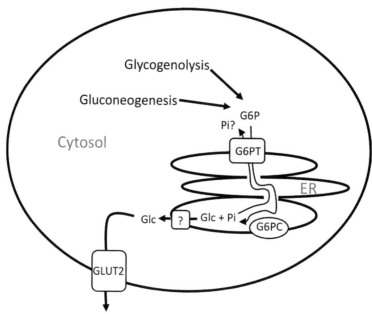

Figure 5.9. Model of glucose-6-phosphatase in liver. G6P, glucose-6-phosphate; G6PT, transporter for G6P and possibly Pi; G6PC, catalytic unit of glucose-6-phosphatase; Glc, glucose; Pi, inorganic phosphate.

components have been proposed for this complex, the existence of only two components has been demonstrated, a catalytic subunit and a glucose-6-P transporter (G6PT) (Cappello et al. 2018).

5.3.4.1 Catalytic Isozymes

Three catalytic unit isozymes have been identified (reviewed in Hutton and O'Brien 2009). G6PC (previously termed G6Pase-alpha) is expressed in liver, kidney and intestine. G6PC2 is expressed in pancreas. G6PC3 (previously termed both G6Pase-beta and ubiquitously expressed glucose-6-phosphatase catalytic subunit-related protein, UGRP) is predominantly expressed in brain, muscle, and kidney.

G6PC. Mouse (Shelly et al. 1993) and human (Lei et al. 1993) G6PC was cloned and biochemically characterized. The gene consists of 5 exons and encodes a protein consisting of 357 amino acids. Sequence alignment coupled with mutational analysis indicates that Arg83, His119, and His176 contribute to the active site of G6PC (Lei et al. 1995, Pan et al. 1998). A number of putative transcription factor binding sites were found in the 5' promoter region of human and rat G6PC including c/EBP, HNF-1, HNF 3, HNF 4, HNF 5, AP-1, CRE, IRE, and GRE (Argaud et al. 1996, Schmoll et al. 1996). G6PC gene expression is regulated by cAMP, glucocorticoids, insulin, glucose, and fructose 2,6, bisphosphate (reviewed in Nordlie et al. 1999, van Schaftingen and Gerin 2002). G6PC mRNA levels were reduced by insulin and increased by dexamethasone and cAMP in FAO hepatoma cells (Lange et al. 1994). When cells were incubated with both insulin and dexamethasone, insulin was dominant. In rat liver, levels of G6PC message and activity were low in the fed and refed state, conditions where insulin levels are elevated (Argaud et al. 1996). These effects of insulin are partially mimicked by an activator of AMPK (Lochhead et al. 2000) and involve PKB and Forkhead (Schmoll et al. 2000). In addition, an AMPK antagonist partially counteracted the inhibitory effects of insulin in hepatoma cells in the basal state (Mues et al. 2009). If cells were stimulated with dexamethasone and cAMP, the effect of insulin was resistant to the AMPK antagonist. G6PC gene expression and enzymatic activity in diabetic rat liver correlated with blood glucose levels and normalization of blood glucose level normalized G6PC mRNA levels (Argaud et al. 1997, Liu et al. 1994, Massillon et al. 1996). cAMP also regulates G6PC activity through a post-translation mechanism that depends on glucose-6-P transport (Soty et al. 2016). Cyclic AMP-responsive element binding protein, hepatocyte specific (CREBH), a circadian transcriptional regulator, interacts with peroxisome proliferator-activated receptor α (PPARα) to regulate the rhythmic expression of G6PC (Kim et al. 2017). The small noncoding RNA miR22-b has been shown to inhibit G6PC expression in human hepatic cells (Ramirez et al. 2013). Several metabolites inhibit the activity of G6PC including fructose-1-phosphate, a proline metabolite, alpha-ketoglutarate, fatty acids, phosphoinositides, and chloride ion (reviewed in Nordlie et al. 1999).

Deficiency of G6PC results in Von Gierke disease (GSD1a). The inability to catalyze the hydrolysis of glucose-6-P leads to fasting hypoglycemia, hypertriglyceridemia, lactic acidosis, hyperuricemia, and hepatomegaly with elevated hepatocyte glycogen levels (Chou et al. 2015). One mouse model disrupted for G6PC exhibited a similar phenotype to the human disease, with the exception of elevated lactate levels (Lei et al. 1996), while another model recapitulated all of the metabolic hallmarks observed in humans with GSD1 (Peng et al. 2009). Liver-specific disruption of G6PC resulted in defective hepatic autophagy, dysregulation of hepatic lipid metabolism, and enhanced hepatic glycolysis and hexose monophosphate shunt (Cho et al. 2018, 2017). Mithieux and co-workers also generated a liver-specific knockout of G6PC (Mutel et al. 2011) as well as intestine-specific and kidney-specific knockouts (reviewed in Rajas et al. 2015).

G6PC2. G6PC2 has 50 percent amino acid sequence identity with G6PC (Arden et al. 1999). It has been controversial as to whether this isozyme hydrolyzes glucose-6-P or not, but supportive evidence is building (Chou et al. 2018, Hutton and O'Brien 2009). Mice with G6PC2 disrupted had a small decrease in blood glucose level (Boortz et al. 2017, Wang et al. 2007). The study's authors suggest that this phenotype may be due to enhanced islet responsiveness at fasting blood glucose levels. This could be due to the glucose phosphorylation catalyzed by glucokinase being unopposed by glucose-6-P hydrolysis normally catalyzed by G6PC2. In addition, glucose-6-P hydrolysis (Pound et al. 2013) and glucose cycling (Wall et al. 2015) were abolished in islets from G6PC2 knockout mice. G6PC2 promoter activity and expression is induced by dexamethasone in isolated human islets (Boortz et al. 2016a,b). Genome wide association studies have indicated that G6PC2 genetic variants are associated with insulin secretion, fasting blood glucose levels, and risk for type 2 diabetes (Hu et al. 2010, Shi et al. 2017).

G6PC3. G6PC3 has a 36 percent amino acid identity with G6PC (Martin et al. 2002). Kinetically, G6PC3 and G6PC have a similar Km for glucose-6-P, but the Vmax is six-fold higher for G6PC (Shieh et al. 2003). Examination of the human G6PC3 promoter revealed that the segment from –455 to –3 relative of the transcriptional start site conferred the highest activity (Bennett et al. 2011). A region of the promoter containing two adjacent enhancer-boxes conferred sensitivity to glucose, with activity correlating with glucose levels between 1 and 5.5 mM. Hypoxia increased G6PC3 levels in brain and heart (Zeng et al. 2016). In addition, coronin 3 reduces expression of G6PC3 in HepG2 cells (Gao et al. 2017).

In humans, deficiency of G6PC3 (GSD-I-related syndrome) results in severe neutropenia and neutrophil dysfunction without the metabolic disturbances associated with G6PC1 deficiency (Banka and Newman 2013). At least 33 different mutations in *G6PC3* have been identified (Lin et al. 2015). Mice disrupted for *G6PC3* had ~ 50 percent reduction in hydrolysis of glucose-6-P in brain (Wang et al. 2006). Liver glycogen content and blood glucose levels were normal, though glucagon levels were elevated in female mice deficient in G6PC3. An even greater reduction (85 percent) in brain glucose-6-P hydrolysis, along with neutropenia, was reported for another G6PC3 knockout mouse (Cheung et al. 2007).

5.3.4.2 *Glucose-6-Phosphate Transporter (G6PT)*

G6PT is ubiquitously expressed and transports glucose-6-P from the cytosol to the ER lumen, possibly in exchange for inorganic phosphate (Chou and Mansfield 2014). The G6PT protein is encoded by the SLC37A4 gene (Chou et al. 2010b, Hiraiwa et al. 1999). The protein contains ten transmembrane helices that span the ER membrane (Chou and Mansfield 2014, Pan et al. 1999). R28 is an essential residue for transporting glucose-6-P (Pan et al. 2009). Using reconstituted proteoliposomes, Chen et al. (Chen et al. 2008a,b) reported that cytoplasmic glucose-6-P is transported in exchange for inorganic phosphate (Pi). In contrast, Marcolongo et al. (Marcolongo et al. 2012) reported that G6PT does not transport Pi, but rather functions as a glucose-6-P uniporter in rat liver microsomes. Transport of glucose-6-P is impacted by the activity of G6PC, though the mechanism is not clear (Chou et al. 2018, Chou and Mansfield 2014). Hepatic microsomes lacking functional G6PC, but expressing functional G6PT, were observed to have decreased glucose-6-P uptake (Lei et al. 1996). In addition, the relationship between transporter and catalytic unit is demonstrated in cell models where the increase in glucose-6-phosphatase activity by the adenylate cyclase activator forskolin required G6PT activity (Soty et al. 2016).

The G6PT promoter has potential binding sites for HIF-1alpha, aryl hydrocarbon receptor (AhR), and Ahr nuclear translator (ARNT) (Lord-Dufour et al. 2009). Hypoxia induced G6PT expression in mesenchymal stromal cells. G6PT levels were increased in the liver of *db/db* mice as well as in primary hepatocytes incubated with corticosterone. This increase was mitigated or blocked, respectively, with the administration of a glucocorticoid antagonist (Wang et al. 2011).

A deficiency in G6PT leads to GSD-1b (Cappello et al. 2018). To date, 97 different mutations have been identified in the human SLC37A4 gene (Chou et al. 2018). The deficiency in synthesizing glucose from both the gluconeogenic and glycogenolytic pathways leads to several metabolic disturbances including fasting hypoglycemia, along with nephromegaly and hepatomegaly, due to accumulation of glycogen. A mouse model of the deficiency mimics known defects of the human disorder and requires dietary intervention to prevent fatal hypoglycemia which has made long-term studies challenging (Chen et al. 2003). Using an inducible Cre-lox system, G6PT was disrupted in adult mice (Raggi et al. 2018). This strategy reduced liver G6PT levels up to 70 percent. These animals exhibited hepatomegaly with variable accumulation of glycogen in the hepatocytes. Disruption was more effective in kidney with a resulting increase in kidney size due to glycogen accumulation. Fasting blood glucose levels were only modestly reduced as compared to control animals.

5.3.4.3 Alternative Roles of Glucose-6-Phosphatase

Until recently, it was thought that the liver, and to a lesser degree the kidney (Kaneko et al. 2018) and intestine (Penhoat et al. 2014), were the only tissues that could release glucose from glucose-6-P due to the expression of glucose-6-phosphatase being limited to these tissues. However, the discovery of G6PC3 expression in muscle (Shieh et al. 2004) and brain expression of both G6PC (Goh et al. 2006) and G6PC3 (Forsyth et al. 1996, 1993, Ghosh et al. 2005) challenges that paradigm. The physiological significance is not well understood, though in astrocytes the gluconeogenesis pathway has been proposed to serve as an alternative pathway for providing glucose to neurons (Yip et al. 2016).

In addition to the classical hydrolysis of glucose-6-P, glucose-6-phosphatase has been shown to catalyze other reactions including the synthesis of glucose-6-P from glucose and carbamoyl-phosphate (carbamyl-P:glucose phosphotransferase activity). The latter activity was investigated extensively by the late Dr. Robert C. Nordlie and co-workers (reviewed in Foster et al. 1997, Nordlie and Foster 2010). This work resulted in the formulation of what he termed the tuning/retuning model. This model incorporates glucokinase along with both the hydrolytic and synthetic activities of glucose-6-phosphatase to explain how the liver regulates blood glucose levels, especially during cases of hyperglycemia and insulin insufficiency. The physiological significance of this hypothesis is awaiting further study.

5.4 Conclusion

In the 160 years since glycogen was discovered (Bernard 1859, Young 1957), much has been learned about the metabolism of this glucose polymer. This work has largely focused on glycogen metabolism in muscle and liver in relation to glucose homeostasis and exercise. Other tissues have been less well studied, but brain has received renewed interest of late. The use of genetically modified mice has given insight into glycogen's role in various physiological processes, increased understanding of the regulation of glycogen metabolism, and provided models for studying and treating glycogen storage diseases. There is still much to be learned about mammalian glycogen metabolism. The roles of glycogen in the brain are actively under investigation. The mechanism(s) by which astrocytic glycogen benefits neurons is not completely understood nor is the significance of the small amounts of neuronal glycogen. In muscle, liver and brain, the relative contributions of the two pathways for glycogen degradation are unclear. Several of the enzymes involved in glycogen metabolism have multiple isoforms, which in some cases have overlapping expression in a given tissue/cell type. In addition, some of these enzymes are subject to multiple forms of isoform-dependent regulation. Further, the subcellular location of these enzymes and association with glycogen and other proteins adds another layer of complexity. Understanding the integration of these various factors will be the subject of study for years to come.

Acknowledgements

The author thanks Dr. Peter J. Roach for kindly providing comments that have improved this chapter. Studies from the author's laboratory have been funded in part by NIH grant DK078370.

References

Adeva-Andany, M.M., M. Gonzalez-Lucan, C. Donapetry-Garcia, C. Fernandez-Fernandez and E. Ameneiros-Rodriguez. 2016. Glycogen metabolism in humans. BBA Clin. 5: 85–100.

Agius, L. 2015. Role of glycogen phosphorylase in liver glycogen metabolism. Mol. Aspects Med. 46: 34–45.

Agius, L. 2016. Hormonal and metabolite regulation of hepatic glucokinase. Annu. Rev. Nutr. 36: 389–415.

Aiston, S., B. Andersen and L. Agius. 2003. Glucose 6-phosphate regulates hepatic glycogenolysis through inactivation of phosphorylase. Diabetes 52: 1333–1339.

Aiston, S., A. Green, M. Mukhtar and L. Agius. 2004. Glucose 6-phosphate causes translocation of phosphorylase in hepatocytes and inactivates the enzyme synergistically with glucose. Biochem. J. 377: 195–204.

Aiston, S., L.J. Hampson, C. Arden, P.B. Iynedjian and L. Agius. 2006. The role of protein kinase B/Akt in insulin-induced inactivation of phosphorylase in rat hepatocytes. Diabetologia 49: 174–182.

Akman, H.O., A. Raghavan and W.J. Craigen. 2011a. Animal models of glycogen storage disorders. Prog. Mol. Biol. Transl. Sci. 100: 369–388.

Akman, H.O., T. Sheiko, S.K. Tay, M.J. Finegold, S. Dimauro and W.J. Craigen. 2011b. Generation of a novel mouse model that recapitulates early and adult onset glycogenosis type IV. Hum. Mol. Genet. 20: 4430–4439.

Allaman, I., S. Lengacher, P.J. Magistretti and L. Pellerin. 2003. A2B receptor activation promotes glycogen synthesis in astrocytes through modulation of gene expression. Am. J. Physiol. Cell Physiol. 284: C696–704.

Amara, S.G. and A.C. Fontana. 2002. Excitatory amino acid transporters: keeping up with glutamate. Neurochem. Int. 41: 313–318.

Amaral, A.I. 2013. Effects of hypoglycaemia on neuronal metabolism in the adult brain: role of alternative substrates to glucose. J. Inherit. Metab. Dis. 36: 621–634.

Andersen, D.K., C.L. Ruiz and C.F. Burant. 1994. Insulin regulation of hepatic glucose transporter protein is impaired in chronic pancreatitis. Ann. Surg. 219: 679–686, discussion 686–677.

Arden, S.D., T. Zahn, S. Steegers, S. Webb, B. Bergman, R.M. O'Brien et al. 1999. Molecular cloning of a pancreatic islet-specific glucose-6-phosphatase catalytic subunit-related protein. Diabetes 48: 531–542.

Argaud, D., Q. Zhang, W. Pan, S. Maitra, S.J. Pilkis and A.J. Lange. 1996. Regulation of rat liver glucose-6-phosphatase gene expression in different nutritional and hormonal states: gene structure and 5'-flanking sequence. Diabetes 45: 1563–1571.

Argaud, D., T.L. Kirby, C.B. Newgard and A.J. Lange. 1997. Stimulation of glucose-6-phosphatase gene expression by glucose and fructose-2,6-bisphosphate. J. Biol. Chem. 272: 12854–12861.

Armstrong, C.G., M.J. Doherty and P.T. Cohen. 1998. Identification of the separate domains in the hepatic glycogen-targeting subunit of protein phosphatase 1 that interact with phosphorylase a, glycogen and protein phosphatase 1. Biochem. J. 336(Pt 3): 699–704.

Arrizabalaga, O., H.M. Lacerda, A.M. Zubiaga and J.L. Zugaza. 2012. Rac1 protein regulates glycogen phosphorylase activation and controls interleukin (IL)-2-dependent T cell proliferation. J. Biol. Chem. 287: 11878–11890.

Ashe, K.M., K.M. Taylor, Q. Chu, E. Meyers, A. Ellis, V. Jingozyan et al. 2010. Inhibition of glycogen biosynthesis via mTORC1 suppression as an adjunct therapy for Pompe disease. Mol. Genet. Metab. 100: 309–315.

Auer, R.N. 1986. Progress review: hypoglycemic brain damage. Stroke 17: 699–708.

Aynsley-Green, A., D.H. Williamson and R. Gitzelmann. 1977. Hepatic glycogen synthetase deficiency. Definition of syndrome from metabolic and enzyme studies on a 9-year-old girl. Arch. Dis. Child. 52: 573–579.

Azpiazu, I., J. Manchester, A.V. Skurat, P.J. Roach and J.C. Lawrence. 2000. Control of glycogen synthesis is shared between glucose transport and glycogen synthase in skeletal muscle fibers. Am. J. Physiol. Endocrinol. 278: E234–E243.

Bachrach, B.E., D.A. Weinstein, M. Orho-Melander, A. Burgess and J.I. Wolfsdorf. 2002. Glycogen synthase deficiency (glycogen storage disease type 0) presenting with hyperglycemia and glucosuria: report of three new mutations. J. Pediatr. 140: 781–783.

Bak, L.K., A.B. Walls, A. Schousboe and H.S. Waagepetersen. 2018. Astrocytic glycogen metabolism in the healthy and diseased brain. J. Biol. Chem. 293: 7108–7116.

Banka, S. and W.G. Newman. 2013. A clinical and molecular review of ubiquitous glucose-6-phosphatase deficiency caused by G6PC3 mutations. Orphanet. J. Rare Dis. 8: 84.

Bao, S. and W.T. Garvey. 1997. Exercise in transgenic mice overexpressing GLUT4 glucose transporters: effects on substrate metabolism and glycogen regulation. Metabolism. 46: 1349–1357.

Bao, Y., T.L. Dawson, Jr. and Y.T. Chen. 1996. Human glycogen debranching enzyme gene (AGL): complete structural organization and characterization of the 5' flanking region. Genomics 38: 155–165.

Baque, S., J.J. Guinovart and J.C. Ferrer. 1997. Glycogenin, the primer of glycogen synthesis, binds to actin. FEBS Lett. 417: 355–359.

Barbetti, F., M. Rocchi, M. Bossolasco, R. Cordera, P. Sbraccia, P. Finelli et al. 1996. The human skeletal muscle glycogenin gene: cDNA, tissue expression and chromosomal localization. Biochem. Biophys. Res. Commun. 220: 72–77.

Baskaran, S., P.J. Roach, A.A. DePaoli-Roach and T.D. Hurley. 2010. Structural basis for glucose-6-phosphate activation of glycogen synthase. Proc. Natl. Acad. Sci. USA. 107: 17563–17568.

Baskaran, S., V.M. Chikwana, C.J. Contreras, K.D. Davis, W.A. Wilson, A.A. DePaoli-Roach et al. 2011. Multiple glycogen-binding sites in eukaryotic glycogen synthase are required for high catalytic efficiency toward glycogen. J. Biol. Chem. 286: 33999–34006.

Bates, E.J., G.M. Heaton, C. Taylor, J.C. Kernohan and P. Cohen. 1975. Debranching enzyme from rabbit skeletal muscle; evidence for the location of two active centres on a single polypeptide chain. FEBS Lett. 58: 181–185.

Beitner, R. 1985. Glucose-1,6-bisphosphate. pp. 1–27. In: Beitner, R. (ed.). The Regulation of Carbohydrate Metabolism (Boca Raton, FL: CRC Press).

Belanger, M., I. Allaman and P.J. Magistretti. 2011. Brain energy metabolism: focus on astrocyte-neuron metabolic cooperation. Cell. Metab. 14: 724–738.

Belenky, D.M. and E.L. Rosenfeld. 1975. Acid alpha-glucosidase (gamma-amylase) from human liver. Clin. Chim. Acta 60: 397–400.

Bennett, K.A., L. Forsyth and A. Burchell. 2011. Functional analysis of the 5' flanking region of the human G6PC3 gene: regulation of promoter activity by glucose, pyruvate, AMP kinase and the pentose phosphate pathway. Mol. Genet. Metab. 103: 254–261.

Bergstrom, J. and E. Hultman. 1966. Muscle glycogen synthesis after exercise: an enhancing factor localized to the muscle cells in man. Nature 210: 309–310.

Bergstrom, J. and E. Hultman. 1967. A study of the glycogen metabolism during exercise in man. Scand. J. Clin. Lab. Invest. 19: 218–228.

Bergstrom, J., L. Hermansen, E. Hultman and B. Saltin. 1967. Diet, muscle glycogen and physical performance. Acta Physiol. Scand. 71: 140–150.

Bernard, C. 1859. De la matiere glycogene consideree comme condition de developpement de certains tissues chez le foetus avant l'apparition de la fonction glycogenique de foie. C.R. Hebd. Seances Acad. Sci. 48: 673–684.

Bhavnani, B.R. 1983. Ontogeny of some enzymes of glycogen metabolism in rabbit fetal heart, lungs, and liver. Can J. Biochem. Cell. Biol. 61: 191–197.

Boortz, K.A., K.E. Syring, C. Dai, L.D. Pound, J.K. Oeser, D.A. Jacobson et al. 2016a. G6PC2 Modulates fasting blood glucose in male mice in response to stress. Endocrinology 157: 3002–3008.

Boortz, K.A., K.E. Syring, R.A. Lee, C. Dai, J.K. Oeser, O.P. McGuinness et al. 2016b. G6PC2 Modulates the effects of dexamethasone on fasting blood glucose and glucose tolerance. Endocrinology 157: 4133–4145.

Boortz, K.A., K.E. Syring, L.D. Pound, H. Mo, L. Bastarache, J.K. Oeser et al. 2017. Effects of G6pc2 deletion on body weight and cholesterol in mice. J. Mol. Endocrinol. 58: 127–139.

Bouju, S., M.F. Lignon, G. Pietu, M. Le Cunff, J.J. Leger, C. Auffray et al. 1998. Molecular cloning and functional expression of a novel human gene encoding two 41–43 kDa skeletal muscle internal membrane proteins. Biochem. J. 335(Pt 3): 549–556.

Bourbon, J.R., M. Rieutort, M.J. Engle and P.M. Farrell. 1982. Utilization of glycogen for phospholipid synthesis in fetal rat lung. Biochim. Biophys. Acta 712: 382–389.

Bouskila, M., M.F. Hirshman, J. Jensen, L.J. Goodyear and K. Sakamoto. 2008. Insulin promotes glycogen synthesis in the absence of GSK3 phosphorylation in skeletal muscle. Am. J. Physiol. Endocrinol. Metab. 294: E28–35.

Bouskila, M., R.W. Hunter, A.F. Ibrahim, L. Delattre, M. Peggie, J.A. van Diepen et al. 2010. Allosteric regulation of glycogen synthase controls glycogen synthesis in muscle. Cell. Metab. 12: 456–466.

Breckenridge, B.M. and E.J. Crawford. 1961. The quantitative histochemistry of the brain. J. Neurochem. 7: 234–240.

Brehier, A. and S.A. Rooney. 1981. Phosphatidylcholine synthesis and glycogen depletion in fetal mouse lung: developmental changes and the effects of dexamethasone. Exp. Lung Res. 2: 273–287.

Brostrom, C.O., F.L. Hunkeler and E.G. Krebs. 1971. The regulation of skeletal muscle phosphorylase kinase by Ca2+. J. Biol. Chem. 246: 1961–1967.

Brown, A.M. 2004. Brain glycogen re-awakened. J. Neurochem. 89: 537–552.

Brown, A.M. and B.R. Ransom. 2015. Astrocyte glycogen as an emergency fuel under conditions of glucose deprivation or intense neural activity. Metab. Brain Dis. 30: 233–239.

Browner, M.F., K. Nakano, A.G. Bang and R.J. Fletterick. 1989. Human muscle glycogen synthase cDNA sequence: a negatively charged protein with an asymmetric charge distribution. Proc. Natl. Acad. Sci. USA. 86: 1443–1447.

Brozinick, J.T., Jr., B.B. Yaspelkis, 3rd, C.M. Wilson, K.E. Grant, E.M. Gibbs, S.W. Cushman et al. 1996. Glucose transport and GLUT4 protein distribution in skeletal muscle of GLUT4 transgenic mice. Biochem. J. 313(Pt1): 133–140.

Bruckner, G. and D. Biesold. 1981. Histochemistry of glycogen deposition in perinatal rat brain: importance of radial glial cells. J. Neurocytol. 10: 749–757.

Brushia, R.J. and D.A. Walsh. 1999. Phosphorylase kinase: the complexity of its regulation is reflected in the complexity of its structure. Front. Biosci. 4: D618–641.

Burwinkel, B., H.D. Bakker, E. Herschkovitz, S.W. Moses, Y.S. Shin and M.W. Kilimann. 1998a. Mutations in the liver glycogen phosphorylase gene (PYGL) underlying glycogenosis type VI. Am. J. Hum. Genet. 62: 785–791.

Burwinkel, B., S. Shiomi, A. Al Zaben and M.W. Kilimann. 1998b. Liver glycogenosis due to phosphorylase kinase deficiency: PHKG2 gene structure and mutations associated with cirrhosis. Hum. Mol. Genet. 7: 149–154.

Cameron, J.M., V. Levandovskiy, N. MacKay, R. Utgikar, C. Ackerley, D. Chiasson et al. 2009. Identification of a novel mutation in GYS1 (muscle-specific glycogen synthase) resulting in sudden cardiac death, that is diagnosable from skin fibroblasts. Mol. Genet. Metab. 98: 378–382.

Camici, M., A.A. DePaoli-Roach and P.J. Roach. 1982. Rabbit liver glycogen synthase. Susceptibility of the enzyme subunit to proteolysis. J. Biol. Chem. 257: 9898–9901.

Camici, M., A.A. DePaoli-Roach and P.J. Roach. 1984. Rabbit liver glycogen synthase. Purification and comparison of the properties of glucose-6-P-dependent and glucose-6-P-independent forms of the enzyme. J. Biol. Chem. 259: 3429–3434.

Canada, S.E., S.A. Weaver, S.N. Sharpe and B.A. Pederson. 2011. Brain glycogen supercompensation in the mouse after recovery from insulin-induced hypoglycemia. J. Neurosci. Res. 89: 585–591.

Cappello, A.R., R. Curcio, R. Lappano, M. Maggiolini and V. Dolce. 2018. The physiopathological role of the exchangers belonging to the SLC37 Family. Front Chem. 6: 122.

Carlson, K.S., P. Davies, B.T. Smith and M. Post. 1987. Temporal linkage of glycogen and saturated phosphatidylcholine in fetal lung type II cells. Pediatr. Res. 22: 79–82.

Caudwell, F.B. and P. Cohen. 1980. Purification and subunit structure of glycogen-branching enzyme from rabbit skeletal muscle. Eur. J. Biochem. 109: 391–394.

Ceulemans, H. and M. Bollen. 2004. Functional diversity of protein phosphatase-1, a cellular economizer and reset button. Physiol. Rev. 84: 1–39.

Chan, E.M., C.A. Ackerley, H. Lohi, L. Ianzano, M.A. Cortez, P. Shannon et al. 2004. Laforin preferentially binds the neurotoxic starch-like polyglucosans, which form in its absence in progressive myoclonus epilepsy. Hum. Mol. Genet. 13: 1117–1129.

Chen, L.Y., J.J. Shieh, B. Lin, C.J. Pan, J.L. Gao, P.M. Murphy et al. 2003. Impaired glucose homeostasis, neutrophil trafficking and function in mice lacking the glucose-6-phosphate transporter. Hum. Mol. Genet. 12: 2547–2558.

Chen, S.Y., C.J. Pan, S. Lee, W. Peng and J.Y. Chou. 2008a. Functional analysis of mutations in the glucose-6-phosphate transporter that cause glycogen storage disease type Ib. Mol. Genet. Metab. 95: 220–223.

Chen, S.Y., C.J. Pan, K. Nandigama, B.C. Mansfield, S.V. Ambudkar and J.Y. Chou. 2008b. The glucose-6-phosphate transporter is a phosphate-linked antiporter deficient in glycogen storage disease type Ib and Ic. FASEB J. 22: 2206–2213.

Cheng, A., M. Zhang, M.S. Gentry, C.A. Worby, J.E. Dixon and A.R. Saltiel. 2007. A role for AGL ubiquitination in the glycogen storage disorders of Lafora and Cori's disease. Genes Dev. 21: 2399–2409.

Cheung, Y.Y., S.Y. Kim, W.H. Yiu, C.J. Pan, H.S. Jun, R.A. Ruef et al. 2007. Impaired neutrophil activity and increased susceptibility to bacterial infection in mice lacking glucose-6-phosphatase-beta. J. Clin. Invest. 117: 784–793.

Chih, C.P. and E.L. Roberts, Jr. 2003. Energy substrates for neurons during neural activity: a critical review of the astrocyte-neuron lactate shuttle hypothesis. J. Cereb. Blood Flow Metab. 23: 1263–1281.

Chikwana, V.M., M. Khanna, S. Baskaran, V.S. Tagliabracci, C.J. Contreras, A. Depaoli-Roach et al. 2013. Structural basis for 2'-phosphate incorporation into glycogen by glycogen synthase. Proc. Natl. Acad. Sci. USA. 110: 20976–20981.

Cho, J.H., G.Y. Kim, C.J. Pan, J. Anduaga, E.J. Choi, B.C. Mansfield et al. 2017. Downregulation of SIRT1 signaling underlies hepatic autophagy impairment in glycogen storage disease type Ia. PLoS Genet. 13: e1006819.

Cho, J.H., G.Y. Kim, B.C. Mansfield and J.Y. Chou. 2018. Hepatic glucose-6-phosphatase-alpha deficiency leads to metabolic reprogramming in glycogen storage disease type Ia. Biochem. Biophys. Res. Commun. 498: 925–931.

Choeiri, C., W. Staines and C. Messier. 2002. Immunohistochemical localization and quantification of glucose transporters in the mouse brain. Neuroscience 111: 19–34.

Choi, I.Y., E.R. Seaquist and R. Gruetter. 2003. Effect of hypoglycemia on brain glycogen metabolism *in vivo*. J. Neurosci. Res. 72: 25–32.

Chou, H.F., K.H. Chuang, Y.S. Tsai and Y.J. Chen. 2010a. Genistein inhibits glucose and sulphate transport in isolated rat liver lysosomes. Br. J. Nutr. 103: 197–205.

Chou, J.Y., H.S. Jun and B.C. Mansfield. 2010b. Glycogen storage disease type I and G6Pase-beta deficiency: etiology and therapy. Nat. Rev. Endocrinol. 6: 676–688.

Chou, J.Y. and B.C. Mansfield. 2014. The SLC37 family of sugar-phosphate/phosphate exchangers. Curr. Top Membr. 73: 357–382.

Chou, J.Y., H.S. Jun and B.C. Mansfield. 2015. Type I glycogen storage diseases: disorders of the glucose-6-phosphatase/glucose-6-phosphate transporter complexes. J. Inherit. Metab. Dis. 38: 511–519.

Chou, J.Y., J.H. Cho, G.Y. Kim and B.C. Mansfield. 2018. Molecular biology and gene therapy for glycogen storage disease type Ib. J. Inherit. Metab. Dis.

Chown, E.E. 2018. Identification of Therapeutic Targets in a Glycogen Storage Disease Type IV Mouse Model. In Institute of Medical Science (University of Toronto).

Chowrashi, P., B. Mittal, J.M. Sanger and J.W. Sanger. 2002. Amorphin is phosphorylase; phosphorylase is an alpha-actinin-binding protein. Cell Motil. Cytoskeleton 53: 125–135.

Cid, E., D. Cifuentes, S. Baque, J.C. Ferrer and J.J. Guinovart. 2005. Determinants of the nucleocytoplasmic shuttling of muscle glycogen synthase. FEBS J. 272: 3197–3213.

Clark, D.D. and L. Sokoloff. 1999. Circulation and energy metabolism of the brain. pp. 637–670. *In*: Siegel, G.J., B.W. Aranoff, R.W. Albers, S.K. Fisher and M.D. Uhler (eds.). Basic Neurochemistry: Molecular, Cellular and Medical Aspects (Philadelphia: Lippincott).

Clayton, N.P., C.A. Nelson, T. Weeden, K.M. Taylor, R.J. Moreland, R.K. Scheule et al. 2014. Antisense oligonucleotide-mediated suppression of muscle glycogen synthase 1 synthesis as an approach for substrate reduction therapy of pompe disease. Mol. Ther. Nucleic. Acids 3: e206.

Clodfelder-Miller, B., P. De Sarno, A.A. Zmijewska, L. Song and R.S. Jope. 2005. Physiological and pathological changes in glucose regulate brain Akt and glycogen synthase kinase-3. J. Biol. Chem. 280: 39723–39731.

Cloix, J.F. and T. Hevor. 2009. Epilepsy, regulation of brain energy metabolism and neurotransmission. Curr. Med. Chem. 16: 841–853.

Cohen, P. 1980. The role of calcium ions, calmodulin and troponin in the regulation of phosphorylase kinase from rabbit skeletal muscle. Eur. J. Biochem. 111: 563–574.

Compernolle, V., K. Brusselmans, T. Acker, P. Hoet, M. Tjwa, H. Beck et al. 2002. Loss of HIF-2alpha and inhibition of VEGF impair fetal lung maturation, whereas treatment with VEGF prevents fatal respiratory distress in premature mice. Nat. Med. 8: 702–710.

Contreras, C.J., D.M. Segvich, K. Mahalingan, V.M. Chikwana, T.L. Kirley, T.D. Hurley et al. 2016. Incorporation of phosphate into glycogen by glycogen synthase. Arch. Biochem. Biophys. 597: 21–29.

Cori, G.T., S.P. Colowick and C.F. Cori. 1938. The action of nucleotides in the disruptive phosphorylastion of glycogen. J. Biol. Chem. 123: 381–389.

Crerar, M.M., O. Karlsson, R.J. Fletterick and P.K. Hwang. 1995. Chimeric muscle and brain glycogen phosphorylases define protein domains governing isozyme-specific responses to allosteric activation. J. Biol. Chem. 270: 13748–13756.

Czech-Damal, N.U., S.J. Geiseler, M.L. Hoff, R. Schliep, J.M. Ramirez, L.P. Folkow et al. 2014. The role of glycogen, glucose and lactate in neuronal activity during hypoxia in the hooded seal (*Cystophora cristata*) brain. Neuroscience 275: 374–383.

Dalsgaard, M.K., F.F. Madsen, N.H. Secher, H. Laursen and B. Quistorff. 2007. High glycogen levels in the hippocampus of patients with epilepsy. J. Cereb. Blood Flow Metab. 27: 1137–1141.

Danos, A.M., S. Osmanovic and M.J. Brady. 2009. Differential regulation of glycogenolysis by mutant protein phosphatase-1 glycogen-targeting subunits. J. Biol. Chem. 284: 19544–19553.

Daran, J.M., N. Dallies, D. Thines-Sempoux, V. Paquet and J. Francois. 1995. Genetic and biochemical characterization of the UGP1 gene encoding the UDP-glucose pyrophosphorylase from Saccharomyces cerevisiae. Eur. J. Biochem. 233: 520–530.

Daugaard, J.R. and E.A. Richter. 2004. Muscle- and fibre type-specific expression of glucose transporter 4, glycogen synthase and glycogen phosphorylase proteins in human skeletal muscle. Pflugers Arch. 447: 452–456.

Dawes, G.S. 1968. Energy metabolism in foetus and after birth. In Foetal and Neonatal Physiology (Chicago: Year Book Medical Publishers, Inc), pp. 210–222.

Dawes, G.S. and H.J. Shelley. 1968. Physiological aspects of carbohydrate metabolism in the foetus and newborn. pp. 87–121. *In*: Dickens, F., P.J. Randle and W.S. Whelan (eds.). Carbohydrate Metabolism and its Disorders (New York: Academic Press).

de Kremer, R.D., A.P. de Capra, C.D. de Boldini, E. Hliba and I. Givogri. 1990. [Hepatic glycogen synthetase deficiency or glycogen storage disease-zero. Mild phenotype with partial enzymatic defect]. Medicina (B Aires) 50: 299–309.

DeLange, R.J., R.G. Kemp, W.D. Riley, R.A. Cooper and E.G. Krebs. 1968. Activation of skeletal muscle phosphorylase kinase by adenosine triphosphate and adenosine 3',5'-monophosphate. J. Biol. Chem. 243: 2200–2208.

DeLano, W.L. 2002. The PyMOL Molecular Graphics System (DeLano Scientific, Palo Alto, CA).

Delaval, E., E. Moreau, S. Andriamanantsara and J.P. Geloso. 1983. Renal glycogen content and hormonal control of enzymes involved in renal glycogen metabolism. Pediatr. Res. 17: 766–769.

Delibegovic, M., C.G. Armstrong, L. Dobbie, P.W. Watt, A.J. Smith and P.T. Cohen. 2003. Disruption of the striated muscle glycogen targeting subunit PPP1R3A of protein phosphatase 1 leads to increased weight gain, fat deposition, and development of insulin resistance. Diabetes 52: 596–604.

DePaoli-Roach, A.A., V.S. Tagliabracci, D.M. Segvich, C.M. Meyer, J.M. Irimia and P.J. Roach. 2010. Genetic depletion of the malin E3 ubiquitin ligase in mice leads to lafora bodies and the accumulation of insoluble laforin. J. Biol. Chem. 285: 25372–25381.

DePaoli-Roach, A.A., C.J. Contreras, D.M. Segvich, C. Heiss, M. Ishihara, P. Azadi et al. 2015. Glycogen phosphomonoester distribution in mouse models of the progressive myoclonic epilepsy, Lafora disease. J. Biol. Chem. 290: 841–850.

Devos, P. and H.G. Hers. 1974. Glycogen metabolism in the liver of the foetal rat. Biochem. J. 140: 331–340.

Dey, N.B., P. Bounelis, T.A. Fritz, D.M. Bedwell and R.B. Marchase. 1994. The glycosylation of phosphoglucomutase is modulated by carbon source and heat shock in *Saccharomyces cerevisiae*. J. Biol. Chem. 269: 27143–27148.

Dickinson, M.E., A.M. Flenniken, X. Ji, L. Teboul, M.D. Wong, J.K. White et al. 2016. High-throughput discovery of novel developmental phenotypes. Nature 537: 508–514.

Dienel, G.A. and N.F. Cruz. 2016. Aerobic glycolysis during brain activation: adrenergic regulation and influence of norepinephrine on astrocytic metabolism. J. Neurochem. 138: 14–52.

Dienel, G.A. 2017. Lack of appropriate stoichiometry: Strong evidence against an energetically important astrocyte-neuron lactate shuttle in brain. J. Neurosci. 95: 2103–2125.

Dienel, G.A. 2019. Does shuttling of glycogen-derived lactate from astrocytes to neurons take place during neurotransmission and memory consolidation? J. Neurosci. Res.

DiNuzzo, M., B. Maraviglia and F. Giove. 2011. Why does the brain (not) have glycogen? BioEssays 33: 319–326.

DiNuzzo, M., F. Giove, B. Maraviglia and S. Mangia. 2015. Monoaminergic control of cellular glucose utilization by glycogenolysis in neocortex and hippocampus. Neurochem. Res. 40: 2493–2504.

DiNuzzo, M. and A. Schousboe (eds.). 2019. Brain Glycogen Metabolism. (Springer).

Doi, R., K. Oishi and N. Ishida. 2010. CLOCK regulates circadian rhythms of hepatic glycogen synthesis through transcriptional activation of Gys2. J. Biol. Chem. 285: 22114–22121.

Douillard-Guilloux, G., N. Raben, S. Takikita, L. Batista, C. Caillaud and E. Richard. 2008. Modulation of glycogen synthesis by RNA interference: towards a new therapeutic approach for glycogenosis type II. Hum. Mol. Genet. 17: 3876–3886.

Douillard-Guilloux, G., N. Raben, S. Takikita, A. Ferry, A. Vignaud, I. Guillet-Deniau et al. 2010. Restoration of muscle functionality by genetic suppression of glycogen synthesis in a murine model of Pompe disease. Hum. Mol. Genet. 19: 684–696.

Dringen, R. and B. Hamprecht. 1992. Glucose, insulin, and insulin-like growth factor I regulate the glycogen content of astroglia-rich primary cultures. J. Neurochem. 58: 511–517.

Duarte, J.M.N., S.S. Nussbaum, R. Gruetter and A.F. Soares. 2016. Glycogen metabolism measured in the brain of insulin-resistant Goto-Kakizaki rats by 13C magnetic resonance spectroscopy *in vivo*. Diabetologia 59 S119.

Duffy, T.E., S.J. Kohle and R.C. Vannucci. 1975. Carbohydrate and energy metabolism in perinatal rat brain: relation to survival in anoxia. J. Neurochem. 24: 271–276.

Duran, J., M.F. Tevy, M. Garcia-Rocha, J. Calbo, M. Milan and J.J. Guinovart. 2012. Deleterious effects of neuronal accumulation of glycogen in flies and mice. EMBO Mol. Med. 4: 719–729.

Duran, J., I. Saez, A. Gruart, J.J. Guinovart and J.M. Delgado-Garcia. 2013. Impairment in long-term memory formation and learning-dependent synaptic plasticity in mice lacking glycogen synthase in the brain. J. Cereb. Blood Flow Metab. 33: 550–556.

Duran, J., A. Gruart, M. Garcia-Rocha, J.M. Delgado-Garcia and J.J. Guinovart. 2014. Glycogen accumulation underlies neurodegeneration and autophagy impairment in Lafora disease. Hum. Mol. Genet. 23: 3147–3156.

Entman, M.L., S.S. Keslensky, A. Chu and W.B. Van Winkle. 1980. The sarcoplasmic reticulum-glycogenolytic complex in mammalian fast twitch skeletal muscle. Proposed *in vitro* counterpart of the contraction-activated glycogenolytic pool. J. Biol. Chem. 255: 6245–6252.

Ercan-Fang, N., M.C. Gannon, V.L. Rath, J.L. Treadway, M.R. Taylor and F.Q. Nuttall. 2002. Integrated effects of multiple modulators on human liver glycogen phosphorylase a. Am. J. Physiol. Endocrinol. Metab. 283: E29–37.

Farkas, I., T.A. Hardy, M.G. Goebl and P.J. Roach. 1991. Two glycogen synthase isoforms in *Saccharomyces cerevisiae* are coded by distinct genes that are differentially controlled. J. Biol. Chem. 266: 15602–15607.

Farrell, C.L. and W.M. Pardridge. 1991. Blood-brain barrier glucose transporter is asymmetrically distributed on brain capillary endothelial lumenal and ablumenal membranes: an electron microscopic immunogold study. Proc. Natl. Acad. Sci. USA. 88: 5779–5783.

Fern, R. 2015. Ischemic tolerance in pre-myelinated white matter: the role of astrocyte glycogen in brain pathology. J. Cereb. Blood Flow Metab. 35: 951–958.

Fernandez-Elias, V.E., J.F. Ortega, R.K. Nelson and R. Mora-Rodriguez. 2015. Relationship between muscle water and glycogen recovery after prolonged exercise in the heat in humans. Eur. J. Appl. Physiol. 115: 1919–1926.

Fernandez-Novell, J.M., A. Roca, D. Bellido, S. Vilaro and J.J. Guinovart. 1996. Translocation and aggregation of hepatic glycogen synthase during the fasted-to-refed transition in rats. Eur. J. Biochem. 238: 570–575.

Fernandez-Novell, J.M., D. Bellido, S. Vilaro and J.J. Guinovart. 1997. Glucose induces the translocation of glycogen synthase to the cell cortex in rat hepatocytes. Biochem. J. 321(Pt 1): 227–231.

Ferrer, J.C., S. Baque and J.J. Guinovart. 1997. Muscle glycogen synthase translocates from the cell nucleus to the cystosol in response to glucose. FEBS Lett. 415: 249–252.

Forsyth, R., A. Fray, M. Boutelle, M. Fillenz, C. Middleditch and A. Burchell. 1996. A role for astrocytes in glucose delivery to neurons? Dev. Neurosci. 18: 360–370.

Forsyth, R.J., K. Bartlett, A. Burchell, H.M. Scott and J.A. Eyre. 1993. Astrocytic glucose-6-phosphatase and the permeability of brain microsomes to glucose 6-phosphate. Biochem. J. 294(Pt 1): 145–151.

Foster, J.D., B.A. Pederson and R.C. Nordlie. 1997. Glucose-6-phosphatase structure, regulation, and function: an update. Proc. Soc. Exp. Biol. Med. 215: 314–332.

Fredriksson, J., M. Ridderstrale, L. Groop and M. Orho-Melander. 2004. Characterization of the human skeletal muscle glycogen synthase gene (GYS1) promoter. Eur. J. Clin. Invest. 34: 113–121.

Froese, D.S., A. Michaeli, T.J. McCorvie, T. Krojer, M. Sasi, E. Melaev et al. 2015. Structural basis of glycogen branching enzyme deficiency and pharmacologic rescue by rational peptide design. Hum. Mol. Genet. 24: 5667–5676.

Froman, B.E., K.R. Herrick and F.A. Gorin. 1994. Regulation of the rat muscle glycogen phosphorylase-encoding gene during muscle cell development. Gene 149: 245–252.

Froman, B.E., R.C. Tait and F.A. Gorin. 1998. Role of E and CArG boxes in developmental regulation of muscle glycogen phosphorylase promoter during myogenesis. DNA Cell Biol. 17: 105–115.

Fuchs, G., C. Diges, L.A. Kohlstaedt, K.A. Wehner and P. Sarnow. 2011. Proteomic analysis of ribosomes: translational control of mrna populations by glycogen synthase GYS1. J. Mol. Biol. 410: 118–130.

Fueger, P.T., S. Heikkinen, D.P. Bracy, C.M. Malabanan, R.R. Pencek, M. Laakso et al. 2003. Hexokinase II partial knockout impairs exercise-stimulated glucose uptake in oxidative muscles of mice. Am. J. Physiol. Endocrinol. Metab. 285: E958–963.

Gao, Y., L. Li, X. Xing, M. Lin, Y. Zeng, X. Liu et al. 2017. Coronin 3 negatively regulates G6PC3 in HepG2 cells, as identified by labelfree massspectrometry. Mol. Med. Rep. 16: 3407–3414.

Garcia-Rocha, M., A. Roca, N. De La Iglesia, O. Baba, J.M. Fernandez-Novell, J.C. Ferrer et al. 2001. Intracellular distribution of glycogen synthase and glycogen in primary cultured rat hepatocytes. Biochem. J. 357: 17–24.

Gaster, M., A. Handberg, H. Beck-Nielsen and H.D. Schroder. 2000. Glucose transporter expression in human skeletal muscle fibers. Am. J. Physiol. Endocrinol. Metab. 279: E529–538.

Gautron, S., D. Daegelen, F. Mennecier, D. Dubocq, A. Kahn and J.C. Dreyfus. 1987. Molecular mechanisms of McArdle's disease (muscle glycogen phosphorylase deficiency). RNA and DNA analysis. J. Clin. Invest. 79: 275–281.

Gayarre, J., L. Duran-Trio, O. Criado Garcia, C. Aguado, L. Juana-Lopez, I. Crespo et al. 2014. The phosphatase activity of laforin is dispensable to rescue Epm2a-/- mice from Lafora disease. Brain 137: 806–818.

Geddes, J., J.E. Schopman, N.N. Zammitt and B.M. Frier. 2008. Prevalence of impaired awareness of hypoglycaemia in adults with Type 1 diabetes. Diabet. Med. 25: 501–504.

Gentry, M.S., C.A. Worby and J.E. Dixon. 2005. Insights into Lafora disease: malin is an E3 ubiquitin ligase that ubiquitinates and promotes the degradation of laforin. Proc. Natl. Acad. Sci. USA. 102: 8501–8506.

Ghafoory, S., K. Breitkopf-Heinlein, Q. Li, C. Scholl, S. Dooley and S. Wolfl. 2013. Zonation of nitrogen and glucose metabolism gene expression upon acute liver damage in mouse. PLoS One 8: e78262.

Ghosh, A., Y.Y. Cheung, B.C. Mansfield and J.Y. Chou. 2005. Brain contains a functional glucose-6-phosphatase complex capable of endogenous glucose production. J. Biol. Chem. 280: 11114–11119.

Gibbons, B.J., P.J. Roach and T.D. Hurley. 2002. Crystal structure of the autocatalytic initiator of glycogen biosynthesis, glycogenin. J. Mol. Biol. 319: 463–477.

Gibbs, M.E. 2015. Role of Glycogenolysis in Memory and Learning: Regulation by Noradrenaline, Serotonin and ATP. Front. Integr. Neurosci. 9: 70.

Gibbs, M.E., D.G. Anderson and L. Hertz. 2006. Inhibition of glycogenolysis in astrocytes interrupts memory consolidation in young chickens. Glia 54: 214–222.

Gibbs, M.E., H.G. Lloyd, T. Santa and L. Hertz. 2007. Glycogen is a preferred glutamate precursor during learning in 1-day-old chick: biochemical and behavioral evidence. J. Neurosci. Res. 85: 3326–3333.

Gibson, W.B., B. Illingsworth and D.H. Brown. 1971. Studies of glycogen branching enzyme. Preparation and properties of -1,4-glucan- -1,4-glucan 6-glycosyltransferase and its action on the characteristic polysaccharide of the liver of children with Type IV glycogen storage disease. Biochemistry 10: 4253–4262.

Gitzelmann, R., M.A. Spycher, G. Feil, J. Muller, B. Seilnacht, M. Stahl et al. 1996. Liver glycogen synthase deficiency: a rarely diagnosed entity. Eur. J. Pediatr. 155: 561–567.

Glaser, T., K.E. Matthews, J.W. Hudson, P. Seth, D.E. Housman and M.M. Crerar. 1989. Localization of the muscle, liver, and brain glycogen phosphorylase genes on linkage maps of mouse chromosomes 19, 12, and 2, respectively. Genomics 5: 510–521.

Goh, B.H., A. Khan, S. Efendic and N. Portwood. 2006. Expression of glucose-6-phosphatase system genes in murine cortex and hypothalamus. Horm. Metab. Res. 38: 1–7.

Goldberg, N.D. and A.G. O'Toole. 1969. The properties of glycogen synthetase and regulation of glycogen biosynthesis in rat brain. J. Biol. Chem. 244: 3053–3061.

Gonzalez, J.T., C.J. Fuchs, J.A. Betts and L.J. van Loon. 2016. Liver glycogen metabolism during and after prolonged endurance-type exercise. Am. J. Physiol. Endocrinol. Metab. 311: E543–553.

Gordon, R.B., D.H. Brown and B.I. Brown. 1972. Preparation and properties of the glycogen-debranching enzyme from rabbit liver. Biochim. Biophys. Acta 289: 97–107.

Gould, G.W. and G.D. Holman. 1993. The glucose transporter family: structure, function and tissue-specific expression. Biochem. J. 295(Pt 2): 329–341.

Groop, L.C., M. Kankuri, C. Schalin-Jantti, A. Ekstrand, P. Nikula-Ijas, E. Widen et al. 1993. Association between polymorphism of the glycogen synthase gene and non-insulin-dependent diabetes mellitus. N. Engl. J. Med. 328: 10–14.

Gruetter, R. 2003. Glycogen: the forgotten cerebral energy store. J. Neurosci. Res. 74: 179–183.

Guillam, M.T., R. Burcelin and B. Thorens. 1998. Normal hepatic glucose production in the absence of GLUT2 reveals an alternative pathway for glucose release from hepatocytes. Proc. Natl. Acad Sci. USA. 95: 12317–12321.

Guinovart, J.J., A. Salavert, J. Massague, C.J. Ciudad, E. Salsas and E. Itarte. 1979. Glycogen synthase: a new activity ratio assay expressing a high sensitivity to the phosphorylation state. FEBS Lett. 106: 284–288.

Gutierrez-Correa, J., M. Hod, J.V. Passoneau and N. Freinkel. 1991. Glycogen and enzymes of glycogen metabolism in rat embryos and fetal organs. Biol. Neonate 59: 294–302.

Hamai, M., Y. Minokoshi and T. Shimazu. 1999. L-Glutamate and insulin enhance glycogen synthesis in cultured astrocytes from the rat brain through different intracellular mechanisms. J. Neurochem. 73: 400–407.

Hanashiro, I. and P.J. Roach. 2002. Mutations of muscle glycogen synthase that disable activation by glucose 6-phosphate. Arch. Biochem. Biophys. 397: 286–292.

Hansen, P.A., E.A. Gulve, B.A. Marshall, J. Gao, J.E. Pessin, J.O. Holloszy et al. 1995. Skeletal muscle glucose transport and metabolism are enhanced in transgenic mice overexpressing the Glut4 glucose transporter. J. Biol. Chem. 270: 1679–1684.

Hansen, P.A., B.A. Marshall, M. Chen, J.O. Holloszy and M. Mueckler. 2000. Transgenic overexpression of hexokinase II in skeletal muscle does not increase glucose disposal in wild-type or Glut1-overexpressing mice. J. Biol. Chem. 275: 22381–22386.

Hargreaves, M. and E.A. Richter. 1988. Regulation of skeletal muscle glycogenolysis during exercise. Can. J. Sport Sci. 13: 197–203.

Hargreaves, M. 2015. Exercise, muscle, and CHO metabolism. Scand. J. Med. Sci. Sports 25 Suppl 4: 29–33.

Harris, R.C., B. Essen and E. Hultman. 1976. Glycogen phosphorylase activity in biopsy samples and single muscle fibres of musculus quadriceps femoris of man at rest. Scand J. Clin. Lab. Invest. 36: 521–526.

Hawley, J.A., C. Lundby, J.D. Cotter and L.M. Burke. 2018. Maximizing cellular adaptation to endurance exercise in skeletal muscle. Cell Metab. 27: 962–976.

He, C., M.C. Bassik, V. Moresi, K. Sun, Y. Wei, Z. Zou et al. 2012. Exercise-induced BCL2-regulated autophagy is required for muscle glucose homeostasis. Nature 481: 511–515.

Hearris, M.A., K.M. Hammond, J.M. Fell and J.P. Morton. 2018. Regulation of muscle glycogen metabolism during exercise: implications for endurance performance and training adaptations. Nutrients 10.

Heikkinen, S., M. Pietila, M. Halmekyto, S. Suppola, E. Pirinen, S.S. Deeb et al. 1999. Hexokinase II-deficient mice. Prenatal death of homozygotes without disturbances in glucose tolerance in heterozygotes. J. Biol. Chem. 274: 22517–22523.

Hemmings, B.A., D. Yellowlees, J.C. Kernohan and P. Cohen. 1981. Purification of glycogen synthase kinase 3 from rabbit skeletal muscle. Copurification with the activating factor (FA) of the (Mg-ATP) dependent protein phosphatase. Eur. J. Biochem. 119: 443–451.

Hermansen, L., E. Hultman and B. Saltin. 1967. Muscle glycogen during prolonged severe exercise. Acta Physiol. Scand. 71: 129–139.

Herrick, K.R., F.A. Gorin, E.A. Park and R.C. Tait. 1993. Characterization of the 5' flanking region of the gene encoding rat liver glycogen phosphorylase. Gene 126: 203–211.

Herzog, R.I., O. Chan, S. Yu, J. Dziura, E.C. McNay and R.S. Sherwin. 2008. Effect of acute and recurrent hypoglycemia on changes in brain glycogen concentration. Endocrinology 149: 1499–1504.

Hespel, P. and E.A. Richter. 1992. Mechanism linking glycogen concentration and glycogenolytic rate in perfused contracting rat skeletal muscle. Biochem. J. 284(Pt 3): 777–780.

Heyen, C.A., V.S. Tagliabracci, L. Zhai and P.J. Roach. 2009. Characterization of mouse UDP-glucose pyrophosphatase, a Nudix hydrolase encoded by the Nudt14 gene. Biochem. Biophys. Res. Commun. 390: 1414–1418.

Hiraiwa, H., C.J. Pan, B. Lin, S.W. Moses and J.Y. Chou. 1999. Inactivation of the glucose 6-phosphate transporter causes glycogen storage disease type 1b. J. Biol. Chem. 274: 5532–5536.

Hirschhorn, R. and A.J. Reuser. 2001. Glycogen storage disease type II: acid alpha-glucosidase (acid maltase) deficiency. pp. 3389–3420. *In*: Scriver, C.R., A.L. Beaudet, W.S. Sly and D. Valle (eds.). The Metabolic and Molecular Bases of Inherited Disease (New York: McGraw Hill).

Hizukuri, S. and J. Larner. 1964. Studies on Udpg: Alpha-1,4-Glucan Alpha-4-Glucosyltransferase. Vii. Conversion of the Enzyme from Glucose-6-Phosphate-Dependent to Independent Form in Liver. Biochemistry 3: 1783–1788.

Hoefsloot, L.H., M. Hoogeveen-Westerveld, A.J. Reuser and B.A. Oostra. 1990. Characterization of the human lysosomal alpha-glucosidase gene. Biochem. J. 272: 493–497.

Hosokawa, M. and B. Thorens. 2002. Glucose release from GLUT2-null hepatocytes: characterization of a major and a minor pathway. Am. J. Physiol. Endocrinol. Metab. 282: E794–801.

Hruz, T., O. Laule, G. Szabo, F. Wessendorp, S. Bleuler, L. Oertle et al. 2008. Genevestigator v3: a reference expression database for the meta-analysis of transcriptomes. Adv. Bioinformatics 2008: 420747.

Hu, C., R. Zhang, C. Wang, W. Yu, J. Lu, X. Ma et al. 2010. Effects of GCK, GCKR, G6PC2 and MTNR1B variants on glucose metabolism and insulin secretion. PLoS One 5: e11761.

Huang, X., A. Vaag, M. Hansson, J. Weng, E. Laurila and L. Groop. 2000. Impaired insulin-stimulated expression of the glycogen synthase gene in skeletal muscle of type 2 diabetic patients is acquired rather than inherited. J. Clin. Endocrinol. Metab. 85: 1584–1590.

Hudson, J.W., K.L. Hefferon and M.M. Crerar. 1993. Comparative analysis of species-independent, isozyme-specific amino-acid substitutions in mammalian muscle, brain and liver glycogen phosphorylases. Biochim. Biophys. Acta 1164: 197–208.

Hutton, J.C. and R.M. O'Brien. 2009. Glucose-6-phosphatase catalytic subunit gene family. J. Biol. Chem. 284: 29241–29245.

Imagawa, E., H. Osaka, A. Yamashita, M. Shiina, E. Takahashi, H. Sugie et al. 2014. A hemizygous GYG2 mutation and Leigh syndrome: a possible link? Hum. Genet. 133: 225–234.

Inoue, N., T. Iwasa, K. Fukunaga, Y. Matsukado and E. Miyamoto. 1987. Phosphorylation and inactivation of brain glycogen synthase by a multifunctional calmodulin-dependent protein kinase. J. Neurochem. 48: 981–988.

Inoue, N., Y. Matsukado, S. Goto and E. Miyamoto. 1988. Localization of glycogen synthase in brain. J. Neurochem. 50: 400–405.

Irimia, J.M., C.M. Meyer, C.L. Peper, L. Zhai, C.B. Bock, S.F. Previs et al. 2010. Impaired glucose tolerance and predisposition to the fasted state in liver glycogen synthase knock-out mice. J. Biol. Chem. 285: 12851–12861.

Irimia, J.M., C.M. Meyer, D.M. Segvich, S. Surendran, A.A. DePaoli-Roach, N. Morral et al. 2017. Lack of liver glycogen causes hepatic insulin resistance and steatosis in mice. J. Biol. Chem. 292: 10455–10464.

Jaldin-Fincati, J.R., M. Pavarotti, S. Frendo-Cumbo, P.J. Bilan and A. Klip. 2017. Update on GLUT4 Vesicle Traffic: A Cornerstone of Insulin Action. Trends Endocrinol. Metab. 28: 597–611.

Janecek, S. 2002. A motif of a microbial starch-binding domain found in human genethonin. Bioinformatics 18: 1534–1537.

Jensen, T.E. and E.A. Richter. 2012. Regulation of glucose and glycogen metabolism during and after exercise. J. Physiol. 590: 1069–1076.

Jiang, S., B. Heller, V.S. Tagliabracci, L. Zhai, J.M. Irimia, A.A. DePaoli-Roach et al. 2010. Starch binding domain-containing protein 1/genethonin 1 is a novel participant in glycogen metabolism. J. Biol. Chem. 285: 34960–34971.

Jiang, S., C.D. Wells and P.J. Roach. 2011. Starch-binding domain-containing protein 1 (Stbd1) and glycogen metabolism: Identification of the Atg8 family interacting motif (AIM) in Stbd1 required for interaction with GABARAPL1. Biochem. Biophys. Res. Commun. 413: 420–425.

Johnson, L.N. 1992. Glycogen phosphorylase: control by phosphorylation and allosteric effectors. FASEB J. 6: 2274–2282.

Johnson, L.N., N.B. Madsen, J. Mosley and K.S. Wilson. 1974. The crystal structure of phosphorylase beta at 6 A resolution. J. Mol. Biol. 90: 703–717.

Johnson, L.N., S.H. Hu and D. Barford. 1992. Catalytic mechanism of glycogen phosphorylase. Faraday Discuss. 131–142.

Johnson, L.N. and D. Barford. 1993. The effects of phosphorylation on the structure and function of proteins. Annu. Rev. Biophys. Biomol. Struct. 22: 199–232.

Jonas, A.J., P. Conrad and H. Jobe. 1990. Neutral-sugar transport by rat liver lysosomes. Biochem. J. 272: 323–326.

Jorgensen, S.B., J.N. Nielsen, J.B. Birk, G.S. Olsen, B. Viollet, F. Andreelli et al. 2004. The alpha2-5'AMP-activated protein kinase is a site 2 glycogen synthase kinase in skeletal muscle and is responsive to glucose loading. Diabetes 53: 3074–3081.

Kadotani, A., M. Fujimura, T. Nakamura, S. Ohyama, N. Harada, H. Maruki et al. 2007. Metabolic impact of overexpression of liver glycogen synthase with serine-to-alanine substitutions in rat primary hepatocytes. Arch. Biochem. Biophys. 466: 283–289.

Kalamidas, S.A., D.J. Kondomerkos, O.B. Kotoulas and A.C. Hann. 2004. Electron microscopic and biochemical study of the effects of rapamycin on glycogen autophagy in the newborn rat liver. Microsc. Res. Tech. 63: 215–219.

Kaneko, K., M. Soty, C. Zitoun, A. Duchampt, M. Silva, E. Philippe et al. 2018. The role of kidney in the inter-organ coordination of endogenous glucose production during fasting. Mol. Metab. 16: 203–212.

Karim, S., D.H. Adams and P.F. Lalor. 2012. Hepatic expression and cellular distribution of the glucose transporter family. World J. Gastroenterol. 18: 6771–6781.

Karlsson, J., L.O. Nordesjo and B. Saltin. 1974. Muscle glycogen utilization during exercise after physical training. Acta Physiol. Scand 90: 210–217.

Karsli-Uzunbas, G., J.Y. Guo, S. Price, X. Teng, S.V. Laddha, S. Khor et al. 2014. Autophagy is required for glucose homeostasis and lung tumor maintenance. Cancer Discov. 4: 914–927.

Kasapkara, C.S., Z. Aycan, E. Acoglu, S. Senel, M.M. Oguz and S. Ceylaner. 2017. The variable clinical phenotype of three patients with hepatic glycogen synthase deficiency. J. Pediatr. Endocrinol. Metab. 30: 459–462.

Kaslow, H.R. and D.D. Lesikar. 1984. Isozymes of glycogen synthase. FEBS Letters 172: 294–298.

Kaslow, H.R., D.D. Lesikar, D. Antwi and A.W. Tan. 1985. L-type glycogen synthase. Tissue distribution and electrophoretic mobility. J. Biol. Chem. 260: 9953–9956.

Kasvinsky, P.J., N.B. Madsen, J. Sygusch and R.J. Fletterick. 1978a. The regulation of glycogen phosphorylase alpha by nucleotide derivatives. Kinetic and x-ray crystallographic studies. J. Biol. Chem. 253: 3343–3351.

Kasvinsky, P.J., S. Shechosky and R.J. Fletterick. 1978b. Synergistic regulation of phosphorylase a by glucose and caffeine. J. Biol. Chem. 253: 9102–9106.

Kato, K., A. Shimizu, N. Kurobe, M. Takashi and T. Koshikawa. 1989. Human brain-type glycogen phosphorylase: quantitative localization in human tissues determined with an immunoassay system. J. Neurochem. 52: 1425–1432.

Kaur, J. and J. Debnath. 2015. Autophagy at the crossroads of catabolism and anabolism. Nat. Rev. Mol. Cell. Biol. 16: 461–472.

Kelsall, I.R., D. Rosenzweig and P.T. Cohen. 2009. Disruption of the allosteric phosphorylase a regulation of the hepatic glycogen-targeted protein phosphatase 1 improves glucose tolerance *in vivo*. Cell. Signal. 21: 1123–1134.

Kelsall, I.R., M. Voss, S. Munro, D.J. Cuthbertson and P.T. Cohen. 2011. R3F, a novel membrane-associated glycogen targeting subunit of protein phosphatase 1 regulates glycogen synthase in astrocytoma cells in response to glucose and extracellular signals. J. Neurochem. 118: 596–610.

Kim, H., Z. Zheng, P.D. Walker, G. Kapatos and K. Zhang. 2017. CREBH maintains circadian glucose homeostasis by regulating hepatic glycogenolysis and gluconeogenesis. Mol. Cell. Biol. 37.

Knull, H.R. and R.L. Khandelwal. 1982. Glycogen metabolizing enzymes in brain. Neurochem. Res. 7: 1307–1317.

Kohler, L., R. Puertollano and N. Raben. 2018. Pompe Disease: From Basic Science to Therapy. Neurotherapeutics.

Koizumi, J. and H. Shiraishi. 1970a. Glycogen accumulation in dendrites of the rabbit pallidum following trifluoperazine administration. Exp. Brain Res. 11: 387–391.

Koizumi, J. and H. Shiraishi. 1970b. Ultrastructural appearance of glycogen in the hypothalamus of the rabbit following chlorpromazine administration. Exp. Brain Res. 10: 276–282.

Kollberg, G., M. Tulinius, T. Gilljam, I. Ostman-Smith, G. Forsander, P. Jotorp et al. 2007. Cardiomyopathy and exercise intolerance in muscle glycogen storage disease 0. N. Engl. J. Med. 357: 1507–1514.

Komatsu, M., S. Waguri, T. Ueno, J. Iwata, S. Murata, I. Tanida et al. 2005. Impairment of starvation-induced and constitutive autophagy in Atg7-deficient mice. J. Cell Biol. 169: 425–434.

Kondomerkos, D.J., S.A. Kalamidas and O.B. Kotoulas. 2004. An electron microscopic and biochemical study of the effects of glucagon on glycogen autophagy in the liver and heart of newborn rats. Microsc. Res. Tech. 63: 87–93.

Kondomerkos, D.J., S.A. Kalamidas, O.B. Kotoulas and A.C. Hann. 2005. Glycogen autophagy in the liver and heart of newborn rats. The effects of glucagon, adrenalin or rapamycin. Histol. Histopathol. 20: 689–696.

Kong, J., P.N. Shepel, C.P. Holden, M. Mackiewicz, A.I. Pack and J.D. Geiger. 2002. Brain glycogen decreases with increased periods of wakefulness: implications for homeostatic drive to sleep. J. Neurosci. 22: 5581–5587.

Korrodi-Gregorio, L., S.L. Esteves and M. Fardilha. 2014. Protein phosphatase 1 catalytic isoforms: specificity toward interacting proteins. Transl. Res. 164: 366–391.

Kotoulas, O.B., J. Ho, F. Adachi, B.I. Weigensberg and M.J. Phillips. 1971. Fine structural aspects of the mobilization of hepatic glycogen. II. Inhibition of glycogen breakdown. Am. J. Pathol. 63: 23–36.

Kotoulas, O.B. and M.J. Phillips. 1971. Fine structural aspects of the mobilization of hepatic glycogen. I. Acceleration of glycogen breakdown. Am. J. Pathol. 63: 1–22.

Kotoulas, O.B., S.A. Kalamidas and D.J. Kondomerkos. 2004. Glycogen autophagy. Microsc. Res. Tech. 64: 10–20.

Kraulis, P.J. 1991. MOLSCRIPT: a program to produce both detailed and schematic plots of protein structures. J. Appl. Crystallogr. 24: 946–950.

Krebs, E.G. and E.H. Fischer. 1956. The phosphorylase b to a converting enzyme of rabbit skeletal muscle. Biochim. Biophys. Acta 20: 150–157.

Kubisch, C., E.M. Wicklein and T.J. Jentsch. 1998. Molecular diagnosis of McArdle disease: revised genomic structure of the myophosphorylase gene and identification of a novel mutation. Hum. Mutat. 12: 27–32.

Kurland, I.J. and S.J. Pilkis. 1989. Indirect versus direct routes of hepatic glycogen synthesis. FASEB J. 3: 2277–2281.

Laberge, A.M., G.A. Mitchell, G. van de Werve and M. Lambert. 2003. Long-term follow-up of a new case of liver glycogen synthase deficiency. Am. J. Med. Genet. A. 120A: 19–22.

Lafora, G.R. and B. Glueck. 1911. Beitrag zur histopathologie der myoklonischen epilepsie. Z. Gesamte Neurol. Psychiatr 6: 1–14.

Lange, A.J., D. Argaud, M.R. el-Maghrabi, W. Pan, S.R. Maitra and S.J. Pilkis. 1994. Isolation of a cDNA for the catalytic subunit of rat liver glucose-6-phosphatase: regulation of gene expression in FAO hepatoma cells by insulin, dexamethasone and cAMP. Biochem. Biophys. Res. Commun. 201: 302–309.

Languren, G., T. Montiel, A. Julio-Amilpas and L. Massieu. 2013. Neuronal damage and cognitive impairment associated with hypoglycemia: An integrated view. Neurochem. Int. 63: 331–343.

Lawrence, J.C., Jr. 1992. Signal transduction and protein phosphorylation in the regulation of cellular metabolism by insulin. Annu. Rev. Physiol. 54: 177–193.

Lawrence, J.C., Jr. and P.J. Roach. 1997. New insights into the role and mechanism of glycogen synthase activation by insulin. Diabetes 46: 541–547.

Lawrence, J.C., Jr., A.V. Skurat, P.J. Roach, I. Azpiazu and J. Manchester. 1997. Glycogen synthase: activation by insulin and effect of transgenic overexpression in skeletal muscle. Biochem. Soc. Trans. 25: 14–19.

Lebo, R.V., F. Gorin, R.J. Fletterick, F.T. Kao, M.C. Cheung, B.D. Bruce et al. 1984. High-resolution chromosome sorting and DNA spot-blot analysis assign McArdle's syndrome to chromosome 11. Science 225: 57–59.

Lee, J.Y., S. Moon, Y.K. Kim, S.H. Lee, B.S. Lee, M.Y. Park et al. 2017. Genome-based exome sequencing analysis identifies GYG1, DIS3L and DDRGK1 are associated with myocardial infarction in Koreans. J. Genet. 96: 1041–1046.

Lee, Y.C., C.J. Chang, D. Bali, Y.T. Chen and Y.T. Yan. 2011. Glycogen-branching enzyme deficiency leads to abnormal cardiac development: novel insights into glycogen storage disease IV. Hum. Mol. Genet. 20: 455–465.

Lehto, M., M. Stoffel, L. Groop, R. Espinosa, 3rd, M.M. Le Beau and G.I. Bell. 1993. Assignment of the gene encoding glycogen synthase (GYS) to human chromosome 19, band q13.3. Genomics 15: 460–461.

Lei, K.J., L.L. Shelly, C.J. Pan, J.B. Sidbury and J.Y. Chou. 1993. Mutations in the glucose-6-phosphatase gene that cause glycogen storage disease type 1a. Science 262: 580–583.

Lei, K.J., C.J. Pan, J.L. Liu, L.L. Shelly and J.Y. Chou. 1995. Structure-function analysis of human glucose-6-phosphatase, the enzyme deficient in glycogen storage disease type 1a. J. Biol. Chem. 270: 11882–11886.

Lei, K.J., H. Chen, C.J. Pan, J.M. Ward, B. Mosinger, Jr., E.J. Lee et al. 1996. Glucose-6-phosphatase dependent substrate transport in the glycogen storage disease type-1a mouse. Nat. Genet. 13: 203–209.

Leino, R.L., D.Z. Gerhart, A.M. van Bueren, A.L. McCall and L.R. Drewes. 1997. Ultrastructural localization of GLUT 1 and GLUT 3 glucose transporters in rat brain. J. Neurosci. Res. 49: 617–626.

Leke, R. and A. Schousboe. 2016. The Glutamine transporters and their role in the glutamate/GABA-glutamine cycle. Adv. Neurobiol. 13: 223–257.

Leloir, L.F., J.M. Olavarria, S.H. Goldemberg and H. Carminatti. 1959. Biosynthesis of glycogen from uridine diphosphate glucose. Arch. Biochem. Biophys. 81: 508–520.

Lewis, G.M., J. Spencer-Peet and K.M. Stewart. 1963. Infantile hypoglycaemia due to inherited deficiency of glycogen synthetase in liver. Arch. Dis. Child. 38: 40–48.

Lim, J.A., L. Li and N. Raben. 2014. Pompe disease: from pathophysiology to therapy and back again. Front Aging Neurosci. 6: 177.

Lin, S.R., C.J. Pan, B.C. Mansfield and J.Y. Chou. 2015. Functional analysis of mutations in a severe congenital neutropenia syndrome caused by glucose-6-phosphatase-beta deficiency. Mol. Genet. Metab. 114: 41–45.

Liu, Z., E.J. Barrett, A.C. Dalkin, A.D. Zwart and J.Y. Chou. 1994. Effect of acute diabetes on rat hepatic glucose-6-phosphatase activity and its messenger RNA level. Biochem. Biophys. Res. Commun. 205: 680–686.

Lochhead, P.A., I.P. Salt, K.S. Walker, D.G. Hardie and C. Sutherland. 2000. 5-aminoimidazole-4-carboxamide riboside mimics the effects of insulin on the expression of the 2 key gluconeogenic genes PEPCK and glucose-6-phosphatase. Diabetes 49: 896–903.

Lopez-Ramos, J.C., J. Duran, A. Gruart, J.J. Guinovart and J.M. Delgado-Garcia. 2015. Role of brain glycogen in the response to hypoxia and in susceptibility to epilepsy. Front Cell. Neurosci. 9: 431.

Lord-Dufour, S., I.B. Copland, L.C. Levros, Jr., M. Post, A. Das, C. Khosla et al. 2009. Evidence for transcriptional regulation of the glucose-6-phosphate transporter by HIF-1alpha: Targeting G6PT with mumbaistatin analogs in hypoxic mesenchymal stromal cells. Stem Cells 27: 489–497.

Lowry, O.H., D.W. Schulz and J.V. Passonneau. 1967. The kinetics of glycogen phosphorylases from brain and muscle. J. Biol. Chem. 242: 271–280.

Lukacs, C.M., N.G. Oikonomakos, R.L. Crowther, L.N. Hong, R.U. Kammlott, W. Levin et al. 2006. The crystal structure of human muscle glycogen phosphorylase a with bound glucose and AMP: an intermediate conformation with T-state and R-state features. Proteins 63: 1123–1126.

Luo, X., Y. Zhang, X. Ruan, X. Jiang, L. Zhu, X. Wang et al. 2011. Fasting-induced protein phosphatase 1 regulatory subunit contributes to postprandial blood glucose homeostasis via regulation of hepatic glycogenesis. Diabetes 60: 1435–1445.

Lutz, P.L., G.E. Nilsson and H. Prentice. 2003. The Brain without Oxygen: Causes of Failure Molecular and Physiological Mechanisms for Survival. (Dordrecht: Kluwers Press.).

Magistretti, P.J., O. Sorg and J.L. Martin. 1993. Regulation of glycogen metabolism in astrocytes:physiological, pharmacological, and pathological aspects. (San Diego: Academic Press, Inc).

Magistretti, P.J. and L. Pellerin. 1996. Cellular bases of brain energy metabolism and their relevance to functional brain imaging: evidence for a prominent role of astrocytes. Cereb. Cortex. 6: 50–61.

Mahalingan, K.K., S. Baskaran, A.A. DePaoli-Roach, P.J. Roach and T.D. Hurley. 2017. Redox switch for the inhibited state of yeast glycogen synthase mimics regulation by phosphorylation. Biochemistry 56: 179–188.

Maile, C.A., J.R. Hingst, K.K. Mahalingan, A.O. O'Reilly, M.E. Cleasby, J.R. Mickelson et al. 2017. A highly prevalent equine glycogen storage disease is explained by constitutive activation of a mutant glycogen synthase. Biochim. Biophys. Acta 1861: 3388–3398.

Malfatti, E., J. Nilsson, C. Hedberg-Oldfors, A. Hernandez-Lain, F. Michel, C. Dominguez-Gonzalez et al. 2014. A new muscle glycogen storage disease associated with glycogenin-1 deficiency. Ann. Neurol. 76: 891–898.

Mamczur, P., B. Borsuk, J. Paszko, Z. Sas, J. Mozrzymas, J.R. Wisniewski et al. 2015. Astrocyte-neuron crosstalk regulates the expression and subcellular localization of carbohydrate metabolism enzymes. Glia 63: 328–340.

Manchester, J., A.V. Skurat, P. Roach, S.D. Hauschka and J.C. Lawrence, Jr. 1996. Increased glycogen accumulation in transgenic mice overexpressing glycogen synthase in skeletal muscle. Proc. Natl. Acad. Sci. U. S. A. 93: 10707–10711.

Mancini, G.M., C.E. Beerens and F.W. Verheijen. 1990. Glucose transport in lysosomal membrane vesicles. Kinetic demonstration of a carrier for neutral hexoses. J. Biol. Chem. 265: 12380–12387.

Mandard, S., R. Stienstra, P. Escher, N.S. Tan, I. Kim, F.J. Gonzalez et al. 2007. Glycogen synthase 2 is a novel target gene of peroxisome proliferator-activated receptors. Cell. Mol. Life Sci. 64: 1145–1157.

Manz, F., H. Bickel, D. Brodehl, D. Feist, K. Gellissen, B. Gescholl-Bauer et al. 1987. Fanconi-Bickel syndrome. Pediatr. Nephrol. 1: 509–518.

March, R.E., W. Putt, M. Hollyoake, J.H. Ives, J.U. Lovegrove, D.A. Hopkinson et al. 1993. The classical human phosphoglucomutase (PGM1) isozyme polymorphism is generated by intragenic recombination. Proc. Natl. Acad. Sci. USA. 90: 10730–10733.

Marcolongo, P., R. Fulceri, R. Giunti, E. Margittai, G. Banhegyi and A. Benedetti. 2012. The glucose-6-phosphate transport is not mediated by a glucose-6-phosphate/phosphate exchange in liver microsomes. FEBS Lett. 586: 3354–3359.

Martin, C.C., J.K. Oeser, C.A. Svitek, S.I. Hunter, J.C. Hutton and R.M. O'Brien. 2002. Identification and characterization of a human cDNA and gene encoding a ubiquitously expressed glucose-6-phosphatase catalytic subunit-related protein. J. Mol. Endocrinol. 29: 205–222.

Massillon, D., N. Barzilai, W. Chen, M. Hu and L. Rossetti. 1996. Glucose regulates in vivo glucose-6-phosphatase gene expression in the liver of diabetic rats. J. Biol. Chem. 271: 9871–9874.

Mathieu, C., I.L. de la Sierra-Gallay, R. Duval, X. Xu, A. Cocaign, T. Leger et al. 2016a. Insights into brain glycogen metabolism: The structure of human brain glycogen phosphorylase. J. Biol. Chem. 291: 18072–18083.

Mathieu, C., R. Duval, A. Cocaign, E. Petit, L.C. Bui, I. Haddad et al. 2016b. An isozyme-specific redox switch in human brain glycogen phosphorylase modulates its allosteric activation by AMP. J. Biol. Chem. 291: 23842–23853.

Mathieu, C., J.M. Dupret and F. Rodrigues Lima. 2017. The structure of brain glycogen phosphorylase-from allosteric regulation mechanisms to clinical perspectives. FEBS J. 284: 546–554.

Matsui, T., T. Ishikawa, H. Ito, M. Okamoto, K. Inoue, M.C. Lee et al. 2012. Brain glycogen supercompensation following exhaustive exercise. J. Physiol. 590: 607–616.

Matsui, T., S. Soya, M. Okamoto, Y. Ichitani, K. Kawanaka and H. Soya. 2011. Brain glycogen decreases during prolonged exercise. J. Physiol. 589: 3383–3393.

McBride, A. and D.G. Hardie. 2009. AMP-activated protein kinase—a sensor of glycogen as well as AMP and ATP? Acta Physiol. (Oxf) 196: 99–113.

McCue, M.E., S.J. Valberg, M.B. Miller, C. Wade, S. DiMauro, H.O. Akman et al. 2008. Glycogen synthase (GYS1) mutation causes a novel skeletal muscle glycogenosis. Genomics 91: 458–466.

McEwen, B.S. and L.P. Reagan. 2004. Glucose transporter expression in the central nervous system: relationship to synaptic function. Eur. J. Pharmacol. 490: 13–24.

McManus, E.J., K. Sakamoto, L.J. Armit, L. Ronaldson, N. Shpiro, R. Marquez et al. 2005. Role that phosphorylation of GSK3 plays in insulin and Wnt signalling defined by knockin analysis. Embo. J. 24: 1571–1583.

McNay, E.C. and R.S. Sherwin. 2004. Effect of recurrent hypoglycemia on spatial cognition and cognitive metabolism in normal and diabetic rats. Diabetes 53: 418–425.

McNay, E.C., A. Williamson, R.J. McCrimmon and R.S. Sherwin. 2006. Cognitive and neural hippocampal effects of long-term moderate recurrent hypoglycemia. Diabetes 55: 1088–1095.

McNay, E.C. and V.E. Cotero. 2010. Mini-review: impact of recurrent hypoglycemia on cognitive and brain function. Physiol. Behav. 100: 234–238.

Mehta, M.B., S.V. Shewale, R.N. Sequeira, J.S. Millar, N.J. Hand and D.J. Rader. 2017. Hepatic protein phosphatase 1 regulatory subunit 3B (Ppp1r3b) promotes hepatic glycogen synthesis and thereby regulates fasting energy homeostasis. J. Biol. Chem. 292: 10444–10454.

Melendez-Hevia, E., T.G. Waddell and E.D. Shelton. 1993. Optimization of molecular design in the evolution of metabolism: the glycogen molecule. Biochem. J. 295(Pt 2): 477–483.

Melpidou, A.E. and N.G. Oikonomakos. 1983. Effect of glucose-6-P on the catalytic and structural properties of glycogen phosphorylase a. FEBS Lett. 154: 105–110.

Merritt, E.A. and D.J. Bacon. 1997. Raster3D: photorealistic molecular graphics. Methods Enzymol. 277: 505–524.

Meyer, H.E., G.F. Meyer, H. Dirks and L.M. Heilmeyer, Jr. 1990. Localization of phosphoserine residues in the alpha subunit of rabbit skeletal muscle phosphorylase kinase. Eur. J. Biochem. 188: 367–376.

Minassian, B.A. 2001. Lafora's disease: towards a clinical, pathologic, and molecular synthesis. Pediatr. Neurol. 25: 21–29.

Minassian, B.A., J.R. Lee, J.A. Herbrick, J. Huizenga, S. Soder, A.J. Mungall et al. 1998. Mutations in a gene encoding a novel protein tyrosine phosphatase cause progressive myoclonus epilepsy. Nat. Genet. 20: 171–174.

Mitchell, J.B., C.R. Lupica and T.V. Dunwiddie. 1993. Activity-dependent release of endogenous adenosine modulates synaptic responses in the rat hippocampus. J. Neurosci. 13: 3439–3447.

Mizushima, N., A. Yamamoto, M. Matsui, T. Yoshimori and Y. Ohsumi. 2004. *In vivo* analysis of autophagy in response to nutrient starvation using transgenic mice expressing a fluorescent autophagosome marker. Mol. Biol. Cell 15: 1101–1111.

Morgan, H.E. and A. Parmeggiani. 1964. Regulation of Glycogenolysis in Muscle. 3. Control of Muscle Glycogen Phosphorylase Activity. J. Biol. Chem. 239: 2440–2445.

Moslemi, A.R., C. Lindberg, J. Nilsson, H. Tajsharghi, B. Andersson and A. Oldfors. 2010. Glycogenin-1 deficiency and inactivated priming of glycogen synthesis. N. Engl. J. Med. 362: 1203–1210.

Mu, J., A.V. Skurat and P.J. Roach. 1997. Glycogenin-2, a novel self-glucosylating protein involved in liver glycogen biosynthesis. J. Biol. Chem. 272: 27589–27597.

Mues, C., J. Zhou, K.N. Manolopoulos, P. Korsten, D. Schmoll, L.O. Klotz et al. 2009. Regulation of glucose-6-phosphatase gene expression by insulin and metformin. Horm. Metab. Res. 41: 730–735.

Muhic, M., N. Vardjan, H.H. Chowdhury, R. Zorec and M. Kreft. 2015. Insulin and Insulin-like Growth Factor 1 (IGF-1) Modulate cytoplasmic glucose and glycogen levels but not glucose transport across the membrane in astrocytes. J. Biol. Chem. 290: 11167–11176.

Munger, R., E. Temler, D. Jallut, E. Haesler and J.P. Felber. 1993. Correlations of glycogen synthase and phosphorylase activities with glycogen concentration in human muscle biopsies. Evidence for a double-feedback mechanism regulating glycogen synthesis and breakdown. Metabolism 42: 36–43.

Murphy, R.M., H. Xu, H. Latchman, N.T. Larkins, P.R. Gooley and D.I. Stapleton. 2012. Single fiber analyses of glycogen-related proteins reveal their differential association with glycogen in rat skeletal muscle. Am. J. Physiol. Cell Physiol. 303: C1146–1155.

Murphy, R.M., M. Flores-Opazo, B.P. Frankish, A. Garnham, D. Stapleton and M. Hargreaves. 2018. No evidence of direct association between GLUT4 and glycogen in human skeletal muscle. Physiol. Rep. 6: e13917.

Mutel, E., A. Abdul-Wahed, N. Ramamonjisoa, A. Stefanutti, I. Houberdon, S. Cavassila et al. 2011. Targeted deletion of liver glucose-6 phosphatase mimics glycogen storage disease type 1a including development of multiple adenomas. J. Hepatol. 54: 529–537.

Nadeau, O.W., J.D. Fontes and G.M. Carlson. 2018. The regulation of glycogenolysis in the brain. J. Biol. Chem. 293: 7099–7107.

Nakayama, A., K. Yamamoto and S. Tabata. 2001. Identification of the catalytic residues of bifunctional glycogen debranching enzyme. J. Biol. Chem. 276: 28824–28828.

Nathan, J.D., P.D. Zdankiewicz, J. Wang, S.A. Spector, G. Aspelund, B.P. Jena et al. 2001. Impaired hepatocyte glucose transport protein (GLUT2) internalization in chronic pancreatitis. Pancreas 22: 172–178.

Newgard, C.B., K. Nakano, P.K. Hwang and R.J. Fletterick. 1986. Sequence analysis of the cDNA encoding human liver glycogen phosphorylase reveals tissue-specific codon usage. Proc. Natl. Acad. Sci. USA. 83: 8132–8136.

Newgard, C.B., R.J. Fletterick, L.A. Anderson and R.V. Lebo. 1987. The polymorphic locus for glycogen storage disease VI (liver glycogen phosphorylase) maps to chromosome 14. Am. J. Hum. Genet. 40: 351–364.

Newgard, C.B., D.R. Littman, C. van Genderen, M. Smith and R.J. Fletterick. 1988. Human brain glycogen phosphorylase. Cloning, sequence analysis, chromosomal mapping, tissue expression, and comparison with the human liver and muscle isozymes. J. Biol. Chem. 263: 3850–3857.

Newgard, C.B., P.K. Hwang and R.J. Fletterick. 1989. The family of glycogen phosphorylases: structure and function. Crit. Rev. Biochem. Mol. Biol. 24: 69–99.

Newman, L.A., D.L. Korol and P.E. Gold. 2011. Lactate produced by glycogenolysis in astrocytes regulates memory processing. PLoS One 6: e28427.

Nielsen, J., H.C. Holmberg, H.D. Schroder, B. Saltin and N. Ortenblad. 2011. Human skeletal muscle glycogen utilization in exhaustive exercise: role of subcellular localization and fibre type. J. Physiol. 589: 2871–2885.

Nilsson, J., A. Halim, A.R. Moslemi, A. Pedersen, J. Nilsson, G. Larson et al. 2012. Molecular pathogenesis of a new glycogenosis caused by a glycogenin-1 mutation. Biochim. Biophys. Acta 1822: 493–499.

Nielsen, J.N., W. Derave, S. Kristiansen, E. Ralston, T. Ploug and E.A. Richter. 2001. Glycogen synthase localization and activity in rat skeletal muscle is strongly dependent on glycogen content. J. Physiol. 531: 757–769.

Nitschke, F., M.A. Sullivan, P. Wang, X. Zhao, E.E. Chown, A.M. Perri et al. 2017. Abnormal glycogen chain length pattern, not hyperphosphorylation, is critical in Lafora disease. EMBO Mol. Med. 9: 906–917.

Nogales-Gadea, G., T. Pinos, A. Lucia, J. Arenas, Y. Camara, A. Brull et al. 2012. Knock-in mice for the R50X mutation in the PYGM gene present with McArdle disease. Brain 135: 2048–2057.

Nogales-Gadea, G., A. Brull, A. Santalla, A.L. Andreu, J. Arenas, M.A. Martin et al. 2015. McArdle Disease: Update of Reported Mutations and Polymorphisms in the PYGM Gene. Hum. Mutat. 36: 669–678.

Nordlie, R.C. and J.D. Foster. 2010. A retrospective review of the roles of multifunctional glucose-6-phosphatase in blood glucose homeostasis: Genesis of the tuning/retuning hypothesis. Life Sci. 87: 339–349.

Nordlie, R.C., J.D. Foster and A.J. Lange. 1999. Regulation of glucose production by the liver. Annu. Rev. Nutr. 19: 379–406.

Nuttall, F.Q., M.C. Gannon, G. Bai and E.Y. Lee. 1994. Primary structure of human liver glycogen synthase deduced by cDNA cloning. Arch. Biochem. Biophys. 311: 443–449.

Obel, L.F., M.S. Muller, A.B. Walls, H.M. Sickmann, L.K. Bak, H.S. Waagepetersen et al. 2012. Brain glycogen-new perspectives on its metabolic function and regulation at the subcellular level. Front Neuroenergetics 4: 3.

Oe, Y., O. Baba, H. Ashida, K.C. Nakamura and H. Hirase. 2016. Glycogen distribution in the microwave-fixed mouse brain reveals heterogeneous astrocytic patterns. Glia 64: 1532–1545.

Oikonomakos, N.G., L.N. Johnson, K.R. Acharya, D.I. Stuart, D. Barford, J. Hajdu et al. 1987. Pyridoxal phosphate site in glycogen phosphorylase b: structure in native enzyme and in three derivatives with modified cofactors. Biochemistry 26: 8381–8389.

Orhan Akman, H., V. Emmanuele, Y.G. Kurt, B. Kurt, T. Sheiko, S. DiMauro et al. 2015. A novel mouse model that recapitulates adult-onset glycogenosis type 4. Hum. Mol. Genet. 24: 6801–6810.

Orho, M., N.U. Bosshard, N.R. Buist, R. Gitzelmann, A. Aynsley-Green, P. Blumel et al. 1998. Mutations in the liver glycogen synthase gene in children with hypoglycemia due to glycogen storage disease type 0. J. Clin. Investig. 102: 507–515.

Ortenblad, N. and J. Nielsen. 2015. Muscle glycogen and cell function—Location, location, location. Scand. J. Med. Sci. Sports 25 Suppl. 4: 34–40.

Oz, G., N. Tesfaye, A. Kumar, D.K. Deelchand, L.E. Eberly and E.R. Seaquist. 2012. Brain glycogen content and metabolism in subjects with type 1 diabetes and hypoglycemia unawareness. J. Cereb. Blood Flow Metab. 32: 256–263.

Pagliarani, S., S. Lucchiari, G. Ulzi, R. Violano, M. Ripolone, A. Bordoni et al. 2014. Glycogen storage disease type III: A novel Agl knockout mouse model. Biochim. Biophys. Acta 1842: 2318–2328.

Pan, C.J., K.J. Lei, B. Annabi, W. Hemrika and J.Y. Chou. 1998. Transmembrane topology of glucose-6-phosphatase. J. Biol. Chem. 273: 6144–6148.

Pan, C.J., B. Lin and J.Y. Chou. 1999. Transmembrane topology of human glucose 6-phosphate transporter. J. Biol. Chem. 274: 13865–13869.

Pan, C.J., S.Y. Chen, S. Lee and J.Y. Chou. 2009. Structure-function study of the glucose-6-phosphate transporter, an eukaryotic antiporter deficient in glycogen storage disease type Ib. Mol. Genet. Metab. 96: 32–37.

Parker, G.E., B.A. Pederson, M. Obayashi, J.M. Schroeder, R.A. Harris and P.J. Roach. 2006. Gene expression profiling of mice with genetically modified muscle glycogen content. Biochem. J. 395: 137–145.

Parker, G.J., K.C. Lund, R.P. Taylor and D.A. McClain. 2003. Insulin resistance of glycogen synthase mediated by o-linked N-acetylglucosamine. J. Biol. Chem. 278: 10022–10027.

Parrish, R.F., R.J. Uhing and D.J. Graves. 1977. Effect of phosphate analogues on the activity of pyridoxal reconstituted glycogen phosphorylase. Biochemistry 16: 4824–4831.

Patel, S., B.W. Doble, K. MacAulay, E.M. Sinclair, D.J. Drucker and J.R. Woodgett. 2008. Tissue-specific role of glycogen synthase kinase 3beta in glucose homeostasis and insulin action. Mol. Cell Biol. 28: 6314–6328.

Patel, S., K. Macaulay and J.R. Woodgett. 2011. Tissue-specific analysis of glycogen synthase kinase-3alpha (GSK-3alpha) in glucose metabolism: effect of strain variation. PLoS One 6: e15845.

Pearce, N.J., J.R. Arch, J.C. Clapham, M.P. Coghlan, S.L. Corcoran, C.A. Lister et al. 2004. Development of glucose intolerance in male transgenic mice overexpressing human glycogen synthase kinase-3beta on a muscle-specific promoter. Metabolism 53: 1322–1330.

Pederson, B.A., C. Cheng, W.A. Wilson and P.J. Roach. 2000. Regulation of glycogen synthase. Identification of residues involved in regulation by the allosteric ligand glucose-6-P and by phosphorylation. J. Biol. Chem. 275: 27753–27761.

Pederson, B.A., A.G. Csitkovits, R. Simon, J.M. Schroeder, W. Wang, A.V. Skurat et al. 2003. Overexpression of glycogen synthase in mouse muscle results in less branched glycogen. Biochem. Biophys. Res. Commun. 305: 826–830.

Pederson, B.A., H. Chen, J.M. Schroeder, W. Shou, A.A. DePaoli-Roach and P.J. Roach. 2004a. Abnormal cardiac development in the absence of heart glycogen. Mol. Cell Biol. 24: 7179–7187.

Pederson, B.A., W.A. Wilson and P.J. Roach. 2004b. Glycogen synthase sensitivity to glucose-6-P is important for controlling glycogen accumulation in Saccharomyces cerevisiae. J. Biol. Chem. 279: 13764–13768.

Pederson, B.A., C.R. Cope, J.M. Irimia, J.M. Schroeder, B.L. Thurberg, A.A. Depaoli-Roach et al. 2005a. Mice with elevated muscle glycogen stores do not have improved exercise performance. Biochem. Biophys. Res. Commun. 331: 491–496.

Pederson, B.A., C.R. Cope, J.M. Schroeder, M.W. Smith, J.M. Irimia, B.L. Thurberg et al. 2005b. Exercise capacity of mice genetically lacking muscle glycogen synthase: In mice, muscle glycogen is not essential for exercise. J. Biol. Chem. 280: 17260–17265.

Pederson, B.A., J.M. Schroeder, G.E. Parker, M.W. Smith, A.A. Depaoli-Roach and P.J. Roach. 2005c. Glucose metabolism in mice lacking muscle glycogen synthase. Diabetes 54: 3466–3473.

Pederson, B.A., J. Turnbull, J.R. Epp, S.A. Weaver, X. Zhao, N. Pencea et al. 2013. Inhibiting glycogen synthesis prevents Lafora disease in a mouse model. Ann. Neurol. 74: 297–300.

Pellegri, G., C. Rossier, P.J. Magistretti and J.L. Martin. 1996. Cloning, localization and induction of mouse brain glycogen synthase. Brain Res. Mol. Brain Res. 38: 191–199.

Pellerin, L., G. Pellegri, P.G. Bittar, Y. Charnay, C. Bouras, J.L. Martin et al. 1998. Evidence supporting the existence of an activity-dependent astrocyte-neuron lactate shuttle. Dev. Neurosci. 20: 291–299.

Peng, W.T., C.J. Pan, E.J. Lee, H. Westphal and J.Y. Chou. 2009. Generation of mice with a conditional allele for G6pc. Genesis 47: 590–594.

Penhoat, A., L. Fayard, A. Stefanutti, G. Mithieux and F. Rajas. 2014. Intestinal gluconeogenesis is crucial to maintain a physiological fasting glycemia in the absence of hepatic glucose production in mice. Metabolism 63: 104–111.

Perros, P., I.J. Deary, R.J. Sellar, J.J. Best and B.M. Frier. 1997. Brain abnormalities demonstrated by magnetic resonance imaging in adult IDDM patients with and without a history of recurrent severe hypoglycemia. Diabetes Care 20: 1013–1018.

Pescador, N., D. Villar, D. Cifuentes, M. Garcia-Rocha, A. Ortiz-Barahona, S. Vazquez et al. 2010. Hypoxia promotes glycogen accumulation through hypoxia inducible factor (HIF)-mediated induction of glycogen synthase 1. PLoS One 5: e9644.

Petit, J.M., I. Tobler, C. Kopp, F. Morgenthaler, A.A. Borbely and P.J. Magistretti. 2010. Metabolic response of the cerebral cortex following gentle sleep deprivation and modafinil administration. Sleep 33: 901–908.

Petit, J.M., S. Burlet-Godinot, P.J. Magistretti and I. Allaman. 2015. Glycogen metabolism and the homeostatic regulation of sleep. Metab. Brain Dis. 30: 263–279.

Petrie, J.L., Z.H. Al-Oanzi, C. Arden, S.J. Tudhope, J. Mann, J. Kieswich et al. 2013. Glucose induces protein targeting to glycogen in hepatocytes by fructose 2,6-bisphosphate-mediated recruitment of MondoA to the promoter. Mol. Cell Biol. 33: 725–738.

Pfeiffer-Guglielmi, B., B. Fleckenstein, G. Jung and B. Hamprecht. 2003. Immunocytochemical localization of glycogen phosphorylase isozymes in rat nervous tissues by using isozyme-specific antibodies. J. Neurochem. 85: 73–81.

Pfeiffer-Guglielmi, B., B. Dombert, S. Jablonka, V. Hausherr, C. van Thriel, N. Schobel et al. 2014. Axonal and dendritic localization of mRNAs for glycogen-metabolizing enzymes in cultured rodent neurons. BMC Neurosci. 15: 70.

Phelps, C.H. 1972. Barbiturate-induced glycogen accumulation in brain. An electron microscopic study. Brain Res. 39: 225–234.

Phillips, M.J., N.J. Unakar, G. Doornewaard and J.W. Steiner. 1967. Glycogen depletion in the newborn rat liver: an electron microscopic and electron histochemical study. J. Ultrastruct. Res. 18: 142–165.

Pickett-Gies, C.A. and D.A. Walsh. 1986. Phosphorylase Kinase. The Enzymes 17: 395–459.

Poitry-Yamate, C., H. Lei and R. Gruetter. 2009. The rate-limiting step for glucose transport into the hypothalamus is across the blood-hypothalamus interface. J. Neurochem. 109 Suppl. 1: 38–45.

Ponce, E., D.P. Witte, R. Hirschhorn, M.L. Huie and G.A. Grabowski. 1999. Murine acid alpha-glucosidase: cell-specific mRNA differential expression during development and maturation. Am. J. Pathol. 154: 1089–1096.

Postic, C., M. Shiota, K.D. Niswender, T.L. Jetton, Y. Chen, J.M. Moates et al. 1999. Dual roles for glucokinase in glucose homeostasis as determined by liver and pancreatic beta cell-specific gene knock-outs using Cre recombinase. J. Biol. Chem. 274: 305–315.

Pound, L.D., J.K. Oeser, T.P. O'Brien, Y. Wang, C.J. Faulman, P.K. Dadi et al. 2013. G6PC2: a negative regulator of basal glucose-stimulated insulin secretion. Diabetes 62: 1547–1556.

Prats, C., J.A. Cadefau, R. Cusso, K. Qvortrup, J.N. Nielsen, J.F. Wojtaszewski et al. 2005. Phosphorylation-dependent translocation of glycogen synthase to a novel structure during glycogen resynthesis. J. Biol. Chem. 280: 23165–23172.

Prats, C., J.W. Helge, P. Nordby, K. Qvortrup, T. Ploug, F. Dela et al. 2009. Dual regulation of muscle glycogen synthase during exercise by activation and compartmentalization. J. Biol. Chem. 284: 15692–15700.

Prats, C., A. Gomez-Cabello and A.V. Hansen. 2011. Intracellular compartmentalization of skeletal muscle glycogen metabolism and insulin signalling. Exp. Physiol. 96: 385–390.

Preisler, N., P. Laforet, K.L. Madsen, R.S. Hansen, Z. Lukacs, M.C. Orngreen et al. 2012. Fat and carbohydrate metabolism during exercise in late-onset Pompe disease. Mol. Genet. Metab. 107: 462–468.

Preisler, N., R.G. Haller and J. Vissing. 2015. Exercise in muscle glycogen storage diseases. J. Inherit. Metab. Dis. 38: 551–563.

Pursell, N., J. Gierut, W. Zhou, M. Dills, R. Diwanji, M. Gjorgjieva et al. 2018. Inhibition of Glycogen Synthase II with RNAi Prevents Liver Injury in Mouse Models of Glycogen Storage Diseases. Mol. Ther. 26: 1771–1782.

Quinlan, K.G., J.T. Seto, N. Turner, A. Vandebrouck, M. Floetenmeyer, D.G. Macarthur et al. 2010. Alpha-actinin-3 deficiency results in reduced glycogen phosphorylase activity and altered calcium handling in skeletal muscle. Hum. Mol. Genet. 19: 1335–1346.

Raben, N., R.C. Nichols, F. Martiniuk and P.H. Plotz. 1996. A model of mRNA splicing in adult lysosomal storage disease (glycogenosis type II). Hum. Mol. Genet. 5: 995–1000.

Raben, N., K. Nagaraju, E. Lee, P. Kessler, B. Byrne, L. Lee et al. 1998. Targeted disruption of the acid alpha-glucosidase gene in mice causes an illness with critical features of both infantile and adult human glycogen storage disease type II. J. Biol. Chem. 273: 19086–19092.

Raben, N., M. Danon, N. Lu, E. Lee, L. Shliselfeld, A.V. Skurat et al. 2001. Surprises of genetic engineering: a possible model of polyglucosan body disease. Neurology 56: 1739–1745.

Raben, N., A. Roberts and P.H. Plotz. 2007. Role of autophagy in the pathogenesis of Pompe disease. Acta Myol. 26: 45–48.

Raggi, F., A.L. Pissavino, R. Resaz, D. Segalerba, A. Puglisi, C. Vanni et al. 2018. Development and characterization of an inducible mouse model for glycogen storage disease type Ib. J. Inherit. Metab. Dis. 41: 1015–1025.

Rai, A., P.K. Singh, V. Singh, V. Kumar, R. Mishra, A.K. Thakur et al. 2018. Glycogen synthase protects neurons from cytotoxicity of mutant huntingtin by enhancing the autophagy flux. Cell Death Dis. 9: 201.

Rajas, F., J. Clar, A. Gautier-Stein and G. Mithieux. 2015. Lessons from new mouse models of glycogen storage disease type 1a in relation to the time course and organ specificity of the disease. J. Inherit. Metab. Dis. 38: 521–527.

Ramirez, C.M., L. Goedeke, N. Rotllan, J.H. Yoon, D. Cirera-Salinas, J.A. Mattison et al. 2013. MicroRNA 33 regulates glucose metabolism. Mol. Cell Biol. 33: 2891–2902.

Rath, V.L., M. Ammirati, P.K. LeMotte, K.F. Fennell, M.N. Mansour, D.E. Danley et al. 2000. Activation of human liver glycogen phosphorylase by alteration of the secondary structure and packing of the catalytic core. Mol. Cell. 6: 139–148.

Ren, J.M., B.A. Marshall, E.A. Gulve, J. Gao, D.W. Johnson, J.O. Holloszy et al. 1993. Evidence from transgenic mice that glucose transport is rate-limiting for glycogen deposition and glycolysis in skeletal muscle. J. Biol. Chem. 268: 16113–16115.

Rencurel, F., G. Waeber, B. Antoine, F. Rocchiccioli, P. Maulard, J. Girard et al. 1996. Requirement of glucose metabolism for regulation of glucose transporter type 2 (GLUT2) gene expression in liver. Biochem. J. 314(Pt 3): 903–909.

Reynet, C., C.R. Kahn and M.R. Loeken. 1996. Expression of the gene encoding glycogen phosphorylase is elevated in diabetic rat skeletal muscle and is regulated by insulin and cyclic AMP. Diabetologia 39: 183–189.

Reynolds, T.H.t., Y. Pak, T.E. Harris, J. Manchester, E.J. Barrett and J.C. Lawrence, Jr. 2005. Effects of insulin and transgenic overexpression of UDP-glucose pyrophosphorylase on UDP-glucose and glycogen accumulation in skeletal muscle fibers. J. Biol. Chem. 280: 5510–5515.

Richter, E.A. and M. Hargreaves. 2013. Exercise, GLUT4, and skeletal muscle glucose uptake. Physiol. Rev. 93: 993–1017.

Roach, P.J. 1990. Control of glycogen synthase by hierarchal protein phosphorylation. FASEB J. 4: 2961–2968.

Roach, P.J. 1991. Multisite and hierarchal protein phosphorylation. J. Biol. Chem. 266: 14139–14142.

Roach, P.J. and A.V. Skurat. 1997. Self-glucosylating initiator proteins and their role in glycogen biosynthesis. Prog. Nucleic. Acid Res. Mol. Biol. 57: 289–316.

Roach, P.J. and J. Larner. 1977. Covalent phosphorylation in the regulation glycogen synthase activity. Mol. Cell Biochem. 15: 179–200.

Roach, P.J., A.V. Skurat and R.A. Harris. 2001. Regulation of glycogen metabolism. pp. 609–647. *In*: Cherrington, A.D. and L.S. Jefferson (eds.). The Endocrine Pancreas and Regulation of Metabolism (New York: Oxford University Press).

Roach, P.J. 2011. Are there errors in glycogen biosynthesis and is laforin a repair enzyme? FEBS Lett. 585: 3216–3218.

Roach, P.J., A.A. Depaoli-Roach, T.D. Hurley and V.S. Tagliabracci. 2012. Glycogen and its metabolism: some new developments and old themes. Biochem. J. 441: 763–787.

Roach, P.J. 2015. Glycogen phosphorylation and Lafora disease. Mol. Aspects Med. 46: 78–84.

Roig-Zamboni, V., B. Cobucci-Ponzano, R. Iacono, M.C. Ferrara, S. Germany, Y. Bourne et al. 2017. Structure of human lysosomal acid alpha-glucosidase-a guide for the treatment of Pompe disease. Nat. Commun. 8: 1111.

Ros, S., M. Garcia-Rocha, J. Dominguez, J.C. Ferrer and J.J. Guinovart. 2009. Control of liver glycogen synthase activity and intracellular distribution by phosphorylation. J. Biol. Chem. 284: 6370–6378.

Ros, S., D. Zafra, J. Valles-Ortega, M. Garcia-Rocha, S. Forrow, J. Dominguez et al. 2010. Hepatic overexpression of a constitutively active form of liver glycogen synthase improves glucose homeostasis. J. Biol. Chem. 285: 37170–37177.

Rosell-Perez, M., C. Villar-Palasi and J. Larner. 1962. Studies on UDPG-glycogen transglucosylase. I. Preparation and differentiation of two activities of UDPG-glycogen transglucosylase from rat skeletal muscle. Biochemistry 1: 763–768.

Rosenfeld, E.L. 1975. Alpha-glucosidases (gamma-amylases) in human and animal organisms. Pathol. Biol. (Paris). 23: 71–84.

Rossi, D.J., J.D. Brady and C. Mohr. 2007. Astrocyte metabolism and signaling during brain ischemia. Nat. Neurosci. 10: 1377–1386.

Rubio-Villena, C., M.A. Garcia-Gimeno and P. Sanz. 2013. Glycogenic activity of R6, a protein phosphatase 1 regulatory subunit, is modulated by the laforin-malin complex. Int. J. Biochem. Cell Biol. 45: 1479–1488.

Ruggero, D., Z.G. Wang and P.P. Pandolfi. 2000. The puzzling multiple lives of PML and its role in the genesis of cancer. Bioessays 22: 827–835.

Rutledge, S.L., J. Atchison, N.U. Bosshard and B. Steinmann. 2001. Case report: liver glycogen synthase deficiency--a cause of ketotic hypoglycemia. Pediatrics 108: 495–497.

Saez, I., J. Duran, C. Sinadinos, A. Beltran, O. Yanes, M.F. Tevy et al. 2014. Neurons have an active glycogen metabolism that contributes to tolerance to hypoxia. J. Cereb. Blood Flow Metab. 34: 945–955.

Sagar, S.M., F.R. Sharp and R.A. Swanson. 1987. The regional distribution of glycogen in rat brain fixed by microwave irradiation. Brain Res. 417: 172–174.

Sakoda, H., M. Fujishiro, J. Fujio, N. Shojima, T. Ogihara, A. Kushiyama et al. 2005. Glycogen debranching enzyme association with beta-subunit regulates AMP-activated protein kinase activity. Am. J. Physiol. Endocrinol. Metab. 289: E474–481.

Santer, R., R. Schneppenheim, A. Dombrowski, H. Gotze, B. Steinmann and J. Schaub. 1997. Mutations in GLUT2, the gene for the liver-type glucose transporter, in patients with Fanconi-Bickel syndrome. Nat. Genet. 17: 324–326.

Schmoll, D., B.B. Allan and A. Burchell. 1996. Cloning and sequencing of the 5' region of the human glucose-6-phosphatase gene: transcriptional regulation by cAMP, insulin and glucocorticoids in H4IIE hepatoma cells. FEBS Lett. 383: 63–66.

Schmoll, D., K.S. Walker, D.R. Alessi, R. Grempler, A. Burchell, S. Guo et al. 2000. Regulation of glucose-6-phosphatase gene expression by protein kinase Balpha and the forkhead transcription factor FKHR. Evidence for insulin response unit-dependent and -independent effects of insulin on promoter activity. J. Biol. Chem. 275: 36324–36333.

Serratosa, J.M. 1999. Idiopathic epilepsies with a complex mode of inheritance. Epilepsia 40 Suppl. 3: 12–16.

Serratosa, J.M., P. Gomez-Garre, M.E. Gallardo, B. Anta, D.B. de Bernabe, D. Lindhout et al. 1999. A novel protein tyrosine phosphatase gene is mutated in progressive myoclonus epilepsy of the Lafora type (EPM2). Hum. Mol. Genet. 8: 345–352.

Shelley, H.J. 1961. Glycogen reserves and their changes at birth and in anoxia. Br. Med. Bull. 17: 137–143.

Shelly, L.L., K.J. Lei, C.J. Pan, S.F. Sakata, S. Ruppert, G. Schutz et al. 1993. Isolation of the gene for murine glucose-6-phosphatase, the enzyme deficient in glycogen storage disease type 1A. J. Biol. Chem. 268: 21482–21485.

Shi, Y., Y. Li, J. Wang, C. Wang, J. Fan, J. Zhao et al. 2017. Meta-analyses of the association of G6PC2 allele variants with elevated fasting glucose and type 2 diabetes. PLoS One 12: e0181232.

Shieh, J.J., C.J. Pan, B.C. Mansfield and J.Y. Chou. 2003. A glucose-6-phosphate hydrolase, widely expressed outside the liver, can explain age-dependent resolution of hypoglycemia in glycogen storage disease type Ia. J. Biol. Chem. 278: 47098–47103.

Shieh, J.J., C.J. Pan, B.C. Mansfield and J.Y. Chou. 2004. A potential new role for muscle in blood glucose homeostasis. J. Biol. Chem. 279: 26215–26219.

Shulman, R.G. and D.L. Rothman. 2001. The "glycogen shunt" in exercising muscle: A role for glycogen in muscle energetics and fatigue. Proc. Natl. Acad. Sci. U. S. A. 98: 457–461.

Sinadinos, C., J. Valles-Ortega, L. Boulan, E. Solsona, M.F. Tevy, M. Marquez et al. 2014. Neuronal glycogen synthesis contributes to physiological aging. Aging Cell 13: 935–945.

Skurat, A.V., Y. Wang and P.J. Roach. 1994. Rabbit skeletal muscle glycogen synthase expressed in COS cells. Identification of regulatory phosphorylation sites. J. Biol. Chem. 269: 25534–25542.

Skurat, A.V. and P.J. Roach. 1995. Phosphorylation of sites 3a and 3b (Ser640 and Ser644) in the control of rabbit muscle glycogen synthase. J. Biol. Chem. 270: 12491–12497.

Skurat, A.V., A.D. Dietrich and P.J. Roach. 2006. Interaction between glycogenin and glycogen synthase. Arch. Biochem. Biophys. 456: 93–97.

Smith, C.H., N.E. Brown and J. Larner. 1971. Molecular characteristics of the totally dependent and independent forms of glycogen synthase of rabbit skeletal muscle. II. Some chemical characteristics of the enzyme protein and of its change on interconversion. Biochim. Biophys. Acta 242: 81–88.

Smith, D. and S.A. Amiel. 2002. Hypoglycaemia unawareness and the brain. Diabetologia 45: 949–958.

Smythe, C. and P. Cohen. 1991. The discovery of glycogenin and the priming mechanism for glycogen biogenesis. Eur. J. Biochem. 200: 625–631.

Sommerfield, A.J., I.J. Deary, V. McAulay and B.M. Frier. 2003a. Moderate hypoglycemia impairs multiple memory functions in healthy adults. Neuropsychology 17: 125–132.

Sommerfield, A.J., I.J. Deary, V. McAulay and B.M. Frier. 2003b. Short-term, delayed, and working memory are impaired during hypoglycemia in individuals with type 1 diabetes. Diabetes Care 26: 390–396.

Sorg, O. and P.J. Magistretti. 1991. Characterization of the glycogenolysis elicited by vasoactive intestinal peptide, noradrenaline and adenosine in primary cultures of mouse cerebral cortical astrocytes. Brain Res. 563: 227–233.

Soty, M., J. Chilloux, F. Delalande, C. Zitoun, F. Bertile, G. Mithieux et al. 2016. Post-translational regulation of the glucose-6-phosphatase complex by cyclic adenosine monophosphate is a crucial determinant of endogenous glucose production and is controlled by the glucose-6-phosphate transporter. J. Proteome Res. 15: 1342–1349.

Spiegel, R., J. Mahamid, M. Orho-Melander, D. Miron and Y. Horovitz. 2007. The variable clinical phenotype of liver glycogen synthase deficiency. J. Pediatr. Endocrinol. Metab. 20: 1339–1342.

Sprang, S., R. Fletterick, M. Stern, D. Yang, N. Madsen and J. Sturtevant. 1982. Analysis of an allosteric binding site: the nucleoside inhibitor site of phosphorylase alpha. Biochemistry 21: 2036–2048.

Stalmans, W., M. Laloux and H.G. Hers. 1974. The interaction of liver phosphorylase a with glucose and AMP. Eur. J. Biochem. 49: 415–427.

Stapleton, D., C. Nelson, K. Parsawar, D. McClain, R. Gilbert-Wilson, E. Barker et al. 2010. Analysis of hepatic glycogen-associated proteins. Proteomics 10: 2320–2329.

Stern, S. and J.D. Biggers. 1968. Enzymatic estimation of glycogen in the cleaving mouse embryo. J. Exp. Zool. 168: 61–66.

Stojkovic, T., J. Vissing, F. Petit, M. Piraud, M.C. Orngreen, G. Andersen et al. 2009. Muscle glycogenosis due to phosphoglucomutase 1 deficiency. N. Engl. J. Med. 361: 425–427.

Striano, P., F. Zara, J. Turnbull, J.M. Girard, C.A. Ackerley, M. Cervasio et al. 2008. Typical progression of myoclonic epilepsy of the Lafora type: a case report. Nat. Clin. Pract. Neurol. 4: 106–111.

Suh, S.W., J.P. Bergher, C.M. Anderson, J.L. Treadway, K. Fosgerau and R.A. Swanson. 2007a. Astrocyte glycogen sustains neuronal activity during hypoglycemia: studies with the glycogen phosphorylase inhibitor CP-316,819 ([R-R*,S*]-5-Chloro-N-[2-hydroxy-3-(methoxymethylamino)-3-oxo-1-(phenylmet hyl)propyl]-1H-indole-2-carboxamide). J. Pharmacol. Exp. Ther. 321: 45–50.

Suh, S.W., A.M. Hamby and R.A. Swanson. 2007b. Hypoglycemia, brain energetics, and hypoglycemic neuronal death. Glia 55: 1280–1286.

Sullivan, M.A., S. Nitschke, E.P. Skwara, P. Wang, X. Zhao, X.S. Pan et al. 2019. Skeletal muscle glycogen chain length correlates with insolubility in mouse models of polyglucosan-associated neurodegenerative diseases. Cell Rep. 27: 1334–1344 e1336.

Sun, T., H. Yi, C. Yang, P.S. Kishnani and B. Sun. 2016. Starch binding domain-containing protein 1 plays a dominant role in glycogen transport to lysosomes in liver. J. Biol. Chem. 291: 16479–16484.

Suzuki, A., S.A. Stern, O. Bozdagi, G.W. Huntley, R.H. Walker, P.J. Magistretti et al. 2011. Astrocyte-neuron lactate transport is required for long-term memory formation. Cell 144: 810–823.

Swanson, R.A. and D.W. Choi. 1993. Glial glycogen stores affect neuronal survival during glucose deprivation *in vitro*. J. Cereb. Blood Flow Metab. 13: 162–169.

Szymanska, E., D. Rokicki, U. Watrobinska, E. Ciara, P. Halat, R. Ploski et al. 2015. Pediatric patient with hyperketotic hypoglycemia diagnosed with glycogen synthase deficiency due to the novel homozygous mutation in GYS2. Mol. Genet. Metab. Rep. 4: 83–86.

Taegtmeyer, H. 2004. Glycogen in the heart—an expanded view. J. Mol. Cell Cardiol. 37: 7–10.

Tagliabracci, V.S., J.M. Girard, D. Segvich, C. Meyer, J. Turnbull, X. Zhao et al. 2008. Abnormal metabolism of glycogen phosphate as a cause for Lafora disease. J. Biol. Chem. 283: 33816–33825.

Tagliabracci, V.S., J. Turnbull, W. Wang, J.M. Girard, X. Zhao, A.V. Skurat et al. 2007. Laforin is a glycogen phosphatase, deficiency of which leads to elevated phosphorylation of glycogen *in vivo*. Proc. Natl. Acad. Sci. USA. 104: 19262–19266.

Tagliabracci, V.S., C. Heiss, C. Karthik, C.J. Contreras, J. Glushka, M. Ishihara et al. 2011. Phosphate incorporation during glycogen synthesis and Lafora disease. Cell Metab. 13: 274–282.

Taylor, R., T.B. Price, L.D. Katz, R.G. Shulman and G.I. Shulman. 1993. Direct measurement of change in muscle glycogen concentration after a mixed meal in normal subjects. Am. J. Physiol. 265: E224–229.

Taylor, R., I. Magnusson, D.L. Rothman, G.W. Cline, A. Caumo, C. Cobelli et al. 1996. Direct assessment of liver glycogen storage by 13C nuclear magnetic resonance spectroscopy and regulation of glucose homeostasis after a mixed meal in normal subjects. J. Clin. Invest. 97: 126–132.

Tegtmeyer, L.C., S. Rust, M. van Scherpenzeel, B.G. Ng, M.E. Losfeld, S. Timal et al. 2014. Multiple phenotypes in phosphoglucomutase 1 deficiency. N. Engl. J. Med. 370: 533–542.

Testoni, G., J. Duran, M. Garcia-Rocha, F. Vilaplana, A.L. Serrano, D. Sebastian et al. 2017. Lack of glycogenin causes glycogen accumulation and muscle function impairment. Cell Metab. 26: 256–266 e254.

Tkacs, N.C., Y. Pan, R. Raghupathi, A.A. Dunn-Meynell and B.E. Levin. 2005. Cortical Fluoro-Jade staining and blunted adrenomedullary response to hypoglycemia after noncoma hypoglycemia in rats. J. Cereb. Blood Flow Metab. 25: 1645–1655.

Trefts, E., A.S. Williams and D.H. Wasserman. 2015. Exercise and the Regulation of Hepatic Metabolism. Prog. Mol. Biol. Transl. Sci. 135: 203–225.

Turnbull, J., A.A. DePaoli-Roach, X. Zhao, M.A. Cortez, N. Pencea, E. Tiberia et al. 2011. PTG depletion removes Lafora bodies and rescues the fatal epilepsy of Lafora disease. PLoS Genet. 7: e1002037.

Turnbull, J., J.R. Epp, D. Goldsmith, X. Zhao, N. Pencea, P. Wang et al. 2014. PTG protein depletion rescues malin-deficient Lafora disease in mouse. Ann. Neurol. 75: 442–446.

Turnbull, J., E. Tiberia, P. Striano, P. Genton, S. Carpenter, C.A. Ackerley et al. 2016. Lafora disease. Epileptic Disord. 18: 38–62.

Tye, L.M. and A.F. Burton. 1980. Glycogen deposition in fetal mouse tissues and the effect of dexamethasone. Biol. Neonate 38: 265–269.

Udoh, U.S., T.M. Swain, A.N. Filiano, K.L. Gamble, M.E. Young and S.M. Bailey. 2015. Chronic ethanol consumption disrupts diurnal rhythms of hepatic glycogen metabolism in mice. Am. J. Physiol. Gastrointest Liver Physiol. 308: G964–974.

van Schaftingen, E. and I. Gerin. 2002. The glucose-6-phosphatase system. Biochem. J. 362: 513–532.

Vannucci, S.J., F. Maher and I.A. Simpson. 1997. Glucose transporter proteins in brain: delivery of glucose to neurons and glia. Glia 21: 2–21.

Velho, G., K.F. Petersen, G. Perseghin, J.H. Hwang, D.L. Rothman, M.E. Pueyo et al. 1996. Impaired hepatic glycogen synthesis in glucokinase-deficient (MODY-2) subjects. J. Clin. Invest. 98: 1755–1761.

Vilchez, D., S. Ros, D. Cifuentes, L. Pujadas, J. Valles, B. Garcia-Fojeda et al. 2007. Mechanism suppressing glycogen synthesis in neurons and its demise in progressive myoclonus epilepsy. Nat. Neurosci. 10: 1407–1413.

Villar-Palasi, C. and J. Larner. 1958. A uridine coenzyme-linked pathway of glycogen synthesis in muscle. Biochim. Biophys. Acta 30: 449.

Villar-Palasi, C. and J.J. Guinovart. 1997. The role of glucose 6-phosphate in the control of glycogen synthase. Faseb. J. 11: 544–558.

von Wilamowitz-Moellendorff, A., R.W. Hunter, M. Garcia-Rocha, L. Kang, I. Lopez-Soldado, L. Lantier et al. 2013. Glucose-6-phosphate-mediated activation of liver glycogen synthase plays a key role in hepatic glycogen synthesis. Diabetes 62: 4070–4082.

Wall, M.L., L.D. Pound, I. Trenary, R.M. O'Brien and J.D. Young. 2015. Novel stable isotope analyses demonstrate significant rates of glucose cycling in mouse pancreatic islets. Diabetes 64: 2129–2137.

Walls, A.B., C.M. Heimburger, S.D. Bouman, A. Schousboe and H.S. Waagepetersen. 2009. Robust glycogen shunt activity in astrocytes: Effects of glutamatergic and adrenergic agents. Neuroscience 158: 284–292.

Wang, J., J.A. Stuckey, M.J. Wishart and J.E. Dixon. 2002. A unique carbohydrate binding domain targets the lafora disease phosphatase to glycogen. J. Biol. Chem. 277: 2377–2380.

Wang, W. and P.J. Roach. 2004. Glycogen and related polysaccharides inhibit the laforin dual-specificity protein phosphatase. Biochem. Biophys. Res. Commun. 325: 726–730.

Wang, Y., J.K. Oeser, C. Yang, S. Sarkar, S.I. Hackl, A.H. Hasty et al. 2006. Deletion of the gene encoding the ubiquitously expressed glucose-6-phosphatase catalytic subunit-related protein (UGRP)/glucose-6-phosphatase catalytic subunit-beta results in lowered plasma cholesterol and elevated glucagon. J. Biol. Chem. 281: 39982–39989.

Wang, Y., C.C. Martin, J.K. Oeser, S. Sarkar, O.P. McGuinness, J.C. Hutton et al. 2007. Deletion of the gene encoding the islet-specific glucose-6-phosphatase catalytic subunit-related protein autoantigen results in a mild metabolic phenotype. Diabetologia 50: 774–778.

Wang, Y., Y. Nakagawa, L. Liu, W. Wang, X. Ren, A. Anghel et al. 2011. Tissue-specific dysregulation of hexose-6-phosphate dehydrogenase and glucose-6-phosphate transporter production in db/db mice as a model of type 2 diabetes. Diabetologia 54: 440–450.

Wang, Z., W.A. Wilson, M.A. Fujino and P.J. Roach. 2001. Antagonistic controls of autophagy and glycogen accumulation by Snf1p, the yeast homolog of AMP-activated protein kinase, and the cyclin-dependent kinase Pho85p. Mol. Cell. Biol. 21: 5742–5752.

Warren, R.E. and B.M. Frier. 2005. Hypoglycaemia and cognitive function. Diabetes Obes. Metab. 7: 493–503.

Watanabe, Y., Y. Makino and K. Omichi. 2006. Activation of 4-alpha-glucanotransferase activity of porcine liver glycogen debranching enzyme with cyclodextrins. J. Biochem. 140: 135–140.

Weinstein, D.A., C.E. Correia, A.C. Saunders and J.I. Wolfsdorf. 2006. Hepatic glycogen synthase deficiency: an infrequently recognized cause of ketotic hypoglycemia. Mol. Genet. Metab. 87: 284–288.

Westphal, S.A. and F.Q. Nuttall. 1992. Comparative characterization of human and rat liver glycogen synthase. Arch. Biochem. Biophys. 292: 479–486.

Wilson, J.E. 2003. Isozymes of mammalian hexokinase: structure, subcellular localization and metabolic function. J. Exp. Biol. 206: 2049–2057.

Wilson, W.A., P.J. Roach, M. Montero, E. Baroja-Fernandez, F.J. Munoz, G. Eydallin et al. 2010. Regulation of glycogen metabolism in yeast and bacteria. FEMS Microbiol. Rev. 34: 952–985.

Wisselaar, H.A., M.A. Kroos, M.M. Hermans, J. van Beeumen and A.J. Reuser. 1993. Structural and functional changes of lysosomal acid alpha-glucosidase during intracellular transport and maturation. J. Biol. Chem. 268: 2223–2231.

Won, S.J., B.G. Jang, B.H. Yoo, M. Sohn, M.W. Lee, B.Y. Choi et al. 2012a. Prevention of acute/severe hypoglycemia-induced neuron death by lactate administration. J. Cereb. Blood Flow Metab. 32: 1086–1096.

Won, S.J., J.H. Kim, B.H. Yoo, M. Sohn, T.M. Kauppinen, M.S. Park et al. 2012b. Prevention of hypoglycemia-induced neuronal death by minocycline. J. Neuroinflammation. 9: 225.

Won, S.J., B.H. Yoo, T.M. Kauppinen, B.Y. Choi, J.H. Kim, B.G. Jang et al. 2012c. Recurrent/moderate hypoglycemia induces hippocampal dendritic injury, microglial activation, and cognitive impairment in diabetic rats. J. Neuroinflammation. 9: 182.

Woodgett, J.R. 2005. Recent advances in the protein kinase B signaling pathway. Curr. Opin. Cell Biol. 17: 150–157.

Wyss, M.T., R. Jolivet, A. Buck, P.J. Magistretti and B. Weber. 2011. *In vivo* evidence for lactate as a neuronal energy source. J. Neurosci. 31: 7477–7485.

Xirouchaki, C.E., S.P. Mangiafico, K. Bate, Z. Ruan, A.M. Huang, B.W. Tedjosiswoyo et al. 2016. Impaired glucose metabolism and exercise capacity with muscle-specific glycogen synthase 1 (gys1) deletion in adult mice. Mol. Metab. 5: 221–232.

Xu, J., D. Song, Z. Xue, L. Gu, L. Hertz and L. Peng. 2013. Requirement of glycogenolysis for uptake of increased extracellular K+ in astrocytes: potential implications for K+ homeostasis and glycogen usage in brain. Neurochem. Res. 38: 472–485.

Yamamoto, E., Y. Watanabe, Y. Makino and K. Omichi. 2009. Inspection of the activator binding site for 4-alpha-glucanotransferase in porcine liver glycogen debranching enzyme with fluorogenic dextrins. J. Biochem. 145: 585–590.

Yan, B., J. Heus, N. Lu, R.C. Nichols, N. Raben and P.H. Plotz. 2001. Transcriptional regulation of the human acid alpha-glucosidase gene. Identification of a repressor element and its transcription factors Hes-1 and YY1. J. Biol. Chem. 276: 1789–1793.

Yang-Feng, T.L., K. Zheng, J. Yu, B.Z. Yang, Y.T. Chen and F.T. Kao. 1992. Assignment of the human glycogen debrancher gene to chromosome 1p21. Genomics 13: 931–934.

Yang, B.Z., J.H. Ding, J.J. Enghild, Y. Bao and Y.T. Chen. 1992. Molecular cloning and nucleotide sequence of cDNA encoding human muscle glycogen debranching enzyme. J. Biol. Chem. 267: 9294–9299.

Yip, J., X. Geng, J. Shen and Y. Ding. 2016. Cerebral Gluconeogenesis and Diseases. Front Pharmacol. 7: 521.

Yip, V., M.E. Pusateri, J. Carter, I.A. Rose and O.H. Lowry. 1988. Distribution of the glucose-1,6-bisphosphate system in brain and retina. J. Neurochem. 50: 594–602.

Young, F.G. 1957. Claude Bernard and the discovery of glycogen, a century of retrospect. Br. Med. J. 1: 1431–1437.

Yu, Q. and X. Zheng. 2012. The crystal structure of human UDP-glucose pyrophosphorylase reveals a latch effect that influences enzymatic activity. Biochem. J. 442: 283–291.

Zani, F., L. Breasson, B. Becattini, A. Vukolic, J.P. Montani, U. Albrecht et al. 2013. PER2 promotes glucose storage to liver glycogen during feeding and acute fasting by inducing Gys2 PTG and G L expression. Mol. Metab. 2: 292–305.

Zeng, Y., Y. Lv, L. Tao, J. Ma, H. Zhang, H. Xu et al. 2016. G6PC3, ALDOA and CS induction accompanies mir-122 down-regulation in the mechanical asphyxia and can serve as hypoxia biomarkers. Oncotarget 7: 74526–74536.

Zeqiraj, E., X. Tang, R.W. Hunter, M. Garcia-Rocha, A. Judd, M. Deak et al. 2014. Structural basis for the recruitment of glycogen synthase by glycogenin. Proc. Natl. Acad. Sci. USA. 111: E2831–2840.

Zeqiraj, E. and F. Sicheri. 2015. Getting a handle on glycogen synthase—Its interaction with glycogenin. Mol. Aspects Med. 46: 63–69.

Zhai, L., J. Mu, H. Zong, A.A. DePaoli-Roach and P.J. Roach. 2000. Structure and chromosomal localization of the human glycogenin-2 gene GYG2. Gene 242: 229–235.

Zhai, L., C.S. Choi, J. Irimia-Dominguez, A.C. McGuire, S. Kim, C.B. Bock et al. 2007. Enhanced insulin sensitivity and energy expenditure in PPP1R3C (PTG) deleted mice. Diabetes 56: A62.

Zhai, L., L. Feng, L. Xia, H. Yin and S. Xiang. 2016. Crystal structure of glycogen debranching enzyme and insights into its catalysis and disease-causing mutations. Nat. Commun. 7: 11229.

Zhang, T., S. Wang, Y. Lin, W. Xu, D. Ye, Y. Xiong et al. 2012. Acetylation negatively regulates glycogen phosphorylase by recruiting protein phosphatase 1. Cell. Metab. 15: 75–87.

Zhang, W.M., M.F. Browner, R.J. Fletterick, A.A. DePaoli-Roach and P.J. Roach. 1989. Primary structure of rabbit skeletal muscle glycogen synthase deduced from cDNA clones. FASEB J. 3: 2532–2536.

Zhang, Y., J. Gu, L. Wang, Z. Zhao, Y. Pan and Y. Chen. 2017. Ablation of PPP1R3G reduces glycogen deposition and mitigates high-fat diet induced obesity. Mol. Cell. Endocrinol. 439: 133–140.

Zhao, H., M. Tang, M. Liu and L. Chen. 2018. Glycophagy: An emerging target in pathology. Clin. Chim. Acta 484: 298–303.

Zhao, S., W. Xu, W. Jiang, W. Yu, Y. Lin, T. Zhang et al. 2010. Regulation of cellular metabolism by protein lysine acetylation. Science 327: 1000–1004.

CHAPTER 6

The Pathologies of a Dysfunctional Glycogen Metabolism

Mitchell A. Sullivan,[1] *Berge A. Minassian*[2] *and Felix Nitschke*[3,]*

6.1 Introduction to Glycogen

Glycogen is a branched polymer of glucose, acting as an important energy reserve for many animal tissues, including liver, muscle, kidney, brain, adipose tissue and heart (Roach et al. 2012). While the specific role of glycogen is tissue-dependant, one common function in all cases is to store glucose in an organized macromolecule, capable of releasing the glucose when required.

In the majority of tissues, an individual glycogen molecule, termed β-particle, will reach sizes of $\approx$ 50,000 glucose molecules, with molecular weights of $\approx 10^7$ Da. In transmission electron microscope (TEM) images, β-particles are typically measured to be 20–30 nm in diameter (Ryu et al. 2009), with a hydrodynamic diameter being reported from size exclusion chromatography of 40–50 nm (Sullivan et al. 2014). In the liver (Drochmans 1962) and heart (Besford et al. 2012), these β-particles, via a currently unknown mechanism, form agglomerates known as α-particles. These particles can reach sizes of 200–300 nm in diameter (Ryu et al. 2009) (under TEM) and molecular weights of $\approx 10^8$ Da (Sullivan et al. 2010).

In glycogen, chains are comprised of glucose residues joined together via α-(1→ 4) glycosidic linkages, with these chains being connected together at branch points, linked via α-(1→ 6) glycosidic bonds. The average chain length of glycogen in eukaryotes has been found to be $\approx$ 10–14 residues (Manners 1991, Wang and Wise 2011). However, a preparation of glycogen molecules comprises a disperse distribution of chain lengths ranging from 3 to over 35 glucosyl units (Nitschke et al. 2017, O'Shea and Morell 1996, Sullivan et al. 2011). The formation of branch points is a critical factor in forming a relatively compact molecule that is water soluble, as longer unbranched glucan chains form double helices and precipitate (Gidley and Bulpin 1987).

While it was initially believed that the presence of phosphate observed in glycogen preparations was due to a contaminant (Roach et al. 2012), it has been conclusively shown to be covalently attached to glycogen at the C2, C3 and C6 position of glucosyl residues (Fontana 1980, Nitschke et al. 2013, Tagliabracci et al. 2011). The phosphate content of skeletal muscle in mice has been

[1] The University of Queensland, Mater Research Institute, Translational Research Institute, Glycation and Diabetes Woolloongabba, Brisbane, Queensland, Australia, mitchell.sullivan@mater.uq.edu.au
[2] University of Texas Southwestern, Department of Pediatrics, Dallas, TX 75390-9063, USA, Email: berge.minassian@utsouthwestern.edu
[3] University of Texas Southwestern, Medical Center Departments of Pediatrics and Biochemistry, Dallas, TX, 75390, USA.
* Corresponding author: felix.nitschke@utsouthwestern.edu

reported to be ≈ 1 per 1500 glucose units (Tagliabracci et al. 2008) or ≈ 1 per 2000 in mouse liver glycogen (Tagliabracci et al. 2007).

6.1.1 The Roles of Glycogen

The formation of glycogen allows storage of extensive quantities of glucose without exposing the cell to the extreme osmotic pressure that would accommodate an equivalent quantity of free glucose. In addition, the incorporation and release of glucose can be tightly regulated via the complex interaction of various enzymes involved in glycogen metabolism.

The specific role of glycogen depends on the tissue of synthesis. The predominant role of liver-glycogen is to act as a blood-glucose buffer. During food digestion, there is a subsequent rise in blood glucose, which is sensed by insulin-producing pancreatic β-cells, resulting in the release of insulin (Schmitz et al. 2008). Insulin stimulates cellular uptake of glucose from the blood and promotes inhibition of gluconeogenesis (Claus and Pilkis 1976) as well as glycogenolysis (Marks and Botelho 1986). During fasting or moderate levels of exercise, glucagon is able to stimulate glycogenolysis in the liver, allowing the glucose levels in the blood to remain relatively stable (Hems and Whitton 1980). Skeletal muscle plays an important role in the maintenance of glucose homeostasis by removing excess glucose from the blood in an insulin-dependent fashion (Saltiel and Kahn 2001). Unlike in liver, however, glucose derived from glycogenolysis in muscle is not released into the blood but instead provides the tissue with an energy source for the high demands of muscle activation (Carnagarin et al. 2015). Glycogen is also metabolically important in the heart. There is evidence that it is an important source of energy during cardiogenesis in the course of embryonic development (Pederson et al. 2004). In the adult heart, glycogen is an important energy source during ischemia but also appears to have a role during aerobic conditions (Depre et al. 1999). In contrast to the liver and skeletal muscle, cardiac glycogen content is increased during fasting, with the fatty acids acting as the main energy source and inhibiting glycolysis more than glucose uptake. The resulting increase in intracellular glucose levels may lead to elevated glycogen synthesis (Schneider et al. 1991). The function of glycogen in the brain is also becoming increasingly clear: during embryonic development glycogen is found in both neurons and astrocytes. By contrast, in adults it has been shown to be mainly present in astrocytes (Cataldo and Broadwell 1986). While blood glucose acts as the main energy source for the brain, glycogen of the brain has been shown to be important for specific purposes, such as learning (Duran et al. 2013). Furthermore, it is essential under stress conditions such as hypoglycaemia (Brown et al. 2003). While it was generally believed that neurons do not have an active glycogen metabolism, it has been demonstrated that the small amounts of glycogen present in neurons protect against hypoxia (Saez et al. 2014).

6.1.2 Metabolic Pathways Feeding into Glycogen

Glycogen synthesis requires the availability of intracellular glucosyl residues, which usually enter cells in the form of glucose. The uptake (and potential release) of glucose is tissue- and/or cell-type-dependant, with various glucose transporters (e.g., GLUTs) being expressed differently in the glycogen metabolizing tissues. GLUTs facilitate glucose permeation across cell membranes along a concentration gradient but differ in kinetic properties.

GLUT2 is the predominant transporter expressed in the liver. However, it is also present in the kidneys and in the pancreas where it has a role in glucose sensing for adequate insulin secretion (Mueckler et al. 1994, Thorens et al. 1988, Tiedge and Lenzen 1991). The GLUT2-mediated rate of glucose permeation is largely dependent on the glucose concentration in the blood (Im et al. 2005).

Skeletal muscle mainly transports glucose via the action of GLUT1 and GLUT4. GLUT1 is constitutively positioned on the plasma membrane and results in basal glucose transport that occurs independently of insulin (Ciaraldi et al. 2005). GLUT1 is also the predominant transporter for glucose crossing the blood-brain barrier (Morgello et al. 1995). Unlike GLUT1 and GLUT2, GLUT4 requires insulin-mediated signalling to stimulate translocation to the plasma membrane. When muscle cells exercise, delivery of GLUT4 to the cell surface is also favoured. Consequently, glucose is taken up by this transporter (Bryant et al. 2002). Similar to skeletal muscle, the heart's main glucose transporters are GLUT1 and GLUT4 (Shao and Tian 2016). In both tissues, the majority of glucose transport is mediated by GLUT1 during embryonic development and by GLUT4 after birth (Montessuit and Thorburn 1999).

Neuronal cells predominantly express GLUT3, which, by its high affinity to glucose but low permeation capacity, maintains a relatively constant influx of glucose which is largely independent of the glucose level in the blood (Simpson et al. 2008).

While GLUT1, GLUT2 and GLUT4 account for a large proportion of glucose transport, there are at least 14 different GLUTs with varying levels of expression across different tissue types (Mueckler and Thorens 2013). Once in the cell, hexokinase, in an essentially irreversible reaction, converts glucose to glucose 6-phosphate (G6P), which then feeds into various pathways, including glycolysis and glycogen synthesis. The latter requires the conversion to glucose 1-phosphate (G1P) mediated by phosphoglucomutase. This enzyme likewise permits the utilization of G1P, the major product of glycogen degradation by phosphorylase, for catalytic pathways such as glycolysis.

6.1.3 Glycogen Synthesis

The main substrate for glycogen synthesis is uridine diphosphate glucose (UDP-glucose) formed from G1P and uridine triphosphate by UDP-glucose pyrophosphorylase (Bhagavan and Ha 2011). Glycogen synthesis is initiated by the self-glucosylation of a glycogenin dimer, transferring the glucosyl unit of UDP-glucose to a tyrosine residue (in humans, amino acid residue 195). Subsequently, glycogenin is able to synthesize a chain of ≈ 10 glucose residues, which acts as a primer for further glycogen synthesis by glycogen synthase (GYS) and glycogen branching enzyme (GBE1) (Hurley et al. 2006, Issoglio et al. 2012, Lomako et al. 1988). In humans, there are two isoforms of glycogenin, glycogenin-1 and glycogenin-2. Glycogenin-1 is widely expressed, such as in skeletal muscle, the heart and the brain, while glycogenin-2 is predominantly expressed in liver, but also in the heart and pancreas (Barbetti et al. 1996, Mu et al. 1997). In mice, only one isoform of glycogenin was found (Roach et al. 2012). Surprisingly, it has been shown that a lack of glycogenin in mice results in an over-accumulation of glycogen, instead of the expected lack of glycogen (Testoni et al. 2017). A similar phenomenon has been observed in humans that lack glycogenin-1 (Malfatti et al. 2014). Thus, glycogen synthesis seems generally possible in the absence of glycogenin. This is supported by the complete knock-out of two glycogenin isoforms in yeast, which results in a small number of colonies possessing glycogen. This low number of colonies can be largely increased by an additional knock-out of one enzyme of the trehalose pathway, trehalose-6-phosphate synthase. The alternative mode of glycogen synthesis may be less efficient (Torija et al. 2005), and glycogenin appears to be important for normal glycogen metabolism in mammals (Testoni et al. 2017, Malfatti et al. 2014): (1) glycogen from mice lacking glycogenin has abnormal structural properties and (2) mice and patients lacking the enzyme show impaired muscle functionality.

GYS, which is able to interact with the C-terminal domain of glycogenin (Skurat et al. 2006), proceeds to add more glucose residues to the growing chain of the glycogenin-containing glycogen primer, again using UDP-glucose as the donor (Nielsen and Richter 2003). In humans, there are two GYS isoforms. While GYS1 is found in many tissues including skeletal muscle, heart, and brain, expression of GYS2 seems to be restricted to the liver (Uhlen et al. 2015). GYS, being the rate determining enzyme in glycogen synthesis, is reversibly modified by the phosphorylation of at least 9 serine residues, with covalently bound phosphates at some of these sites decreasing GS activity. Some kinases that have been shown to phosphorylate GYS include glycogens synthase kinase 3 (GSK-3), phosphorylase kinase (PhK), protein kinase A (PKA), protein kinase C (PKC) and AMP-activated protein kinase (AMPK) (Roach 1990, Roach and Larner 1977). Irrespective of the GYS phosphorylation state, full activity can be restored by exposing to a sufficiently concentrated allosteric activator, G6P (Hunter et al. 2015).

As GYS elongates chains, GBE1 is able to cut the growing chain and to attach the chain released via an α-$(1 \rightarrow 6)$ linkage, creating a branch point. The chain lengths transferred are predominantly 7 glucose units, coming from a chain with a minimum degree of polymerisation of 11 (Brown and Brown 1966b, Verhue and Hers 1966). The combination of chain elongation and branching creates a highly branched, roughly spherical glycogen molecule. Eventually the growing molecule will reach a maximum chain density where steric constraints limit further synthesis, termed crowding, resulting in a fully formed β-particle (Deng et al. 2015, Melendez et al. 1998). The mechanism for the agglomeration of β-particles into α-particles, as seen in the liver and heart, is currently unknown. However, there is increasing evidence that a protein "glue" is involved (Sullivan et al. 2015, 2012).

6.1.4 Glycogen Degradation

Glycogen degradation occurs mainly in the cytosol but the lysosome also appears to be involved (Hirschhorn and Reuser 2000). In the cytosol, glycogen phosphorylase (GP) cleaves the terminal glucose residues of glycogen-related chains utilizing orthophosphate (P_i) and releasing glucose-1-phosphate, which is subsequently converted to G6P via the action of phosphoglucomutase. In humans, there are three GP isoforms: PYGL, PYGM and PYGB (liver, muscle and brain, respectively).

Similar to GYS in the glycogen synthetic pathway, GP is regulated by phosphorylation, allosteric effectors, and changes in the redox potential of the cell (Mathieu et al. 2016, Newgard et al. 1989). However, the effect of each regulatory factor on the GP activity depends on the specific isoform. The phosphorylated (a form of) muscle and brain GP is generally active but inhibited by allosteric effectors, such as glucose, G6P and ATP. By contrast, the liver GPa is inactivated by allosteric binding of glucose; however, this inhibition is counteracted by AMP (Kasvinsky et al. 1978). The non-phosphorylated (b form of) muscle and brain GP (but not liver GP) is activated by adenosine monophosphate (AMP) (Agius 2015, Muller et al. 2015). Only the brain GPb is reversibly deactivated by reactive oxygen species, even in the presence of AMP (Mathieu et al. 2016). Activation by GP phosphorylation is mediated by phosphorylase kinase (Phk), which itself is regulated by a cyclic-AMP-dependent protein kinase. This kinase acts in response to hormonal regulation by glucagon and adrenalin (Agius 2015).

The glucosyl residue removing action of glycogen phosphorylase is stalled when a chain contains only four glucose units. Further degradation requires the action of glycogen debranching enzyme (AGL) which, in the indirect debranching process that is common in mammals, possesses two enzymatic activities (Walker and Whelan 1960). Here, AGL's glucosyltransferase is able to remove three glucose units and reattach them to the end of another branch. The second enzymatic activity of AGL cleaves the remaining α-$(1 \rightarrow 6)$ bond, releasing the final glucosyl residue in the form of glucose (Zhai et al. 2016). Debranching is essential for continued GP-mediated glycogen degradation (Walker and Whelan 1960) (Fig. 6.1).

Thus, the cytosolic glycogenolysis breaks down the macromolecule, producing mainly G1P, which is easily converted to G6P. The latter can be further metabolized in both oxidative and glycolytic metabolic pathways. In the liver, glucose 6-phosphatase is able to dephosphorylate G6P, allowing the release of glucose into the bloodstream.

The lysosome also plays an important role in glycogen metabolism. In rats, apparently $\approx 10\%$ of the liver glycogen is located in lysosomes (Geddes and Stratton 1977). Similarly, approximately 5% of the skeletal muscle glycogen is degraded by the lysosomal path (Calder and Geddes 1989). The lysosomal degradation of glycogen is mediated by acid α-glucosidase (GAA), converting glycogen to glucose. The mechanism of glycogen transport to the lysosome is not well understood; however, autophagy appears to be an important component, with this specific pathway being termed "glycophagy" (Jiang et al. 2010, Raben et al. 2010). In the liver, Stbd1 (Starch Binding Domain-containing Protein 1) seems to be essential for transporting glycogen to lysosomes but this does not appear to be the case in skeletal muscle or heart (Sun et al. 2016).

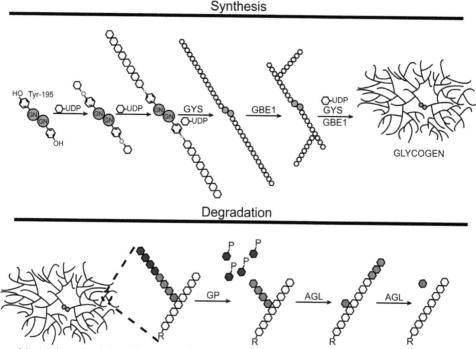

Figure 6.1. A schematic of the main enzymes involved in cytosolic glycogen metabolism. Glycogenin (GN) initiates glycogen synthesis with glycosylation at Tyr-195. Glycogenin can then facilitate the addition of more glucose units, making an initial chain. Glycogen synthase (GYS) then elongates chains with new branch points being introduced by glycogen branching enzyme (GBE1). During degradation, glycogen phosphorylase (GP) removes glucose from the terminal ends of chains, releasing glucose-1-phosphate. When a chain reaches a length of 4 units, glycogen debranching enzyme (AGL) transfers 3 glucose units to the end of another branch, followed by the removal of the final glucose unit.

6.2 Glycogen Storage Diseases

As glycogen metabolism comprises many enzymatic reactions and some carbohydrate translocations (see above), it is affected by a wide variety of genetic disorders. Many of them were historically termed 'glycogen storage disease' and numbered according to the chronological order in which they were described. Some diseases have only recently been placed in the context of GSDs (but not officially assigned a GSD number) when the clinical phenotype of those diseases was associated with changes in glycogen metabolism. All changes in glycogen metabolism are the defining criterion for using the term GSD in this chapter. The severity of the clinical phenotype and the disease symptoms vary greatly among the GSDs, mostly depending on the affected organ and the degree to which the cellular metabolism is impaired by the particular enzyme deficiency. An extensive summary of the GSDs is given in a recent review (Adeva-Andany et al. 2016). Figure 6.2 summarizes the different GSDs, the affected protein(s), and the phenotypical effects on glucose and glycogen metabolism.

6.2.1 Categories of GSDs

The various GSDs can be grouped into a number of different categories, depending on which particular aspect of glycogen metabolism is impaired.

A significant proportion of GSDs result from a deficiency in the capability of metabolising glucose for energy production via glycolysis. These GSDs often result in exercise intolerance, as glycogen's predominant role in skeletal muscle is to supply glucose for glycolysis when energy is required. The GSDs VII, X, XI (LDH A), XII, and XIII all fall into this category. While termed glycogen storage diseases, a more appropriate label for at least a part of this category might be glucose utilization

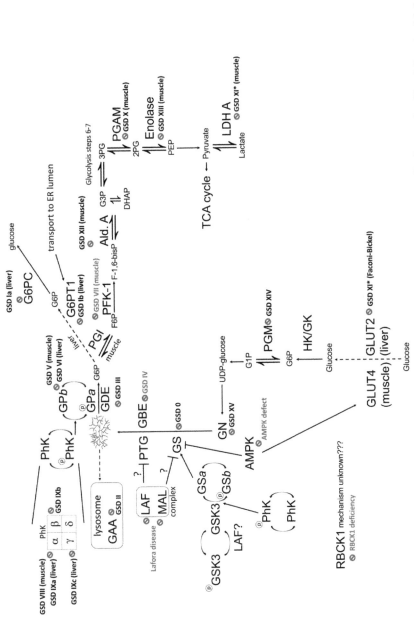

Figure 6.2. Various GSDs and glycogen metabolism. GSDs resulting in longer glycogen chain-length are depicted in gray. Following are the abbreviations in alphabetical order: 2PG (2-phosphoglycerate); 3PG (3-phosphoglycerate); Ald. A (aldolase A); AMPK (5' adenosine monophosphate-activated protein kinase); DHAP (dihydroxyacetone phosphate); F-1,6-bisP (fructose 1,6-bisphosphate); F6P (fructose 6-phosphate); G1P (glucose 1-phosphate); G6P (glucose 6-phosphate); G6PC (glucose 6-phosphatase catalytic subunit); G6PT1 (glucose 6-phosphate transporter 1); GAA (acid a-glucosidase); GBE (glycogen branching enzyme); GDE (glycogen debranching enzyme); GK (glucokinase); GLUT2 (glucose transporter 2); GLUT4 (glucose transporter 4); GN (glycogenin); GP (glycogen phosphorylase); GPa (glycogen phosphorylase active); GPb (glycogen phosphorylase inactive); GS (glycogen synthase); GSa (glycogen synthase active); GSb (glycogen synthase inactive); GSK3 (glycogen synthase kinase 3); HK (hexokinase); LAF (laforin); LHD A (lactate dehydrogenase A); MAL (malin); PEP (phosphoenolpyruvate); PFK1 (phosphofructokinase 1); PGI (phosphoglucose isomerase); PGAM (phosphoglycerate mutase); PGM (phosphoglucomutase); PhK (phosphorylase kinase); TCA (tricarboxylic acid); UDP (uridine diphosphate). Broken arrows represent transport.

Table 6.1. List of GSDs – Effects on Glycogen and Clinical Symptoms.

Disease	Affected protein	Phenotype regarding glycogen	Symptoms
GSD 0	Glycogen synthase: either GYS1 (muscle isoform) or GYS2 (liver isoform)	Reduction in muscle or liver glycogen synthesis.	Exercise intolerance Cardiac arrest
GSD I (von Gierke's disease)	Glucose 6-phosphatase Catalytic-1 (G6PC1) GSD Ia, Glucose 6-phosphate translocase (G6PT) GSD Ib	Glycogen accumulation in the liver, kidney and small intestine.	Hypoglycemia Enlarged liver/kidneys Osteoporosis Gout Kidney disease
GSD II (Pompe disease)	Lysosomal acid α-glucosidase (GAA)	Glycogen accumulation in lysosomes.	Exercise intolerance Enlarged heart Cardiac arrest
GSD III (Cori-Forbes disease)	Glycogen debranching enzyme (AGL)	Accumulation of malformed glycogen in the liver heart and skeletal muscle.	Hypoglycemia Hyperlipidemia Enlarged liver Immunodeficiency
GSD IV 1. Anderson disease	Glycogen branching enzyme (GBE1) reduction	Accumulation of polyglucosan bodies in glycogen producing tissues.	Hypotonia Exercise intolerance Enlarged liver
2. Adult polyglucosan body disease	Glycogen branching enzyme (GBE1) less severe reduction of activity	Accumulation of polyglucosan bodies in glycogen producing tissues.	Adult onset form (less severe) Exercise intolerance
GSD V (McArdle disease)	Skeletal muscle isoform of glycogen phosphorylase (PYGM)	Inability to breakdown glycogen in skeletal muscle.	Exercise intolerance Breakdown of muscle tissue
GSD VI (Hers disease)	Liver isoform of glycogen phosphorylase (PYGL)	Inability to breakdown glycogen in the liver.	Enlarged liver Hypoglycemia
GSD VII (Tarui disease)	Skeletal muscle isoform of Phosphofructokinase-1 (PFKM)	Glycogen accumulation in skeletal muscle and exercise intolerance. Polyglucosan bodies have been reported in some cases.	Exercise intolerance Breakdown of muscle tissue
GSD IXa	α subunit of the liver isoform of glycogen phosphorylase kinase (PHKA2)	Impaired glycogenolysis in the liver.	Enlarged liver hypoglycemia
GSD IXb	β subunit of the muscle and liver isoforms of glycogen phosphorylase kinase (PHKB)	Impaired glycogenolysis in the skeletal muscle and liver.	Enlarged liver hypoglycemia
GSD IXc	γ subunit of the liver isoform of glycogen phosphorylase kinase (PHKG2)	Impaired glycogenolysis in the liver.	Enlarged liver hypoglycemia
GSD IXd	α subunit of the skeletal muscle isoform of glycogen phosphorylase kinase (PHKA1)	Decreased ability to degrade glycogen in skeletal muscle.	Exercise intolerance Breakdown of muscle tissue
GSD X	Skeletal muscle phosphoglycerate mutase (PGAM2)	Relatively normal skeletal muscle glycogen stores.	Exercise intolerance Breakdown of muscle tissue
GSD XI (Fanconi-Bickel disease)[1]:	Glucose transporter-2 (GLUT2)	Large accumulations of glycogen in the liver and kidney.	Enlarged liver Kidney dysfunction Hypoglycaemia
GSDXI[1]	Lactate dehydrogenase A (LDHA)		Exercise intolerance Breakdown of muscle tissue

Table 6.1 contd. ...

...Table 6.1 contd.

Disease	Affected protein	Phenotype regarding glycogen	Symptoms
GSD XII	Aldolase A (ALDOA)	Absence of glycogen in skeletal muscle.	Exercise intolerance
GSD XIII	Enolase-3 or β-enolase (ENO3)	Glycogen levels remain relatively normal.	Exercise intolerance
GSD XIV	Phosphoglucomutase-1 (PGM1)	Inability to properly utilize glucose.	Exercise intolerance Hepatopathy
GSD XV	Glycogenin-1 (GYG1)	Muscle glycogen depletion and glycogen accumulation in cardiac tissue.	Muscle weakness
Lafora disease	Laforin (EPM2A) or malin (EPM2B) or PR/SET domain 8 (PRDM8)	Polyglucosan accumulations in most glycogen containing tissues.	Myoclonus Epilepsy Intellectual decline
RBCK1 disease	RANBP2-Type And C3HC4-Type Zinc Finger Containing 1 (RBCK1)	Polyglucosan accumulation in the heart and skeletal muscle.	Immunodeficiency Exercise intolerance Cardiomyopathy
AMPK defect	AMP-dependent protein kinase (AMPK)	Polyglucosan accumulation in the heart.	Cardiomyopathy

[1] There are two different GSDs that have been called GSD XI

diseases, or glycolysis diseases, as in some of these diseases there is no evidence for qualitative and/ or quantitative changes of glycogen in the affected organs.

GSD I can be placed in its own unique category, and predominantly involves the liver and kidneys. When the blood glucose level is low, glycogen-derived glucose has to be released into the bloodstream from organs such as liver and, therefore, glucose monophosphate has to be converted to free glucose. Due to a lack of the functional glucose phosphate phosphatase, GSD Ia and Ib are characterized by the inability to convert G6P to free glucose which can lead to hypoglycaemia (Chou et al. 2010). Consequently, G6P accumulates, which is an allosteric activator of GS and an inhibitor of GP. Therefore, glycogen is accumulated in the liver, kidney and intestinal mucosa (Chou and Mansfield 2008, Kishnani et al. 2014, Mahmoud et al. 2017).

Another distinct category of GSDs (GSD 0, GSD XI [Faconi-Bickel], GSD XIV and GSDXV) is characterized by a non-functional step within the pathways feeding into glycogen synthesis. These GSDs arise either due to: (1) an impairment in the processes involved in the uptake of glucose or its conversion to UDP-glucose, the main substrate of glycogen synthesis (GSD XI and GSD XIV); (2) the initiation of glycogen synthesis (GSD XV); or (3) the glycogen synthase transferring a glucosyl moiety from UDP-glucose into glycogen (GSD 0).

A number of GSDs (GSD III, V, VI, and $GSDIX_{a-d}$) have a decreased capacity to degrade glycogen in the cytosol. GSD III results in a decreased ability to degrade glycogen as a result of an impairment of glycogen debranching, a prerequisite for continued GP-mediated degradation. The accumulating glycogen in GSD III contains many short branches, which cannot be degraded by GP but require a functional AGL (see Fig. 6.1). GSD V and VI possess a lower level of the GP protein. $GSDIX_{a,b,c,d}$ have a lower activity of the key activator of GP, PhK (Beauchamp et al. 2007). GSD II also results in an impairment of glycogen degradation; however, it is specific to lysosomal glycogenolysis.

The final category of GSDs, which this chapter will mainly focus on, includes those that are characterized by an accumulation of abnormal (i.e., insoluble) glycogen, termed polyglucosan, which aggregates and forms so-called polyglucosan bodies (PBs) in various tissues from patients of GSD IV, Lafora disease, RBCK1 deficiency, with AMPK defect and phosphofructokinase deficiency. In Lafora disease, it is established that PBs are the driver of the disease progression (Nitschke et al. 2018).

6.2.1.1 Features of Polyglucosan Bodies

PBs have been identified using various methods. Their characteristic fibrillar morphology and the cell encompassing size of PBs have made them easy to identify under the electron microscope. These bodies are seen in Lafora disease (Lohi and Minassian 2005), GSD IV (both APBD (Peress et al. 1979, Robertson et al. 1998) and Andersen's (Guerra et al. 1986)), RBCK1 deficiency (Nilsson et al. 2013), in GSD VII (Hays et al. 1981), and in patients with AMPK defects (Arad et al. 2002). An example of what a PB looks like under the electron microscope is given in Fig. 6.3C. Another characteristic of PBs is their relative resistance to α-amylase (historically termed diastase), an enzyme capable of degrading normal glycogen. By pre-treating embedded tissue with diastase, and then staining for carbohydrate using periodic acid-Schiff (PAS), the PBs (stained magenta) can be visualized (an example in Fig. 6.3D). Early studies in GSDIV, VII, and LD showed that iodine spectra of the patient glycogen resembled that of amylopectin (Boudouresques et al. 1978, Hays et al. 1981, Peress et al. 1979), the major constituent of plant starch. Though amylopectin, like glycogen, is a glucose polymer it is water-insoluble. This implies that glycogen and amylopectin have distinct structural characteristics that cause this profound difference in water-solubility. One of these characteristics is chain length distribution.

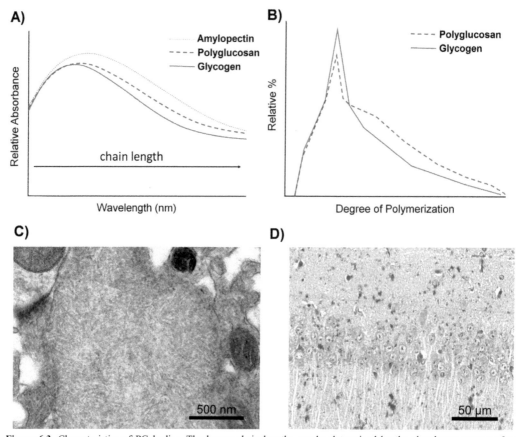

Figure 6.3. Characteristics of PG bodies. The longer chain lengths can be determined by the absorbance spectra after reacting with iodine **(A)**, or by comparing chain length distributions obtained by HPAEC-PAD or FACE **(B)**; PBs appear fibrillar under the EM. This image is from the brain of a laforin KO mouse **(C)**; PBs are resistant to diastase treatment and stain magenta with PAS. This image is from the hippocampus of a malin KO mouse **(D)**.

6.2.1.2 Evaluation of Glycogen Chain Lengths

Early attempts to evaluate glycogen chain lengths, which are directly associated with the degree of branching, involved the complexing of glycogen with iodine and measuring absorbance across a range of wavelengths (Krisman 1962). The wavelength at which there is an absorbance maximum (l_{max}) would shift to longer wavelengths as the average chain lengths increase, with glycogen having a l_{max} at $\approx$ 460 nm and the longer-chain amylopectin having a l_{max} of $\approx$ 520 nm. Using this method, it could be shown that glycogen from laforin-deficient mice (Tagliabracci et al. 2008), malin-deficient mice (Valles-Ortega et al. 2011), patients with GSD IV (Peress et al. 1979) and phosphofructokinase-deficient humans (Hays et al. 1981) have absorbance spectra more similar to amylopectin, indicating chain lengths longer than healthy glycogen. A schematic showing the different spectra seen in healthy glycogen, polyglucosan glycogen and amylopectin is given in Fig. 6.3A. While this method was able to reveal general trends on the relative differences between carbohydrate branching frequencies, the use of iodine absorption spectra is unable to reveal the distribution of chain lengths.

A method that greatly improves the amount of detail that can be obtained on the glycogen chain lengths is High Performance Anion Exchange Chromatography with Pulsed Amperometric Detection (HPAEC-PAD). This method, which involves the enzymatic debranching of glycogen using isoamylase, results in a distribution of chain lengths, with a resolution capable of a complete separation of chains only differing in length by one glucose unit. This method allows the comparison between different biological samples, with the relative abundance of each length between samples being quantitatively comparable. A schematic representing the difference between healthy glycogen and glycogen from PB accumulating tissue is seen in Fig. 6.3B. It is important to note that the sensitivity of PAD detection is dependent on the chain length, with larger chains resulting in a higher signal per chain (Koch et al. 1998). The obtained data therefore does not reveal the true distributions. Distributions between different samples, however, can still be quantitatively compared for each chain length. This makes this method very useful for the purpose of comparing healthy glycogen to that from various GSDs (Nitschke et al. 2017, 2013), as relative differences generally convey enough information to test hypotheses related to the glycogen CLDs. This limitation with HPAEC-PAD becomes more important when it is essential to know the relative abundance of chains within a distribution, for example when trying to model glycogen biosynthesis from the CLDs. For this purpose, fluorophore-assisted carbohydrate electrophoresis (FACE) is advantageous (Deng et al. 2015, Sullivan et al. 2011). Here, each chain is covalently linked with a fluorophore, the detection of which is not affected by the length of the chain.

6.2.1.3 The Role of Chain-Length in Maintaining Glycogen Solubility

A useful way to understand how the degree of branching and the branching pattern affects the solubility of glycogen is to consider the solubility of similar polymers of glucose, starch (amylose and amylopectin) and phytoglycogen. Glycogen, starch and phytoglycogen all consist of chains of glucose monomers connected via α-(1$\rightarrow$4) linkages, which have branch points via α-(1$\rightarrow$6) linkages. Despite the fundamental similarities with regards to the covalent linkages in these branched polysaccharides, glycogen and phytoglycogen are largely water-soluble while amylose and amylopectin usually form semi-crystalline (water-insoluble) starch granules. When examining the reasons for this, there are two main factors to consider: (1) the chain-length distribution and (2) the spatial organization of the branches within the different polyglucans. The chain length distribution refers to the relative quantity of chains of a particular length of attached glucose units.

For example, amylopectin, the major component of starch, has an average chain length of $\approx$ 25 (Manners 1991, Thompson 2000) and exterior chain of, on average, $\approx$ 15 glucose units. Importantly, the organization of branch points in amylopectin molecules appears to be non-random, with branch points occurring in clusters. This clustering gives rise to areas essentially free of branch points wherein the formation of double helices is more likely as chains are aligned close to each other (Bertoft et al. 1999, Thompson 2000). Amylose, which is typically less abundant than amylopectin in starch granules, is a mixture of linear chains and those with only few branches. While amylose chain lengths cover a wide range, they are on average much longer than those in amylopectin with chain lengths reaching even over 10,000 glucose units. These large chains are also able to form helices (Putaux et al. 2008). The formation of helices facilitates tight packing of glucan chains, which largely excludes water and decreases the solubility of polyglucans. Note that technically it is hard to determine the native solubility of amylose as *in vivo* amylose does not exist free of amylopectin. In fact, it has been shown in *Chlamydomonas reinhardtii* that amylose has its origin in amylopectin (van de Wal et al. 1998). Nevertheless, amylose and amylopectin can be physically separated by size-exclusion chromatography, but that requires solubilisation of the hydro-insoluble native starch by heat treatment (Lemos et al. 2019). So-called high-amylose starches possess a largely altered amylopectin structure enriched with long chains but depleted of short ones due to absent or genetically altered branching enzymes. These changes impact the crystal structure of the starch and tend to make high-amylose starch more resistant to enzymatic degradation (Wang et al. 2017).

Conversely, while glycogen has an average chain-length of ≈ 13 units (reports ranging from 10–18 units) (Manners 1991), it has been shown that the outer chains of the molecule are on average only ≈ 6–8 glucose units long (Akai et al. 1971) and the organization of branch points is described as generally random. It has been shown that the minimum length for purified malto-oligosaccharides to crystallize is 10 glucose units; however, in the presence of longer chains, malto-oligosaccharides of 6 units can co-crystallise (Gidley and Bulpin 1987). Using molecular models, it has also been shown that even with adjacent branch points, at least 2 glucose units per chain cannot be a part of the helices (Umeki and Kainuma 1981). Therefore, in glycogen it is much less likely that at any point in time a sufficiently high number of outer chains are present that are long enough to align and form crystalline double helices. Phytoglycogen has been shown to have an arrangement of branch points similar to glycogen, with an average chain-length of ≈ 11–12 units (Huang and Yao 2011). One study compared the chain-length distribution and solubility of a range of starch mutants, finding that there is indeed a strong correlation, with a greater number of shorter branches leading to more soluble polysaccharides (Pfister et al. 2014). There are a number of GSDs that result in a population of glycogen molecules with longer chain, resulting in the build-up of insoluble PBs. The differences between the average chain length and solubility of glycogen, phytoglycogen, PBs, amylopectin and amylose are given in Fig. 6.4.

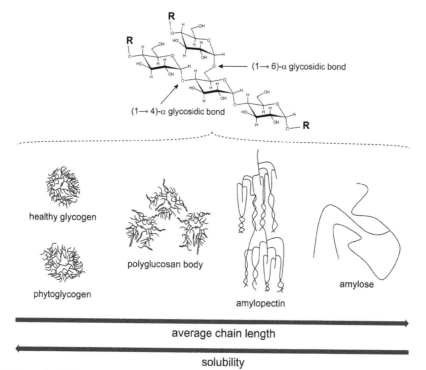

Figure 6.4. Schematic of different types of polyglucans. While glycogen, phytoglycogen, polyglucosan bodies, amylopectin and amylose have the same covalent building blocks, their average chain length and solubilities vary significantly.

6.2.2 Lafora Disease

Lafora disease (LD, OMIM #254780), a rare neurodegenerative disease that results in progressive myoclonus epilepsy, is generally manifested during adolescence and usually ends up being fatal within a decade of the first epileptic episode. This disease characteristically involves the accumulation of neurotoxic Lafora bodies (LBs), insoluble glycogen-derived particles that have longer chain lengths and are hyperphosphorylated. These LBs are present in the brain, skeletal muscle, heart, liver and skin, with the relatively less invasive skin biopsy historically being chosen as the diagnostic method (Carpenter and Karpati 1981, Minassian 2001). Currently, however, targeted genetic testing is the gold standard to confirm diagnosis as it is the least invasive and readily available as well as less expensive method (Nitschke et al. 2018).

Loss-of-function mutations in the genes *EPM2A* (encoding laforin) and *EPM2B* (encoding malin) cause this disease, with over 100 distinct mutations being discovered in over 200 independent families (Singh and Ganesh 2009). The number of documented mutations continues to increase (Lesca et al. 2010, Poyrazoglu et al. 2015). While deficiencies in laforin and malin account for the majority of LD cases, there were also cases discovered in which neither *EPM2A* nor *EPM2B* are mutated (Chan et al. 2004b). It has since been discovered that mutations in *PRDM8* (PR/SET Domain 8) lead to an early-onset form of LD with LBs absent in skin biopsy but present in that of muscle (Turnbull et al. 2012).

The clinical outcomes of mutations in either *EPM2A* or *EPM2B* are very similar; however, some studies have reported a later onset or slower disease progression when *EPM2B* is affected (Baykan et al. 2005, Gomez-Abad et al. 2005). One study reported a similar slow progression in some patients but a rapid progression in another patient with mutations in *EPM2B* (Franceschetti et al. 2006). Recently, a novel *EPM2A* mutation (p.N163D) was found in a patient with a slow progression form of the disease (Garcia-Gimeno et al. 2018). Studies in mouse models of LD reported that muscle glycogen from $Epm2a^{-/-}$ mice had higher levels of phosphorylation than that from $Epm2b^{-/-}$ mice (DePaoli-Roach et al. 2015), which is not surprising given laforin's glycogen phosphatase activity (Tagliabracci et al. 2007). It should be noted that a knockout of malin still leads to significantly higher levels of glycogen phosphorylation compared to the wild-type (DePaoli-Roach et al. 2015, Tiberia et al. 2012), which was hypothesized to be due to a decrease in the amount of available laforin which becomes occluded with the insoluble Lafora bodies (Tiberia et al. 2012). More recent data shows that the lower levels of phosphorylation in $Epm2b^{-/-}$ muscle glycogen, compared to that from $Epm2a^{-/-}$ mice, is due to normal (i.e., wildtype-like) levels of phosphorylation in the soluble fraction of the glycogen, as demonstrated by measuring glycogen C6 phosphate. The amount of C6 phosphorylation in the insoluble glycogen fractions was similarly high in both $Epm2b^{-/-}$ and $Epm2a^{-/-}$ mice, resulting in lower C6 phosphate levels of the total glycogen in the $Epm2b^{-/-}$ mice (Sullivan et al. 2019).

At present, there is no curative treatment for Lafora disease, with therapies being limited to alleviating the myclonus epilepsy using antiepileptic drugs. Some of the treatments used include administration of valproate, levetiracetam, barbital, zonisamide, piracetam and benzodiazeprines (Striano et al. 2008).

6.2.2.1 Laforin

Laforin is a 331 amino acid protein encoded by *EPM2A*, a 130-kb gene located on chromosome 6q24 (Gentry et al. 2013). Alternative splicing of this four-exon gene may lead to additional isoforms (Dubey et al. 2012). Laforin contains an N-terminal carbohydrate-binding module (CBM), similar to other enzymes involved in starch and glycogen metabolism. This CBM is expected to target laforin, and any other laforin-interacting proteins to glycogen molecules (Emanuelle et al. 2016, Wang et al. 2002). In addition to the CBM, laforin contains a C-terminal dual-specificity phosphatase domain (DSP), which is common in the large family of protein phosphatases capable of dephosphorylating phosphoproteins at the amino acid residues, threonine/serine and tyrosine. Laforin has been shown to be capable of interacting with an expanding number of proteins (Gentry et al. 2013). These include

GSK3β, three regulatory subunits of protein phosphatase 1 (PTG (Fernandez-Sanchez et al. 2003), GL (Worby et al. 2006), R6 (Worby et al. 2008)), glycogen synthase (Worby et al. 2006), AMPKα/β (Solaz-Fuster et al. 2008), TAU (Puri et al. 2009), HIRIP5 (Ganesh et al. 2003), EPM2AIP1 (Ianzano et al. 2003) and malin (Fernandez-Sanchez et al. 2003). The significance of some of these interactions is controversial, as many of the experiments involved the overexpression of both laforin and the putative target protein, an interaction being reported when co-immunoprecipitation occurred. Because these methods do not examine the proteins at biologically relevant concentrations under representative environments, it is possible that some of these interactions are not replicated *in vivo*. It remains that laforin's biological role is likely complex and multidimensional (Gentry et al. 2013).

Adding to the complexity, laforin has also been shown to belong to the class of glucan phosphatases (Worby et al. 2006). Being able to remove the small amounts of phosphate attached to glycogen (Tagliabracci et al. 2007), it was postulated that the cause for LD was an abnormality in the removal of this phosphate. It has since been shown, however, that the introduction of phosphatase-inactive laforin into Lafora mice is sufficient to rescue LD, despite the glycogen still maintaining high phosphate levels (Gayarre et al. 2014, Nitschke et al. 2017).

One interaction with laforin, now supported by a large number of studies, is that with malin (Fernandez-Sanchez et al. 2003, Lohi et al. 2005, Sanchez-Martin et al. 2015, Solaz-Fuster et al. 2008, Vilchez et al. 2007, Worby et al. 2008). A recently described *EPM2A* mutation (p.N163D), associated with a slow progression form of Lafora disease, did not affect laforin's phosphatase activity, but evidently its interaction with previously identified binding partner, including malin, was impaired (Garcia-Gimeno et al. 2018).

6.2.2.2 Malin

Malin, encoded by *EPM2B* (also called *NHLRC1*), is a 395 amino acid protein containing an N-terminal RING finger motif followed by 6 NHL domains. This protein is encoded by a single-exon gene located on chromosome 6p22.3. Malin's RING finger domain predicts an E3 ubiquitin ligase functionality (Chan et al. 2003, Freemont 2000). Similar to the search for laforin's substrate for phosphatase activity, research efforts aimed to discover the putative substrates that malin ubiquitinated. It was quickly confirmed that malin was indeed an E3 ubiquitin ligase and is capable of self-ubiquitination (common with E3 ligases under certain experimental conditions) (Lohi et al. 2005). It was also reported that malin ubiquitinates laforin, promoting its degradation (Gentry et al. 2005).

Malin was then shown to be capable of ubiquitinating glycogen debranching enzyme (AGL) (Cheng et al. 2007), GS (Vilchez et al. 2007) and PTG (protein targeting to glycogen), an important scaffold protein that targets several enzymes to glycogen (Worby et al. 2008). The malin-laforin complex has also been shown to ubiquitinate AMPK (Moreno et al. 2010), pyruvate kinase 1 (PKM1) and pyruvate kinase 2 (PKM2) (Viana et al. 2015). Because most of these malin-interacting proteins were discovered in over-expression systems, again it has been questioned whether these proteins interact at physiological concentrations. Likewise, the exact fate of proteins ubiquitinated by the laforin-malin complex remains controversial. As the type of ubiquitin linkage strongly correlates with its function (Herrmann et al. 2007), investigations were aiming to determine the type of ubiquitin linkages on putative substrates of the laforin-malin complex. Evidence for incorporation of predominantly K63-linked polyubiquitin chains (Moreno et al. 2010, Roma-Mateo et al. 2011, Rubio-Villena et al. 2013) supports the idea that protein targets of the laforin-malin complex are subject to autophagic inclusion and degradation (Deshaies and Joazeiro 2009, Nathan et al. 2013, Tan et al. 2008). These results are challenged by the finding that malin-dependent target degradation can be counteracted by inhibitors of the proteasome (Vilchez et al. 2007, Worby et al. 2008) and by one publication that also demonstrates the incorporation of K48-linked polyubiquitin which suggests proteasomal targeting (Sharma et al. 2012). Thus, while it is still unclear whether targets of the laforin-malin complex are subjected to the cytosolic proteasome or to autophagy (or both), it seems very likely that malin targets are subjected to degradation. Studies showing that PTG, AGL and GS levels are not elevated in either *Epm2a^-/-* or *Epm2b^-/-* mice have led to scepticism that these proteins are ubiquitinated by malin (DePaoli-Roach et al. 2010). However, a recent model proposes that the action of the laforin-malin complex is locally confined to only a small fraction of glycogen molecules, where PTG and GS are targeted to decrease chain elongation and thereby the risk of precipitation of these specific glycogen molecules. The model includes the possibility that local changes of PTG and GS may not be detectable in whole tissue lysates (Sullivan et al. 2017) (see below).

6.2.2.3 PRDM8

While the majority of LD patients have a mutation in either *EPM2A* (laforin) or *EPM2B* (malin), a third locus was predicted after a family had members present with a pathology similar to LD, despite having no mutations in these two genes (Chan et al. 2004b). This early-onset form of Lafora disease was subsequently shown to result from a mutation, c.781T > C (p.F261L) on chromosome 4q21.21, in a gene with an unknown function called *PRDM8* (PR/SET Domain 8) (Turnbull et al. 2012). Despite the earlier onset, the mutations discovered in *PRDM8* led to a more protracted progression compared to typical LD, with the patients living into their 30s. The symptoms are also not identical, with the associated dysarthria not usually occurring in LD and the psychosis involving prolonged agitation and screaming, also a unique feature of this early-onset variety.

In cell culture, overexpressed and epitope-tagged PRDM8 was shown to be located exclusively and diffusely in the nucleus. Co-overexpression of PRDM8 with laforin or malin resulted in relocation of both proteins from the cytoplasm to the nucleus, resulting in a co-localisation with PRDM8. It was also shown that these three proteins formed a complex that was more stable when all three were present. Experiments with the mutated PRDM8 (p.F261L) showed that nuclear translocation and co-localisation still occurred; however, the co-localisation foci size was significantly larger as compared to when wild-type PRDM8 was used. The authors suggested that PRDM8 regulates the cytosolic quantities of laforin and malin, and that the p.F262L mutation results in a gain-of-function that removes too much laforin and malin from the cytoplasm and leads to pathologically low cytosolic activities of these proteins (Turnbull et al. 2012).

A naturally occurring canine model of Lafora disease has a mutation in *Epm2b*, leading to an almost complete lack of expression and function of malin (Hajek et al. 2016, Lohi and Minassian 2005). Mouse models include laforin (Ganesh et al. 2002) and malin knockouts (DePaoli-Roach et al. 2010, Turnbull et al. 2010). Laforin- and malin-KO mice form polyglucosan bodies (here termed Lafora bodies, LBs) in neurons, spinal cord, muscle, liver and heart, but the disease-related symptoms are much less severe than in humans. There is, however, a phenotype in both mouse models described by spontaneous myoclonus and a greater susceptibility to hippocampal seizures when injected with kainic acid (Pederson et al. 2013, Turnbull et al. 2011, 2014, Valles-Ortega et al. 2011).

6.2.3 Glycogen Storage Disease IV

Glycogen branching enzyme (GBE1) is responsible for creating new branch points in a growing glycogen molecule (Fig. 6.1). Without this enzyme, it is expected that long branches are synthesized by GYS, which can form double helices that exclude water and promote the molecule to become insoluble. As discussed above, the phenomenon of insoluble double helices is also described for amylopectin.

GBE1 is located on chromosome 3p14 (Thon et al. 1993), mutations of which can lead to a variety of different diseases, including Andersen disease (OMIM 232500) and the later-onset adult polyglucosan body disease (APBD) (OMIM 263570). These diseases are rare and account for 3% of all GSDs (Mochel et al. 2012).

All forms of GSD IV exhibit the accumulation of amylase-resistant PBs, similar to those described in Lafora disease. These can form in the liver, skeletal muscle, heart, brain, spinal cord, peripheral nerve and skin (Schochet et al. 1970) and can be observed by electron microscopy as large deposits

(Peress et al. 1979). As observed in mouse models of Lafora disease, GSD IV glycogen consists of longer chains (Brown and Brown 1966a, Mercier and Whelan 1973). However, according to recent studies on muscle of the APBD mouse model, the abnormal glycogen structure seems to be restricted to the insoluble glycogen, while the soluble glycogen is indistinguishable from that of wild type. Unlike in mouse models of LD, neither the soluble nor the insoluble glycogen appears to contain elevated levels of covalently bound phosphate (Sullivan et al. 2019).

The age of onset and symptoms of these diseases are affected by the residual amount of GBE1 activity, which greatly depends on the specific mutation a patient harbors. The variety of existing *GBE1* mutations explains the clinically heterogeneous nature of GSD IV, including a large variation in both symptoms and age of onset (Akman et al. 2015, Mochel et al. 2012).

The total absence of glycogen branching activity results in an infantile onset of GSD IV (Andersen's disease), which often leads to childhood death (Andersen 1956). Foetal onset has even been observed, resulting in perinatal or neonatal death (Akman et al. 2006, Janecke et al. 2004). One of the main causes of fatality in infantile GSD IV is the severe cirrhosis leading to liver failure, usually between 3 and 5 years of age (Andersen 1956). In addition to the damaging effects of polyglucosan bodies in the liver, cardiomyopathy and neuromuscular dysfunction are also often associated with the disease (Brown and Brown 1989). Orthotopic liver transplantation has been successful in a number of cases, allowing survival of patients far beyond 5 years of age. In most of these cases, cardiomyopathy and myopathy did not develop after the transplantation (Davis and Weinstein 2008). While a total lack of branching enzyme activity generally leads to these severe early-onset forms of GSD IV, there have also been a few cases of people surviving into adulthood, with a less severe, non-progressive variant (Dhawan et al. 1994, McConkie-Rosell et al. 1996).

The later-onset APBD causes significantly different clinical outcomes compared to the more severe Andersen's disease. The characteristic symptoms of APBD include motor-neuron dysfunction, distal sensory loss (especially in the lower extremities), neurogenic bladder and cognitive decline (Cafferty et al. 1991). A combination of peripheral neuropathy and myelopathy are responsible for the motor and sensory pathologies, with polyglucosan bodies accumulating in the central and peripheral nervous system (Bigio et al. 1997, Cafferty et al. 1991, Peress et al. 1979). The majority of APBD patients are homozygous for a *GBE1* mutation encoding the p.Y329S variant, resulting in GBE1 activity levels reduced to 10–20% of normal (Mochel et al. 2012).

As outlined in detail (Akman et al. 2011a), the first naturally occurring animal model of GSD IV was found in Norwegian forest cats (Fyfe et al. 1992). Another naturally occurring model was found in the American Quarterhorse (Ward et al. 2004).

A mouse model of Andersen's disease was created by deleting exon 7 of *Gbe1*, leading to an unstable protein. The mutated GBE protein lacks the α-amylase domain that is responsible for the first step of creating a branching point in glycogen, i.e., cleaving a glycogen chain. Similar to the typical clinical outcome from Andersen's disease, the pups died at or soon after birth without a liver transplant (Akman et al. 2011b). To model the later-onset APBD, a p.Y329S mutation was introduced, the most common mutation found in APBD patients. This mutation results in reduced GBE activity as compared to wild type: ≈ 20% in the heart and brain, ≈ 37% in the liver, and ≈ 16% in the muscle. In all of these organs, large accumulations of polyglucosan material can be found (Akman et al. 2015). Interfering with the transcription of *Gbe1* by insertion of a neo cassette into intron 7 had a similar effect like the p.Y329S mutation: reduced GBE activity (≈ 16% in brain, ≈ 11% in brain, ≈ 13% in liver and ≈ 15% in muscle) and polyglucosan accumulation in all of these tissues (Akman et al. 2011b).

6.2.4 RBCK1 deficiency

A deficiency in RBCK1 (RBCC protein interacting with PKC1) leads to a disease termed polyglucosan body myopathy 1 with or without immunodeficiency (OMIM #615895). *RBCK1* is located on chromosome 20p13, encoding for a 58 kDa protein that contains an ubiquitin-like domain (UBL) and a RING-Inbetween RING-RING (RBR), characteristic of E3 ubiquitin ligases (Kirisako et al. 2006).

RBCK1, also often called HOIL-1 (heme-oxidized IRP2 ubiquitin ligase-1) (Yamanaka et al. 2003), has been reported to be involved in a wide variety of biological processes. As thoroughly reviewed (Elton et al. 2015), the role of RBCK1 in inflammation (Boisson et al. 2012, Gerlach et al. 2011, Ombrello et al. 2012), organ development (Landgraf et al. 2010), cancer prevention (Queisser et al. 2014), iron metabolism (Yamanaka et al. 2003) and the immune system (Boisson et al. 2012, Gerlach et al. 2011, Ombrello et al. 2012) has been investigated with great progress within the last 2 decades. Although RBCK1 has been clearly demonstrated to be involved in glycogen metabolism, the precise function remains unknown (Boisson et al. 2012). A mutation in the C-terminal end of RBCK1, containing the E3 ligase RBR domain, leads to an accumulation of amylase-resistant polyglucosan bodies. In this case, they are reported to be predominantly in the heart and in skeletal muscle. Patients lacking a functional C-terminus develop cardiomyopathy, often needing cardiac transplants, as well as muscle weakness (Nilsson et al. 2013). Mutations in the N-terminus appear to be more detrimental to many of the other functions of RBCK1, including its involvement in the linear ubiquitin chain assembly complex LUBAC, which is involved in the immune system (Fiil and Gyrd-Hansen 2014), inflammatory system (Boisson et al. 2015) and in cell proliferation/apoptotic pathways (Ikeda et al. 2011).

A mouse deficient in RBCK1 was constructed by adding a neo resistance gene, replacing exon 7 and part of exon 8 in the gene (Tokunaga et al. 2009). These mice contain PG deposits in cardiac tissue (MacDuff et al. 2015), similar to those seen in human patients (Nilsson et al. 2013). In addition to accumulating PG bodies, RBCK1 KO mice showed immunodeficiency (MacDuff et al. 2015) and exhibited enhanced apoptosis in hepatocytes (Tokunaga et al. 2009).

The mechanism that explains why RBCK1 deficiency leads to PG accumulation requires further investigation and may uncover a new layer of regulation in glycogen metabolism.

6.2.5 Phosphofructokinase Deficiency

Phosphofuctokinase (PFK1) deficiency (GSD VII or Tarui disease), OMIM 232800, results from a mutation in the gene *PFKM* (muscle isoform), located on chromosome 12q13. GSD VII is an autosomal recessive disorder, with the most common mutation being the c.237-1G > A splice acceptor variant, which results in an in-frame deletion of exon 5 (Raben et al. 1993). This mutation is most commonly found in the Ashkenazi Jewish population, along with a c.20003delC mutation that results in a frameshift and premature stop codon (Sherman et al. 1994). These two mutations make up ≈94% of all mutant alleles (Hedberg-Oldfors and Oldfors 2015). The heterogeneous age of onset and symptoms of PFK1 deficiency has led to the suggestion that the phenotypes could be divided into 5 subtypes (Vora et al. 1983); however, the classical syndrome results in exercise intolerance, rhabdomyolysis and haemolytic anaemia, with a number of studies reporting an accumulation of polyglucosan bodies in muscle fibres. These bodies had the characteristics of being relatively resistant to diastase digestion, having fibrillary morphology and longer chains, as demonstrated with the iodine absorption spectra (Hays et al. 1981, Malfatti et al. 2012, Nakajima et al. 2002). PFK1 (in the muscle a homotetramer of the muscle isoform of PFK1) catalyses the conversion of fructose-6-phosphate to fructose-1,6-biphosphate, an early step in the glycolytic pathway (Fig. 6.2). Without PFK1 carrying out this phosphorylation, there is an accumulation of F6P and G6P (Nakajima et al. 2002). A plausible mechanism for the polyglucosan body accumulation is that the elevated levels of G6P, an allosteric activator of glycogen synthase, leads to an imbalance between GS and GBE, resulting in glycogen with longer chains and an increased propensity to become insoluble (Hays et al. 1981, Nakajima et al. 2002).

A PFK1-deficient mouse model was generated by replacing exon 3 with a neo cassette, resulting in the absence of *PFK1* transcript (Garcia et al. 2009). The lack of PFK1 resulted in a phenotype of high occurrence of premature death, impaired exercise endurance, and glycogen accumulation in both the heart and skeletal muscle. As seen in human patients, there was an increase in G6P levels, which could explain the increased levels of glycogen. The glycogen chain length was not analysed.

The mouse and human *PFKM* genes are under the control of two promoters, a distal and a proximal promoter (Yamasaki et al. 1991). A mouse model containing a disrupting tag near the distal promoter of *PFKM* has been shown to have 50–75% loss in skeletal muscle PFK1 protein but an almost complete depletion of the muscle isoform of PFK1 in brain. These data support the proposed two-promoter system of the gene, with ubiquitous use of the distal promoter and additional use of the proximal promoter selectively in muscle (Richard et al. 2007).

Phosphofructokinase deficiency has been reported in a canine, showing progressive myopathy and the accumulation of polyglucosan bodies in muscle fibres (Harvey et al. 1990).

6.2.6 Mechanisms of Polyglucosan Formation

One feature common in Lafora disease, GSD IV (Andersen and APBD) and GSD VII (Phosphofuctokinase deficiency) is the accumulation of polyglucosan bodies that consist of glycogen molecules with longer chain lengths, are resistant to diastase digestion and have fibrillar morphology. A closer inspection of these diseases reveals that despite the similarities, there are some important differences (Table 6.1).

Both GSD IV and GSD VII appear to be explained by a general imbalance between GS and GBE activity, here termed 'Polyglucosan Body Disease (PGBD) Class I'. In GSD IV, this imbalance is caused by a greatly reduced amount of GBE1 activity. In GSD VII, when phosphofructokinase is deficient, the higher levels of G6P, a potent allosteric activator of GS, could lead to an over-activation of GYS, the glycogen chain-elongating enzyme. In both of these cases, the increased elongation:branching ratio promotes the formation of glycogen with longer chains. Genetic defects in subunits of AMPK, which lead to chronic activation of AMPK, have been associated with the formation of PBs in heart (Arad et al. 2002). In mouse models, it could be shown that acute activation of AMPK stimulates glycogen synthesis through allosteric activation of GS via increasing glucose uptake and the subsequent rise in intracellular G6P. Therefore, this disease can also be regarded as a PGBD Class I, as defects in AMPK may lead to an imbalance of GS:GBE activity (Hunter et al. 2011).

The term 'PGBD Class II' is introduced here to describe pathologies where PG bodies form without a detectable imbalance in GS and GBE activity. For example, in a mouse model of Lafora disease, it has been shown that in the brain there were no obvious differences in the GS and GBE activities (Wang et al. 2007). In malin KO mice, no differences were observed in Western blots in the protein levels of GS, PTG or GDE in muscle or brain (DePaoli-Roach et al. 2010). There appeared to be a paradox, with some studies showing that malin ubiquitinates PTG (Worby et al. 2008) and GS (Vilchez et al. 2007). This apparent paradox is resolved if the laforin-malin complex is assumed to only ubiquitinate PTG and GS at molecules that are at an elevated risk of becoming insoluble and eventually forming polyglucosan bodies (Sullivan et al. 2017). It has been recently shown that soluble wild-type glycogen consists of glycogen molecules that differ in their degree of average chain length and solubility properties (Sullivan et al. 2019). The observation that laforin binds preferentially to the longer chained potato starch and polyglucosan bodies, compared to glycogen, provides a possible mechanism for how the laforin-malin complex could target longer glycogen chains, which would

Table 6.2. Characteristics of Polyglucosan Bodies in Various GSDs Glycogen chain length and glycogen phosphate in bodies as compared to that in the respective glycogen in WT. GS to GBE ratio as compared to that in the respective WT tissue. Symbols: increased (↑), equivalent (=), present (✓), absent (X), not determined (?).

	Glycogen Chain Length	Glycogen Phosphate	Diastase Resistant	Fibrillar Morphology	GS to GBE Ratio	White Matter	Grey Matter
GSD IV	↑	=	✓	✓	↑	✓	✓
GSD VII	↑	?	✓	✓	↑	?	?
Lafora Disease	↑	↑	✓	✓	=	X	✓
RBCK1 deficiency	?	?	✓	✓	?	?	?

be at a higher risk of precipitation (Chan et al. 2004a). At these sites, the local levels of GS and PTG may be decreased by the action of the laforin-malin complex to increase the relative branching frequency and prevent particular glycogen molecules from precipitating. However, the average GS and PTG levels across the tissue may not have changed in a measureable way (Sullivan et al. 2017).

One difference between PGBD class I and II is the location of the PG bodies. In GSD IV, these bodies are scattered throughout the brain fairly uniformly in both areas of high white and grey matter density. In Class II, the PG bodies are found mainly in areas dense in grey matter. This includes the spinal cord, where bodies are found scattered ubiquitously in GSD IV but are only found in the grey matter of Lafora disease. Unfortunately, this analysis has not yet been performed for GSD VII; however, the prediction would be that it is more similar to GSD IV, based on a similar mechanism of PG body formation.

PGBD class II also has higher levels of glycogen phosphate, while there is evidence GSD IV does not (Sullivan et al. 2019). There does not appear to be research on whether GSD VII glycogen has elevated covalently bound phosphate to glycogen. Assuming a similar mechanism of polyglucosan body formation in GSD IV and VII, both with an imbalance between GS and GBE activity, it is possible that GSD IV glycogen contains normal levels of glycogen phosphate as well.

Acknowledgements

This work was supported in part by the National Institute of Neurological Disorders and Stroke of the NIH under award number P01 NS097197. Mitchell Sullivan is supported by a Mater Research McGuckin Early Career Fellowship, the University of Queensland's Amplify Initiative, Mater Foundation, Equity Trustees and the L G McCallam Est and George Weaber Trusts.

References

Adeva-Andany, M.M., M. Gonzalez-Lucan, C. Donapetry-Garcia, C. Fernandez-Fernandez and E. Ameneiros-Rodriguez. 2016. Glycogen metabolism in humans. Biochim. Biophys. Acta Clin. 5: 85–100.

Agius, L. 2015. Role of glycogen phosphorylase in liver glycogen metabolism. Mol. Aspects Med. 46: 34–45.

Akai, H., K. Yokobaya, A. Misaki and T. Harada. 1971. Complete hydrolysis of branching linkages in glycogen by Pseudomonas isoamylase-distribution of linear chains. Biochim. Biophys. Acta 237: 422–429.

Akman, H.O., C. Karadimas, Y. Gyftodimou, M. Grigoriadou, H. Kokotas, A. Konstantinidou et al. 2006. Prenatal diagnosis of glycogen storage disease type IV. Prenatal Diagn. 26: 951–955.

Akman, H.O., A. Raghavan and W.J. Craigen. 2011a. Animal models of glycogen storage disorders. pp. 369–388. *In:* Chang, K.T. and K.-T. Min (eds.). Prog. Mol. Biol. Transl. Sci. (Academic Press).

Akman, H.O., T. Sheiko, A. Raghavan, M. Finegold and W.J. Craigen. 2011b. Generation of a novel mouse model that recapitulates early and adult early and adult onset glycogenosis type 4. J. Inherit. Metab. Dis. 34: S173.

Akman, H.O., V. Emmanuele, Y.G. Kurt, B. Kurt, T. Sheiko, S. DiMauro et al. 2015. A novel mouse model that recapitulates adult-onset glycogenosis type 4. Hum. Mol. Genet. 24: 6801–6810.

Andersen, D.H. 1956. Familial cirrhosis of the liver with storage of abnormal glycogen. Lab. Invest. 5: 11–20.

Arad, M., D.W. Benson, A.R. Perez-Atayde, W.J. McKenna, E.A. Sparks, R.J. Kanter et al. 2002. Constitutively active AMP kinase mutations cause glycogen storage disease mimicking hypertrophic cardiomyopathy. J. Clin. Invest. 109: 357–362.

Barbetti, F., M. Rocchi, M. Bossolasco, R. Cordera, P. Sbraccia, P. Finelli et al. 1996. The human skeletal muscle glycogenin gene: cDNA, tissue expression, and chromosomal localization. Biochem. Biophys. Res. Commun. 220: 72–77.

Baykan, B., P. Striano, S. Gianotti, N. Bebek, E. Gennaro, C. Gurses et al. 2005. Late-onset and slow-progressing lafora disease in four siblings with EPM2B mutation. Epilepsia 46: 1695–1697.

Beauchamp, N.J., A. Dalton, U. Ramaswami, H. Niinikoski, K. Mention, P. Kenny et al. 2007. Glycogen storage disease type IX: High variability in clinical phenotype. Mol. Genet. Metab. 92: 88–99.

Bertoft, E., Q. Zhu, H. Andtfolk and M. Jungner. 1999. Structural heterogeneity in waxy-rice starch. Carbohydr. Polym. 38: 349–359.

Besford, Q.A., M.A. Sullivan, L. Zheng, R.G. Gilbert, D. Stapleton and A. Gray-Weale. 2012. The structure of cardiac glycogen in healthy mice. Int. J. Biol. Macromol. 51: 887–891.

Bhagavan, N.V. and C.-E. Ha. 2011. Carbohydrate metabolism II: Gluconeogenesis, glycogen synthesis and breakdown, and alternative pathways. pp. 151–168. *In*: Bhagavan, N.V. and C.-E. (eds.). Essentials of Medical Biochemistry (San Diego: Academic Press).

Bigio, E.H., M.F. Weiner, F.J. Bonte and C.L. White. 1997. Familial dementia due to adult polyglucosan body disease. Clin. Neuropathol. 16: 227–234.

Boisson, B., E. Laplantine, C. Prando, S. Giliani, E. Israelsson, Z. Xu et al. 2012. Immunodeficiency, autoinflammation and amylopectinosis in humans with inherited HOIL-1 and LUBAC deficiency. Nat. Immunol. 13: 1178–1186.

Boisson, B., E. Laplantine, K. Dobbs, A. Cobat, N. Tarantino, M. Hazen et al. 2015. Human HOIP and LUBAC deficiency underlies autoinflammation, immunodeficiency, amylopectinosis, and lymphangiectasia. J. Exp. Med. 212: 939–951.

Boudouresques, J., J. Roger, R. Khalil, J.F. Pellissier, A. Ali Cherif, B. Tafani et al. 1978. 2 familial cases of Lafora disease. Clinical, electroencephalographic and pathologic study. Rev. Neurol. (Paris) 134: 523–540.

Brown, A.M., S.B. Tekkok and B.R. Ransom. 2003. Glycogen regulation and functional role in mouse white matter. J. Physiol. 549: 501–512.

Brown, B.I., and D.H. Brown. 1966a. Lack of an alpha-1,4-glucan: alpha-1,4-glucan 6-glycosyl transferase in a case of type IV glycogenosis. Proc. Natl. Acad. Sci. U. S. A. 56: 725–729.

Brown, B.I. and D.H. Brown. 1989. Branching enzyme-activity of cultured amniocytes and chorionic villi—prenatal testing for type IV glycogen storage disease. Am. J. Hum. Genet. 44: 378–381.

Brown, D.H. and B.I. Brown. 1966b. Action of a muscle branching enzyme on polysaccharides enlarged from UDP ^{14}C glucose. Biochim. Biophys. Acta 130: 263–266.

Bryant, N.J., R. Govers and D.E. James. 2002. Regulated transport of the glucose transporter glut4. Nat. Rev. Mol. Cell Biol. 3: 267–277.

Cafferty, M.S., R.E. Lovelace, A.P. Hays, S. Servidei, S. Dimauro and L.P. Rowland. 1991. Polyglucosan body disease. Muscle Nerve 14: 102–107.

Calder, P.C. and R. Geddes. 1989. Rat skeletal-muscle lysosomes contain glycogen. Int. J. Biochem. 21: 561–567.

Carnagarin, R., A.M. Dharmarajan and C.R. Dass. 2015. Molecular aspects of glucose homeostasis in skeletal muscle—a focus on the molecular mechanisms of insulin resistance. Mol. Cell. Endocrinol. 417: 52–62.

Carpenter, S. and G. Karpati. 1981. Sweat gland duct cells in Lafora disease: Diagnosis by skin biopsy. Neurology 31: 1564–1568.

Cataldo, A.M. and R.D. Broadwell. 1986. Cytochemical identification of cerebral glycogen and glucose-6-phosphatase activity under normal and experimental conditions. 1. Neurons and glia. J. Electron Microsc. Tech. 3: 413–437.

Chan, E.M., E.J. Young, L. Ianzano, I. Munteanu, X.C. Zhao, C.C. Christopoulos et al. 2003. Mutations in NHLRC1 cause progressive myoclonus epilepsy. Nat. Genet. 35: 125–127.

Chan, E.M., C.A. Ackerley, H. Lohi, L. Ianzano, M.A. Cortez, P. Shannon et al. 2004a. Laforin preferentially binds the neurotoxic starch-like polyglucosans, which form in its absence in progressive myoclonus epilepsy. Hum. Mol. Genet. 13: 1117–1129.

Chan, E.M., S. Omer, M. Ahmed, L.R. Bridges, C. Bennett, S.W. Scherer et al. 2004b. Progressive myoclonus epilepsy with polyglucosans (Lafora disease)—Evidence for a third locus. Neurology 63: 565–567.

Cheng, A., M. Zhang, M.S. Gentry, C.A. Worby, J.E. Dixon and A.R. Saltiel. 2007. A role for AGL ubiquitination in the glycogen storage disorders of Lafora and Cori's disease. Genes Dev. 21: 2399–2409.

Chou, J.Y. and B.C. Mansfield. 2008. Mutations in the glucose-6-phosphatase-a (G6PC) gene that cause type Ia glycogen storage disease. Hum. Mutat. 29: 921–930.

Chou, J.Y., H.S. Jun and B.C. Mansfield. 2010. Glycogen storage disease type I and G6Pase-beta deficiency: etiology and therapy. Nat. Rev. Endocrinol. 6: 676–688.

Ciaraldi, T.P., S. Mudaliar, A. Barzin, J.A. Macievic, S.V. Edelman, K.S. Park et al. 2005. Skeletal muscle GLUT1 transporter protein expression and basal leg glucose uptake are reduced in type 2 diabetes. J. Clin. Endocrinol. Metab. 90: 352–358.

Claus, T.H. and S.J. Pilkis. 1976. Regulation by insulin of gluconeogenesis in isolated rat hepatocytes. Biochim. Biophys. Acta 421: 246–262.

Davis, M.K. and D.A. Weinstein. 2008. Liver transplantation in children with glycogen storage disease: Controversies and evaluation of the risk/benefit of this procedure. Pediatr. Transplant. 12: 137–145.

Deng, B., M.A. Sullivan, A.C. Wu, J.L. Li, C. Chen and R.G. Gilbert. 2015. The mechanism for stopping chain and total-molecule growth in complex branched polymers, exemplified by glycogen. Biomacromolecules 16: 1870–1872.

DePaoli-Roach, A.A., V.S. Tagliabracci, D.M. Segvich, C.M. Meyer, J.M. Irimia and P.J. Roach. 2010. Genetic depletion of the Malin E3 Ubiquitin Ligase in mice leads to Lafora Bodies and the accumulation of insoluble laforin. J. Biol. Chem. 285: 25372–25381.

DePaoli-Roach, A.A., C.J. Contreras, D.M. Segvich, C. Heiss, M. Ishihara, P. Azadi et al. 2015. Glycogen phosphomonoester distribution in mouse models of the progressive myoclonic epilepsy, Lafora Disease. J. Biol. Chem. 290: 841–850.

Depre, C., J.L.J. Vanoverschelde and H. Taegtmeyer. 1999. Glucose for the heart. Circulation 99: 578–588.

Deshaies, R.J. and C.A.P. Joazeiro. 2009. RING Domain E3 Ubiquitin Ligases. Annu. Rev. Biochem. 78: 399–434.

Dhawan, A., K.C. Tan, B. Portmann and A.P. Mowat. 1994. Glycogenosis type-IV liver-transplant at 12 years. Arch. Dis. Child. 71: 450–451.

Drochmans, P. 1962. Study under the electron microscope of negative colourings of particulate glycogen. J. Ultrastruct. Res. 6: 141–163.

Dubey, D., R. Parihar and S. Ganesh. 2012. Identification and characterization of novel splice variants of the human EPM2A gene mutated in Lafora progressive myoclonus epilepsy. Genomics 99: 36–43.

Duran, J., I. Saez, A. Gruart, J.J. Guinovart and J.M. Delgado-Garcia. 2013. Impairment in long-term memory formation and learning-dependent synaptic plasticity in mice lacking glycogen synthase in the brain. J. Cereb. Blood Flow Metab. 33: 550–556.

Elton, L., I. Carpentier, K. Verhelst, J. Staal and R. Beyaert. 2015. The multifaceted role of the E3 ubiquitin ligase HOIL-1: beyond linear ubiquitination. Immunol. Rev. 266: 208–221.

Emanuelle, S., M.K. Brewer, D.A. Meekins and M.S. Gentry. 2016. Unique carbohydrate binding platforms employed by the glucan phosphatases. Cell. Mol. Life Sci. 73: 2765–2778.

Fernandez-Sanchez, M.E., O. Criado-Garcia, K.E. Heath, B. Garcia-Fojeda, I. Medrano-Fernandez, P. Gomez-Garre et al. 2003. Laforin, the dual-phosphatase responsible for Lafora disease, interacts with R5 (PTG), a regulatory subunit of protein phosphatase-1 that enhances glycogen accumulation. Hum. Mol. Genet. 12: 3161–3171.

Fiil, B.K. and M. Gyrd-Hansen. 2014. Met1-linked ubiquitination in immune signalling. FEBS J. 281: 4337–4350.

Fontana, J.D. 1980. Presence of phosphate in glycogen. FEBS Lett. 109: 85–92.

Franceschetti, S., A. Gambardella, L. Canafoglia, P. Striano, H. Lohi, E. Gennaro et al. 2006. Clinical and genetic findings in 26 Italian patients with Lafora disease. Epilepsia 47: 640–643.

Freemont, P.S. 2000. Ubiquitination: RING for destruction? Curr. Biol. 10: R84–R87.

Fyfe, J.C., U. Giger, T.J. Vanwinkle, M.E. Haskins, S.A. Steinberg, P. Wang et al. 1992. Glycogen-storage-disease type-IV – inherent deficiency of branching enzyme-activity in cats. Pediatr. Res. 32: 719–725.

Ganesh, S., A.V. Delgado-Escueta, T. Sakamoto, M.R. Avila, J. Machado-Salas, Y. Hoshii et al. 2002. Targeted disruption of the Epm2a gene causes formation of Lafora inclusion bodies, neurodegeneration, ataxia, myoclonus epilepsy and impaired behavioral response in mice. Hum. Mol. Genet. 11: 1251–1262.

Ganesh, S., N. Tsurutani, T. Suzuki, K. Ueda, K.L. Agarwala, H. Osada et al. 2003. The Lafora disease gene product laforin interacts with HIRIP5, a phylogenetically conserved protein containing a NIfU-like domain. Hum. Mol. Genet. 12: 2359–2368.

Garcia, M., A. Pujol, A. Ruzo, E. Riu, J. Ruberte, A. Arbos et al. 2009. Phosphofructo-1-kinase deficiency leads to a severe cardiac and hematological disorder in addition to skeletal muscle glycogenosis. PLOS Genet. 5.

Garcia-Gimeno, M.A., P.N. Rodilla-Ramirez, R. Viana, X. Salas-Puig, M.K. Brewer, M.S. Gentry et al. 2018. A novel EPM2A mutation yields a slow progression form of Lafora disease. Epilepsy Res. 145: 169–177.

Gayarre, J., L. Duran-Trio, O.C. Garcia, C. Aguado, L. Juana-Lopez, I. Crespo et al. 2014. The phosphatase activity of laforin is dispensable to rescue Epm2a(-/-) mice from Lafora disease. Brain 137: 806–818.

Geddes, R. and G.C. Stratton. 1977. Influence of lysosomes on glycogen-metabolism. Biochem. J 163: 193–200.

Gentry, M.S., C.A. Worby and J.E. Dixon. 2005. Insights into Lafora disease: Malin is an E3 ubiquitin ligase that ubiquitinates and promotes the degradation of laforin. Proc. Natl. Acad. Sci. U. S. A. 102: 8501–8506.

Gentry, M.S., C. Roma-Mateo and P. Sanz. 2013. Laforin, a protein with many faces: glucan phosphatase, adapter protein, et alii. FEBS J. 280: 525–537.

Gerlach, B., S.M. Cordier, A.C. Schmukle, C.H. Emmerich, E. Rieser, T.L. Haas et al. 2011. Linear ubiquitination prevents inflammation and regulates immune signalling. Nature 471: 591–596.

Getty-Kaushik, L., J.C. Viereck, J.M. Goodman, Z.F. Guo, N.K. LeBrasseur, A.M.T. Richard et al. 2010. Mice deficient in phosphofructokinase-M have greatly decreased fat stores. Obesity 18: 434–440.

Gidley, M.J. and P.V. Bulpin. 1987. Crystallisation of malto-oligosaccharides as models of the crystalline forms of starch: Minimum chain-length requirement for the formation of double-helices. Carbohydr. Res. 161: 291–300.

Gomez-Abad, C., P. Gomez-Garre, E. Gutierrez-Delicado, S. Saygi, R. Michelucci, C.A. Tassinari et al. 2005. Lafora disease due to EPM2B mutations—A clinical and genetic study. Neurology 64: 982–986.

Guerra, A.S., O.P. Vandiggelen, F. Carneiro, R.M. Tsou, S. Simoes and N.T. Santos. 1986. A juvenile variant of glycogenosis-IV (Andersen disease). Eur. J. Pediatr. 145: 179–181.

Hajek, I., F. Kettner, V. Simerdova, C. Rusbridge, P. Wang, B.A. Minassian et al. 2016. NHLRC1 repeat expansion in two beagles with Lafora disease. J. Small Anim. Pract. 57: 650–652.

Harvey, J.W., M.B.C. Mays, K.E. Gropp and F.J. Denaro. 1990. Polysaccharide storage myopathy in canine phosphofructokinase deficiency (type VII glycogen storage disease). Vet. Pathol. 27: 1–8.

Hays, A.P., M. Hallett, J. Delfs, J. Morris, A. Sotrel, M.M. Shevchuk et al. 1981. Muscle phosphofructokinase deficiency - Abnormal polysaccharide in a case of late-onset myopathy. Neurology 31: 1077–1086.

Hazard, B., X.Q. Zhang, P. Colasuonno, C. Uauy, D.M. Beckles and J. Dubcovsky. 2012. Induced mutations in the Starch Branching Enzyme II (SBEII) genes increase amylose and resistant starch content in Durum Wheat. Crop Sci. 52: 1754–1766.

Hedberg-Oldfors, C. and A. Oldfors. 2015. Polyglucosan storage myopathies. Mol. Aspects Med. 46: 85–100.

Hems, D.A. and P.D. Whitton. 1980. Control of hepatic glycogenolysis. Physiol. Rev. 60: 1–50.

Herrmann, J., L.O. Lerman and A. Lerman. 2007. Ubiquitin and ubiquitin-like proteins in protein regulation. Circ. Res. 100: 1276–1291.

Hirschhorn, R. and A.J. Reuser. 2000. Glycogen storage disease type II: acid α-glucosidase (acid maltase) deficiency. pp. 3389–3420. *In:* Beaudet, A.L., W.S. Sly and D. Valle (eds.). The Metabolic and Molecular Basis of Inherited Disease. C.R. Scriver (McGraw-Hill).

Huang, L. and Y. Yao. 2011. Particulate structure of phytoglycogen nanoparticles probed using amyloglucosidase. Carbohydr. Polym. 83: 1665–1671.

Hunter, R.W., J.T. Treebak, J.F. Wojtaszewski and K. Sakamoto. 2011. Molecular mechanism by which AMP-activated protein kinase activation promotes glycogen accumulation in muscle. Diabetes 60: 766–774.

Hunter, R.W., E. Zeqiraj, N. Morrice, F. Sicheri and K. Sakamoto. 2015. Expression and purification of functional human glycogen synthase-1:glycogenin-1 complex in insect cells. Protein Expr. Purif. 108: 23–29.

Hurley, T.D., C. Walls, J.R. Bennett, P.J. Roach and M. Wang. 2006. Direct detection of glycogenin reaction products during glycogen initiation. Biochem. Biophys. Res. Commun. 348: 374–378.

Ianzano, L., X.C. Zhao, B.A. Minassian and S.W. Scherer. 2003. Identification of a novel protein interacting with laforin, the EPM2A progressive myoclonus epilepsy gene product. Genomics 81: 579–587.

Ikeda, F., Y.L. Deribe, S.S. Skanland, B. Stieglitz, C. Grabbe, M. Franz-Wachtel et al. 2011. SHARPIN forms a linear ubiquitin ligase complex regulating NF-kappa B activity and apoptosis. Nature 471: 637–U120.

Im, S.-S., S.-Y. Kang, S.-Y. Kim, H.-i. Kim, J.-W. Kim, K.-S. Kim et al. 2005. Glucose-stimulated upregulation of GLUT2 gene Is mediated by sterol response element-binding protein-1c in the hepatocytes. Diabetes 54: 1684.

Issoglio, F.M., M.E. Carrizo, J.M. Romero and J.A. Curtino. 2012. Mechanisms of monomeric and dimeric glycogenin autoglucosylation. J. Biol. Chem. 287: 1955–1961.

Janecke, A.R., S. Dertinger, U.P. Ketelsen, L. Bereuter, B. Simma, T. Muller et al. 2004. Neonatal type IV glycogen storage disease associated with "null" mutations in glycogen branching enzyme 1. J. Pediatr. 145: 705–709.

Jiang, S.X., B. Heller, V.S. Tagliabracci, L.M. Zhai, J.M. Irimia, A.A. DePaoli-Roach et al. 2010. Starch binding domain-containing protein 1/genethonin 1 Is a novel participant in glycogen metabolism. J. Biol. Chem. 285: 34960–34971.

Kasvinsky, P.J., S. Shechosky and R.J. Fletterick. 1978. Synergistic regulation of phosphorylase a by glucose and caffeine. J. Biol. Chem. 253: 9102–9106.

Kirisako, T., K. Kamei, S. Murata, M. Kato, H. Fukumoto, M. Kanie et al. 2006. A ubiquitin ligase complex assembles linear polyubiquitin chains. EMBO J. 25: 4877–4887.

Kishnani, P.S., S.L. Austin, J.E. Abdenur, P. Arn, D.S. Bali, A. Boney et al. 2014. Diagnosis and management of glycogen storage disease type I: a practice guideline of the American College of Medical Genetics and Genomics. Genet. Med. 16: e1.

Koch, K., R. Andersson and P. Aman. 1998. Quantitative analysis of amylopectin unit chains by means of high-performance anion-exchange chromatography with pulsed amperometric detection. J. Chromatogr. A 800: 199–206.

Krisman, C.R. 1962. A method for colorimetric estimation of glycogen with iodine. Anal. Biochem. 4: 17–23.

Landgraf, K., F. Bollig, M.-O. Trowe, B. Besenbeck, C. Ebert, D. Kruspe et al. 2010. Sipl1 and Rbck1 are novel Eya1-binding proteins with a role in craniofacial development. Mol. Cell. Biol. 30: 5764–5775.

Lemos, P.V.F., L.S. Barbosa, I.G. Ramos, R.E. Coelho and J.I. Druzian. 2019. Characterization of amylose and amylopectin fractions separated from potato, banana, corn, and cassava starches. Int. J. Biol. Macromol. 132: 32–42.

Lesca, G., N. Boutry-Kryza, B. de Toffol, M. Milh, D. Steschenko, M. Lemesle-Martin et al. 2010. Novel mutations in EPM2A and NHLRC1 widen the spectrum of Lafora disease. Epilepsia 51: 1691–1698.

Lohi, H.T. and B.A. Minassian. 2005. Starch-like polyglucosan formation in neuronal dendrites in the Lafora form of human epilepsy: a theory of pathogenesis. Biologia (Bratisl.) 60: 123–129.

Lohi, H., L. Ianzano, X.C. Zhao, E.M. Chan, J. Turnbull, S.W. Scherer et al. 2005. Novel glycogen synthase kinase 3 and ubiquitination pathways in progressive myoclonus epilepsy. Hum. Mol. Genet. 14: 2727–2736.

Lomako, J., W.M. Lomako and W.J. Whelan. 1988. A self-glucosylating protein is the primer for rabbit muscle glycogen biosynthesis. FASEB J. 2: 3097–3103.

MacDuff, D.A., T.A. Reese, J.M. Kimmey, L.A. Weiss, C. Song, X. Zhang et al. 2015. Phenotypic complementation of genetic immunodeficiency by chronic herpesvirus infection. Elife 4.

Mahmoud, S.K., A. Khorrami, M. Rafeey, R. Ghergherehchi and M.D. Sima. 2017. Molecular analysis of glycogen storage disease type Ia in Iranian Azeri Turks: identification of a novel mutation. Journal of Genetics 96: 19–23.

Malfatti, E., N. Birouk, N.B. Romero, M. Piraud, F.M. Petit, J.Y. Hogrel et al. 2012. Juvenile-onset permanent weakness in muscle phosphofructokinase deficiency. J. Neurol. Sci. 316: 173–177.

Malfatti, E., J. Nilsson, C. Hedberg-Oldfors, A. Hernandez-Lain, F. Michel, C. Dominguez-Gonzalez et al. 2014. A new muscle glycogen storage disease associated with glycogenin-1 deficiency. Ann. Neurol. 76: 891–898.

Manners, D.J. 1991. Recent developments in our understanding of glycogen structure. Carbohydr. Polym. 16: 37–82.

Marks, J.S. and L.H.P. Botelho. 1986. Synergistic inhibition of glucagon-induced effects on hepatic glucose-metabolism in the presence of insulin and a cAMP antagonist. J. Biol. Chem. 261: 5895–5899.

Mathieu, C., R. Duval, A. Cocaign, E. Petit, L.-C. Bui, I. Haddad et al. 2016. An isozyme-specific redox switch in human brain glycogen phosphorylase modulates its allosteric activation by AMP. J. Biol. Chem. 291: 23842–23853.

McConkie-Rosell, A., C. Wilson, D.A. Piccoli, J. Boyle, T. DeClue, P. Kishnani et al. 1996. Clinical and laboratory findings in four patients with the non-progressive hepatic form of type IV glycogen storage disease. J. Inherit. Metab. Dis. 19: 51–58.

Melendez, R., E. Melendez-Hevia, F. Mas, J. Mach and M. Cascante. 1998. Physical constraints in the synthesis of glycogen that influence its structural homogeneity: A two-dimensional approach. Biophys. J. 75: 106–114.

Mercier, C. and W.J. Whelan. 1973. Further characterization of glycogen from type-IV glycogen-storage disease. Eur. J. Biochem. 40: 221–223.

Minassian, B.A. 2001. Lafora's disease: Towards a clinical, pathologic, and molecular synthesis. Pediatr. Neurol. 25: 21–29.

Mochel, F., R. Schiffmann, M.E. Steenweg, H.O. Akman, M. Wallace, F. Sedel et al. 2012. Adult polyglucosan body disease: Natural history and key magnetic resonance imaging findings. Ann. Neurol. 72: 433–441.

Montessuit, C. and A. Thorburn. 1999. Transcriptional activation of the glucose transporter GLUT1 in ventricular cardiac myocytes by hypertrophic agonists. J. Biol. Chem. 274: 9006–9012.

Moreno, D., M.C. Towler, D.G. Hardie, E. Knecht and P. Sanz. 2010. The laforin-malin complex, involved in Lafora disease, promotes the incorporation of K63-linked ubiquitin chains into AMP-activated protein kinase beta subunits. Mol. Biol. Cell 21: 2578–2588.

Morgello, S., R.R. Uson, E.J. Schwartz and R.S. Haber. 1995. The human blood-brain-barrier glucose-transporter (GLUT1) is a glucose-transporter of gray-matter astrocytes. Glia 14: 43–54.

Mu, J., A.V. Skurat and P.J. Roach. 1997. Glycogenin-2, a novel self-glucosylating protein involved in liver glycogen biosynthesis. J. Biol. Chem. 272: 27589–27597.

Mueckler, M., M. Kruse, M. Strube, A.C. Riggs, K.C. Chiu and M.A. Permutt. 1994. A mutation in the Glut2 glucose-transporter gene of a diabetic patient abolishes transport activity. J. Biol. Chem. 269: 17765–17767.

Mueckler, M. and B. Thorens. 2013. The SLC2 (GLUT) family of membrane transporters. Mol. Aspects Med. 34: 121–138.

Muller, M.S., S.E. Pedersen, A.B. Walls, H.S. Waagepetersen and L.K. Bak. 2015. Isoform-selective regulation of glycogen phosphorylase by energy deprivation and phosphorylation in astrocytes. Glia 63: 154–162.

Nakajima, H., N. Raben, T. Hamaguchi and T. Yamasaki. 2002. Phosphofructokinase deficiency; past, present and future. Curr. Mol. Med. 2: 197–212.

Nathan, J.A., H.T. Kim, L. Ting, S.P. Gygi and A.L. Goldberg. 2013. Why do cell proteins linked to K63-polyubiquitin chains not associate with proteasomes? EMBO J. 32: 552–565.

Newgard, C.B., P.K. Hwang and R.J. Fletterick. 1989. The family of glycogen phosphorylases: structure and function. Crit. Rev. Biochem. Mol. Biol. 24: 69–99.

Nielsen, J.N. and E.A. Richter. 2003. Regulation of glycogen synthase in skeletal muscle during exercise. Acta Physiol. Scand. 178: 309–319.

Nilsson, J., B. Schoser, P. Laforet, O. Kalev, C. Lindberg, N.B. Romero et al. 2013. Polyglucosan body myopathy caused by defective ubiquitin ligase RBCK1. Ann. Neurol. 74: 914–919.

Nitschke, F., P. Wang, P. Schmieder, J.-M. Girard, D.E. Awrey, T. Wang et al. 2013. Hyperphosphorylation of glucosyl C6 carbons and altered structure of glycogen in the neurodegenerative epilepsy Lafora disease. Cell Metab. 17: 756–767.

Nitschke, F., M.A. Sullivan, P. Wang, X. Zhao, E.E. Chown, A.M. Perri et al. 2017. Abnormal glycogen chain length pattern, not hyperphosphorylation, is critical in Lafora disease. EMBO Mol. Med.: e201707608.

Nitschke, F., S.J. Ahonen, S. Nitschke, S. Mitra and B.A. Minassian. 2018. Lafora disease—from pathogenesis to treatment strategies. Nat. Rev. Neurol. 14: 606–617.

O'Shea, M.G. and M.K. Morell. 1996. High resolution slab gel electrophoresis of 8-amino-1,3, 6-pyrenetrisulfonic acid (APTS) tagged oligosaccharides using a DNA sequencer. Electrophoresis 17: 681–686.

Ombrello, M.J., D.L. Kastner and J.D. Milner. 2012. HOIL and water: the two faces of HOIL-1 deficiency. Nat. Immunol. 13: 1133–1135.

Pederson, B.A., H.Y. Chen, J.M. Schroeder, W.N. Shou, A.A. DePaoli-Roach and P.J. Roach. 2004. Abnormal cardiac development in the absence of heart glycogen. Mol. Cell. Biol. 24: 7179–7187.

Pederson, B.A., J. Turnbull, J.R. Epp, S.A. Weaver, X.C. Zhao, N. Pencea et al. 2013. Inhibiting glycogen synthesis prevents Lafora disease in a mouse model. Ann. Neurol. 74: 297–300.

Peress, N.S., S. Dimauro and V.A. Roxburgh. 1979. Adult polysaccharidosis—clinico-pathological, ultrastructural, and biochemical features. Arch. Neurol. 36: 840–845.

Pfister, B., K.J. Lu, S. Eicke, R. Feil, J.E. Lunn, S. Streb et al. 2014. Genetic evidence that chain length and branch point distributions are linked determinants of starch granule formation in Arabidopsis. Plant Physiol. 165: 1457–1474.

Poyrazoglu, H.G., E. Karaca, H. Per, H. Gumus, H. Onay, M. Canpolat et al. 2015. Three patients with lafora disease: different clinical presentations and a novel mutation. J. Child Neurol. 30: 777–781.

Puri, R., T. Suzuki, K. Yamakawa and S. Ganesh. 2009. Hyperphosphorylation and aggregation of Tau in laforin-deficient mice, an animal model for Lafora disease. J. Biol. Chem. 284: 22657–22663.

Putaux, J.L., M.B. Cardoso, D. Dupeyre, M. Morin, A. Nulac and Y. Hu. 2008. Single crystals of V-amylose Inclusion complexes. Macromol. Symp. 273: 1–8.

Queisser, M.A., L.A. Dada, N. Deiss-Yehiely, M. Angulo, G. Zhou, F.M. Kouri et al. 2014. HOIL-1L functions as the PKC zeta ubiquitin ligase to promote lung tumor growth. Am. J. Respir. Crit. Care Med. 190: 688–698.

Raben, N., J. Sherman, F. Miller, H. Mena and P. Plotz. 1993. A 5' splice junction mutation leading to exon deletion in an Ashkenazic Jewish family with phosphofructokinase deficiency (Tarui disease). J. Biol. Chem. 268: 4963–4967.

Raben, N., C. Schreiner, R. Baum, S. Takikita, S.G. Xu, T. Xie et al. 2010. Suppression of autophagy permits successful enzyme replacement therapy in a lysosomal storage disorder-murine Pompe disease. Autophagy 6: 1078–1089.

Richard, A.-M.T., D.-L. Webb, J.M. Goodman, V. Schultz, J.N. Flanagan, L. Getty-Kaushik et al. 2007. Tissue-dependent loss of phosphofructokinase-M in mice with interrupted activity of the distal promoter: impairment in insulin secretion. Am. J. Physiol. Endocrinol. Metab. 293: E794–E801.

Roach, P.J. and J. Larner. 1977. Covalent phosphorylation in the regulation of glycogen synthase activity. Mol. Cell. Biochem. 15: 179–200.

Roach, P.J. 1990. Control of glycogen synthase by hierarchal protein phosphorylation. FASEB J. 4: 2961–2968.

Roach, P.J., A.A. DePaoli-Roach, T.D. Hurley and V.S. Tagliabracci. 2012. Glycogen and its metabolism: some new developments and old themes. Biochem. J 441: 763–787.

Robertson, N.P., S. Wharton, J. Anderson and N.J. Scolding. 1998. Adult polyglucosan body disease associated with an extrapyramidal syndrome. J. Neurol. Neurosurg. Psychiatry 65: 788–790.

Roma-Mateo, C., D. Moreno, S. Vernia, T. Rubio, T.M. Bridges, M.S. Gentry et al. 2011. Lafora disease E3-ubiquitin ligase malin is related to TRIM32 at both the phylogenetic and functional level. BMC Evol. Biol. 11: 225–225.

Rubio-Villena, C., M.A. Garcia-Gimeno and P. Sanz. 2013. Glycogenic activity of R6, a protein phosphatase 1 regulatory subunit, is modulated by the laforin-malin complex. Int. J. Biochem. Cell Biol. 45: 1479–1488.

Ryu, J.-H., J. Drain, J.H. Kim, S. McGee, A. Gray-Weale, L. Waddington et al. 2009. Comparative structural analyses of purified glycogen particles from rat liver, human skeletal muscle and commercial preparations. Int. J. Biol. Macromol. 45: 478–482.

Saez, I., J. Duran, C. Sinadinos, A. Beltran, O. Yanes, M.F. Tevy et al. 2014. Neurons have an active glycogen metabolism that contributes to tolerance to hypoxia. J. Cereb. Blood Flow Metab. 34: 945–955.

Saltiel, A.R. and C.R. Kahn. 2001. Insulin signalling and the regulation of glucose and lipid metabolism. Nature 414: 799–806.

Sanchez-Martin, P., C. Roma-Mateo, R. Viana and P. Sanz. 2015. Ubiquitin conjugating enzyme E2-N and sequestosome-1 (p62) are components of the ubiquitination process mediated by the malin-laforin E3-ubiquitin ligase complex. Int. J. Biochem. Cell Biol. 69: 204–214.

Schneider, C.A., V.T.B. Nguyen and H. Taegtmeyer. 1991. Feeding and fasting determine postischemic glucose-utilization in isolated working rat hearts. Am. J. Physiol. 260: H542–H548.

Schmitz, O., J. Rungby, L. Edge and C.B. Juhl. 2008. On high-frequency insulin oscillations. Ageing Res. Rev. 7: 301–305.

Schochet, S.S., Jr., W.F. McCormick and H. Zellweger. 1970. Type IV glycogenosis (amylopectinosis). Light and electron microscopic observations. Arch. Pathol. 90: 354–363.

Shao, D. and R. Tian. 2016. Glucose transporters in cardiac metabolism and hypertrophy. Compr. Physiol. 6: 331–351.

Sharma, J., S. Mulherkar, D. Mukherjee and N.R. Jana. 2012. Malin regulates Wnt signaling pathway through degradation of dishevelled2. J. Biol. Chem. 287: 6830–6839.

Sherman, J.B., N. Raben, C. Nicastri, Z. Argov, H. Nakajima, E.M. Adams et al. 1994. Common mutations in the phosphofructokinase-M gene in Ashkhenazi Jewish patients with glycogenesis-VII—and their popupation frequency. Am. J. Hum. Genet. 55: 305–313.

Simpson, I.A., D. Dwyer, D. Malide, K.H. Moley, A. Travis, and S.J. Vannucci. 2008. The facilitative glucose transporter GLUT3: 20 years of distinction. Am. J. Physiol. Endocrinol. Metab. 295: E242–E253.

Singh, S. and S. Ganesh. 2009. Lafora progressive myoclonus epilepsy: A meta-analysis of reported mutations in the first decade following the discovery of the EPM2A and NHLRC1 genes. Hum. Mutat. 30: 715–723.

Skurat, A.V., A.D. Dietrich and P.J. Roach. 2006. Interaction between glycogenin and glycogen synthase. Arch. Biochem. Biophys. 456: 93–97.

Solaz-Fuster, M.C., J.V. Gimeno-Alcaniz, S. Ros, M.E. Fernandez-Sanchez, B. Garcia-Fojeda, O.C. Garcia et al. 2008. Regulation of glycogen synthesis by the laforin-malin complex is modulated by the AMP-activated protein kinase pathway. Hum. Mol. Genet. 17: 667–678.

Striano, P., F. Zara, J. Turnbull, J.M. Girard, C.A. Ackerley, M. Cervasio et al. 2008. Typical progression of myoclonic epilepsy of the Lafora type: a case report. Nat. Clin. Pract. Neurol. 4: 106–111.

Sullivan, M.A., F. Vilaplana, R.A. Cave, D.I. Stapleton, A.A. Gray-Weale and R.G. Gilbert. 2010. Nature of alpha and beta particles in glycogen using molecular size distributions. Biomacromolecules 11: 1094–1100.

Sullivan, M.A., J. Li, C. Li, F. Vilaplana, L. Zheng, D. Stapleton et al. 2011. Molecular structural differences between type-2-diabetic and healthy glycogen. Biomacromolecules 12: 1983–1986.

Sullivan, M.A., M.J. O'Connor, F. Umana, E. Roura, K. Jack, D.I. Stapleton et al. 2012. Molecular insights into glycogen alpha-particle formation. Biomacromolecules 13: 3805–3813.

Sullivan, M.A., P.O. Powell, T. Witt, F. Vilaplana, E. Roura and R.G. Gilbert. 2014. Improving size-exclusion chromatography for glycogen. J. Chromatography A 1332: 21–29.

Sullivan, M.A., S.H. Li, S.T.N. Aroney, B. Deng, C. Li, E. Roura et al. 2015. A rapid extraction method for glycogen from formalin-fixed liver. Carbohydr. Polym. 118: 9–15.

Sullivan, M.A., S. Nitschke, M. Steup, B.A. Minassian and F. Nitschke. 2017. Pathogenesis of Lafora disease: Transition of soluble glycogen to insoluble polyglucosan. Int. J. Mol. Sci. 18: 1743.

Sullivan, M.A., S. Nitschke, E.P. Skwara, P. Wang, X. Zhao, X.S. Pan et al. 2019. Skeletal muscle glycogen chain length correlates with insolubility in polyglucosan body associated neurodegenerative disease. Cell Rep. 27: 1334–1344.

Sun, T., H.Q. Yi, C.Y. Yang, P.S. Kishnani and B.D. Sun. 2016. Starch binding domain-containing protein 1 plays a dominant role in glycogen transport to lysosomes in liver. J. Biol. Chem. 291: 16479–16484.

Tagliabracci, V.S., J.M. Girard, D. Segvich, C. Meyer, J. Turnbull, X.C. Zhao et al. 2008. Abnormal metabolism of glycogen phosphate as a cause for Lafora disease. J. Biol. Chem. 283: 33816–33825.

Tagliabracci, V.S., J. Turnbull, W. Wang, J.M. Girard, X. Zhao, A.V. Skurat et al. 2007. Laforin is a glycogen phosphatase, deficiency of which leads to elevated phosphorylation of glycogen *in vivo*. Proc. Natl. Acad. Sci. U. S. A. 104: 19262–19266.

Tagliabracci, V.S., C. Heiss, C. Karthik, C.J. Contreras, J. Glushka, M. Ishihara et al. 2011. Phosphate incorporation during glycogen synthesis and Lafora disease. Cell Metab. 13: 274–282.

Tan, J.M.M., E.S.P. Wong, D.S. Kirkpatrick, O. Pletnikova, H.S. Ko, S.P. Tay et al. 2008. Lysine 63-linked ubiquitination promotes the formation and autophagic clearance of protein inclusions associated with neurodegenerative diseases. Hum. Mol. Genet. 17: 431–439.

Testoni, G., J. Duran, M. Garcia-Rocha, F. Vilaplana, A.L. Serrano, D. Sebastian et al. 2017. Lack of glycogenin causes glycogen accumulation and muscle function impairment. Cell Metab. 26: 256–266.

Thompson, D.B. 2000. On the non-random nature of amylopectin branching. Carbohydr. Polym. 43: 223–229.

Thon, V.J., M. Khalil and J.F. Cannon. 1993. Isolation of human glycogen branching enzyme cDNAs by screening complementation in yeast. J. Biol. Chem. 268: 7509–7513.

Thorens, B., H.K. Sarkar, H.R. Kaback and H.F. Lodish. 1988. Cloning and functional expression in bacteria of a novel glucose transporter present in liver, intestine, kidney, and beta-pancreatic islet cells. Cell 55: 281–290.

Tiberia, E., J. Turnbull, T. Wang, A. Ruggieri, X.C. Zhao, N. Pencea et al. 2012. Increased laforin and laforin binding to glycogen underlie Lafora body formation in Malin-deficient Lafora Disease. J. Biol. Chem. 287: 25650–25659.

Tiedge, M. and S. Lenzen. 1991. Regulation of glucokinase and Glut-2 glucose-transporter gene-expression in pancreatic B-cells. Biochem. J. 279: 899–901.

Tokunaga, F., S. Sakata, Y. Saeki, Y. Satomi, T. Kirisako, K. Kamei et al. 2009. Involvement of linear polyubiquitylation of NEMO in NF-kappa B activation. Nat. Cell Biol. 11: 123–U140.

Torija, M.J., M. Novo, A. Lemassu, W. Wilson, P.J. Roach, J. François et al. 2005. Glycogen synthesis in the absence of glycogenin in the yeast Saccharomyces cerevisiae. FEBS Lett. 579: 3999–4004.

Turnbull, J., P. Wang, J.M. Girard, A. Ruggieri, T.J. Wang, A.G. Draginov et al. 2010. Glycogen hyperphosphorylation underlies lafora body formation. Ann. Neurol. 68: 925–933.

Turnbull, J., A.A. DePaoli-Roach, X. Zhao, M.A. Cortez, N. Pencea, E. Tiberia et al. 2011. PTG depletion removes Lafora bodies and rescues the fatal epilepsy of Lafora disease. PLOS Genet. 7: e1002037.

Turnbull, J., J.M. Girard, H. Lohi, E.M. Chan, P.X. Wang, E. Tiberia et al. 2012. Early-onset Lafora body disease. Brain 135: 2684–2698.

Turnbull, J., J.R. Epp, D. Goldsmith, X. Zhao, N. Pencea, P. Wang et al. 2014. PTG protein depletion rescues malin-deficient Lafora disease in mouse. Ann. Neurol. 75: 442–446.

Uhlen, M., L. Fagerberg, B.M. Hallstrom, C. Lindskog, P. Oksvold, A. Mardinoglu et al. 2015. Tissue-based map of the human proteome. Science 347.

Umeki, K. and K. Kainuma. 1981. Fine-structure of Nageli amylodextrin obtained by acid treatment of defatted waxy-maize starch—structural evidence to support the double-helix hypothesis. Carbohydr. Res. 96: 143–159.

Valles-Ortega, J., J. Duran, M. Garcia-Rocha, C. Bosch, I. Saez, L. Pujadas et al. 2011. Neurodegeneration and functional impairments associated with glycogen synthase accumulation in a mouse model of Lafora disease. EMBO Mol. Med. 3: 667–681.

Verhue, W. and H.G. Hers. 1966. A study of the reaction catalyzed by the liver branching enzyme. Biochem. J. 99: 222.

Viana, R., P. Lujan and P. Sanz. 2015. The laforin/malin E3-ubiquitin ligase complex ubiquitinates pyruvate kinase M1/M2. BMC Biochem. 16.

Vilchez, D., S. Ros, D. Cifuentes, L. Pujadas, J. Valles, B. Garcia-Fojeda et al. 2007. Mechanism suppressing glycogen synthesis in neurons and its demise in progressive myoclonus epilepsy. Nat. Neurosci. 10: 1407–1413.

Vora, S., M. Davidson, C. Seaman, A.F. Miranda, N.A. Noble, K.R. Tanaka et al. 1983. Heterogeneity of the molecular lesions in inherited phosphofructokinase deficiency. J. Clin. Invest. 72: 1995–2006.

van de Wal, M., C. D'Hulst, J.P. Vincken, A. Buleon, R. Visser and S. Ball. 1998. Amylose is synthesized *in vitro* by extension of and cleavage from amylopectin. J. Biol. Chem. 273: 22232–22240.

Walker, G.J. and W.J. Whelan. 1960. The mechanism of carbohydrase action. 8. Structures of the muscle-phosphorylase limit dextrins of glycogen and amylopectin. Biochem. J. 76: 264–268.

Wang, J., P. Hu, Z. Chen, Q. Liu and C. Wei. 2017. Progress in high-amylose cereal crops through inactivation of starch branching enzymes. Front. Plant Sci. 8: 469.

Wang, J.Y., J.A. Stuckey, M.J. Wishart and J.E. Dixon. 2002. A unique carbohydrate binding domain targets the Lafora disease phosphatase to glycogen. J. Biol. Chem. 277: 2377–2380.

Wang, L. and M. Wise. 2011. Glycogen with short average chain length enhances bacterial durability. Sci. Nat. 98: 719–729.

Wang, W., H. Lohi, A.V. Skurat, A.A. DePaoli-Roach, B.A. Minassian and P.J. Roach. 2007. Glycogen metabolism in tissues from a mouse model of Lafora disease. Arch. Biochem. Biophys. 457: 264–269.

Ward, T.L., S.J. Valberg, D.L. Adelson, C.A. Abbey, M.M. Binns and J.R. Mickelson. 2004. Glycogen branching enzyme (GBE1) mutation causing equine glycogen storage disease IV. Mamm. Genome 15: 570–577.

Worby, C.A., M.S. Gentry and J.E. Dixon. 2006. Laforin, a dual specificity phosphatase that dephosphorylates complex carbohydrates. J. Biol. Chem. 281: 30412–30418.

Worby, C.A., M.S. Gentry and J.E. Dixon. 2008. Malin decreases glycogen accumulation by promoting the degradation of protein targeting to glycogen (PTG). J. Biol. Chem. 283: 4069–4076.

Yamanaka, K., H. Ishikawa, Y. Megumi, F. Tokunaga, M. Kanie, T.A. Rouault et al. 2003. Identification of the ubiquitin-protein ligase that recognizes oxidized IRP2. Nat. Cell Biol. 5: 336–340.

Yamasaki, T., H. Nakajima, N. Kono, k. Hotta, K. Yamada, E. Imai et al. 1991. Structure of the entire human muscle phosphofructokinase-encoding gene: a two-promoter system. Gene 104: 277–282.

Zhai, L., L. Feng, L. Xia, H. Yin and S. Xiang. 2016. Crystal structure of glycogen debranching enzyme and insights into its catalysis and disease-causing mutations. Nat. Commun. 7: 11229.

CHAPTER 7

Reversible Phosphorylation in Glycogen and Starch

Katherine J. Donohue,[1] *Andrea Kuchtova,*[1] *Craig W. Vander Kooi*[2,*]
and *Matthew S. Gentry*[3,*]

7.1 Introduction

Organisms have complex and sophisticated systems for harvesting, storing, and mobilizing energy in the form of carbohydrates. Energy is primarily stored in carbohydrates in the form of the polysaccharides glycogen and starch. While polysaccharide synthesis and degradation has been studied for many decades, regulation of carbohydrate storage via reversible phosphorylation by dikinase and phosphatase enzymes has only recently been identified. The discovery of these enzymes is intimately linked to the discovery of aberrantly formed glycogen and starch in humans and plants, respectively. These aberrant forms of the polysaccharides contain atypical levels of phosphate and abnormal architecture, and have revealed key roles for reversible phosphorylation in glycogen and starch synthesis and metabolism. This chapter reviews the discovery and enzymology of the glucan phosphatase and dikinase enzymes involved in phosphorylation-mediated regulation of storage carbohydrates and how they function in carbohydrate metabolism. These findings are placed in the context of outstanding questions regarding the biology and utilization of reversible carbohydrate phosphorylation.

7.2 Glycogen

Glycogen is a storage carbohydrate found in humans, animals, fungi, and bacteria. The molecule was first observed in the 1850s in the liver of a dog (Bernard 1857). Initial studies defined the composition of glycogen and revealed glycogen molecules in muscle that decreased in size after periods of exercise, suggesting that glycogen played some role in storing and mobilizing carbohydrates for the organism (Young 1957). More than a century after its discovery, the modern molecular picture of glycogen synthesis and the structure of the molecule began to come into view. In 1985, the Whelan

[1] University of Kentucky College of Medicine, Department of Molecular and Cellular Biochemistry, Center for Structural Biology, Lexington, KY 40536, United States.
[2] University of Kentucky College of Medicine, Department of Molecular and Cellular Biochemistry, Center for Structural Biology, Lafora Epilepsy Cure Initiative, Lexington, KY 40536, United States.
[3] University of Kentucky College of Medicine, Department of Molecular and Cellular Biochemistry, Center for Structural Biology, Lafora Epilepsy Cure Initiative, Epilepsy & Brain Metabolism Alliance, Lexington, KY 40536, United States.
* Corresponding authors: craig.vanderkooi@uky.edu; matthew.gentry@uky.edu

group demonstrated that each glycogen molecule is primed by the scaffold enzyme glycogenin, which enzymatically primes the glycogen synthesis (Whelan 1986). The glycogen granule is then built by glycogen synthase and other enzymes using a combination of linear -1,4- and branched -1,6-linkages between glucose moieties (Calder 1991, Pitcher et al. 1987, Lomako et al. 1988) (Fig. 7.1A). Each glycogen molecule consists of ~ 20,000–50,000 glucose moieties, with branch points occurring on average every 10–12 glucose units (Melendez-Hevia et al. 1993, Shearer and Graham 2004, Prats et al. 2018). This arrangement creates the round, globular shape of glycogen that is observed via electron microscopy (Fig. 7.1B). Depending on the tissue type, the morphology of glycogen can range from small oblong spheres in muscle to round spherical structures in brain tissue.

7.2.1 Covalent Phosphate in Glycogen and Lafora Bodies

The presence of phosphate in glycogen was first described in 1980 (Fontana 1980). Initially considered an anomaly or contaminant, significant research on glycogen phosphorylation was not stimulated until critical findings regarding increased glycogen phosphorylation were discovered in the study of Lafora's Disease (LD) (Yokoi et al. 1968, Sakai et al. 1970, Worby et al. 2006, Gentry et al., 2007, Tagliabracci et al. 2007, 2008, Tagliabracci 2011, Turnbull et al. 2016, Gentry et al. 2018; see also Chapter 3 of this volume). LD is both a progressive myoclonic epilepsy and a glycogen storage disease that presents during a child's early teen years with seizures and neurodegeneration. The disease rapidly progresses in severity until the child dies, typically ten years after the onset of the first symptoms (Hicks 2011, Turnbull et al. 2016, Gentry et al. 2018). The histological hallmark of LD is Periodic Acid-Schiff (PAS) positive staining of cytoplasmic inclusions known as Lafora Bodies (LBs) (Lafora 1911, Turnbull et al. 2016, Gentry et al. 2018). Upon analysis, it was discovered that LBs are aberrant glycogen-like aggregates containing higher phosphate levels compared to wild type glycogen (Yokoi et al. 1968, Sakai et al. 1970, Tagliabracci et al. 2007, Tagliabracci 2011) (Fig. 7.1C). Subsequent studies of LBs revealed that phosphate groups can be attached to the C2, C3, or C6 position on a glucose molecule (Nitschke et al. 2013). Further study of normal glycogen revealed the same sites of phosphorylation, although not present at the high levels seen in LBs, and variation between the level of phosphorylation of glycogen from different tissues (Depaoli-Roach et al. 2015, Nitschke et al. 2013, 2017) (Fig. 7.1A).

A series of seemingly unrelated events provided an intriguing connection between animal and plant researchers that continues to result in synergistic scientific advances. In the 1960s, Dr. James Austin, a researcher who focused on glycogen storage diseases, noticed similarities between the structure of plant starch and LBs from LD patients. He presented his work at the 1967 American Association of Neuropathologists meeting, and a colleague concluded, "I think that Doctor Austin has demonstrated that a devious and atavistic quirk leads to the accumulation of vegetable matter in the brain, causing the human body to lapse into a vegetative state. Does Dr. Austin have any information regarding the state of enzymes concerned with the anabolism of the glucose polymers?" (Yokoi 1967). Dr. Austin replied, "In answer to Dr. Zeman, it [the Lafora body] does resemble a vegetable starch. It is conceivable that this is a catabolic enzyme defect as with some other glycogenoses" (Yokoi 1967). Years later, this account inspired researchers to explore the starch field to inform their studies on LBs. In doing so, they realized what scientists in the plant world had known for several decades—covalent modification by phosphorylation can play an important role in regulating the properties of carbohydrate storage molecules.

7.3 Starch

The first report on the nature of starch came when Leeuwenhoek observed it via microscopy as discrete granules in 1716—more than a century before the discovery of glycogen (Fig. 7.1D). The phosphorylation of starch was first proposed in 1897 (Coehn 1897), and just as with the study of

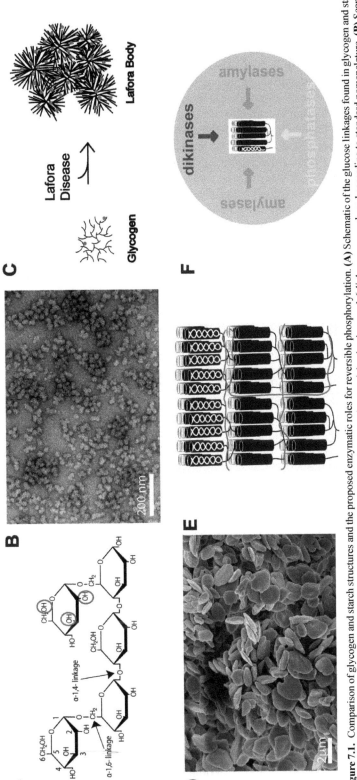

Figure 7.1. Comparison of glycogen and starch structures and the proposed enzymatic roles for reversible phosphorylation. (**A**) Schematic of the glucose linkages found in glycogen and starch. Grey circles depict hydroxyls that can be phosphorylated. The carbons of the glucose moiety participating in an α-1,6-linkage are numbered according to standard nomenclature. (**B**) Scanning electron micrograph of glycogen particles isolated from rabbit muscle (provided by Jean-Luc Putaux). (**C**) Glycogen is a water-soluble branched glucose polymer with limited covalently attached phosphate (grey balls). Mutations in *EPM2A* or *EPM2B* result in Lafora disease (LD) and lead to glycogen that is hyperphosphorylated and has disrupted glucose chain branching. This abnormal, less-soluble glycogen then aggregates to form the Lafora bodies that cause LD. (**D**) Illustration of amylopectin (black) and amylose (grey) and how the structures combine to form a layered assembly within starch granules. (**E**) Scanning electron micrograph of *Arabidopsis* starch granules. (**F**) Schematic overview of the enzymes involved in starch degradation and the cyclic action of reversible starch phosphorylation.

glycogen, starch phosphate was initially considered to be a metabolic mistake or contaminant. Further research, however, suggested that phosphate addition was functional and regulated (reviewed in Blennow 2015). It was observed that phosphate groups were attached via monoester bonds at the C6 and C3 positions in the amylopectin fraction of starch (Hizukuri et al. 1970, Tabata and Hizukuri 1971, Bay-Smidt et al. 1994, Blennow et al. 1998).

Starch is primarily composed of two distinct glucose polymers: amylopectin and amylose, which form semi-crystalline, water insoluble starch granules with a lamellar structure (Fig. 7.1E). Highly organized amylopectin is the major component of starch and constitutes 65–85% of the starch granule, depending on the plant tissue and species (Hoover et al. 2001). Amylopectin is composed of α-1,4-glucans with α-1,6-linked branches, similar to glycogen (Fig. 7.1A). Unlike glycogen, the amylopectin in starch contains clustered branch points and longer glucose chain lengths, with α-1,6-branches on average every 20–25 glucose units, allowing double helix formation that is necessary for crystalline lamellae (Gallant et al. 1997, Tester et al. 2004, Tang et al. 2006, Seetharaman and Bertoft 2012, Streb and Zeeman 2012). The glucan chains of the crystalline lamellae adopt left-handed double helices stabilized by hydrogen bonding with six glucose units per turn, resulting in exclusion of water (Sarko and Wu 1978, Sinnot 2007). Amylose is primarily a linear chain of α-1,4-glucan and is predominantly found in the amorphous regions of the granule (Buléon et al. 1998, Hancock and Tarbet 2000, Ball et al. 2011, Bertoft 2015). High-amylose starches are produced by plants that have a reduced level of the starch branching enzyme activity and actually possess a strongly modified amylopectin (see Chapter 10 of this volume).

Starch serves as the energy cache in plants in the form of both transitory (assimilatory) and storage (reserve) starch (Zeeman et al. 2010, Cenci et al. 2014, Pfister and Zeeman 2016, see also Chapters 4, 8, 9, and 10 of this volume). Transitory starch is produced in photosynthesis-competent chloroplasts, while reserve starch is synthesized in non-green plastids and the process is unrelated to photosynthesis. Transitory starch is the major storage site for fixed carbon that is produced by photosynthesis during the day. At the beginning of the dark period, transitory starch metabolism switches from synthesis to degradation. While ideal for energy storage density, the insoluble crystalline structure of starch presents a barrier to degradation via amylases, the enzymes responsible for starch degradation. Therefore, for efficient degradation, solubilization of the granule surface is needed to make it accessible to amylases. Genetic screens using *Arabidopsis* discovered specific mutant lines that yielded a Starch EXcess (SEX) phenotype of transitory starch, where starch granule degradation during the night was dramatically reduced (Caspar 1991, Streb and Zeeman 2012). In these starch excess plants, the phosphate level was either significantly higher or significantly lower from the wildtype starch phosphorylation levels, suggesting that both adding phosphate to and removing phosphate from starch played an important role in starch degradation (Ritte et al. 2002, 2005, Baunsgaard et al. 2005, Kotting et al. 2005, Kerk et al. 2006, Niittyla et al. 2006, Comparot-Moss et al. 2010, Santelia et al. 2011). As discussed below, it is currently thought that reversible starch phosphorylation plays an important role in the control of starch synthesis and degradation. In degradation, this process involves the cyclic action of starch dikinases that phosphorylate starch, amylases that degrade the starch, and phosphatases that dephosphorylate starch (Fig. 7.1F).

7.4 Discovery and Enzymology of the Starch Phosphatases

After the discovery of phosphate in starch, researchers sought to identify enzymes responsible for regulating starch phosphorylation. A genetic screen described an *Arabidopsis* mutant line that has higher starch content than wild-type plants throughout the diurnal cycle (Zeeman and Rees 1999). They named the At3g52180 gene product Starch EXcess4 (SEX4). It was subsequently demonstrated that SEX4 possesses a dual specificity phosphatase domain near its N-terminus (Fordham-Skelton et al. 2002). This finding was followed by the identification of a carbohydrate binding module (CBM48, as defined by CAZy) (Lombard et al. 2014)) near the C-terminus by multiple groups, along with a chloroplast targeting peptide at the N-terminus and a carboxy-terminal motif (CT) that is required to stabilize the protein and promote enzymatic activity (Kerk et al. 2006, Nittyla et al. 2006, Vander Kooi et al. 2010; Fig. 7.2A). Subsequent work has demonstrated that SEX4 utilizes an integrated DSP-CBM glucan-binding platform to bind, position, and dephosphorylate starch (Meekins et al. 2014, 2016, Gentry et al. 2016, Emmanuelle et al. 2016) (Fig. 7.2A). These unique structural features allow SEX4 to dephosphorylate both the C6- and C3-positions with a preference for the C6-position (Fig. 7.2B).

Bioinformatic searches revealed two additional putative phosphatases in plants: like-SEX4-one (LSF1) and like-SEX4-two (LSF2) (Comparot-Moss et al. 2010, Santelia et al. 2011). LSF1 contains a similar domain structure to SEX4, with an additional PSD95-Dlg1-Zo1 (PDZ) protein interaction domain between the N-terminal chloroplast targeting peptide and the DSP (Comparot-Moss et al. 2010). LSF2, however, lacks a CBM and so its function was unclear (Fig. 7.2C). LSF2 was later shown

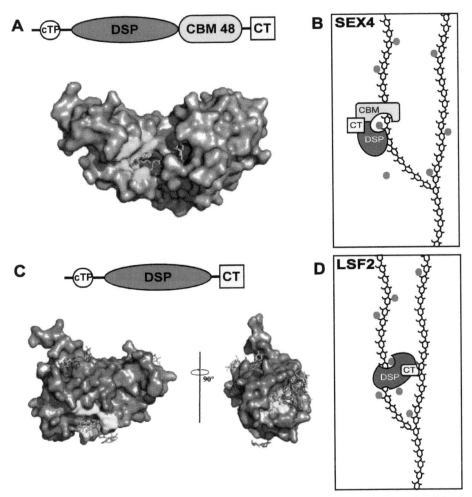

Figure 7.2. Glucan phosphatases that participate in reversible starch phosphorylation. **(A)** A schematic of the domains present in SEX4, along with a surface diagram of the SEX4 x-ray crystal structure bound to phosphate and maltoheptaose (PDB: 4PYH). The carbohydrate binding site in the carbohydrate binding module (CBM) is colored cyan, the active site in the dual specificity phosphatase (DSP) domain is in red, and the glucan is in gold. The C-terminal (CT) motif is shown in tan. The chloroplast targeting peptide (cTP) was removed for crystallization experiments. **(B)** An illustration depicting putative binding interactions of SEX4 with amylopectin. **(C)** A schematic of the domains present in LSF2, along with a surface diagram of the LSF2 x-ray crystal structure bound to phosphate and maltohexaose (PDB: 4KYR). The LSF2 carbohydrate surface binding sites (SBS) are in green, the active site of the phosphatase domain is in red, and the glucan is in gold. The C-terminal (CT) motif is shown in tan. The chloroplast targeting peptide (cTP) was removed for crystallization experiments. **(D)** An illustration depicting possible binding interactions of LSF2 with amylopectin.

to act as a starch phosphatase, dephosphorylating the C3 position (Santelia et al. 2011). Structural studies of LSF2 revealed that it utilizes Surface Binding Sites (SBS) to interact with starch in place of a CBM (Meekins et al. 2013, 2016) (Fig. 7.2C). SBS are formed by non-contiguous aromatic amino acids that interact with carbohydrates. These surface binding sites allow LSF2 to interact with glucan chains in amylopectin, facilitating C3 phosphate removal (Fig. 7.2D).

Interestingly, when *lsf2* knock-out plants were generated a starch excess phenotype was not observed suggesting compensation, presumably by SEX4. Indeed, double knock-out *sex4/lsf2* plants display a starch excess phenotype more severe than *sex4* knock-out alone (Santelia et al. 2011). The current data suggest that SEX4 and LSF2 are primarily responsible for removing phosphate groups from starch.

7.4.1 The Role of SEX4 and LSF2 in Reversible Starch Phosphorylation

Upon identifying these mechanisms, researchers then sought to uncover why reversible phosphorylation of starch was critical for the plant life cycle. Starch is an insoluble storage carbohydrate molecule that requires an initiation step whereby the starch surface undergoes significant structural changes for degradation to begin. The starch excess phenotypes of *sex4* and *lsf2* mutant plants in *Arabidopsis* and other plants led researchers to hypothesize that these two phosphatases are critical in the starch degradation pathway (Kotting et al. 2009, Nittyla et al. 2006, Santelia et al. 2011). Indeed, research suggests that phosphorylation at the surface of the starch granule increases solvent accessibility and allows for effective energy mobilization from starch (Smith 2012). The action of phosphatases is thought to be of particular importance for processive starch degradation by exo-acting hydrolases such as the β-amylases (Blennow and Engelsen 2010, Takeda and Hizikuri 1981, Silver et al. 2014).

7.5 Enzymes Involved in Glycogen Dephosphorylation

The molecular insights from plant phosphatases have directly contributed to understanding of glycogen dephosphorylation and the current exciting new research on LD. These novel molecular insights commenced with the identification of recessive mutations in two genes that cause LD. The first gene identified was *Epilepsy Progressive Myoclonus 2A* (*EPM2A*) that encodes the phosphatase laforin (Minassian et al. 1998, Serratosa et al. 1999). The second gene, *EPM2B*, was identified in 2003 and was later shown to encode for an E3 ubiquitin ligase named malin (Chan et al. 2003, Gentry et al. 2005). Research into the enzymology of these proteins sparked new interest in the role of glycogen phosphorylation.

7.5.1 Discovery and Enzymology of a Glycogen Phosphatase

Laforin is a bi-modular protein that contains a carbohydrate binding module (CBM20) at its N-terminus and a highly conserved dual specificity phosphatase domain (DSP) at its C-terminus (Ganesh et al. 2000, Wang et al. 2002) (Fig. 7.3A). Laforin binds glycogen and co-localizes with it in cell culture via the CBM, and was subsequently demonstrated to dephosphorylate glycogen at C2, C3, and C6 positions (Wang et al. 2002, Worby et al. 2006, Gentry et al. 2007, Tagliabracci 2011, Nitschke 2013, Roach 2015). To date, no other mammalian phosphatase has been identified that contains a CBM, suggesting that laforin could play a unique role in glycogen dephosphorylation (Gentry et al. 2016, 2009). Laforin exists as an antiparallel homodimer (CBM-DSP-DSP-CBM), forming a coordinated tetramodular architecture (Raththagala et al. 2015) (Fig. 7.3A). There is a large shared interface between the CBM and DSP domains. Mutations that disrupt this interface lead to LD pathology, suggesting that the interaction between the two domains is critical for laforin functionality (Raththagala et al. 2015, Gentry et al. 2016). Laforin was interrogated by a number of biophysical techniques to define its tetramodular architecture and gain insights into how the laforin dimer binds glucans. Laforin

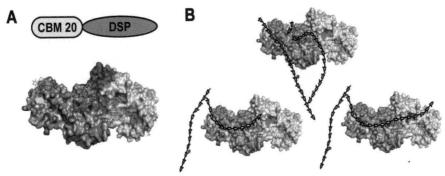

Figure 7.3. The glucan phosphatase laforin. **(A)** A schematic of the domains present in laforin, along with a surface diagram of the laforin x-ray crystal structure bound to phosphate and maltohexaose (PDB: 4RKK). Laforin is a dimer both in solution and in the x-ray structure. Each laforin monomer is represented as a different shade of grey. The residues that bind carbohydrates in the carbohydrate binding module (CBM) are in cyan and the dual specificity phosphatase (DSP) active site is in red. **(B)** An illustration depicting possible binding interactions between laforin and glycogen.

was preferentially stabilized when incubated with longer glucan chains compared to shorter ones, demonstrating a preference for binding to longer glucan chains (Raththagala et al. 2015). This result suggests that laforin could target glycogen with longer than normal chain lengths and dephosphorylate these chains. Because laforin functions as a dimer, it is possible that the two DSP domains function collaboratively to associate with the longer chains and target them for dephosphorylation (Fig. 7.3B). Laforin also possesses a scaffold role and coordinates binding with multiple other proteins (reviewed in Gentry et al. 2013). Therefore, it is also possible that once dimeric laforin binds to glycogen, one molecule could dephosphorylate glycogen while the second acts as a scaffold for other proteins needed for glycogen regulation.

7.5.2 The Role of Laforin in Glycogen Architecture

The critical function of laforin has initiated investigations that were not only related to LD, but also aimed to identify the role of phosphorylation in glycogen metabolism and its regulation. Several studies have demonstrated that physiological glycogen phosphorylation correlates with changes in glycogen architecture (Nitschke et al. 2013, Irimia et al. 2015, Nitschke et al. 2017). Additionally, glycogen phosphate levels vary depending on tissue type (Irimia et al. 2015, Nitschke et al. 2017). There is a distinct difference in glycogen morphology between liver, muscle, and brain glycogen that may be related to the different degree of glycogen phosphorylation in these tissues (Simpson and Macleod 1927, Shearer and Graham 2002). At one end of the spectrum there is liver glycogen: phosphorylation levels are low and the branching is highly ordered, displaying uniform chain lengths of around 15 glucose moieties. The glucose cache stored in liver glycogen structure is easily and rapidly accessed during times of metabolic need (Chasiotis et al. 1982, Tagliabracci et al. 2007, Nitschke et al. 2013). In muscle tissues, glycogen phosphorylation levels are higher, the molecules are more highly branched, and the glucose cache is not as rapidly accessible (Wang et al. 2002, Turnbull 2010).

In LBs, the architecture of the glucan is significantly altered. A correlation was defined between the increase in phosphate content and aberrant branching in LBs versus normal glycogen: as phosphate content increases, so does the linear glucose chain length (Nitschke et al. 2013). Branch points, which normally occur every 10–12 moieties, instead occur every 15–20+ moieties (Tagliabracci et al. 2008, Nitschke et al. 2013). The longer chains of glucosyl residues alter the architecture of the macromolecule by an unknown mechanism. As a consequence, the altered architecture may result in LBs that are 2–10 µm in diameter compared to the average 20–40 nm diameter of normal brain glycogen (Gentry et al. 2018). A breakthrough in the LD field was the discovery from multiple labs that LBs are the pathogenic disease agent in LD (Turnbull et al. 2011, Pederson et al. 2013, Turnbull et al. 2014, Duran et al. 2014, Gentry et al. 2018).

LD mouse models have yielded additional insights into the link between phosphorylation and morphology (reviewed in Roach 2015, Gentry et al. 2013, 2018). In laforin knock-out mice, the phosphate content in muscle glycogen increases 4-fold, and the phosphate content in liver glycogen increases by 40% (Tagliabracci et al. 2007). In muscle tissue of laforin knock-out mice, the glycogen formed immediately after exercise displays aberrant branching and phosphorylation patterns, and is catabolized more slowly (Irimia et al. 2015). These data led to the hypothesis that glycogen phosphorylation may indicate the age of the glycogen molecule (Irimia et al. 2015).

Despite the advances in understanding laforin structure and enzymology, there are significant outstanding questions regarding how laforin may regulate glycogen architecture. A current hypothesis asserts that phosphate is incorporated into glycogen at the C2- and C3-positions by glycogen synthase as an error during glycogen synthesis (Roach 2015). According to this model, appropriate levels of glycogen phosphate promotes glycogen solubility and proper branching (Roach 2015, Tagliabracci et al. 2007, 2011, Nitschke et al. 2013). Without functional laforin, excess phosphate groups are not removed from glycogen and branching decreases, leading to longer glucose chains and aggregated LBs (Tagliabracci et al. 2011, Roach 2015). There are multiple unknowns in this area of research and it is an active area of research in multiple labs. Strikingly, overexpression of a catalytic dead version of laforin in the laforin knockout mouse rescued the LB phenotype and glycogen structural studies suggest that chain length and not phosphorylation is the most critical component (Gayarre et al. 2014, Nitschke et al. 2017, Sullivan et al. 2017, 2019). These data suggest that the phosphatase activity of laforin is not necessary in preventing LBs. However, structural and biophysical data suggest that catalytically dead laforin possesses unique features that could promote clearance of LBs (Raththagala et al. 2015). These incongruent data demonstrate the exciting nature of fundamental questions that remain to be answered and are currently being explored.

The varying levels of glycogen phosphorylation in different tissues and metabolic states, and the necessity of laforin to remove phosphate groups suggests that a mechanism beyond a metabolic error is responsible for glycogen phosphorylation (Chikwana et al. 2013, Gentry et al. 2018). However, no glycogen kinase or dikinase has been identified to date and the above studies suggest glycogen synthase incorporates only C2 and C3 phosphate into the glycogen molecule. All work to date acknowledges that no obvious mechanism can account for C6 phosphorylation (Contreras et al. 2016, Depaoli-Roach et al. 2015). Recently, researchers have again turned to the world of plants to catalyze our thinking about glucan phosphorylation.

7.6 Glucan Phosphorylation by Dikinases

While phosphorylation of glycogen remains an active topic of research, significant advances have been made in our understanding of starch phosphorylation. The concentration of phosphate in starch varies depending on plant origin. For example, 0.1% of the glucose residues are phosphorylated in *Arabidopsis* leaf starch compared to 1% in potato tuber starch and 0.03% in cereal starch (Blennow et al. 2000, Carciofi et al. 2011). The majority of starch phosphate monoester bonds are found in the amylopectin portion of the starch granule (Hizukuri et al. 1970, Takeda and Hizikuri 1981). In principle, each carbon atom of a glucose residue connected to a hydroxyl group could be modified with a phosphate group. However, starch phosphorylation occurs specifically at the C3- and C6-positions, with C6-phosphorylation accounting for approximately 70% of the total phosphorylation (Hizukuri et al. 1970, Tabata and Hizukuri 1971).

7.6.1 Discovery and Enzymology of Glucan Dikinases

The first gene encoding an α-glucan, water dikinase (GWD) was originally described in the 1990s *Arabidopsis SEX* screens (Caspar et al. 1991). In 1998, the gene of a starch binding protein from potato was cloned and named R1 (Lorberth et al. 1998). Reduced levels of R1 resulted in decreased starch phosphorylation (Lorberth et al. 1998). It was subsequently demonstrated that the *sex1* gene in *Arabidopsis* encodes the R1 protein, which is now known as GWD (Yu et al. 2001). GWD orthologs have been identified in species ranging from land plants to single-cell green algae (Ball et al. 2011, 2015).

GWD is a large multi-domain protein that consists of a chloroplast targeting peptide (cTP), tandem CBM45 carbohydrate binding domains, a catalytic histidine domain (HIS-domain), and a nucleotide binding domain (NBD) (Fig. 7.4A). The HIS-domain coupled with an NBD is a highly conserved domain structure and has similarities to pyruvate, phosphate dikinase, which phosphorylates pyruvate and is a key enzyme in photosynthetic carbon fixation for C4 plants (Edwards et al. 1985, Herzberg et al. 1996, Nakanishi et al. 2005, Jiang et al. 2016, Minges et al. 2017). Mechanistically, GWD phosphorylates glucose residues in starch by transferring the β-phosphate of ATP first to the HIS-domain, forming a phospho-enzyme intermediate, and then transferring the β-phosphate exclusively to the C6-position of a glucose residue on the amylopectin chain (Ritte et al. 2002, Baunsgaard et al. 2005). The γ-phosphate from ATP is transferred to water (Ritte et al. 2002, 2006, Hejazi et al. 2012, Skeffington et al. 2014 Mahlow et al. 2014, Hejazi et al. 2009, Mikkelsen et al. 2004) (Fig. 7.4A). Functionally, GWD has a high-affinity for crystalline malto-oligosaccharides (Ritte et al. 2002). Glucan phosphorylation disrupts the crystalline structure of amylopectin and increases the oligosaccharide chain solvent exposure at the starch surface to allow for degradation by amylases. Consistent with this proposed mechanism, deficient levels of GWD in *Arabidopsis* (the *sex1* mutant) lead to a severe starch excess phenotype with decreased levels of starch phosphate (Lorberth et al. 1998, Yu et al. 2001). In *Arabidopsis* a second GWD isoform, GWD2, has been described but it is located in the cytosol with no access to chloroplast starch and its function remains to be determined (Glaring et al. 2007, Pirone et al. 2017).

Subsequently, another starch dikinase was identified in *Arabidopsis* and named phosphoglucan, water dikinase (PWD or GWD3) (Baunsgaard et al. 2005, Kötting et al. 2005, Ritte et al. 2006). PWD has a similar architecture to GWD (Fig. 7.4B). The most significant structural difference between GWD and PWD is that PWD has a single CBM20 domain at its N-terminus (Mikkelsen et al. 2006, Glaring et al. 2011). PWD requires a C6-phosphorylated glucan substrate and it exclusively phosphorylates the C3-position of glucose residues in amylopectin chains (Baunsgaard et al. 2005, Kötting et al. 2005, Ritte et al. 2006). These findings were further supported when PWD mutants showed decreased levels of C3-phosphate and displayed a mild starch excess phenotype (Baunsgaard et al. 2005, Kötting et al. 2005).

7.6.2 The Role of Glucan Dikinases in Starch Phosphorylation

Based on their distinct enzymology, it is now thought that the two dikinases act sequentially in starch degradation with GWD first phosphorylating the C6-position followed by PWD localizing to the phosphorylated region and phosphorylating the C3-position (Fig. 7.4C). Once the granule surface is phosphorylated, amylases can hydrolyze the solvent exposed glucan chains. The activity of β-amylases (BAM1) and isoamylases (ISA3) is significantly increased by phosphorylation at the granule surface by GWD (Edner et al. 2007). The β-amylases hydrolyze the -1,4-linkages, but require isoamylases to remove the -1,6-branchpoints. The current degradation model hypothesizes that amylase activity is hindered once they reach a phosphate monoester linkage, therefore requiring the glucan phosphatases SEX4 and LSF2 to remove the phosphate groups and continue cyclic degradation at the granule surface (Figs. 7.4C and 7.1F). Once a layer is degraded, then GWD initiates C6-phosphorylation of the next layer and the cycle repeats until degradation is complete or halted.

While there is clearly an important role for glucan dikinases in starch degradation, recent studies suggest that starch degradation may not be the only role for starch phosphorylation. While GWD reduction results in a starch excess phenotype, overexpression of GWD in barley and *Arabidopsis* did not accelerate starch degradation (Carciofi et al. 2011, Skeffington et al. 2014). Starch Synthase 1 transfers more glucosyl residues to starch that have been pre-phosphorylated by GWD *in vitro*

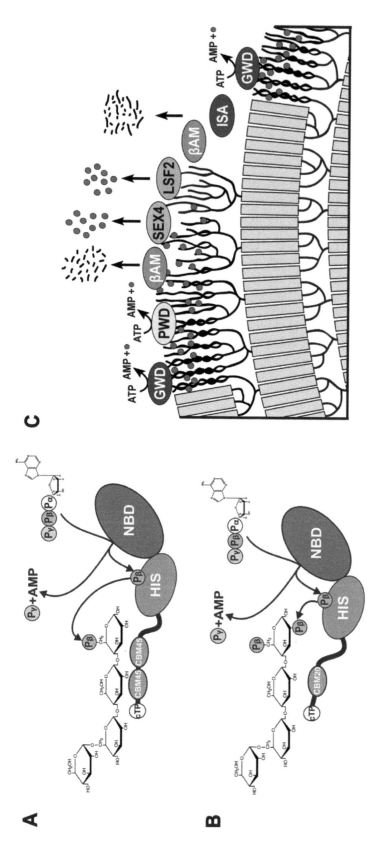

Figure 7.4. Glucan dikinases that participate in reversible starch phosphorylation. (**A**) A schematic representing the mechanism of GWD and the mechanism whereby GWD transfers the β-phosphate of ATP to a C6-glucose hydroxyl of glucose. (**B**) A schematic of PWD and the mechanism whereby PWD transfers the β-phosphate of ATP to a C3-glucose hydroxyl that has been pre-phosphorylated at the C6-position. (**C**) An illustration depicting transient starch degradation. Grey boxes represent glucose helices. The glucan dikinases GWD and PWD phosphorylate (grey circles) the outer glucans to facilitate glucan solvent accessibility. β-amylases (BAM) and isoamylase (ISA) release glucose and maltose. SEX4 and LSF2 dephosphorylate surface glucans so the process can continue and repeat on the next layer (Emanuelle et al. 2016). Copyright © 2016 Cellular and Molecular Life Sciences. Used with permission.

(Delvallé et al. 2005, Mahlow et al. 2014). These results led several groups to suggest that while the levels of starch phosphorylation are dramatically higher during the degradation process, there may be a role for phosphorylation during starch synthesis (Blennow and Engelsen 2010, Zeeman et al. 2010, Skeffington et al. 2014). It is possible that phosphorylation during synthesis could govern granule structure or contribute to signaling events for starch turnover (Mahlow et al. 2014, Blennow et al. 2015). The suppression of GWD in potato tubers results in starch granules with modified structure (i.e., cracks and fissures), lower percentage of amylopectin, and a higher amylose content (Viksø-Nielsen et al. 2001, Glaring et al. 2006), further supporting the assertion that phosphorylation may play an important role in synthesis.

7.6.3 Starch Morphology and Phosphate Levels

While there are clear links between reversible glucan phosphorylation and development of a starch excess phenotype, there are intriguing differences in starch morphology that indicate a more complex mechanism regulates starch turnover. Indeed, the *sex1-8 Arabidopsis* GWD T-DNA line has five-fold higher leaf starch content (Yu et al. 2001, Kötting et al. 2005, Mahlow et al. 2014). Biophysical analysis of the granules revealed that the granule surface of mature *sex1-8* mutant starch contains a greater abundance of short soluble glucan chains compared to WT. When wild type GWD was incubated with mutant and wild-type starch, GWD activity was 30% lower when incubated with the mutant starch (Mahlow et al. 2014). Transmission electron microscopy revealed that mature leaves of the *sex1-8* mutants contain more starch granules per chloroplast than WT plants, and these starch granules have a greater size distribution and are irregularly shaped compared to disc-shaped WT granules (Mahlow et al. 2014). Thus, the lack of GWD results in alterations of the amount, granule number, size variability, and surface properties of starch (Mahlow et al. 2014). A similar difference in starch granule composition of *sex4* mutant starch has also been demonstrated, with mutant starch granules having a higher content of phosphorylated linear chains (Kötting et al. 2009). These results support the hypothesis that the phosphatases act on the starch granule after GWD and PWD disassociate the glucan helices to generate linear phosphorylated chains.

Although many of these observations were made using the *Arabidopsis* or potato plant models, similar results were obtained in different starch-producing organisms. GWD was overexpressed in the barley endosperm, leading to a 10-fold increase of phosphate content and a more "potato-like" starch structure with granules exhibiting a more porous surface and irregular shape (Carciofi et al. 2011). Knockdown of GWD expression in the wheat endosperm led to a corresponding decrease in starch phosphate content, although it did not lead to an elevation in plant starch content (Ral et al. 2012). Additionally, a transgenic cassava plant with a redox-insensitive form of potato GWD produced starch granules with elevated phosphate levels. While the change in phosphate levels did not appear to impact functionality of the starch granule, it did increase the swelling power of the granule and the paste clarity of the starch (Wang et al. 2018). From these studies, it is clear that starch phosphorylation plays an important role in the formation of correct granule structure and impacts the physical properties of the granule.

7.7 Future Directions in the Study of Reversible Phosphorylation of Glycogen and Starch

Although originally thought to be a contaminant, it is now clear that covalent phosphate attachment to glycogen and starch occurs naturally and even at low levels plays an important role in governing the morphology and regulation of these polyglucans (reviewed in Blennow 2015). The enzymes responsible for regulating phosphate content in polyglucans have crucial roles in maintaining normal granule structure and facilitating proper metabolism.

Starch is globally important in agriculture and immediately significant in human and animal dietary needs. Phosphorylation of starch directly affects granule hydration, crystallinity, freeze-thaw stability, viscosity and transparency, properties that are central to many commercial applications (Muhrbeck and Eliasson 1991, Blennow et al. 2001, Blennow 2015). Studying the impact of overexpression and reduction of GWD activity on starch biophysical properties has already highlighted a variety of potential applications. Downregulating the activity of GWD in transgenic ryegrass (*Lolium perenne*), alfalfa (*Medicago sativa*) and clover (*Trifolium repens*) (Zeeman et al. 2010) produces an easily digestible carbohydrate in fodder crops; this is also true for maize plants used for the production of silage (Frohberg and Bauerlein 2006). Controlling starch content in crops is important for inhibiting the generation of reducing sugars, glucose and fructose upon storage (cold-sweetening) which leads to product quality deterioration and spoiling. Downregulating GWD in potato tubers reduces cold-sweetening (Lorberth et al. 1998) and limits acrylamide production during frying, preventing undesirable dark coloring (Rommens et al. 2007).

By contrast, increasing degradation of starch has a potential value for crops that are grown for bioethanol production and other industrial applications like adhesive manufacturing (Lanahan et al. 2006). As researchers uncover how to manipulate the starch synthesis and degradation process, it may be possible to engineer plants as more efficient carbon sinks and increase their value as starting material for bioethanol production (Weise et al. 2011). The ability to manipulate phosphate content in starch impacts the clarity of the starch granule, which is an important attribute to control in the manufacturing of thickening agents (Wang et al. 2018). A better understanding of the mechanism for reversible phosphorylation of starch and novel methods for phosphate modulation will give scientists the ability to design starches with ideal properties for their nutritional or industrial applications.

Aberrant glycogen shares key characteristics with starch. Therefore, the ability to control starch degradation may open new therapeutic avenues for the treatment of LD and other glycogen storage diseases. As we gain further insight into the mechanisms of phosphorylation in starch synthesis and degradation, that knowledge can shed light on the mysteries surrounding glycogen architecture and regulation of glycogen phosphorylation. Research thus far suggests a link between phosphate content, chain lengths, and branching patterns, and this correlation could explain the formation of aberrant glycogen in LD patients. The change in glycogen structure and the subsequent impact on enzyme accessibility may prove to be an important mechanism for metabolism regulation with broad implications for both physiology and disease.

Although first identified in humans, laforin was subsequently shown to be present in all vertebrates. Surprisingly, laforin-like phosphatases were also identified in two invertebrates, members of the nematode phylum, and several parasitic protists suggesting novel mechanisms with relevance to human health (Gentry et al. 2007, Gentry and Pace 2009, Kuchtova et al. 2018). Bioinformatics searches show that these laforin-like phosphatases share less than 35% sequence identity with human laforin. However, the presence of a laforin-like phosphatase in parasites opens the intriguing possibility that reversible glucan phosphorylation could be integral to the parasites' lifecycle. Given the divergence between human laforin and the parasitic laforin-like enzymes, one could target these enzymes as a novel therapeutic alternative.

Cumulatively, recent findings from diverse kingdoms of life suggest a broad role for reversible phosphorylation. The diverse biological roles and enzymatic actions of the glucan dikinases and glucan phosphatases highlight the current exciting opportunities for understanding, ameliorating, and utilizing glucan phosphorylation in multiple biotechnology and medical applications. It is clear that as researchers continue to elucidate the mechanism of reversible phosphorylation, cross-fertilization between researches studying plant and animal models, and perhaps those studying parasites, will continue to be critical.

Acknowledgements

This work was supported by grant funding from NIH R01 NS070899, P01 NS097197, and NSF 1817417 to M.S.G. and NSF 1808304 to C.W.V.K. K.J.D was supported by NSF 1355438. The authors would also like to thank Dr. Jean-Luc Putaux (French National Centre for Scientific Research) for the use of his electron microscopy image of muscle glycogen and members of the Gentry and Vander Kooi labs for helpful discussions.

References

Ball, S., C. Colleoni, U. Cenci, J.N. Raj and C. Tritaux. 2011. The evolution of glycogen and starch metabolism in eukaryotes gives molecular clues to understand the establishment of plastid endosymbiosis. J. Exp. Bot. 62(6): 1775–1801.

Ball, S., C. Colleoni and M.C. Arias. 2015. Starch: Metabolism and Structure. pp. 93–158. Springer Japan, Tokyo.

Baunsgaard, L., H. Lutken, R. Mikkelsen, M.A. Glaring, T.T. Pham and A. Blennow. 2005. A novel isoform of glucan, water dikinase phosphorylates pre-phosphorylated α-glucans and is involved in starch degradation in Arabidopsis. Plant. J. 41(4): 595–605.

Bay-Smidt, A.M., B. Wischmann, C.E. Olsen and T.H. Nielsen. 1994. Starch bound phosphate in potato as studied by a simple method for determination of organic phosphate and (31)P-NMR. Starch/Staerke 46(5): 167–172.

Bernard, C. 1857. Sur le mécanisme physiologique de la formation du sucre dans le foie. Paris: C R Acad Sci. 44: 578–586.

Bertoft, E. 2015. Fine structure of amylopectin. pp. 3–40. *In*: Nakamura, Y. (ed.). Starch: Metabolism and Structure. Springer Japan, Tokyo.

Blennow, A., M.A. Bay-Schmidt, B. Wischmann, C.E. Olsen and B.L. Møller. 1998. The degree of starch phosphorylation is related to chain length distribution of neutral and phosphorylated chains of amylopectin. Carbohydr. Res. 307: 45–54.

Blennow, A., A.M. Bay-Smit, C.E. Olsen and B.L. Moller. 2000. The distribution of covalently bound phosphate in the starch granule in relation to starch crystallinity. Int. J. Bio. Macromol. 27(3): 211–218.

Blennow, A., A.M. Bay-Smidt and R. Bauer. 2001. Amylopectin aggregation as a function of starch phosphate content studied by size exclusion chromatography and on-line refractive index and light scattering. Int. J. Bio. Macromol. 28(5): 409–420.

Blennow, A., T.H. Nielsen, L. Baunsgaard, R. Mikkelsen and S.B. Engelsen. 2002. Starch phosphorylation: a new front line in starch research. Trends Plant Sci. 7(10): 445–450.

Blennow, A. and S.B. Engelsen. 2010. Helix-breaking news: fighting crystalline starch energy deposits in the cell. Trends in Plant Sci. 15(4): 236–240.

Blennow, A. and B. Svensson. 2010. Dynamics of starch granule biogenesis—the role of redox-regulated enzymes and low-affinity carbohydrate-binding modules. Biocatal. Biotransform. 28(1): 3–9.

Blennow, A. 2015. Phosphorylation of the Starch Granule. pp. 399–424. *In*: Nakamura, Y. (ed.). Starch: Metabolism and Structure. Springer Japan, Tokyo.

Buléon, A., P. Colonna, V. Planchot and S. Ball. 1998. Starch granules: structure and biosynthesis. Int. J. Bio. Macromol. 23(2): 85–112.

Calder, P.C. 1991. Glycogen structure and biogenesis. Int. J. Biochem. 23(2): 1335–1352.

Carciofi, M., S.S. Shaik, S.L. Jensen, A. Blennow, J.T. Svensson, E. Vincze and K.H. Hebelstrup. 2011 Hyperphosphorylation of cereal starch. J. Cereal Sci. 54(3): 339–346.

Caspar, T., T.P. Lin, G. Kakefuda, L. Benbow, J. Preiss and C. Somerville. 1991. Mutants of Arabidopsis with Altered Regulation of Starch Degradation. Plant. Physiol. 95(4): 1181.

Cenci, B., F. Nitschke, M. Steup, B.A. Minassian, C. Colleoni and S.G. Ball. 2014 Transition from glycogen to starch metabolism in Archaeplastida. Trends Plant Sci. 19(1): 18–28.

Chan, E.M., D.E. Bulman, A.D. Paterson, J. Turnbull, E. Andermann, F. Andermann et al. 2003. Genetic mapping of a new Lafora progressive myoclonus epilepsy locus (EPM2B) on 6p22. J. Med. Genet. 40(9): 671–675.

Chasiotis, D., K. Sahlin and E. Hultman. 1982. Regulation of glycogenolysis in human muscle at rest and during exercise. J. Appl. Physiol. Respir. Environ. Exerc. Physiol. 53(3): 708–715.

Chikwana, V.M., M. Khanna, S. Baskaran, V.S. Tagliabracci, C.J. Contreras, A. DePaoli-Roach et al. 2013. Structural basis for 2'-phosphate incorporation into glycogen by glycogen synthase. Proc. Natl. Acad. Sci. USA. 110(52): 20976–20981.

Coehn, A. 1897. Über elektrische Wanderung von Kolloiden. Z. Electrochem. 4: 63–67.

Comparot-Moss, S., O. Kötting, M. Stettler, C. Edner, A. Graf, S.E. Weise et al. 2010. A putative phosphatase, LSF1, is required for normal starch turnover in Arabidopsis leaves. Plant Physiol. 152: 685–697.

Contreras, C.J., D.M. Segvich, K. Mahalingan, V.M. Chikwana, T.L. Kirley, T.D. Hurley et al. 2016. Incorporation of phosphate into glycogen by glycogen synthase. Arch. Biochem. BioPhys. 597: 21–29.

Delvallé, D., S. Dumez, F. Wattebled, I. Roldán, V. Planchot, P. Berbezy et al. 2005. Soluble starch synthase I: a major determinant for the synthesis of amylopectin in *Arabidopsis thaliana* leaves. Plant J. 43: 398–412.

DePaoli-Roach, A.A., C.J. Contreras, D.M. Segvich, C. Heiss, M. Ishihara, P. Azadi and P.J. Roach. 2015. Glycogen phosphomonoester distribution in mouse models of the progressive myoclonic epilepsy, Lafora disease. J. Biol. Chem. 290(2): 841–850.

Dukhande, V.V., D.M. Rogers, C. Roma-Mateo, J. Donderis, A. Marina, A.O. Taylor et al. 2011. Laforin, a dual specificity phosphatase involved in Lafora disease, is present mainly as monomeric form with full phosphatase activity. PLoS One. 6(8): e24040.

Duran, J., A. Gruart, M. Garcia-Rocha, J.M. Delgado-Garcia and J.J. Guinovart. 2014. Glycogen accumulation underlies neurodegeneration and autophagy impairment in Lafora disease. Hum. Mol. Genet. 23(12): 3147–3156.

Edner, C., J. Li, T. Albrecht, S. Mahlow, M. Hejazi, H. Hussain et al. 2007. Glucan, water dikinase activity stimulates breakdown of starch granules by plastidial β-amylases. Plant Phys. 145(1): 17–28.

Edwards, G.E., H. Nakamoto, J.N. Burnell and M.D. Hatch. 1985. Pyruvate, Pi Dikinase and NADP-Malate Dehydrogenase in C4 Photosynthesis: Properties and Mechanism of Light/Dark Regulation. Annu. Rev. Plant Physiol. 36(1): 255–286.

Evans, H.J. and H.G. Wood. 1968. The mechanism of the pyruvate, phosphate dikinase reaction. Proc. Natl. Acad. Sci. USA. 61(4): 1448–1453.

Fettke, J., M. Hejazi, J. Smirnova, E. Hochel, M. Stage and M. Steup. 2009. Eukaryotic starch degradation: integration of plastidial and cytosolic pathways. J. Exp. Biol. 60(10): 2907–2922.

Fontana, J.D. 1980. The presence of phosphate in glycogen. FEBS Lett. 109(1): 85–92.

Fordham-Skelton, A.P., P. Chilley, V. Lumbreras, S. Reignoux, T.R. Fenton, C.C. Dahm et al. 2002. A novel higher plant protein tyrosine phosphatase interacts with SNF1-related protein kinases via a KIS (kinase interaction sequence) domain. Plant J. 29(6): 705–715.

Frohberg, C. and M. Bauerlein. 2006. Method for generating maize plants with an increased leaf starch content, and their use for making maize silage: Publication number: 20060150278.

Gallant, D.J., B. Bouchet and P.M. Baldwin. 1997. Microscopy of starch: Evidence of a new level of granule organization. Carbohydr. Polym. 32: 177–191.

Ganesh, S., K.L. Agarwala, K. Ueda, T. Akagi, K. Shoda, T. Usui et al. 2000. Laforin, defective in the progressive myoclonus epilepsy of Lafora type, is a dual-specificity phosphatase associated with polyribosomes. Hum. Mol. Genet. 9(15): 2251–2261.

Ganesh, S., A.V. Delgado-Escueta, T. Sakamoto, M.R. Avila, J. Machado-Salas, Y. Hoshii et al. 2002. Targeted disruption of the Epm2a gene causes formation of Lafora inclusion bodies, neurodegeneration, ataxia, myoclonus epilepsy and impaired behavioral response in mice. Hum. Mol. Genet. 11(11): 1251–1262.

Gayarre, J., L. Duran-Trio, O. Criado Garcia, C. Aguado, L. Juana-Lopez, I. Crespo et al. 2014. The phosphatase activity of laforin is dispensable to rescue Epm2a-/- mice from Lafora disease. Brain. 137(3): 806–818.

Gentry, M.S., C.A. Worby and J.E. Dixon. 2005. Insights into Lafora disease: malin is an E3 ubiquitin ligase that ubiquitinates and promotes the degradation of laforin. Proc. Natl. Acad. Sci. USA. 102(24): 8501–8506.

Gentry, M.S., R.H. Dowen 3rd, C.A. Worby, S. Mattoo, J.R. Ecker and J.E. Dixon. 2007. The phosphatase laforin crosses evolutionary boundaries and links carbohydrate metabolism to neuronal disease. J. Cell Biol. 178(3): 477–488.

Gentry, M.S., J.E. Dixon and C.A. Worby. 2009. Lafora disease: insights into neurodegeneration from plant metabolism. Trends Biochem. Sci. 34(12): 628–639.

Gentry, M.S. and R.M. Pace. 2009. Conservation of the glucan phosphatase laforin is linked to rates of molecular evolution and the glucan metabolism of the organism. BMC Evol. Biol. 9: 138.

Gentry, M.S., C. Roma-Mateo and P. Sanz. 2013. Laforin, a protein with many faces: glucan phosphatase, adapter protein, et alii. FEBS J. 280(2): 525–537.

Gentry, M.S., M.K. Brewer and C.W. Vander Kooi. 2016. Structural biology of glucan phosphatases from humans to plants. Curr. Opin. Struct. Biol. 40: 62–69.

Gentry, M.S., J.J. Guinovart, B.A. Minassian, P.J. Roach and J.M. Serratosa. 2018. Lafora disease offers a unique window into neuronal glycogen metabolism. J. Biol. Chem. 293(19): 7117–7125.

Glaring, M.A., C.B. Koch and A. Blennow. 2006. Genotype-specific spatial distribution of starch molecules in the starch granule: a combined CLSM and SEM approach. Biomacromolecules 7(8): 2310–2320.

Glaring, M.A., A. Zygadlo, D. Thorneycroft, A. Schulz, S.M. Smith, A. Blennow et al. 2007. An extra-plastidial alpha-glucan, water dikinase from Arabidopsis phosphorylates amylopectin *in vitro* and is not necessary for transient starch degradation. J. Exp. Bot. 58(14): 3949–3960.

Glaring, M.A., M.J. Baumann, M.A. Hachem, H. Nakai, N. Nakai, D. Santelia et al. 2011. Starch-binding domains in the CBM45 family—low-affinity domains from glucan, water dikinase and α-amylase involved in plastidial starch metabolism. FEBS J. 278(7): 1175–1185.

Hancock, R.D. and B.J. Tarbet. 2000. The other double helix—the fascinating chemistry of starch. J. Chem. Educ. 77(8): 988.

Hejazi, M., J. Fettke, S. Haebel, C. Edner, O. Paris, C. Frohberg et al. 2008. Glucan, water dikinase phosphorylates crystalline maltodextrins and thereby initiates solubilization. Plant J. 55(2): 323–334.

Hejazi, M., J. Fettke, O. Paris and M. Steup. 2009. The two plastidial starch-related dikinases sequentially phosphorylate glucosyl residues at the surface of both the a- and b-type allomorphs of crystallized maltodextrins but the mode of action differs. Plant Phys. 150(2): 962.

Hejazi, M., M. Steup and J. Fettke. 2012. The plastidial glucan, water dikinase (GWD) catalyses multiple phosphotransfer reactions. FEBS J. 279(11): 1953–1966.

Herzberg, O., C.C. Chen, G. Kapadia, M. McGuire, L.J. Carroll, S.J. Noh et al. 1996. Swiveling-domain mechanism for enzymatic phosphotransfer between remote reaction sites. Proc. Natl. Acad. Sci. USA. 93(7): 2652–2657.

Hicks, J., E. Wartchow and G. Mierau. 2011. Glycogen storage diseases: a brief review and update on clinical features, genetic abnormalities, pathologic features, and treatment. Ultrastruct. Pathol. 35(5): 183–196.

Hizukuri, S., S. Tabata, Kagoshima, and Z. Nikuni. 1970. Studies on Starch Phosphate Part 1. Estimation of glucose-6-phosphate residues in starch and the presence of other bound phosphate(s). Starch. 22: 338–343.

Hoover, R. 2001. Composition, molecular structure, and physicochemical properties of tuber and root starches: a review. Carbohydr. Polym. 45(3): 253–267.

Irimia, J.M., V.S. Tagliabracci, C.M. Meyer, D.M. Segvich, A.A. DePaoli-Roach and P.J. Roach. 2015. Muscle glycogen remodeling and glycogen phosphate metabolism following exhaustive exercise of wild type and laforin knockout mice. J. Biol. Chem. 290(37): 22686–22698.

Jiang, L., Y.B. Chen, J. Zheng, Z. Chen, Y. Liu, Y. Tao et al. 2016. Structural basis of reversible phosphorylation by maize pyruvate orthophosphate dikinase regulatory protein. Plant Phys. 170(2): 732–741.

Kerk, D., T.R. Conley, F.A. Rodriguez, H.T. Tran, M. Nimick, D.G. Muench et al. 2006. A chloroplast-localized dual-specificity protein phosphatase in Arabidopsis contains a phylogenetically dispersed and ancient carbohydrate-binding domain, which binds the polysaccharide starch. Plant J. 46(3): 400–413.

Kotting, O., K. Pusch, A. Tiessen, P. Geigenberger, M. Steup and G. Ritte. 2005. Identification of a novel enzyme required for starch metabolism in Arabidopsis leaves. The phosphoglucan, water dikinase. Plant Physiol. 137(1): 242–252.

Kotting, O., D. Santelia, C. Edner, S. Eicke, T. Marthaler and M.S. Gentry et al. 2009. Starch-Excess4 is a laforin-like Phosphoglucan phosphatase required for starch degradation in Arabidopsis thaliana. Plant Cell. 21(1): 334–346.

Kuchtová, A., M.S. Gentry and Š. Janeček. 2018. The unique evolution of the carbohydrate-binding module CBM20 in laforin. FEBS Lett. 592: 586–598.

Lafora, G.R. and B. Glueck. 1911. Beitrag zur Histopathologie der myoklonischen Epilepsie. Zeitschrift für die gesamte Neurologie und Psychiatrie 6: 1–14.

Lanahan, M.B., S.S. Basu, C.J. Batie, W. Chen, J. Craig and M. Kinkema. 2006. Self-processing plants and plant parts: US patent number: 7102057B2.

Lomako, J., W.M. Lomako and W.J. Whelan. 1988. A self-glucosylating protein is the primer for rabbit muscle glycogen biosynthesis. FASEB J. 2(15): 3097–3103.

Lombard, V., H.G. Ramulu, E. Drula, P.M. Coutinho and B. Henrissat. 2014. The carbohydrate-active enzymes database (CAZy) in 2013. Nucleic Acids Res. 42(Database issue): D490–D495.

Lorberth, R., G. Ritte, L. Willmitzer and J. Kossmann. 1998. Inhibition of a starch-granule–bound protein leads to modified starch and repression of cold sweetening. Nat. Biotechnol. 16: 473–477.

Mahlow, S., M. Hejazi, F. Kuhnert, A. Garz, H. Brust, O. Baumann et al. 2014. Phosphorylation of transitory starch by alpha-glucan, water dikinase during starch turnover affects the surface properties and morphology of starch granules. New Phytol. 203(2): 495–507.

Meekins, D.A., H.F. Guo, S. Husodo, B.C. Paasch, T.M. Bridges, D. Santelia et al. 2013. Structure of the Arabidopsis glucan phosphatase like sex four2 reveals a unique mechanism for starch dephosphorylation. Plant Cell. 25(6): 2302–2314.

Meekins, D.A., M. Raththagala, S. Husodo, C.J. White, H.F. Guo, O. Kötting et al. 2014. Phosphoglucan-bound structure of starch phosphatase Starch Excess4 reveals the mechanism for C6 specificity. Proc. Natl. Acad. Sci. USA. 111(20): 7272–7277.

Meekins, D.A., M. Raththagala, K.D. Auger, B.D. Turner, D. Santelia, O. Kotting et al. 2015. Mechanistic insights into glucan phosphatase activity against polyglucan substrates. J. Biol. Chem. 290(38): 23361–23370.

Meekins, D.A., C.W. Vander Kooi and M.S. Gentry. 2016. Structural mechanisms of plant glucan phosphatases in starch metabolism. FEBS J. 283(13): 2427–2447.

Melendez-Hevia, E., T.G. Waddell and E.D. Shelton. 1993. Optimization of molecular design in the evolution of metabolism: the glycogen molecule. Biochem. J. 295(2): 477–483.

Mikkelsen, R., L. Baunsgaard and A. Blennow. 2004. Functional characterization of alpha-glucan,water dikinase, the starch phosphorylating enzyme. Biochem. J. 377(2): 525–532.

Mikkelsen, R., K.E. Mutenda, A. Mant, P. Schürmann and A. Blennow. 2005. α-Glucan, water dikinase (GWD): A plastidic enzyme with redox-regulated and coordinated catalytic activity and binding affinity. Proc. Natl. Acad. Sci. USA. 102(5): 1785–1790.

Mikkelsen, R., K. Suszkiewicz and A. Belnnow. 2006. A novel type carbohydrate-binding module identified in α-glucan, water dikinases is specific for regulated plastidial starch metabolism. Biochem. 45(14): 4674–4682.

Minges, A., D. Ciupka, C. Winkler, A. Hoppner, H. Gohlke and G. Groth. 2017. Structural intermediates and directionality of the swiveling motion of Pyruvate Phosphate Dikinase. Sci. Rep. 7: 45389.

Minassian, B.A., J.R. Lee, J.A. Herbrick, J. Huizenga, S. Soder, A.J. Mungall et al. 1998. Mutations in a gene encoding a novel protein tyrosine phosphatase cause progressive myoclonus epilepsy. Nat. Genet. 20: 171–174.

Muhrbeck, P. and A.C. Eliasson. 1987. Influence of the naturally occurring phosphate esters on the crystallinity of potato starch. J. Sci. Food Agric. 55(1): 13–18.

Nakianishi, T., T. Nakatsu, M. Matsouka, K. Sakata and H. Kato. 2005. Crystal structures of pyruvate phosphate dikinase from maize revealed an alternative conformation in the swiveling-domain motion. Biochemistry 44(4): 1136–1144.

Niittylä, T., S. Comparot-Moss, W.L. Lue, G. Messerli, M. Trevisan, M.D.J. Seymour et al. 2006. Similar protein phosphatases control starch metabolism in plants and glycogen metabolism in mammals. J. Biol. Chem. 281(17): 11815–11818.

Nitschke, F., P. Wang, P. Schmieder, J.M. Girard, D.E. Awrey, T. Wang et al. 2013. Hyperphosphorylation of glucosyl C6 carbons and altered structure of glycogen in the neurodegenerative epilepsy Lafora disease. Cell Metab. 17(5): 756–767.

Nitschke, F., M.A. Sullivan, P. Wang, X. Zhao, E.E. Chown, A.M. Parri et al. 2017. Abnormal glycogen chain length pattern, not hyperphosphorylation, is critical in Lafora disease. EMBO Mol. Med. 9(7): 906–917.

Pederson, B.A., J. Turnbull, J.R. Epp, S.A. Weaver, X. Zhao, N. Pencea et al. 2013. Inhibiting glycogen synthesis prevents Lafora disease in a mouse model. Ann. Neurol. 74(2): 297–300.

Pfister, B. and S.C. Zeeman. 2016. Formation of starch in plant cells. Cell Mol. Life Sci. 73: 2781–2807.

Pitcher, J., C. Smythe and P. Cohen. 1988. Glycogenin is the priming glucosyltransferase required for the initiation of glycogen biogenesis in rabbit skeletal muscle. Eur. J. Biochem. 176(2): 391–395.

Pirone, C., L. Gurrieri, I. Gaiba, A. Adamiano, F. Valle, P. Trost et al. 2017. The analysis of the different functions of starch-phosphorylating enzymes during the development of *Arabidopsis thaliana* plants discloses an unexpected role for cytosolic isoform GWD2. Physiol. Plant. 160(4): 447–457.

Posternak, T. 1935. Sur le phosphore des amidons. HCA. 18: 1351–1369.

Prats, C., T.E. Graham and J. Shearer. 2018. The dynamic life of the glycogen granule. J. Biol. Chem. 293(19): 7089–7098.

Ral, J.P., A.F. Bowerman, Z. Li, X. Sirault, R. Furbank, J.R. Pritchard et al. 2012. Down-regulation of Glucan, Water-Dikinase activity in wheat endosperm increases vegetative biomass and yield. Plant Biotechnol. 10(7): 871–882.

Raththagala, M., M.K. Brewer, M.W. Parker, A.R. Sherwood, B.K. Wong, S. Hsu et al. 2015. Structural mechanism of laforin function in glycogen dephosphorylation and lafora disease. Mol Cell. 57(2): 261–272.

Ritte, G., J.R. Lloyd, N. Eckermann, A. Rottmann, J. Kossmann and M. Steup. 2002. The starch-related R1 protein is an α-glucan, water dikinase. Proc. Natl. Acad. Sci. USA. 99(10): 7166–7171.

Ritte, G., A. Scharf, N. Eckermann, S. Haebel and M. Steup. 2004. Phosphorylation of transitory starch is increased during degradation. Plant Phys. 135(4): 2068.

Ritte, G., M. Heydenreich, S. Mahlow, S. Haebel, O. Kötting and M. Steup. 2006. Phosphorylation of C6- and C3-positions of glucosyl residues in starch is catalysed by distinct dikinases. FEBS Lett. 580(20): 4872–4876.

Roach, P.J. 2015. Glycogen phosphorylation and Lafora disease. Mol. Aspects Med. 46: 78–84.

Roach, P. and S. Zeeman. 2015 Glycogen and Starch. *In*: [ed.]. Encyclopedia of cell biology. Elsevier Inc.

Rommens, C.M., M.A. Haring, K. Swords, H.V. Davies and W.R. Belknap. 2007. The intragenic approach as a new extension to traditional plant breeding. Trends Plant Sci. 12(9): 397–403.

Romá-Mateo, C., P. Sanz and M.S. Gentry. 2012. Deciphering the role of malin in the Lafora progressive myoclonus epilepsy. IUBMB life 64(10): 801–808.

Rubio-Villena, C., R. Viana, J. Bonet, M.A. Garcia-Gimeno, M. Casado, M. Heredia et al. 2018. Astrocytes: new players in progressive myoclonus epilepsy of Lafora type. Hum. Mol. Genet. 27(7): 1290–1300.

Sakai, M., J. Austin, F. Witmer and L. Trueb. 1970. Studies in myoclonus epilepsy (Lafora body form): II. Polyglucosans in the systemic deposits of myoclonus epilepsy and in corpora amylacea. Neurology 20: 160–176.

Sarko, A. and H.C.H. Wu. 1978. The crystal structures of A-, B- and C-polymorphs of amylose and starch. Starch. 30(3): 73–78.

Santelia, D., O. Kotting, D. Seung, M. Schubert, M. Thalmann, S. Bischof et al. 2011. The phosphoglucan phosphatase like sex Four2 dephosphorylates starch at the C3-position in Arabidopsis. Plant Cell. 23(11): 4096–4111.

Scialdone, A., S.T. Mugford, D. Feike, A. Skeffington, P. Borrill, A. Graf et al. 2013 Arabidopsis plants perform arithmetic division to prevent starvation at night. eLife. 2: e00669–e00669.

Serratosa, J.M., P. Gómez-Garre, M.E. Gallardo, B. Anta, D.B.V. de Bernabé, D. Lindhout et al. 1999. A novel protein tyrosine phosphatase gene is mutated in progressive myoclonus epilepsy of the lafora type (EPM2). Hum. Mol. Genet. 8(2): 345–352.

Seetharaman, K. and E. Bertoft. 2012. Perspectives on the history of research on starch. Part 1: On the linkages in starch. Starch/Starke 64: 677–682.

Shearer, J. and T.E. Graham. 2002. New perspectives on the storage and organization of muscle glycogen. Can. J. Appl. Physiol. 27(2): 179–203.

Shearer, J. and T.E. Graham. 2004. Novel aspects of skeletal muscle glycogen and its regulation during rest and exercise. Exerc. Sport Sci. Rev. 32(3): 120–126.

Silver, D.M., O. Kötting and G.B.G. Moorhead. 2014. Phosphoglucan phosphatase function sheds light on starch degradation. Trends Plant Sci. 19(7): 471–478.

Sinnott, M.L. 2007. Carbohydrate chemistry and biochemistry: structure and mechanism. Royal Society of Chemistry: London, UK.

Simpson, W.W. and J.J. Macleod. 1927. The immediate products of post-mortem glycogenolysis in mammalian muscle and liver. J. Physiol. 64(3): 255–266.

Skeffington, A.W., A. Graf, Z. Duxbury, W. Gruissem and A.M. Smith. 2014. Glucan, water dikinase exerts little control over starch degradation in arabidopsis leaves at night. Plant Phys. 165(2): 866–879.

Smith, A.M. 2012. Starch in the Arabidopsis plant. Starch. 64(6): 421–434.

Solaz-Fuster, M.C., J.V. Gimeno-Alcaniz, S. Ros, M.E. Fernandez-Sanchez, B. Garcia-Fojeda, O. Criado Garcia et al. 2008. Regulation of glycogen synthesis by the laforin-malin complex is modulated by the AMP-activated protein kinase pathway. Hum. Mol. Genet. 17(5): 667–678.

Sorokina, O., F. Corellou, D. Dauvillee, A. Sorokin, I. Goryanin, S. Ball et al. 2011. Microarray data can predict diurnal changes of starch content in the picoalga Ostreococcus. BMC Syst. Biol. 5(1): 36.

Streb, S. and S.C. Zeeman. 2012. Starch metabolism in arabidopsis. *In:* The Arabidopsis Book. American Society of Plant Biologists. 10: e0160.

Sullivan, M.A., S. Nitschke, M. Steup, B.A. Minassian and F. Nitschke. 2017. Pathogenesis of Lafora disease: Transition of soluble glycogen to insoluble polyglucosan. Int. J. Mol. Sci. 18(8): E1743.

Sullivan, M.A., S. Nitschke, E.P. Skwara, P. Wang, X. Zhao, X.S. Pan et al. 2019. Skeletal muscle glycogen chain length correlates with insolubility in mouse models of polyglucosan-associated neurodegenerative diseases. Cell Rep. 27(5): 1334–1344.

Tabata, S. and S. Hizukuri. 1971. Studies on starch phosphate. Part 2. isolation of glucose 3-phosphate and maltose phosphate by acid hydrolysis of potato starch. Starch. 23: 267–272.

Tagliabracci, V.S., J. Turnbull, W. Wang, J.M. Girard, X. Zhao, A.V. Skurat et al. 2007. Laforin is a glycogen phosphatase, deficiency of which leads to elevated phosphorylation of glycogen *in vivo*. Proc. Natl. Acad. Sci. USA. 104(49): 19262–19266.

Tagliabracci. V.S., J.M. Girard, D. Segvich, C. Meyer, J. Turnbull, X. Zhao et al. 2008. Abnormal metabolism of glycogen phosphate as a cause of Lafora disease. J. Biol. Chem. 283(49): 33816–33825.

Tagliabracci, V.S., C. Heiss, C. Karthik, C.J. Contreras, J. Glushka, M. Ishihara et al. 2011. Phosphate incorporation during glycogen synthesis and Lafora disease. Cell Metab. 13(3): 274–282.

Tang, H., T. Mitsunaga and Y. Kawamura. 2006. Molecular arrangement in blocklets and starch granule architecture. Carbohydr Polym. 63(4): 555–560.

Tester, R.F., J. Karkalas and X. Qi. 2004. Starch—composition, fine structure and architecture. J. Cereal Sci. 39(2): 151–165.

Turnbull, J., P. Wang, J.M. Girard, A. Ruggieri, T.J. Wang, A.G. Draginov et al. 2010. Glycogen hyperphosphorylation underlies lafora body formation. Ann. Neurol. 68(6): 925–933.

Turnbull, J., A.A. Depaoli-Roach, X. Zhao, M.A. Cortez, N. Pencea, E. Tiberia et al. 2011. PTG depletion removes Lafora bodies and rescues the fatal epilepsy of Lafora disease. PLoS Genet. 7(4): e1002037.

Turnbull, J., J.R. Epp, D. Goldsmith, X. Zhao, N. Pencea, P. Wang et al. 2014. PTG protein depletion rescues malin-deficient Lafora disease in mouse. Ann. Neurol. 75(3): 442–446.

Turnbull, J., E. Tiberia, P. Striano, P. Genton, S. Carpenter, C.A. Ackerley et al. 2016. Lafora disease. Epileptic Disorders 18: S38–S62.

Vander Kooi, C.W., A.O. Taylor, R.M. Pace, D.A. Meekins, H.F. Guo, Y. Kim et al. 2010. Structural basis for glucan phosphatase activity of Starch Excess4. PNAS 107(35): 15379–15384.

Visko-Nielsen, A., A. Blennow, K. Jorgensen, K.H. Kristensen, A. Jensen and B.L. Moller. 2001. Structural, Physicochemical, and pasting properties of starches from potato plants with repressed r 1-gene. Biomacromolecules 2(3): 836–843.

Wang, J., J.A. Stuckey, M.J. Wishart and J.E. Dixon. 2002. A unique carbohydrate binding domain targets the lafora disease phosphatase to glycogen. J. Biol. Chem. 277(4): 2377–2380.

Wang, W., C.E. Hostettler, F.F. Damberger, J. Kossmann, J.R. Lloyd and S.C. Zeeman. 2018. Modification of cassava root starch phosphorylation enhances starch functional properties. Front Plant Sci. 9: 1562.

Weise, S.E., T.D. Sharkey and K.J. van Wijk. 2011. The role of transitory starch in C3, CAM, and C4 metabolism and opportunities for engineering leaf starch accumulation. J. Exp. Bot. 62(9): 3109–3118.

Whelan, W.J. 1986. The initiation of glycogen synthesis. BioEssays 5(3): 136–140.

Whelan, W.J. 1998. Pride and prejudice: The discovery of the primer for glycogen synthesis. Protein Sci. 7(9): 2038–2041.

Worby, C.A., M.S. Gentry and J.E. Dixon. 2006. Laforin, a dual specificity phosphatase that dephosphorylates complex carbohydrates. J. Biol. Chem. 281(41): 30412–30418.

Worby, C.A., M.S. Gentry and J.E. Dixon. 2008. Malin decreases glycogen accumulation by promoting the degradation of protein targeting to glycogen (PTG). J. Biol. Chem. 283(7): 4069–4076.

Xu, X., X.F. Huang, R.G. Visser and L.M. Trindade. 2017. Engineering potato starch with a higher phosphate content. PloS one. 12(1): e0169610.

Yokoi, S. 1967. Isolation and characterization of lafora bodies in two cases of myoclonus epilepsy. J. Neuropathol. Exp. Neurol. 26(1): 25–127.

Yokoi, S., J. Austin, F. Witmer and M. Sakai. 1968. Studies in myoclonus epilepsy (Lafora body form). Isolation and preliminary characterization of Lafora bodies in two cases. Arch. Neurol. 19(1): 15–33.

Young, F.G. 1957. Claude Bernard and the discovery of glycogen; a century of retrospect. BMJ. 1(5033): 1431–1437.

Yu, T.S., H. Kofler, R.E. Häusler, D. Hille, U.I. Flügge, S.C. Zeeman et al. 2001. The Arabidopsis Sex1 mutant is defective in the R1 protein, a general regulator of starch degradation in plants, and not in the chloroplast hexose transporter. Plant Cell. 13(8): 1907–1918.

Zeeman, S.C., F. Northrop, A.M. Smith and T. Rees. 1998. A starch-accumulating mutant of Arabidopsis thaliana deficient in a chloroplastic starch-hydrolysing enzyme. Plant J. 15(3): 357–365.

Zeeman, S.C. and T.A. Rees. 1999. Changes in carbohydrate metabolism and assimilate export in starch-excess mutants of Arabidopsis. Plant Cell Environ. 22(11): 1445–1453.

Zeeman, S.C., J. Kossmann–and A.M. Smith. 2010. Starch: its metabolism, evolution, and biotechnological modification in plants. Annu. Rev. Plant Biol. 61: 209–234.

CHAPTER 8

Starch Granules and their Glucan Components

Eric Bertoft

8.1 Introduction

Starch is produced in plants and many microorganisms as discrete granules containing two major glucan components, namely amylose and amylopectin. Amylose is a mixture of strictly linear [(1,4)-linkages] and slightly branched α-D-glucans, i.e., also possessing a few α-(1,6)-linkages (Takeda et al. 1987b). In contrast, amylopectin contains comparatively short chains of α-(1,4)-linked D-glucosyl units that are connected by α-(1,6)-linkages. These linkages account for about 5% of the total interglucose bonds and, therefore, amylopectin is extensively branched (Manners 1989). In most natural starches, amylopectin represents the major component and amylose makes up 15 ~ 35% of the granules. In addition, many starch granules also contain smaller amounts of so-called intermediate materials with structures that appear to be in between amylose and amylopectin (Banks et al. 1974, Gérard et al. 2002, Han et al. 2017, Lloyd et al. 1996, Schwall et al. 2000, Wang and White 1994, Whistler and Doane 1961, Yano et al. 1985). Besides the polyglucans, starch granules also contain small amounts, generally below one or up to a few per cent (w/w), of lipids (Hoover 2001, Morrison et al. 1993) and proteins (Goering and DeHaas 1974, Han and Hamaker 2002, Hoover 2001, Wang et al. 2013), together with trace elements (Blennow et al. 2005). Phosphorus is relatively abundant in root and tuber starches, in which phosphate groups are esterified at C-6 and C-3 positions of some of the glucosyl units of amylopectin (Blennow et al. 1998a, Hizukuri et al. 1970). Potato starch is particularly rich in phosphate esters, containing about 20–40 nmol phosphate/mg starch (Blennow et al. 1998b, Glaring et al. 2006, Wikman et al. 2014). However, the phosphate content of shoti (*Curcuma zedoaria*) starch is even up to three times higher (Blennow et al. 1998b, Glaring et al. 2006, Vamadevan et al. 2018). Amylopectin from other starches have lower phosphate ester contents. In cereal starches, the phosphate content is comparatively high. In these plants, however, only a minor phosphate part is esterified to amylopectin, but it mostly exists as lysophospholipids associated with amylose (Morrison et al. 1993, 1984).

Two major types of starch are distinguished: transient (assimilatory) starch is produced in chloroplasts as a result of photosynthesis during daylight and used by the plant in the night (Zeeman et al. 2010). Storage (reserve) starch is produced in special organelles called amyloplasts in storage organs (e.g. roots and tubers or the endosperm of seeds) (Tetlow 2011, Tetlow and Emes 2017, Zeeman et al. 2010). Whereas transient starch mostly finds great interest for research purposes, storage starches is of high importance for nutritional purposes. Most starch for food consumption comes from cereals, such as maize, rice, and wheat, as well as from roots and tubers, such as cassava (tapioca), sweet potato, and potato. Starch is also used for a wide range of non-food purposes for paper production,

Bertoft Solutions, 20960 Turku, Finland, Email: eric.bertoft@abo.fi in textile industries, as encapsulation materials and biodegradable plastics, to mention a few (Apinan et al. 2007, Autio et al. 1992, Chen and Jane 1995, Santacesaria et al. 1994, Shamekh et al. 2002, Wang and Copeland 2015, Zhao et al. 1996, Zhu 2015).

This chapter deals with the molecular structure of the starch polysaccharide components. The chapter also examines how these polyglucans contribute to the structure of the starch granules. Interestingly, the morphology of the starch granules are largely plant (and tissue) specific, whereas their internal architecture to a major extent is common to all granules regardless of their origin (French 1972, Jenkins et al. 1993, Gallant et al. 1997).

8.2 Carbohydrate Components of Starch

8.2.1 Amylopectin

Amylopectin is the major component of practically all starch granules. Amylopectin molecules are considered to be among the largest found in nature. Weight-average molecular weight (M_w) is in the order of 10^7–10^8 Da (Aberle et al. 1994, Franco et al. 2002, Lelievre et al. 1986, Perez-Rea et al. 2015), but number-average values (M_n) are around 10^5–10^6 (Hizukuri et al. 1989, McIntyre et al. 2013, Shibanuma et al. 1994) and the polydispersity index expressed as M_w/M_n is large (Stacy and Foster 1957). Takeda et al. (2003) analyzed the size-distribution of amylopectin preparations from a range of plants and found that, generally, amylopectin exists as three size-fractions with DP_n 13,400–26,500, 4400–8400, and 700–2100, of which the largest molecules predominate both by weight (83–90%) and by mole (43–63%). The amylopectin molecule consists of a large number of comparatively short chains with average chain lengths (CL) in the order of 17 ~ 23 residues. Chains in root and tuber amylopectin tend to be longer than in cereals (Table 8.1). Generally, about 5–6% of the interglucose linkages are branching points.

Table 8.1. Some Chain Ratios and Chain Segment Lengths in Amylopectin from Selected Plants.

Source	A:B	S:L	CL	ECL	TICL	ICL
Barley	1.0	17.9	17.6	11.5	12.3	5.1
Wheat	1.4	16.2	17.7	12.3	12.7	4.4
Maize	1.1	9.9	19.7	13.1	12.6	5.6
Rice	1.0	14.2	16.9	10.7	12.4	5.2
Sweet potato	1.1	10.1	19.6	12.8	14.0	5.8
Cassava	1.3	11.0	18.8	12.4	14.6	5.3
Smooth pea	1.4	7.3	22.0	14.1	–	6.9
Potato	1.2	6.3	23.1	14.1	19.9	8.0

A:B = molar ratio of A- to B-chains; S:L = molar ratio of S- to L-chains; CL = average chain length; ECL = average external CL; TICL = average total internal CL; ICL = average internal CL. Values represent a collection from the literature (Bertoft et al. 2008, 1993, Biliaderis et al. 1981, Goldstein et al. 2017a, Kalinga et al. 2013, Matheson 1990, Zhu et al. 2013a, 2011c).

8.2.1.1 Categories of Chains and Chain Segments of Amylopectin

In 1952, Peat et al. defined three major categories of chains in amylopectin: A-chains are connected to another chain through a α-(1,6)-linkage at the reducing-end side and do not carry other chains; B-chains are connected to other chains in the similar way but are also carrying other chains, either A- or other B-chains; the sole C-chain carries the free reducing end of the macromolecule but is otherwise similar to the B-chains (Fig. 8.1). Beside these categories, different chain segments are also distinguished. External chains are defined as chain segments that extend from the outermost branch point to the non-reducing end of the chain (Manners 1989). Whereas every B-chain contains one external segment, the entire A-chains are external. Internal chains are defined as the segments

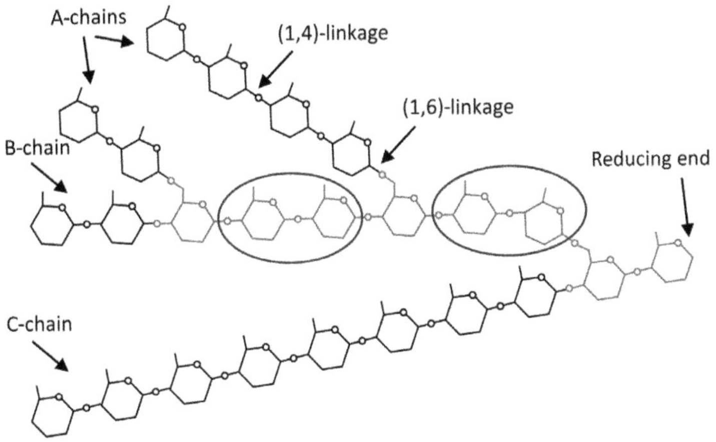

Figure 8.1. Basic chain types and chain segments in starch molecular components. External segments are shown in black, total internal chain segments in gray, and internal chain segments are encircled.

between branches (excluding the branched residues) and exist therefore only in B-chains (Manners 1989). The total internal chain segment refers to the total length of the internal part of a B-chain, i.e., from the outermost branch to the reducing-end side of the chain including all branched residues along the chain (Fig. 8.1) (Bertoft 1991). The core chain was defined in a similar way but excludes the outermost branch point (Yun and Matheson 1993). Typical average external and internal chain lengths are given in Table 8.1.

The chains of amylopectin can be isolated by using the enzymes isoamylase and/or pullulanase, which specifically hydrolyze the (1,6)-linkages. The size-distribution of the chains, often called the unit chain profile, can be measured by gel-permeation chromatography (GPC) (Akai et al. 1971, Lee et al. 1968, Mercier 1973) or high-performance size-exclusion chromatography (HPSEC) (Fredriksson et al. 1998, Hanashiro et al. 2002, Yao et al. 2005). Higher resolution of the chains, especially of shorter chains with DP up to ~ 60, is obtained by fluorophore-assisted carbohydrate electrophoresis (FACE) (Morell et al. 1998, O'Shea et al. 1998, Yao et al. 2005) or alternatively by high-performance anion-exchange chromatography (HPAEC) (Hanashiro et al. 1996, Koch et al. 1998, Koizumi et al. 1991, Wong and Jane 1997). These latter techniques have shown that the shortest chains in amylopectin have DP 6 (Koizumi et al. 1991). The shortest chains with DP 6–8 were called "fingerprint" A_{fp}-chains, due to their "fingerprint" profile with a characteristic pattern for different samples (Fig. 8.2) (Bertoft 2004). Amylopectin chains are typically distinguished as two major groups, short and long chains, with a division at about DP 36 (Bertoft et al. 2008, Hanashiro et al. 1996, Takeda et al. 2003). Short chains are mixtures of A-chains and short B-chains (BS-chains), whereas long chains are believed to belong to B-chains (BL-chains, Fig. 8.2), although in some starches tiny amounts of long A-chains also might exist (Bertoft 2004, Bertoft et al. 2008). The molar ratio of short:long chains (S:L) is generally higher in cereal starches than in root and tuber starches (Table 8.1).

Hizukuri (1986) suggested that A-chains generally are the shortest chains with DP 6 ~ 16 and he named BS-chains as B1-chains with DP up to ~ 36. BL-chains were subdivided into B2- (DP 36 ~ 60) and B3-chains (DP 60 ~ 80), based on a periodicity in chain length of DP 27 ~ 28 in different samples. Eventually, even longer chains (B4) are found in trace amounts (Hizukuri 1986). A decade later, Hanashiro et al. (1996) compared the unit chain profile of 11 samples and found another, shorter periodicity of only 12 residues. They named the fractions as fa (DP 6–12), fb_1 (13–24), fb_2 (25–36), and fb_3 (DP > 36) (at DP > 36 there was no periodicity). They suggested that these fractions may correspond to A-, B1-, B2-, and B3-chains, respectively, and the vast majority of researchers have adopted this nomenclature ever since. This is unfortunate, however, because the definitions of the chain categories were different from the original definitions and has caused a lot of confusion, especially with regards to the structural models of amylopectin, as we shall see later on. It has, in

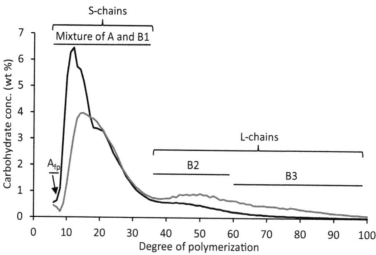

Figure 8.2. Examples of the unit chain profile of amylopectin from oat (type 1 structure, black line) and lesser yam (type 4 structure, gray line) obtained by HPAEC. Short chains (S) and long chains (L) and different sub-categories of chains are indicated.

fact, been shown that the number of A-chains in amylopectin is larger than fraction fa and, therefore, chains with DP 6–12 do not account for all A-chains (Bertoft et al. 2008). Moreover, Witt and Gilbert (2014) argued that there is no reason to believe that A-chains generally are the shortest chains in amylopectin. Actually, it is likely that chains as short as DP 9, or maybe even DP 8, are mixtures of A- and B-chains (Vamadevan and Bertoft 2018).

Besides chains with DP up to 100 or somewhat more, some plants have amylopectin with very long chains. These super-long chains (or extra-long chains) have DP up to 1000 or more and are thereby similar in length to chains normally found in amylose. Super-long chains were first reported in rice, especially in *indica* varieties (Takeda et al. 1987a), but have since then also been reported in many other starches, such as barley (Schulman et al. 1995, Takeda et al. 1999), wheat (Shibanuma et al. 1994), maize (Takeda et al. 1988), and cassava (Charoenkul et al. 2006, Laohaphatanaleart et al. 2009). Super-long chains are very few by number, but on a weight basis they may represent as much as ~ 14% of the amylopectin molecule (Takeda et al. 1987a). Most super-long chains appear to be slightly branched (Hanashiro et al. 2005). Interestingly, super-long chains are not found in amylose-free starches (known as "waxy" starches) and it was shown that granular-bound starch synthase I, an enzyme needed for amylose synthesis, is responsible for the appearance of super-long chains as well (Hanashiro et al. 2008).

8.2.1.2 Internal Chain Distribution

The size-distribution of the internal chains has been analyzed after removal of the external chain segments with exo-acting enzymes to form limit dextrins, in which the internal part of the molecule, together with all branches, remains intact. Two enzymes are useful for this purpose, namely β-amylase and phosphorylase *a*. β-Amylase cleaves maltosyl units from the non-reducing ends of the chains, but it cannot by-pass (1,6)-linkages. The enzyme reduces the length of A-chains to either maltosyl or maltotriosyl units depending on if the chains contain an even or odd number of glucose residues, whereas the external segments of B-chains are reduced into maltosyl or glucosyl units, respectively (Summer and French 1956). The molar amount of maltose and maltotriose obtained after debranching of the β-limit dextrin corresponds to the amount of A-chains in the sample (Fig. 8.3). Phosphorylase *a* (from rabbit muscle) produces glucose 1-phosphate (by phosphorolysis instead of hydrolysis) by attack at the non-reducing ends (Hestrin 1949). In the resultant so-called φ-limit dextrin, all A-chains remain as maltotetraosyl units, regardless of whether the original A-chain had an even or odd number

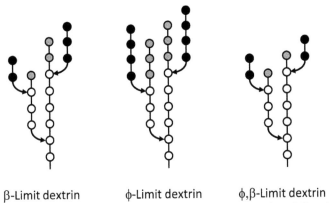

β-Limit dextrin   φ-Limit dextrin   φ,β-Limit dextrin

Figure 8.3. The principal external structure of limit dextrins obtained by the action of β-amylase (β-limit dextrin), phosphorylase *a* (φ-limit dextrin), or the successive treatment with phosphorylase *a* and β-amylase (φ,β-limit dextrin). Branches are shown as arrows and the glucose residues (O) of A-chains (black), external segments of B-chains (gray) and internal segments of B-chains (white) are indicated.

of residues (Walker and Whelan 1960) and the external segments of B-chains remain as maltotriosyl units (Fig. 8.3) (Bertoft 1989b, 2004). If β-amylase is added to the φ-limit dextrin, an additional maltose is cleaved from each chain and a φ,β-limit dextrin is obtained, in which all A-chains have DP 2 (maltose) and all external segments of B-chains are left as a single glucosyl unit (Fig. 8.3) (Bertoft 2004). All three types of limit dextrins can be used to study the internal chain distribution of amylopectin and to measure the ratio of A:B chains (Table 8.1).

Like the unit chain profile, the internal chain distribution of amylopectin also possesses two major groups of chains that represent the internal part of the BS- and BL-chains with the division at DP 23 ~ 27 depending on the sample (Fig. 8.4). BL-chains are distinguished as B2- and B3-chains, but these are shorter than in the original unit chain profile because the external segments have been removed. The internal BS-chains are also distinguished as two groups (Bertoft et al. 2008, Klucinec and Thompson 2002, Shi and Seib 1995): a minor group of chains with DP 3–7 were named "fingerprint" B_{fp}-chains analogous to A_{fp}-chains because of their size-distribution profile, which is characteristic of starches in different plants (Bertoft 2004). A major group of BS-chains (BS_{major}) have DP 8 ~ 25,

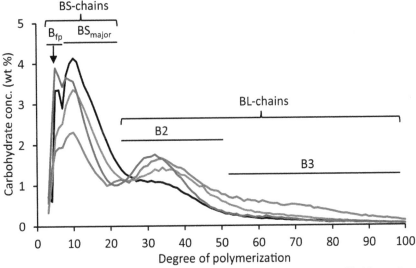

Figure 8.4. Internal chain profile from four different structural types of amylopectin, exemplified by oat (type 1, black), rice (type 2, blue), mung bean (type 3, green), and lesser yam (type 4, red), obtained by debranching of the φ,β-limit dextrin of the amylopectin and analysis by HPAEC. Short (BS) and long B-chains (BL) and different sub-categories are indicated.

depending on the sample (Bertoft et al. 2008). It should be noted that the two subgroups of BS-chains are only distinguished in the internal profile, not in the original unit chain profile.

Amylopectin from a broad range of plants (17 samples) were divided into four structural types based on the characteristics of their internal chain distributions (Fig. 8.4) (Bertoft et al. 2008). Type 1 amylopectin was found in some cereals, such as barley, rye and oat. This structure has characteristically very low content of B2-chains and B3-chains are almost absent. Instead, BS-chains are found in large number and this is due to a broad size-distribution of the BS_{major}-chains up to DP ~ 28, which makes the division between BS- and BL-chains unclear in the chromatograms. Type 2 amylopectin, which is found in cereals such as millets (Annor et al. 2014) and rice (Gayin et al. 2016a), but also in for example sago starch (Bertoft et al. 2008), has more B2-chains and a narrower range of BS_{major}-chains up to DP 23 ~ 24 so that a clear groove between long and short chains is seen in the chromatograms (Fig. 8.4). These starches, particularly those in cereals, have lot of B_{fp}-chains that give rise to a clear peak in the chromatograms at DP 5 or 6. Wheat amylopectin appears to have a structure in between types 1 and 2 (Kalinga et al. 2013). Type 3 amylopectin, represented by for example cassava and arrowroot starch (Bertoft et al. 2008), has comparatively little B_{fp}-chains, but somewhat more BL-chains than type 2 starches. Sweet potato amylopectin was reported to have a structure in between type 2 and 3 (Zhu et al. 2011c). Type 4 amylopectin has been described in some other tuber and root starches (potato, edible canna and lesser yam) and has characteristically more B3-chains than the other types (Bertoft et al. 2008). From the above, it is clear that a precise definition of the different structural types of amylopectin cannot be made and more intermediate structures, or eventually even new structural types, might be identified as more samples are investigated. For example, starch-related mutations in plants may not always follow the respective wild-type scheme. The *dull 1* mutation in maize rendered its amylopectin structure more close to type 1, instead of type 2 for the wild-type starch (Zhu et al. 2013a). Nevertheless, these initial findings suggest a possible systematic organization of the structure of amylopectin from diverse plants, which would aid in the understanding of amylopectin structure.

8.2.1.3 Branched Units

The final structure of amylopectin depends on the organization of the unit chains within the macromolecule (Thompson 2000). To understand this organization, endo-acting enzymes that cleave the internal chains between the branches have been utilized to enable an isolation and structural characterization of smaller branched units inside the molecule (Bender et al. 1982, Bertoft 1986, 1989b, Bertoft and Spoof 1989, Finch and Sebesta 1992). Only few enzymes have been tested for this purpose, of which the α-amylase of *Bacillus amyloliquefaciens* by far have been used most and is described in this section. The enzyme contains nine subsites around its catalytically active site (Robyt and French 1963). The reaction is effective only when all subsites are filled with glucosyl units and, therefore, mostly internal chains with nine or more units are cleaved at a high rate during the initial stages of the hydrolysis of amylopectin (Robyt and French 1963). Such long internal chains are believed to be found between groups of short chains known as clusters (Hizukuri 1986). The clusters are, however, slowly further degraded by the enzyme by attack at shorter internal chains with lengths of approximately 5 ~ 8 residues (Zhu et al. 2011a). These short internal chain segments are found between even shorter, yet branched chains inside the clusters and are known as building blocks. Those constitute the final, resistant α-limit dextrins and are considered to be the actual branched units in the macromolecule (Bertoft et al. 2010, 1999, Zhu et al. 2011a). The apparent clusters may therefore more correctly be considered as clusters of α-amylase-resistant building blocks, rather than clusters of chains.

8.2.1.4 Clusters of Building Blocks

Clusters are isolated from amylopectin by interrupting the reaction with the α-amylase of *B. amyloliquefaciens* when the initial, fast reaction changes into a slower rate (Laohaphatanaleart et al. 2010, Zhu et al. 2011b). Besides attack at the longer internal chains, the enzyme also shorten the external chains, giving rise to a complex mixture of linear dextrins with DP centred around 6–7 (Bertoft 1989a, Laohaphatanaleart et al. 2010, Robyt and French 1963). To remove the remaining external chain stubs, which have diverse lengths after the reaction, the isolated dextrins are further treated with exo-enzymes to produce the β- or φ,β-limit dextrins of the clusters (Bertoft 2007b, Bertoft et al. 2011b, Laohaphatanaleart et al. 2010). The clusters obtained in this manner generally have very large size-distributions ranging from approximately DP 15 to 1000 (Bertoft et al. 2012a). Average DP-values of the cluster preparations are given in Table 8.2. Generally, isolated clusters from type 1 and 2 amylopectin tend to be larger than those isolated from type 3 amylopectin, whereas type 4 amylopectin has the smallest clusters (Bertoft et al. 2012a).

The internal chain distribution of the clusters (Fig. 8.5) is obtained by debranching the clusters with isoamylase and pullulanase. The average number of chains (NC) in clusters depends on the size of the clusters (Table 8.2). Large clusters in type 1 and 2 amylopectin have between 11 ~ 19 chains. The largest clusters have been isolated from barley (Bertoft et al. 2011b, Goldstein et al. 2017a). NC in type 4 amylopectin is only between 5 ~ 9 (Bertoft 2007b, Bertoft et al. 2012a). The chains in clusters correspond mostly to the lengths of the short BS-chains originally found in amylopectin together with the A-chains, which in the limit dextrins are reduced into DP 2 or 3. The average internal chain length (ICL) in clusters is about 3~6, which is shorter than in total amylopectin (Table 8.1) and suggests that the longest internal chains are absent in the isolated clusters.

Because the longer internal chains of amylopectin have been cleaved by the α-amylase in order to cut out the clusters from the macromolecule, the number of BL-chains in clusters is largely reduced and new short chains are formed (Fig. 8.5). Therefore, many chains in the clusters (that are obtained by treatment with α-amylase) are not directly comparable with the chain categories defined for amylopectin. Consequently, chain categories in clusters were given small letters instead of capital symbols and defined, besides a-chains, as b0- (DP 4–6), b1- (DP 7–18), b2- (DP 19–27), and b3-chains (DP > 27) (Bertoft et al. 2012a). b0-chains are generally produced in large number by the α-amylase, and longer chains are produced in decreasing number with a pattern that appears universal regardless of the starch source (Bertoft et al. 2012a). b2-chains have lengths that correspond to the groove found between BS- and BL-chains in the limit dextrins of amylopectin, and they are especially well distinguished in clusters from type 4 amylopectins (Fig. 8.5). However, even if clusters characteristically consist of short chains (DP approximately < 23), long chains also remain in isolated clusters in the form of b3- and b2-chains. The average number of long chains per cluster varies between 0.4 and 1.3 chains depending on the starch source. Thus, the numbers suggest that,

Table 8.2. Structural Properties of Limit Dextrins of Clusters of Building Blocks Isolated from Amylopectin.

Source	AP structure	DP	NC	b2+b3-chains	CL	ICL
Barley	1	123.1	19.5	1.3	6.3	4.0
Wheat	1-2	84.2	14.2	0.8	5.8	3.6
Maize	2	68.2	12.5	0.6	5.4	3.3
Rice	2	81.0	12.0	0.6	6.8	4.7
Sweet potato	2-3	79.0	12.2	0.4	6.5	4.5
Cassava	3	70.2	10.0	0.9	7.0	4.6
Waxy potato	4	48.6	6.2	1.0	7.9	5.4

AP structure = internal structural type of amylopectin; DP = average degree of polymerization; NC = average number of chains per cluster; b2+b3-chains = average number of long chains per cluster; CL = average chain length; ICL = average internal CL. Values represent a collection from the literature (Bertoft 2007b, Gayin et al. 2016b, Goldstein et al. 2017a, Kalinga et al. 2014, Laohaphatanaleart et al. 2010, Zhu et al. 2013b, 2011b).

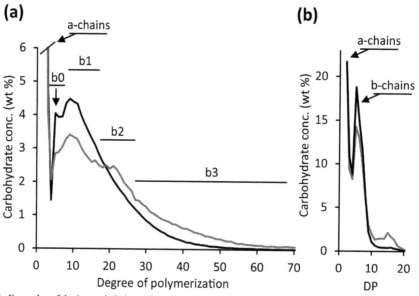

Figure 8.5. Examples of the internal chain profile of clusters of building blocks (a) and the isolated building blocks (b) from amylopectin of oat (type 1 structure, black line) and lesser yam (type 4 structure, gray line) obtained by HPAEC. Different chain categories in the clusters, which are given in small letters instead of capital symbols, are shown. a-chains (unsubstituted chains) are obtained as a maltose peak after debranching of the φ,β-limit dextrin, and b-chains (substituted chains) have DP ≥ 4. Subcategories of b-chains are b0- (DP 4–6), b1- (DP 7–18), b2- (DP 19–27), and b3-chains (DP > 27), whereas chains at DP 3 are mixtures of a- and b-chains in unknown proportions.

for example, roughly every second cluster in rice or sweet potato contains a long chain, whereas approximately every cluster in potato possesses a long chain (Table 8.2).

8.2.1.5 Building Blocks

The smallest branched unit in amylopectin is the building block. Building blocks have been isolated by an extensive hydrolysis with the α-amylase of *B. amyloliquefaciens* from a range of cluster preparations of diverse plants (Bertoft 2007a, Bertoft et al. 2011a, 2012a, 2010, Gayin et al. 2016b, Kong et al. 2009, Peymanpour et al. 2016, Wikman et al. 2011, Zhu et al. 2013b, 2011a), including transient (assimilatory) starch from the leaves of *Arabidopsis* (Zhu et al. 2015). Building blocks are divided into different groups (2–6) based on their NC (Bertoft et al. 2012b). The smallest building blocks have only two chains (i.e., one branch) with DP 5–9. The structure, shown in Fig. 8.6, is thus very simple as it consists of only one a-chain and one b-chain with short chain stubs surrounding the branched residue (French et al. 1972, Umeki and Yamamoto 1972a, 1975b). The lengths of the chain stubs depend on the action pattern of the α-amylase, rather than the structure of the building block, and in practice the blocks can be considered as having the same principal structure (i.e., a single branched dextrin) regardless of the DP (Umeki and Yamamoto 1972b). Group 3 building blocks have three chains and DP 10–14. The structure of these dextrins is of two principal types with either two a-chains and one b-chain (so-called Staudinger configuration) or one a-chain and two b-chains (Haworth configuration). The structure of group 4 building blocks with four chains (DP 15–19) becomes increasingly more complex with mixed configurations (examples are shown in Fig. 8.6) (Umeki and Yamamoto 1975a). Group 5 building blocks have DP 20 ~ 35 and have average NC about 6, i.e., they contain between 5 – 7 chains, whereas group 6 with DP > 35 have at average NC 10 ~ 11, and were not isolated in more pure form (Bertoft et al. 2012b). The average ICL in building blocks is only around 1.5 ~ 2.2 (Bertoft et al. 2012b). Very detailed structural analyses of α-limit dextrins from waxy rice amylopectin were performed by Umeki and Yamamoto (1975b), who showed

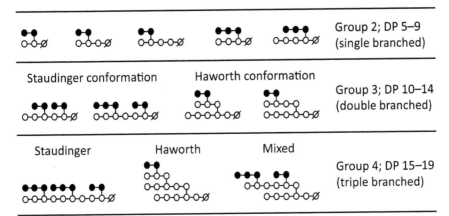

	Group 2; DP 5–9 (single branched)
Staudinger conformation Haworth conformation	Group 3; DP 10–14 (double branched)
Staudinger Haworth Mixed	Group 4; DP 15–19 (triple branched)

Figure 8.6. Examples of building blocks from groups 2, 3 and 4 formed by the α-amylase of *B. amyloliquefaciens* and successive treatment with β-amylase. The structures are primarily based on the work of Umeki and Yamamoto (1975a,b). The glucose residues in a-chains (•), and b-chains (O), reducing ends (Ø), and (1,6)-linkages (|) are indicated.

that the shortest internal chain segment between branches consists of one glucose residue, whereas branches immediately adjacent to each other do not exist.

The composition of the groups of building blocks in clusters appears to be surprisingly similar in all plants. Thus, group 2 building blocks is most common, constituting approximately 45 ~ 60% by number of all the blocks. Group 3 building blocks is second most abundant (roughly 25 ~ 35%) and larger blocks are found in decreasing numbers. Group 6 building blocks are generally found in only small numbers (≤ 5%) (Bertoft et al. 2012a). Building blocks are interconnected through inter-block segments with chain lengths (IB-CL) 5–8. IB-CL is generally shortest in type 1 amylopectin and increases through types 2 and 3 to type 4 amylopectin, in which it is 7–8 (Bertoft et al. 2012a). The chains in building blocks are very short and correspond mostly to b0-chains in the cluster unit chains profile (Fig. 8.5). Somewhat more of b1-chains are found in larger blocks (Bertoft et al. 2012b). The nomenclature of b-chains in clusters suggests how many inter-block segments the chain likely is involved in (Bertoft et al. 2012a). Thus, b0-chains have no inter-block segments and are found completely inside building blocks, b1-chains have one inter-block segment (and thus interconnect two building blocks), b2-chains have two segments, etc. (Fig. 8.7). As group 2 building blocks are most common, it follows that most branching points in clusters are separated by 5–8 glucose residues of inter-block segments, whereas groups of more frequently branched chains become increasingly rare with their size.

(a)   **(b)**

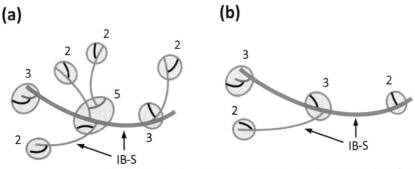

Figure 8.7. Hypothetical structure of φ,β-limit dextrins of clusters of building blocks from amylopectin of structural type 1 (a) and type 4 (b). The building blocks are encircled and numbers indicate blocks of different structural groups. a-chains are symbolized by black lines, b0-chains are blue, b1-chains are green, and b2-chains are bold red and are probably parts of the backbone of amylopectin, whereas b1-chains mostly form side-chains to the backbone. IB-S indicates some of the inter-block segments, which are somewhat longer in (b) than in (a). Side-chains (green) to the backbone are more frequent in (a) than in (b).

Even though the composition of groups of building blocks is very similar in different plants, it is interesting to notice that the exact structure of the building blocks might be different. Thus, cereal amylopectins appear to have more building blocks with the Haworth configuration, whereas the Staudinger configuration predominates in roots and tubers (Bertoft et al. 2012b). However, it remains unclear if—and how—this affects the properties of starch.

8.3 Amylose

8.3.1 Molecular Structure

Amylose consists of long chains of D-glucosyl units with a broad size-distribution. The degree of polymerization (DP) ranges from a few hundred to several thousand. Generally, the amylose in cereal starches has lower average DP values than in roots and tubers (Table 8.3). Most starches contain both linear and branched amylose molecules. The molar fraction of branched amylose varies greatly between plants from about 17 to 70%. The linear amylose molecules can be removed from a mixture of linear and branched amylose by treatment with β-amylase, whereas the internal part of the branched amylose remains as limit dextrin. The DP of branched amylose in rice was calculated to be larger than its linear counterpart (Hizukuri et al. 1989). This is also suggested from the fact that DP of the β-limit dextrin is often similar or even higher than that of the original mixture (Takeda et al. 1999, Yoshimoto et al. 2000).

The size-distribution of the chains can be measured after debranching of the molecules with isoamylase or pullulanase. It was shown, however, that some branches are resistant to the enzymes (possibly because of sterical reasons), which complicates the estimation of the chain lengths (Hanashiro et al. 2013). Moreover, the lengths of the chains in branched amyloses are difficult to measure because there are no straightforward methods to separate intact branched amylose from the linear counterpart or to distinguish from which molecules the chains derive after debranching. Nevertheless, the unit chain profile of the remaining branched limit dextrins after β-amylolysis has been analyzed

Table 8.3. Content and Properties of Amylose in Selected Plants.

Source	AM (%)	LAM (%)	DP_n	MF_B (%)	β-Limit dextrin		
					DP_n	CL_n	NC
Barley	22–30	5.0–9.6	940–1850	17–35	1110–1440	104–190	3.7–13.8
Wheat	25–34	4.4–6.8	570–2100	26–44	700–1430	50–94	12.9–20.7
Maize	21–32	4.1–6.1	790–1100	44–48	850	140–160	5.3–5.4
Rice	8–29	4.1–6.7	1100	31–43	890–1030	105–115	8.5–9.0
Smooth pea	28–35	0.0	770–1400	–	–	–	–
Sweet potato	15–24	–	3025–4100	70	1770	130	13.6
Cassava	17–26	–	1540–3642	42	1970	115	17.1
Potato	18–36	0.0	2110–8025	38	–	–	–

AM = amylose content of starch granules; LAM = lipid-amylose complexes in starch granules; DP_n = number-average degree of polymerization; MF_B = molar fraction of branched molecules; CL_n = number-average chain length; NC = number of chains per molecule. Values represent a collection from the literature (Andersson et al. 1999, Ao and Jane 2007, Biliaderis et al. 1981, Franco et al. 2002, Fredriksson et al. 1998, Gérard et al. 2001, Glaring et al. 2006, Greenwood and Thomson 1962a,b, Hoover 2001, Ishiguro et al. 2000, Jane et al. 1999, Li and Yeh 2001, Manion et al. 2011, Morrison and Laignelet 1983, Morrison et al. 1984, Nakazawa and Wang 2003, Orford et al. 1987, Ren 2017, Schulman et al. 1995, Shibanuma et al. 1994, Srichuwong et al. 2005, Takeda et al. 1989a, 1987b, 1992a, Takeda and Preiss 1993, Takeda et al. 1984, 1988, 1999, Yoshimoto et al. 2002, 2000, Zhang and Oates 1999).

and the average number of chains (NC) in branched molecules was estimated to be between 4 ~ 21 (Table 8.3). Large amylose molecules in maize were shown to contain larger fractions of branched molecules and more branches per molecule compared to small amylose molecules (Takeda et al. 1992b). It appears that the chains in branched amylose at average are shorter than the linear amylose, except for the smallest amylose molecules, which have chain lengths similar to the large molecules (Takeda et al. 1992b). A substantial number of the chains in branched amylose are as short as DP 6 ~ 50, which corresponds to the chain lengths generally found in amylopectin (Shibanuma et al. 1994, Takeda et al. 1990, 1993). Larger branched amylose molecules contain more of short chains than small molecules (Takeda et al. 1992b). The size-distribution profile of these short chains is, however, different from that in amylopectin and thus they are not organized as in amylopectin (Hanashiro et al. 2013). Even if the number of short chains may be large, their contribution to the total weight of the branched molecule is very low and, therefore, these chains normally escape detection (Takeda et al. 1993).

8.3.2 Helical Structure

When starch granules are boiled in water, their semi-crystalline structure is destroyed and the molecular components are solubilized. Amylose is inherently unstable in water solution, however. At higher pH, the molecule is a random coil or forms partial helical conformations (Banks and Greenwood 1975). At neutral pH, segments of the molecules form left-handed double-helices with six residues per turn in each strand and a pitch of 2.1 nm (Imberty and Pérez 1988). The double-helical segments crystallize into a structure known as B-type allomorph, which precipitates from the solution (Gidley 1989, Jane and Robyt 1984). This process is referred to as retrogradation. The structure of the crystalline hexagonal unit cell is well characterized and has the dimensions $a = b = 1.85$ nm and $c = 1.04$ nm. Each unit contains 36 water molecules (Imberty and Pérez 1988).

Amylose is also well known for its ability to form inclusion complexes with a range of ligand molecules. These complexes are single α-glucan helices with the ligand in the central cavity of the helix and they crystallize into so-called V-type allomorphs. The structure of the single helix depends on the ligand and it consists typically of 6–8 glucose residues per turn with increasing diameter, but almost constant pitch length of ~ 0.8 nm (Buléon et al. 1998a, French 1979, French and Zobel 1967, Godet et al. 1993, Hinkle and Zobel 1968). Hydrated helices are denoted as V_h-helices and the anhydrous counterparts are V_a-helices (Rappenecker and Zugenmaier 1981, Zobel et al. 1967). Typical ligand molecules are fatty acids, alcohols such as 1-butanol, and flavour substances (Buléon et al. 1998a, French and Zobel 1967, Hinkle and Zobel 1968, Jane and Robyt 1984, Nuessli et al. 1997). A fraction of the amylose in cereal starch granules is complexed with free fatty acids and lysophospholipids (Table 8.3) (Biais et al. 2006, Buléon et al. 1998a, LeBail et al. 1999, Morrison et al. 1993, 1984, 1986). The lipid-complexed amylose is abbreviated LAM to distinguish it from the lipid-free fraction FAM. The most familiar inclusion complex is that with iodine (Rundle and French 1943), which forms an intense blue colour. The colour is commonly used to measure the (apparent) content of amylose in starch (Chrastil 1987, Kaufman et al. 2015, Knutson and Grove 1994, Zhu et al. 2008). The intensity of the colour, as well as the wavelength maximum, depends on the chain length and therefore amylose stains dark blue or black, whereas amylopectin stains brownish-red (Bailey and Whelan 1961, Banks et al. 1971). In some starches, however, amylopectin also contains long chains that stain similarly to amylose, which overestimates the true amylose content (Takeda et al. 1989b). On the other hand, if the starch contains LAM, the iodine binding is weak and the amylose content is underestimated, unless the starch is defatted prior to amylose staining (Morrison and Laignelet 1983).

8.4 Intermediate Materials

Besides amylose and amylopectin, many starches have been shown to contain intermediate materials. These are especially abundant in mutant plants, in which the biosynthesis of starch has been altered. Starches with an increased amount of (apparent) amylose are often rich in intermediate materials (Banks et al. 1974, Colonna and Mercier 1984, Li et al. 2008, Matheson and Welsh 1988, Perera et al. 2001, Schwall et al. 2000). Most frequently, plants producing such starches have a low activity of starch branching enzyme IIb (amylose-extender mutation), but also other mutations give rise to increased (apparent) amylose content (Bertoft and Seetharaman 2012, Nakamura 2018). However, as already mentioned, the measurement of amylose is not straightforward and many methods are based on the inclusion complexes formed between amylose and iodine, which gives rise to the characteristic blue colour that can be quantified colorimetrically or by potentiometric means. As intermediate materials have chain length characteristics in between amylopectin and amylose, these measurements might overestimate the true amylose content due to the complexation of iodine with the intermediate material.

The true structural nature of intermediate materials has been proved to be difficult to analyze. This is mostly due to the fact that the isolation of intermediate materials from "true" amylose and amylopectin is challenging, as there are no definite structural definitions for the different molecules. Several methods have been proposed for isolation of intermediate materials (Adkins and Greenwood 1969, Colonna and Mercier 1984, Klucinec and Thompson 1998, Li et al. 2008, Whistler and Doane 1961, Yoon and Lim 2003), and possibly they all result in slightly different constitutions in the preparations and thus, diverse pictures of the structures of the isolated material are obtained. In addition, the same genetic mutation in diverse inbred backgrounds of the same plant species was shown to give starches with different structures (Boyer and Liu 1985). It is therefore apparent that not only any given mutation of starch synthase or branching enzymes specific genes but also other, so far unknown, alterations in the plant genome has important influence on the final starch structure.

Intermediate materials were isolated from maize starches with different mutations by precipitation of intermediate material and amylose in a mixture of 1-butanol and isoamyl alcohol. The mixture was then treated with pure 1-butanol, whereby only the latter precipitated and intermediate material was retained in solution (Klucinec and Thompson 1998). This intermediate material, from both normal and *ae* maize starch, had longer chains than normal amylopectin, but was otherwise quite similar to the amylopectin fraction. However, the "normal" amylopectin also contained longer chains than in non-mutant maize (Klucinec and Thompson 1998). Later, it was shown that the amylopectin in *ae* maize starch in fact contained two fractions with high and low molecular weight, of which the former was similar to amylopectin in non-mutant maize starch and the latter had longer chains. Moreover, the clusters of the amylopectin in the high-amylose starches were smaller and had a lower branching density than in the non-mutant counterpart (Peymanpour et al. 2016). The intermediate material fraction in both regular and *ae* maize starch was shown to possess super-long chains, and internal chain lengths appeared longer than in the amylopectin fraction, whereas the external chain length was indifferent (Han et al. 2017).

Wang et al. (1993) removed the amylose fraction from several mutants of maize (*ae, dull 1* [*du1*], *brittle 1* [*bt1*], *ae du1,* and *ae bt1*) of the Oh43 inbred line by precipitation in 1-butanol and then fractionated the remaining starch into intermediate material and amylopectin by GPC. They found 15.2–22.5 wt% intermediate material in all samples except *bt1*, in which there was only 1.8%. The classification of chains (see above) was similar in intermediate material and in amylopectin, but only in *ae* and *ae du1* the intermediate material possessed clearly lower proportion of short chains, and thus relatively more of the long chains, suggesting a lower degree of branching (Wang et al. 1993). On the other hand, Bertoft et al. (2000) found that intermediate material in the double and triple waxy mutants *du wx* and *ae du wx* of maize starch from the inbred line Ia453 was partially resistant to attack by the α-amylase of *B. amyloliquefaciens* and suggested that it contained a more regularly branched structure with too short internal segments to allow α-amylase action.

Pea starch, especially wrinkled pea starch, has also been subject to the structural analyses of intermediate materials. Different mutations give rise to the wrinkled character (Kooistra 1962, Lloyd et al. 1996, Smith et al. 1989) and can possibly contribute to different opinions about the nature of this fraction. Banks et al. (1974) found that the intermediate material consists of very short amylose chains mixed with normal amylopectin, whereas others described it as being amylopectin-like, but with much longer chains (Biliaderis et al. 1981, Boyer et al. 1980, Matheson 1990, Potter et al. 1953). Colonna and Mercier (1984) found that the intermediate material was similar to amylopectin, but with lower molecular weight and different proportions of long and short chains, and this view was largely supported by Bertoft et al. (1993b).

Intermediate materials in oats were described by Wang and White (1994), who found that the material was similar to amylopectin in structure but had longer unit chain lengths. Paton (1979) found a fraction in oat starch with sizes intermediate to that of amylopectin and amylose. On the other hand, Manelius and Bertoft (1996) could not confirm the existence of intermediate materials in oat starch. Waduge et al. (2014) found intermediate materials in developing wheat starch granules based on their iodine binding properties, which suggested comparatively long iodine-complexing segments in the molecules.

A special kind of intermediate material is (hydrosoluble) phytoglycogen, which has been described in isoamylase deficient maize, sorghum and rice (Boyer and Liu 1983, Inouchi et al. 1983, Wong et al. 2003). The mutation is known as *sugary-1*. Phytoglycogen is extremely branched compared to amylopectin: about 9% of the glucosyl units are branched (Scheffler et al. 2010, Shin et al. 2008). The molecule has a unit chain profile that resembles that of glycogen, i.e., a unimodal size-distribution lacking long B-chains as opposed to the bi- or polymodal distribution of amylopectin (Gunja-Smith et al. 1970, Inouchi et al. 1987, Palmer et al. 1983, Shin et al. 2008, Wong et al. 2003). Phytoglycogen particles from maize exhibit spheroidal shape with diameters between 30–100 nm (Putaux et al. 1999). Interestingly, this water-soluble molecule is not found in the granular starch of the plant, which suggests that debranching is a part of the normal starch biosynthetic process (James et al. 1995). The fact that phytoglycogen does not crystallize suggests that partial debranching of amylopectin by isoamylase during starch biosynthesis, also called 'trimming', is a prerequisite in order to form starch granules (Ball et al. 1996, Mouille et al. 1996).

8.5 Starch Granules

The native form of starch exists as discrete water-insoluble granules having a semi-crystalline structure, i.e., they contain both crystalline and amorphous parts. The relative crystallinity range from 17 to about 50% in different plants, and granules that lack amylose are generally slightly more crystalline than the amylose-containing counterparts (Bertoft et al. 2008, Buléon et al. 1998a, Ren 2017, Srichuwong et al. 2005, Witt et al. 2012). Starch granules have a wide diversity of shapes and sizes but are usually defined by the plant species and the tissue in which they are formed (Jane et al. 1994, Li and Yeh 2001, Ren 2017, Srichuwong et al. 2005). Very small granules with diameters of < 1 μm to a few μm are found in, for example, amaranth (Linsberger-Martin et al. 2012, Wilhelm et al. 2002) and quinoa (Linsberger-Martin et al. 2012, Steffolani et al. 2013). Rice has small granules that form groups known as compound granules (Ratnayake and Jackson 2007). Barley, wheat, rye and triticale have two distinct size-groups (Fredriksson et al. 1998), of which the large granules (known as A-granules) have lenticular shape and diameters around 20 μm, whereas the smaller, round or irregular B-granules are 3 ~ 10 μm (Ao and Jane 2007, Kim and Huber 2008). Maize has both round and polyhedral granules (Banks et al. 1974). The round to oval granules in potato tubers have a large size-distribution up to > 100 μm in diameter (Gomand et al. 2010). The surface of wheat granules is comparatively smooth and shows protrusions of about 10–50 nm in diameter when studied by atomic force microscopy (AFM). Potato granules have larger protrusions with diameters about 50–300 nm (Baldwin et al. 1998). Many granules, for example in wheat, barley, maize and sorghum, have pores or channels that penetrate into the interior of the granules (Fannon et al. 1992, 1993, Glaring et al. 2006, Huber and BeMiller 1997). These channels were shown to contain proteins and phospholipids (Han et al. 2005, Kim and Huber 2008, Naguleswaran et al. 2011). Other granules, for example from potato tubers, lack channels and appear to be comparatively resistant to penetration by enzymes or chemicals (Fannon et al. 1992).

The starch granule is kept intact without covalent bindings between the polyglucan constituents. Amylopectin is the principal component of the granules that contributes to the granular structure, which is shown by the fact that amylose-free granules (waxy granules) generally retain the size and shape of the normal, amylose-containing counterparts. So-called high amylose-containing granules, on the other hand, are often deformed and obtain elongated or irregular shapes (Banks et al. 1974, Glaring et al. 2006, Yano et al. 1985). For long, it was therefore believed that granular starch could not exist in the absence of amylopectin. However, recently, a mutant barley starch with < 1% amylopectin was engineered having granular starch, albeit with deformed, irregular shapes (Goldstein et al. 2016). The exact location of amylose in the granules remains a matter of debate. By pealing the starch granules from potato tubers layer-by-layer in concentrated $CaCl_2$ solution, Jane et al. (1993) found that amylose is more confined to the peripheral layers of the granules. They also found that the amylopectin component at the core of the granules had longer BL-chains than in the peripheral parts. On the other hand, Blennow et al. (2003), who used confocal laser scanning microscopy to analyze potato starch granules, found the opposite, i.e., amylose tends to be more concentrated in the interior of the granules. A similar conclusion was made by Buléon et al. (2014), who analysed phosphorus by synchrotron X-ray microfluorescence mapping.

8.5.1 Internal Architecture of Starch Granules

Amylose is generally believed to exist in a disorganized, amorphous state inside the starch granules, though in the so-called high-amylose starches some amylose molecules might form double-helices and partly crystallize into the B-allomorph pattern and/or crystallize together with amylopectin chains (Gérard et al. 2002). Amylose is known to form V-type inclusion complexes with lysophospholipids in cereal starches (Morrison et al. 1993). Amylopectin forms a semi-crystalline "skeleton" with the short chains organized in a radial fashion having the non-reducing ends directed towards the surface (French 1972). This is shown by the fact that the granules are birefringent possessing a Maltese cross seen in cross-polarized light in a microscope (Fig. 8.8). The cross extends from the so-called hilum, which in some granules is situated at the centre of the granules, while in others it is eccentric. Rings, or shells, known as "growth rings", surround the hilum and are observable in an optical microscope or by scanning electron microscopy, especially after a gentle treatment of the granules in dilute acid or with α-amylase (Goldstein et al. 2017b, Pilling and Smith 2003). However, the granules of lesser yam, and possibly also shoti starch, were shown to possess "layers" rather than rings (Vamadevan et al. 2018). Also, protrusions of the giant granules of *Phajus grandifolius* appear to have more layer-like structures (Chanzy et al. 2006). The rings, or layers, are thought to represent semi-crystalline structures with a thickness roughly about 100 μm embedded in an amorphous background (Jenkins et al. 1993). The rings are generally thinner towards the periphery of the granules. "Growth rings" in cereals were believed to be the result of diurnal photosynthesis of plants synthesizing the crystalline areas during daytime (Buttrose 1962). Wheat and barley grown under constant light were reported to possess granules devoid of the rings (Buttrose 1960, 1962, Sande-Bakhuyzen 1926). However, a later investigation with barley could not confirm the older results, as granules grown under constant light retained the rings and, in addition, the relative crystallinity of the granules was found to be slightly lower than when growing in diurnal light conditions (Goldstein et al. 2017b). Growth rings also remain in starch granules from potato tubers grown under either constant light or constant dark conditions (Buttrose 1962, Pilling and Smith 2003).

Diverse microscopic techniques have revealed boll-like structures called blocklets in the "growth rings" (Fig. 8.8), the nature of which remains obscure (Atkin et al. 1998, Baker et al. 2001, Gallant

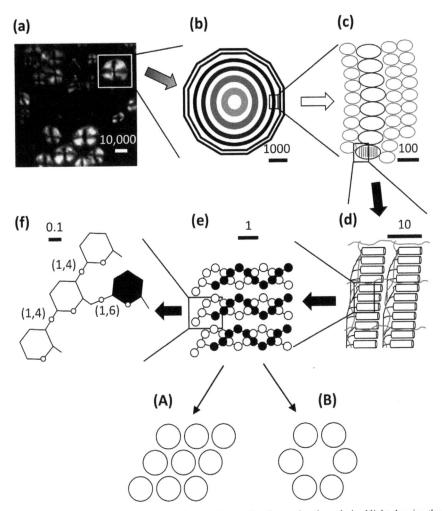

Figure 8.8. From granules to glucosyl units. (a) Maize starch granules observed under polarized light showing the "Maltese cross". (b) A hypothetical granule with growth rings extending from the hilum. (c) Blocklets in semi-crystalline (black) and amorphous (grey) rings. (d) Crystalline and amorphous lamellae formed by double helices (cylinders) and branched segments of amylopectin (black lines), respectively. Amylose molecules (red lines) are interspersed among the amylopectin molecules. (e) Three double-helices of amylopectin. Each double-helix consists of two polyglucosyl chains, in which the glucosyl units are symbolized by white and black circles, respectively. The double helices form either A- or B-type allomorphic crystals (A and B, respectively, in which the circles symbolize the double-helices seen from the edge). (f) Glucosyl units showing α-(1,4)- and α-(1,6)-linkages at the base of the double-helix. The bar scale (in nm) is only approximate to give an impression of the size dimensions. Reproduced from Agronomy 7(3) (2017): 56, doi 10.3390/agronomy7030056.

et al. 1997, Huang et al. 2014, Ohtani et al. 2000, Perez-Herrera et al. 2017, Ridout et al. 2002). The blocklets, which have diameters in the order of 10–500 nm (Baker et al. 2001, Dang and Copeland 2003), were found in both crystalline and amorphous rings (also described as dark, hard bands and bright, soft bands) and were proposed to consist of amylopectin and embedded in an amylose matrix (Gallant et al. 1997, Ridout et al. 2006). Tang et al. (2006) suggested that amylose might also be involved in the blocklet structure and proposed that hard shells have "normal" blocklets, whereas soft shells have "defective" blocklets. Blocklets in maize and potato starch granules were found to swell and merge upon hydration during *in situ* AFM observations (Park et al. 2011).

Investigations by small-angle X-ray scattering (SAXS) have revealed a characteristic repeat-distance of 9 ~ 10 nm, which is universal for all granules regardless of their botanical origin (Cameron and Donald 1992, Daniels and Donald 2003, Jenkins et al. 1993, Kiseleva et al. 2005, Sanderson et al. 2006, Vermeylen et al. 2004, Wang et al. 2012). This repeat-distance stems from the organization of amylopectin into alternating amorphous and crystalline lamellae that forms stacks within the semi-crystalline rings (Fig. 8.8). The crystalline lamellae are formed by the short chains of amylopectin and have a thickness of 4 ~ 6 nm (Andreev et al. 1999, Genkina et al. 2003, 2007, Kiseleva et al. 2005), corresponding to a chain length of 11 ~ 17 glucosyl units, which is within the range of the external chain length of amylopectin (Table 8.1). The chains form double helices that extend from the outermost branch points (Imberty and Pérez 1989) and crystallize into either an A- or B-allomorph pattern (Fig. 8.8), which is distinguished by wide-angle X-ray diffraction (XRD or WAXS) (Buléon et al. 1987). Therefore, starch granules are generally divided into A- and B-crystalline granules, the former found especially in cereals and the latter in many roots and tubers (Hizukuri 1985), but also in the transient granules of *Arabidopsis* leaves (Wattebled et al. 2008). However, some starches have a mixed XRD pattern known as C-type, and were shown to have both types of crystallites within the same granule (Bogracheva et al. 1998, Buléon et al. 1998b). For example, sweet potato and most legume starches have C-type granules (Gernat et al. 1990, Hizukuri 1985). The A-type crystallinity has been associated with more dry conditions during starch development, whereas B-type starch is formed in wet conditions. The structure of the B-type allomorph in amylopectin is the same as found in retrograded (recrystallized) amyloses, whereas the A-type allomorph has a more compact structure with only 8 water molecules in the crystalline monoclinic unit cell, which has the dimensions $a = 2.083$ nm, $b = 1.145$ nm, and $c = 1.058$ nm (Popov et al. 2009). Amylopectin in B-crystalline starches generally has longer average CL than in A-crystalline starches (Hizukuri 1985).

The amorphous lamellae are believed to consist of the internal parts of the amylopectin molecules, i.e., the internal segments with most of the branches, although some branches are probably scattered into the crystalline lamellae as well, especially in A-crystalline starches (Jane et al. 1997). Amylose is probably also found in the amorphous parts, but it might also traverse the crystalline lamellae causing defects in the crystallites (Genkina et al. 2007). The involvement of amylose in the crystalline lamellae appears to depend on the type of starch as well as the amylose content of the granules (Gérard et al. 2002, Koroteeva et al. 2007, Kozlov et al. 2007a,b). Amylose is also believed to be a major component of the amorphous background, or amorphous "growth rings" (Atkin et al. 1999), though amylopectin is also found there, because the ring structure apparently exists both in amylose-containing and in waxy granules (Atkin et al. 1998, Glaring et al. 2006, Goldstein et al. 2017b, Li et al. 2003).

8.6 Structural Models of Amylopectin

Though the structure of the alternating crystalline and amorphous lamellae in starch granules is well known, it remains a matter of debate exactly how amylopectin contributes to their architecture, and this again depends on the exact molecular structure of the macromolecule itself. Two models of the amylopectin structure are currently under consideration: the cluster model, which represents a traditional view, and the building block backbone model, which is more recent.

8.6.1 The Cluster Model

Several investigators in 1969 and the early 1970s proposed the cluster model independently (French 1972, Nikuni 1969, Robin et al. 1974). The model was based on important findings in the preceding decade, such as the finding of a repeat spacing of approximately 10 nm in starch granules (Sterling 1962) and the bimodal size-distribution of the chains in amylopectin, which was made possible by the discovery of debranching enzymes and the development of gel-permeation chromatography (Gunja-Smith et al. 1970). Analysis of acid-treated starch granules, in which mostly only the crystalline, short chains remain and most of the branches disappear, was also an important contributor (Kainuma and French 1971, 1972, Robin et al. 1974, Umeki and Kainuma 1981). These results suggested that the short chains are found in clusters and are responsible for the structure of the crystalline parts in the granules, whereas the branches are part of the amorphous layers. The Maltese cross shown in polarized light suggested that the amylopectin molecules are oriented radially toward the granule surface (French 1972). Taken together, these results strongly support a model, in which the short chains are clustered together and the long chains stretch through the clusters, thus interconnecting them, and stretch further through the stacks of lamellae, so that the entire macromolecule is directed towards the surface of the granule (Fig. 8.9) (Cameron and Donald 1992).

Later in 1986, Hizukuri refined the model based on the finding of a periodicity of 27–28 residues, as discussed above. He suggested that this periodicity stems from the regular, repeating distance of the lamellae, as the chains would correspond to 9–10 nm in the form of the double-helix (in which

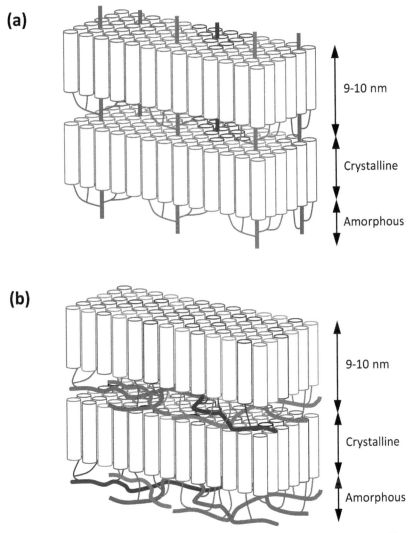

Figure 8.9. Amylopectin as the principal component of the stacks of amorphous and crystalline lamellae in (A-type crystalline) starch granules. (a) Possible arrangement of double-helices (cylinders) in the crystalline lamellae based on the cluster model. Local crystallites are formed by each cluster of chains, and individual amylopectin molecules (symbolized by different colours) are lined up side-by-side and penetrate the stacks of lamellae. Long B-chains are drawn as thick lines. (b) Possible arrangement of the double-helices based on the building block backbone model. Local crystallites constitute cooperative structures formed by double-helices from several individual amylopectin molecules, which form the stacks of lamellae layer-by-layer. Long chains of the backbone (thick lines) extend in diverse directions and are embedded in the amorphous lamellae (but their external segments possibly form double helices with the short chains and penetrate into the crystalline part).

each residue corresponds to a distance of 0.35 nm in the length direction of the axis). Moreover, he suggested that A-chains together with the short B1-chains form the clusters, whereas B2-chains interconnect two clusters, B3-chains three clusters, and so on. At about the same time, Imberty and co-workers (Imberty et al. 1987, Imberty and Pérez 1988) revealed the fine structure of the A- and B-type allomorphs in amylose and showed that short chains in amylopectin can also form these crystallites without being disturbed by the branch point (Imberty and Pérez 1989). On the contrary, they found that the branch linkage actually stabilizes the helix. Finally, the length of the crystalline chains in acid-treated starches closely matches the thickness of the crystalline lamellae in the starch granules (Umeki and Kainuma 1981). Thus, it is beyond doubt that the cluster model nicely explains how the cluster structure of amylopectin fits into what is known about the lamellar structure in the granules.

However, as already discussed above, in 1996 Hanashiro et al. found another, shorter periodicity in chain length of only 12, instead of 27–28 residues. In their discussion, they proposed that both B1- and B2-chains are of shorter lengths than previously suggested, and only B3-chains correspond to long chains (DP > 36), albeit also being shorter than the earlier proposed DP around 70. This, then, was not in accordance with the chain periodicity on which the proposed cluster model rests. The starch research community has largely overseen this obstacle, however.

8.6.2 The Building Block Backbone Model

It should be noticed that the cluster model is based solely on indirect evidences for the existence of the clusters, not on direct evidences. In addition, no definition was proposed for the structural features of the clusters, i.e., how close together should chains be in order to be considered as "clustered"? To prove the structure, it would be necessary to isolate clusters from amylopectin and show that they have the expected structure. The cluster model predicts that clusters consist of an average number of chains that closely corresponds to the ratio of short:long (S:L) chains in amylopectin (Hanashiro et al. 2002, Takeda et al. 2003) because only short chains form the clusters and long chains interconnect them. Actually, the number of chains in isolated clusters should be closer to S:L + 1, because the long chains have to be cleaved into short chains in order to cut them out from the macromolecule. Another distinct feature would be that isolated clusters only contain short chains, because of the same reason. Finally, one would expect that the average internal chain length (ICL) of isolated clusters is considerably shorter than in amylopectin.

As has been discussed already above, α-dextrins have been isolated from amylopectin and structurally characterized (Table 8.2). These dextrins were first believed to be the clusters of chains predicted by the cluster model, but consistent results from a large variety of starch samples could not confirm their structures as being compatible with the cluster model (Bertoft 2007b, Bertoft et al. 2012a, Laohaphatanaleart et al. 2010). That is, isolated clusters contain long chains, the average ICL is only slightly shorter than in amylopectin, and the NC of the clusters is mostly either higher or lower than expected (compare Tables 8.1 and 8.2). In light of these results, another model was therefore invented in agreement with the new experimental facts that were not available at the time of the proposal of the cluster model (Bertoft 2013). The new model is called the building block backbone model, which implies the two important structural motifs that it contains. The new model is also compatible with the side-chain liquid crystalline model proposed by Waigh et al. (2000), who were studying the properties of starch. Further, and importantly, the model is also compatible with all the results in favour of the cluster model discussed above.

The building block backbone model suggests that amylopectin consists of a backbone of mostly the long chains (B2- and B3-chains) of amylopectin (Fig. 8.9); however, some B1-chains might also be included, especially in the more branched type 1 amylopectins (Bertoft et al. 2012a). To this backbone, the short chains are attached. (Note that most other branched polysaccharides also have backbones with shorter branches attached: the exceptions appear to be phytoglycogen and glycogen.) The short chains are organized along the backbone in small, tight groups, namely the building blocks, which are separated by the inter-block segments (DP 5 ~ 8). Occasionally, and without any regularity

(Källman et al. 2013), these segments are longer and constitute the segments that are easily attacked by the α-amylase of *B. amyloliquefaciens* during initial attack. These latter segments have CL in the order of 9 ~ 15 residues (Kong et al. 2009, Laohaphatanaleart et al. 2010, Zhu et al. 2011c). The branched dextrins that are isolated post this initial event are the *clusters of building blocks* (and not clusters of chains, as in the cluster model). Due to the random distribution of inter-block chains with DP ≥ 9, the size-distribution of the isolated clusters of building blocks is broad, as found experimentally (Bertoft et al. 2012a). It is noted that in the backbone model, the long chains mediate the interconnection of building blocks, not clusters of chains as in the traditional model and, therefore, the nomenclature (B2, B3, etc.) does not imply their involvement in cluster interconnection, only their chain lengths. Depending on how many building block segments are arranged along the long chains without interruption by longer internal segments (ICL ≥ 9), the chains attain quite variable lengths in the isolated clusters of building blocks, which explains the existence of remaining long chains (b2- and b3-chains) in the clusters (cf. Fig. 8.7). Clusters from type 4 amylopectins have generally long b-chains with building blocks basically outspread along the backbone. Type 1 amylopectins, however, have more branches along the backbone in the form of b1-chains (Fig. 8.7) (Bertoft et al. 2012a). As the name b1-chains suggests, they have a single inter-block segment and connects to a building block beside the backbone.

The backbone model of amylopectin also fits nicely into the lamellar organization in starch granules. In this case, the backbone, and therefore the entire macromolecule, extends tangentially along the amorphous lamella, whereas the short chains extend into the crystalline lamella and form double-helices, just as in the cluster model (Fig. 8.9). The major difference in the cluster model is the direction of the long chains in the granule: in the cluster model, they are radial, whereas in the new model they are tangential. How, then, are the crystallites formed between the double-helices if there are no clusters of short chains? Ball et al. (1996) proposed that too tightly branched glucopolymers cannot crystallize, and therefore molecules such as phytoglycogen and glycogen are not crystalline, whereas the cluster structure provides groups of chains in optimal organization for crystallization to occur. In this context, it is of interest to discuss the work of O'Sullivan and Pérez (1999), who analyzed the structural conditions that enable two double-helices to align in parallel fashion by using molecular modelling techniques, in which they tested the preferred internal chain lengths that interconnect the two double-helices (segments m and n in Fig. 8.10). They found that, out of an extensive number of tested combinations, only two combinations of segments m and n give rise to a thermodynamically stable A-type allomorph and two combinations give the B-type structure, one of which is common with that giving the A-type structure. A fourth combination (m = n = 7) could give parallel double-helices situated in different clusters within the context of the traditional model (but would appear to be different building blocks in the backbone model). Considering the broad range of internal chains with DP ≥ 3 that amylopectin consists of (cf. Fig. 8.4), it is striking that as few as only two combinations of the internal segments m and n could give rise to either A- or B-crystallites. The fact that the degree of crystallinity is around 40% in waxy starches suggests that the vast majority of the external chains in amylopectin form double-helices and crystallize. Therefore, chains with DP < 5 or DP < 6 should hardly exist in φ,β-limit dextrins of A- or B-crystalline samples, respectively (Fig. 8.10), but they do (Fig. 8.4). Therefore, it appears that most of the chains as depicted in the traditional cluster model are too tightly packed to be able to crystallize.

Another interesting work was done by Putaux et al. (2003), who isolated nanocrystals from acid-treated waxy-maize starch. The nanocrystals consisted of discrete platelets corresponding to the crystalline lamellae in the granules with dimensions suggesting that each platelet contained more than hundred double-helices (Pérez and Bertoft 2010). This strongly suggests that the crystallites in starch granules are co-operative structures composed of double-helices from several amylopectin molecules lining up together, rather than crystallites being formed by clusters of chains within the same molecule. The building block backbone model offers a convenient explanation to this finding (Fig. 8.9): the backbone is a flexible structure that intermingles with backbone segments of neighbouring molecules in the amorphous lamella in such a way that their double-helices can crystallize into either

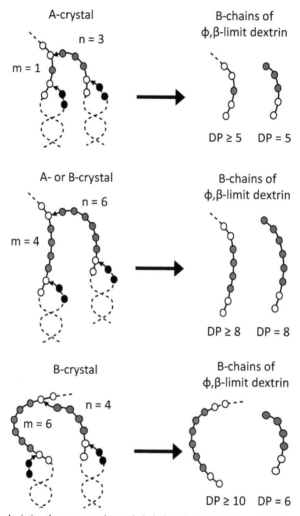

Figure 8.10. A schematic depicting the necessary internal chain length segments (m and n) between two paralleled aligned double-helices forming either A- or B-type crystallites based on the work of O'Sullivan and Pérez (1999), and the expected B-chain lengths obtained by debranching of the φ,β-limit dextrin of the amylopectin component participating in the respective crystallites. m-segments are shown in red, n-segments in blue, branches as arrows, and dotted lines traces the external segments of the double-helices as well as the variable length of the internal part of the B-chain carrying the m-segment. Black circles are glucosyl units of A-chains as they will appear in the limit dextrin.

A- or B-patterns and, thus, intra-molecular internal chain segments between double-helices are less important. Interestingly, this implies that the structure of the backbone in the amorphous lamella to a large extent is indirectly responsible for the packing ability of the double-helices in the crystalline lamella. Vamadevan and co-workers (2013a,b) argued that longer inter-block chain segments give a more flexible backbone that enables more perfect crystalline structure, whereas short segments restrict the parallel alignment of double-helices, resulting in weak crystals with low melting temperature.

The backbone model was also used to explain the swelling of starch granules (Bertoft and Nilsson 2017), which happens when they are heated in water. The swelling would be possible because of the network of backbones in the amorphous lamellae (Fig. 8.9) that allows the lamellae to expand in all directions without the backbones losing contact with each other, much resembling the surface of a balloon. Only when the contact is lost, the granule finally disintegrates. The model was also used to explain retrogradation (recrystallization) of dissolved amylopectin (Vamadevan and Bertoft 2018). External chains form intra- and intermolecular double-helices, the latter being junction zones between molecules. The formation of double-helices is supported by a flexible backbone, the flexibility being higher with longer inter-block segments, which enables a firmer contact between the helices. Double-helices were also proposed to be possible between external chains and longer internal segments in adjacent molecules (Bertoft et al. 2016). Longer internal segments are possibly found in single-helical conformation, as shown by the fact that limit dextrins of waxy starches bind iodine (Chauhan and Seetharaman 2013, Shen et al. 2013).

Finally, the "trimming" hypothesis, presented by Ball in 1995, suggests that partial debranching of amylopectin during biosynthesis is necessary to form clusters with double-helices that are able to crystallize, otherwise they will remain too densely branched to be organized, resulting in phytoglycogen. Likewise, but based on the backbone model, one could argue that the "trimming" is necessary in order to form the backbone, otherwise phytoglycogen, which lacks the backbone (Scheffler et al. 2010), is formed. That is, the model suggests that only a critical density of branches and branch lengths along the backbone allows the organization of chains into the characteristic lamellar structure of starch granules. A too densely branched structure will lack the backbone, whereas a too loosely branched backbone will not give rise to the lamellae and result in intermediate materials and eventually amylose, which might partly crystallize, albeit not in the same ordered fashion shown by the delicate organization of stacks of lamellae in starch granules.

8.7 Conclusions

Starch granules consist primarily of amylose and amylopectin and eventually intermediate materials. Amylose is a linear or slightly branched polymer with mostly long chains, whereas amylopectin is highly branched and has comparatively short chains. The structure of intermediate materials appears to be in between that of amylose and amylopectin. All three types of polymers can therefore be considered as having a basic structural motif in common, namely, a backbone structure (Bertoft 2013): amylopectin in type 1 starches is representative of the most extreme branched structure (besides phytoglycogen), in which a backbone is built from several long B-chains and it carries several shorter or longer branches, whereas type 4 amylopectin has a less branched backbone. Intermediate materials have even less branched backbone with longer chains as well as longer branches attached. Branched amylose has a sparsely branched backbone with few long branch-chains (though short chains also exist) and, finally, linear amylose has only a single, very long chain representing the most extreme, non-branched backbone structure.

Two models exist at present that tend to explain the role of amylopectin as the principal contributor to the lamellar organization of the starch granules. The cluster model suggests that amylopectin molecules align in a radial fashion penetrating the stacks of alternating amorphous and crystalline lamellae, so that each layer consists of clusters of short chains that crystallize and is interconnected to the second layer by long chains. The building block backbone model suggests that each layer consists of a network of amylopectin molecules extending in the tangential direction within the amorphous lamella and with short chains extending into the crystalline lamellae. The crystalline lamellae are cooperative structures formed by double-helices from diverse amylopectin molecules and the stacks of lamellae consist of molecules laid down layer-by-layer.

References

Aberle, T., W. Burchard, W. Vorwerg and S. Radosta. 1994. Conformational contributions of amylose and amylopectin to the structural properties of starches from various sources. Starch/Stärke 46: 329–335.

Adkins, G.K. and C.T. Greenwood. 1969. Studies on starches of high amylose-content. Part X. An improved method for the fractionation of maize and amylomaize starches by complex formation from aqueous dispersion after pretreatment with methyl sulphoxide. Carbohydr. Res. 11: 217–224.

Akai, H., K. Yokobayashi, A. Misaki and T. Harada. 1971. Structural analysis of amylopectin using Pseudomonas isoamylase. Biochim. Biophys. Acta 252: 427–431.

Andersson, L., H. Fredriksson, M. Oscarsson-Bergh, R. Andersson and P. Åman. 1999. Characterisation of starch from inner and peripheral parts of normal and waxy barley kernels. J. Cereal Sci. 30: 165–171.

Andreev, N.R., E.N. Kalistratova, L.A. Wasserman and V.P. Yuryev. 1999. The influence of heating rate and annealing on the melting thermodynamic parameters of some cereal starches in excess water. Starch/Stärke 51: 422–429.

Annor, G.A., M. Marcone, E. Bertoft and K. Seetharaman. 2014. Unit and internal chain profile of millet amylopectin. Cereal Chem. 91: 29–34.

Ao, Z. and J.-l. Jane. 2007. Characterization and modeling of the A- and B-granule starches of wheat, triticale, and barley. Carbohydr. Polym. 67: 46–55.

Apinan, S., I. Yujiro, Y. Hidefumi, F. Takeshi, P. Myllärinen, P. Forssell et al. 2007. Visual observation of hydrolyzed potato starch granules by α-amylase with confocal laser scanning microscopy. Starch/Stärke 59: 543–548.

Atkin, N.J., R.M. Abeysekera, S.L. Cheng and A.W. Robards. 1998. An experimentally-based predictive model for the separation of amylopectin subunits during starch gelatinization. Carbohydr. Polym. 36: 173–192.

Atkin, N.J., S.L. Cheng, R.M. Abeysekera and A.W. Robards. 1999. Localisation of amylose and amylopectin in starch granules using enzyme-gold labelling. Starch/Stärke 51: 163–172.

Autio, K., T. Suortti, A. Hamunen and K. Poutanen. 1992. Microstructural and physicochemical properties of oxidized potato starch for paper coating. Starch/Stärke 44: 393–398.

Bailey, J.M. and W.J. Whelan. 1961. Physical properties of starch. I. Relationship between iodine stain and chain length. J. Biol. Chem. 236: 969–973.

Baker, A.A., M.J. Miles and W. Helbert. 2001. Internal structure of the starch granule revealed by AFM. Carbohydr. Res. 330: 249–256.

Baldwin, P.M., J. Adler, M.C. Davies and C.D. Melia. 1998. High resolution imaging of starch granule surfaces by atomic force microscopy. J. Cereal Sci. 27: 255–265.

Ball, S.G. 1995. Recent views on the biosynthesis of the plant starch granule. Trends Glycosci. Glycotechnol. 7: 405–415.

Ball, S., H.-P. Guan, M. James, A. Myers, P. Keeling, G. Mouille et al. 1996. From glycogen to amylopectin: A model for the biogenesis of the plant starch granule. Cell 86: 349–352.

Banks, W., C.T. Greenwood and K.M. Khan. 1971. The interaction of linear amylose oligomers with iodine. Carbohydr. Res. 17: 25–33.

Banks, W., C.T. Greenwood and D.D. Muir. 1974. Studies on starches of high amylose content. Part 17. A review of current concepts. Starch/Stärke 26: 289–300.

Banks, W. and C.T. Greenwood. 1975. Starch and its components. Edinburgh University Press, Edinburgh.

Bender, H., R. Siebert and A. Stadler-Szöke. 1982. Can cyclodextrin glycosyltransferase be useful for the investigation of the fine structure of amylopectins? Characterisation of highly branched clusters isolated from digests with potato and maize starches. Carbohydr. Res. 110: 245–259.

Bertoft, E. 1986. Hydrolysis of amylopectin by the alpha-amylase of B. subtilis. Carbohydr. Res. 149: 379–387.

Bertoft, E. and L. Spoof. 1989. Fractional precipitation of amylopectin alpha-dextrins using methanol. Carbohydr. Res. 189: 169–180.

Bertoft, E. 1989a. Partial characterisation of amylopectin alpha-dextrins. Carbohydr. Res. 189: 181–193.

Bertoft, E. 1989b. Investigation of the fine structure of amylopectin using alpha- and beta-amylase. Carbohydr. Res. 189: 195–207.

Bertoft, E. 1991. Investigation of the fine structure of alpha-dextrins derived from amylopectin and their relation to the structure of waxy-maize starch. Carbohydr. Res. 212: 229–244.

Bertoft, E., Z. Qin and R. Manelius. 1993a. Studies on the structure of pea starches. Part 3: Amylopectin of smooth pea starch. Starch/Stärke 45: 377–382.

Bertoft, E., Z. Qin and R. Manelius. 1993b. Studies on the structure of pea starches. Part 4: Intermediate material of wrinkled pea starch. Starch/Stärke 45: 420–425.

Bertoft, E., Q. Zhu, H. Andtfolk and M. Jungner. 1999. Structural heterogeneity in waxy-rice starch. Carbohydr. Polym. 38: 349–359.

Bertoft, E., C. Boyer, R. Manelius and A.-K. Åvall. 2000. Observations on the α-amylolysis pattern of some waxy maize starches from the inbred line Ia453. Cereal Chem. 77: 657–664.

Bertoft, E. 2004. On the nature of categories of chains in amylopectin and their connection to the super helix model. Carbohydr. Polym. 57: 211–224.

Bertoft, E. 2007a. Composition of clusters and their arrangement in potato amylopectin. Carbohydr. Polym. 68: 433–446.

Bertoft, E. 2007b. Composition of building blocks in clusters from potato amylopectin. Carbohydr. Polym. 70: 123–136.

Bertoft, E., K. Piyachomkwan, P. Chatakanonda and K. Sriroth. 2008. Internal unit chain composition in amylopectins. Carbohydr. Polym. 74: 527–543.

Bertoft, E., K. Laohaphatanaleart, K. Piyachomkwan and K. Sriroth. 2010. The fine structure of cassava amylopectin. Part 2. Building block structure of clusters. Int. J. Biol. Macromol. 47: 325–335.

Bertoft, E., A. Källman, K. Koch, R. Andersson and P. Åman. 2011a. The building block structure of barley amylopectin. Int. J. Biol. Macromol. 49: 900–909.

Bertoft, E., A. Källman, K. Koch, R. Andersson and P. Åman. 2011b. The cluster structure of barley amylopectins of different genetic backgrounds. Int. J. Biol. Macromol. 49: 441–453.

Bertoft, E. and K. Seetharaman. 2012. Starch structure. pp. 1–27. *In*: Tetlow, I. (ed.). Starch: Origins, Structure and Metabolism. The Society for Experimental Biology, London.

Bertoft, E., K. Koch and P. Åman. 2012a. Structure of building blocks in amylopectins. Carbohydr. Res. 361: 105–113.

Bertoft, E., K. Koch and P. Åman. 2012b. Building block organisation of clusters in amylopectin of different structural types. Int. J. Biol. Macromol. 50: 1212–1223.

Bertoft, E. 2013. On the building block and backbone concepts of amylopectin structure. Cereal Chem. 90: 294–311.

Bertoft, E., G.A. Annor, X. Shen, P. Rumpagaporn, K. Seetharaman and B.R. Hamaker. 2016. Small differences in amylopectin fine structure may explain large functional differences of starch. Carbohydr. Polym. 140: 113–121.

Bertoft, E. and L. Nilsson. 2017. Starch. Analytical and structural aspects. pp. 377–478. *In*: Eliasson, A.-C. (ed.). Carbohydrates in Food. CRC Press, Taylor & Francis Group, Boca Raton.

Biais, B., P. LeBail, P. Robert, B. Pontoire and A. Buléon. 2006. Structural and stoichiometric studies of complexes between aroma compounds and amylose. Polymorphic transitions and quantification in amorphous and crystalline areas. Carbohydr. Polym. 66: 306–315.

Biliaderis, C.G., D.R. Grant and J.R. Vose. 1981. Structural characterization of legume starches. I. Studies on amylose, amylopectin, and beta-limit dextrins. Cereal Chem. 58: 496–502.

Blennow, A., A.M. Bay-Smidt, C.E. Olsen and B. Lindberg-Møller. 1998a. Analysis of starch-bound glucose 3-phosphate and glucose 6-phosphate using controlled acid treatment combined with high-performance anion-exchange chromatography. J. Chromatogr. A 829: 385–391.

Blennow, A., A.M. Bay-Smidt, B. Wischmann, C.E. Olsen and B. Lindberg-Møller. 1998b. The degree of starch phosphorylation is related to the chain length distribution of the neutral and the phosphorylated chains of amylopectin. Carbohydr. Res. 307: 45–54.

Blennow, A., M. Hansen, A. Schulz, K. Jørgensen, A.M. Donald and J. Sanderson. 2003. The molecular deposition of transgenically modified starch in the starch granule as imaged by functional microscopy. J. Struct. Biol. 143: 229–241.

Blennow, A., A.K. Sjöland, R. Andersson and P. Kristiansson. 2005. The distribution of elements in the native starch granule as studied by particle-induced X-ray emission and complementary methods. Anal. Biochem. 347: 327–329.

Bogracheva, T.Y., V.J. Morris, S.G. Ring and C.L. Hedley. 1998. The granular structure of C-type pea starch and its role in gelatinization. Biopolymers 45: 323–332.

Boyer, C.D., P.A. Damewood and G.L. Matters. 1980. Effect of gene dosage at high amylose loci on properties of the amylopectin fractions of the starches. Starch/Stärke 32: 217–222.

Boyer, C.D. and K.-C. Liu. 1983. Starch and water-soluble polysaccharides from sugary endosperm of sorghum. Phytochemistry 22: 2513–2515.

Boyer, C.D. and K.-C. Liu. 1985. The interaction of endosperm genotype and genetic background. Part I. Differences in chromatographic profiles of starches from nonmutant and mutant endosperms. Starch/Stärke 37: 73–79.

Buléon, A., H. Bizot, M.M. Delage and B. Pontoire. 1987. Comparison of X-ray diffraction patterns and sorption properties of the hydrolyzed starches of potato, wrinkled and smooth pea, broad bean and wheat. Carbohydr. Polym. 7: 461–482.

Buléon, A., P. Colonna, V. Planchot and S. Ball. 1998a. Starch granules: structure and biosynthesis. Int. J. Biol. Macromol. 23: 85–112.

Buléon, A., C. Gérard, C. Riekel, R. Vuong and H. Chanzy. 1998b. Details of the crystalline ultrastructure of C-starch granules revealed by synchrotron microfocus mapping. Macromolecules 31: 6605–6610.

Buléon, A., M. Cotte, J.-L. Putaux, C. D'Hulst and J. Susini. 2014. Tracking sulfur and phosphorus within starch granules using synchrotron X-ray microfluorescence mapping. Biochim. Biophys. Acta 1840: 113–119.

Buttrose, M.S. 1960. Submicroscopic development and structure of starch granules in cereal endosperm. J. Ultrastruct. Res. 4: 231–257.

Buttrose, M.S. 1962. The influence of environment on the shell structure of starch granules. J. Cell Biol. 14: 159–167.

Cameron, R.E. and A.M. Donald. 1992. A small-angle X-ray scattering study of the annealing and gelatinization of starch. Polymer 33: 2628–2635.

Carciofi, M., A. Blennow, S.L. Jensen, S.S. Shaik, A. Henriksen, A. Buléon et al. 2012. Concerted suppression of all starch branching enzyme genes in barley produces amylose-only starch granules. BMC Plant Biol. 12: 223–238.

Chanzy, H., J.-L. Putaux, D. Dupeyre, R. Davies, M. Burghammer, S. Montanari et al. 2006. Morphological and structural aspects of the giant starch granules from Phajus grandifolius. J. Struct. Biol. 154: 100–110.

Charoenkul, N., D. Uttapap, W. Pathipanawat and Y. Takeda. 2006. Molecular structure of starches from cassava varieties having different cooked root textures. Starch/Stärke 58: 443–452.

Chauhan, F. and K. Seetharaman. 2013. On the organization of chains in amylopectin. Starch/Stärke 65: 191–199.

Chen, J. and J. Jane. 1995. Effectiveness of granular cold-water-soluble starch as a controlled-release matrix. Cereal Chem. 72: 265–268.

Chrastil, J. 1987. Improved colorimetric determination of amylose in starches or flours. Carbohydr. Res. 159: 154–158.

Colonna, P. and C. Mercier. 1984. Macromolecular structure of wrinkled- and smooth-pea starch components. Carbohydr. Res. 126: 233–247.

Dang, J.M.C. and L. Copeland. 2003. Imaging rice grains using atomic force microscopy. J. Cereal Sci. 37: 165–170.

Daniels, D.R. and A.M. Donald. 2003. An improved model for analyzing the small angle X-ray scattering of starch granules. Biopolymers 69: 165–175.

Fannon, J.E., R.J. Hauber and J.N. BeMiller. 1992. Surface pores of starch granules. Cereal Chem. 69: 284–288.

Fannon, J.E., J.M. Shull and J.N. BeMiller. 1993. Interior channels of starch granules. Cereal Chem. 70: 611–613.

Finch, P. and D.W. Sebesta. 1992. The amylase of Pseudomonas stutzeri as a probe of the structure of amylopectin. Carbohydr. Res. 227: c1–c4.

Franco, C.M.L., K.-S. Wong, S.-h. Yoo and J.-l. Jane. 2002. Structural and functional characteristics of selected soft wheat starches. Cereal Chem. 79: 243–248.

Fredriksson, H., J. Silverio, R. Andersson, A.-C. Eliasson and P. Åman. 1998. The influence of amylose and amylopectin characteristics on gelatinization and retrogradation properties of different starches. Carbohydr. Polym. 35: 119–134.

French, A.D. and H.F. Zobel. 1967. X-ray diffraction of oriented amylose fibers. I. Amylose dimethyl sulfoxide complex. Biopolymers 5: 457–464.

French, A.D. 1979. Allowed and preferred shapes of amylose. Bakers Digest 53: 39–54.

French, D. 1972. Fine structure of starch and its relationship to the organization of starch granules. J. Jpn. Soc. Starch Sci. 19: 8–25.

French, D., E.E. Smith and W.J. Whelan. 1972. The structural analysis and enzymic synthesis of a pentasaccharide alpha-limit dextrin formed from amylopectin by Bacillus subtilis alpha-amylase. Carbohydr. Res. 22: 123–134.

Gallant, D.J., B. Bouchet and P.M. Baldwin. 1997. Microscopy of starch: evidence of a new level of granule organization. Carbohydr. Polym. 32: 177–191.

Gayin, J., E.-S.M. Abdel-Aal, J. Manful and E. Bertoft. 2016a. Unit and internal chain profile of African rice (*Oryza glaberrima*) amylopectin. Carbohydr. Polym. 137: 466–472.

Gayin, J., E.-S.M. Abdel-Aal, M. Marcone, J. Manful and E. Bertoft. 2016b. Structure of clusters and building blocks in amylopectin from African rice accessions. Carbohydr. Polym. 148: 125–133.

Genkina, N.K., T. Noda, G.I. Koltisheva, L.A. Wasserman, R.F. Tester and V.P. Yuryev. 2003. Effects of growth temperature on some structural properties of crystalline lamellae in starches extracted from sweet potatoes (Sunnyred and Ayamurasaki). Starch/Stärke 55: 350–357.

Genkina, N.K., J. Wikman, E. Bertoft and V.P. Yuryev. 2007. Effects of structural imperfection on gelatinization characteristics of amylopectin starches with A- and B-type crystallinity. Biomacromol. 8: 2329–2335.

Gérard, C., C. Barron, P. Colonna and V. Planchot. 2001. Amylose determination in genetically modified starches. Carbohydr. Polym. 44: 19–27.

Gérard, C., P. Colonna, A. Buléon and V. Planchot. 2002. Order in maize mutant starches revealed by mild acid hydolysis. Carbohydr. Polym. 48: 131–141.

Gernat, C., S. Radosta, G. Damaschun and F. Schierbaum. 1990. Supramolecular structure of legume starches revealed by X-ray scattering. Starch/Stärke 42: 175–178.

Gidley, M.J. 1989. Molecular mechanisms underlying amylose aggregation and gelation. Macromolecules 22: 351–358.

Glaring, M.A., C.B. Koch and A. Blennow. 2006. Genotype-specific spatial distribution of starch molecules in the starch granule: A combined CLSM and SEM approach. Biomacromol. 7: 2310–2320.

Godet, M.C., V. Tran, M.M. Delage and A. Buléon. 1993. Molecular modelling of the specific interactions involved in the amylose complexation by fatty acids. Int. J. Biol. Macromol. 15: 11–18.

Goering, K.J. and B. DeHaas. 1974. A comparison of the properties of large- and small-granule starch isolated from several isogenic lines of barley. Cereal Chem. 51: 573–578.

Goldstein, A., G. Annor, J.-L. Putaux, K.H. Hebelstrup, A. Blennow and E. Bertoft. 2016. Impact of full range of amylose contents on the architecture of starch granules. Int. J. Biol. Macromol. 89: 305–318.

Goldstein, A., G. Annor, A. Blennow and E. Bertoft. 2017a. Effect of diurnal photosynthetic activity on the fine structure of amylopectin from normal and waxy barley starch. Int. J. Biol. Macromol. 102: 924–932.

Goldstein, A., G. Annor, V. Vamadevan, I. Tetlow, J.J.K. Kirkensgaard, K. Mortensen et al. 2017b. Influence of diurnal photosynthetic activity on the morphology, structure, and thermal properties of normal and waxy barley starch. Int. J. Biol. Macromol. 98: 188–200.

Gomand, S.V., L. Lamberts, J.L. Derde, H. Goesaert, G.E. Vandeputte, B. Goderis et al. 2010. Structural properties and gelatinisation characteristics of potato and cassava starches and mutants thereof. Food Hydrocoll. 24: 307–317.

Greenwood, C.T. and J. Thomson. 1962a. Physicochemical studies of starches. Part XXIV. The fractionation and characterization of starches of various plant origins. J. Chem. Soc. 222–229.

Greenwood, C.T. and J. Thomson. 1962b. Studies on the biosynthesis of starch granules. 2. The properties of the components of starches from smooth- and wrinkled-seeded peas during growth. Biochem. J. 82: 156–164.

Gunja-Smith, Z., J.J. Marshall, C. Mercier, E.E. Smith and W.J. Whelan. 1970. A revision of the Meyer-Bernfeld model of glycogen and amylopectin. FEBS Lett. 12: 101–104.

Han, W., B. Zhang, J. Li, S. Zhao, M. Niu, C. Jia et al. 2017. Understanding the fine structure of intermediate materials of maize starches. Food Chem. 233: 450–456.

Han, X.-Z. and B.R. Hamaker. 2002. Location of starch granule-associated proteins revealed by confocal laser scanning microscopy. J. Cereal Sci. 35: 109–116.

Han, X.-Z., M. Benmoussa, J.A. Gray, J.N. BeMiller and B.R. Hamaker. 2005. Detection of proteins in starch granule channels. Cereal Chem. 82: 351–355.

Hanashiro, I., J.-i. Abe and S. Hizukuri. 1996. A periodic distribution of chain length of amylopectin as revealed by high-performance anion-exchange chromatography. Carbohydr. Res. 283: 151–159.

Hanashiro, I., M. Tagawa, S. Shibahara, K. Iwata and Y. Takeda. 2002. Examination of molar-based distribution of A, B and C chains of amylopectin by fluorescent labeling with 2-aminopyridine. Carbohydr. Res. 337: 1211–1215.

Hanashiro, I., J.-i. Matsugasako, T. Egashira and Y. Takeda. 2005. Structural characterization of long unit-chains of amylopectin. J. Appl. Glycosci. 52: 233–237.

Hanashiro, I., K. Itoh, Y. Kuratomi, M. Yamazaki, T. Igarashi, J.-i. Matsugasako et al. 2008. Granule-bound starch synthase I is responsible for biosynthesis of extra-long unit chains of amylopectin in rice. Plant Cell Physiol. 49: 925–933.

Hanashiro, I., I. Sakaguchi and H. Yamashita. 2013. Branched structures of rice amylose examined by differential fluorescence detection of side-chain distribution. J. Appl. Glycosci. 60: 79–85.

Hestrin, S. 1949. Action pattern of crystalline muscle phosphorylase. J. Biol. Chem. 179: 943–955.

Hinkle, M.E. and H.F. Zobel. 1968. X-ray diffraction of oriented amylose fibers. III. The structure of amylose-n-butanol complexes. Biopolymers 6: 1119–1128.

Hizukuri, S., S. Tabata and Z. Nikuni. 1970. Studies on starch phosphate. Part 1. Estimation of glucose-6-phosphate residues in starch and the presence of other bound phosphate(s). Starch/Stärke 22: 338–343.

Hizukuri, S. 1985. Relationship between the distribution of the chain length of amylopectin and the crystalline structure of starch granules. Carbohydr. Res. 141: 295–306.

Hizukuri, S. 1986. Polymodal distribution of the chain lengths of amylopectins, and its significance. Carbohydr. Res. 147: 342–347.

Hizukuri, S., Y. Takeda, N. Maruta and B.O. Juliano. 1989. Molecular structure of rice starch. Carbohydr. Res. 189: 227–235.

Hoover, R. 2001. Composition, molecular structure, and physicochemical properties of tuber and root starches: a review. Carbohydr. Polym. 45: 253–267.

Huang, J., N. Wei, H. Li, S. Liu and D. Yang. 2014. Outer shell, inner blocklets, and granule architecture of potato starch. Carbohydr. Polym. 103: 355–358.

Huber, K.C. and J.N. BeMiller. 1997. Visualization of channels and cavities of corn and sorghum starch granules. Cereal Chem. 74: 537–541.

Imberty, A., H. Chanzy, S. Pérez, A. Buléon and V. Tran. 1987. New three-dimensional structure for A-type starch. Macromolecules 20: 2634–2636.

Imberty, A. and S. Pérez. 1988. A revisit to the three-dimensional structure of B-type starch. Biopolymers 27: 1205–1221.

Imberty, A. and S. Pérez. 1989. Conformational analysis and molecular modelling of the branching point of amylopectin. Int. J. Biol. Macromol. 11: 177–185.

Inouchi, N., D.V. Glover, T. Takaya and H. Fuwa. 1983. Development changes in fine structure of starches of several endosperm mutants of maize. Starch/Stärke 35: 371–376.

Inouchi, N., D.V. Glover and H. Fuwa. 1987. Chain length distribution of amylopectins of several single mutants and the normal counterpart, and sugary-1 phytoglycogen in maize (*Zea mays* L.). Starch/Stärke 39: 259–266.

Ishiguro, K., T. Noda, K. Kitahara and O. Yamakawa. 2000. Retrogradation of sweetpotato starch. Starch/Stärke 52: 13–17.

James, M.G., D.S. Robertson and A.M. Myers. 1995. Characterization of the maize gene sugary1, a determinant of starch composition in kernels. Plant Cell 7: 417–429.

Jane, J., Y.Y. Chen, L.F. Lee, A.E. McPherson, K.S. Wong, M. Radosavljevic et al. 1999. Effects of amylopectin branch chain length and amylose content on gelatinization and pasting properties of starch. Cereal Chem. 76: 629–637.

Jane, J.-L. and J.F. Robyt. 1984. Structure studies of amylose-V complexes and retrograded amylose by action of alpha-amylases, and a new method for preparing amylodextrins. Carbohydr. Res. 132: 105–118.

Jane, J.-l. and J.J. Shen. 1993. Internal structure of the potato starch granule revealed by chemical gelatinization. Carbohydr. Res. 247: 279–290.

Jane, J.-l., T. Kasemsuwan, S. Leas, H. Zobel and J.F. Robyt. 1994. Anthology of starch granule morphology by scanning electron microscopy. Starch/Stärke 46: 121–129.

Jane, J.-l., K.-s. Wong and A.E. McPherson. 1997. Branch-structure difference in starches of A- and B-type X-ray patterns revealed by their Naegeli dextrins. Carbohydr. Res. 300: 219–227.

Jenkins, P.J., R.E. Cameron and A.M. Donald. 1993. A universal feature in the structure of starch granules from different botanical sources. Starch/Stärke 45: 417–420.

Kainuma, K. and D. French. 1971. Nägeli amylodextrin and its relationship to starch granules structure. I. Preparation and properties of amylodextrins from various starch types. Biopolymers 10: 1673–1680.

Kainuma, K. and D. French. 1972. Naegeli amylodextrin and its relationship to starch granule structure. II. Role of water in crystallization of B-starch. Biopolymers 11: 2241–2250.

Kalinga, D.N., R. Waduge, Q. Liu, R.Y. Yada, E. Bertoft and K. Seetharaman. 2013. On the differences in granular architechture and starch structure between pericarp and endosperm wheat starches. Starch/Stärke 65: 791–800.

Kalinga, D.N., E. Bertoft, I. Tetlow and K. Seetharaman. 2014. Structure of clusters and building blocks in amylopectin from developing wheat endosperm. Carbohydr. Polym. 112: 325–333.

Källman, A., E. Bertoft, K. Koch, P. Åman and R. Andersson. 2013. On the interconnection of clusters and building blocks in barley amylopectin. Int. J. Biol. Macromol. 55: 75–82.

Kaufman, R.C., J.D. Wilson, S.R. Bean, T.J. Herald and Y.C. Shi. 2015. Development of a 96-well plate iodine assay for amylose content determination. Carbohydr. Polym. 115: 444–447.

Kim, H.-S. and K.C. Huber. 2008. Channels within soft wheat starch A- and B-type granules. J. Cereal Sci. 48: 159–172.

Kiseleva, V.I., A.V. Krivandin, J. Fornal, W. Blaszczak, T. Jelinski and V.P. Yuryev. 2005. Annealing of normal and mutant wheat starches. LM, SEM, DSC, and SAXS studies. Carbohydr. Res. 340: 75–83.

Klucinec, J.D. and D.B. Thompson. 1998. Fractionation of high-amylose maize starches by differential alcohol precipitation and chromatography of the fractions. Cereal Chem. 75: 887–896.

Klucinec, J.D. and D.B. Thompson. 2002. Structure of amylopectins from ae-containing maize starches. Cereal Chem. 79: 19–23.

Knutson, C.A. and M.J. Grove. 1994. Rapid method for estimation of amylose in maize starches. Cereal Chem. 71: 469–471.

Koch, K., R. Andersson and P. Åman. 1998. Quantitative analysis of amylopectin unit chains by means of high-performance anion-exchange chromatography with pulsed amperometric detection. J. Chromatogr. A 800: 199–206.

Koizumi, K., M. Fukuda and S. Hizukuri. 1991. Estimation of the distributions of chain length of amylopectins by high-performance liquid chromatography with pulsed amperometric detection. J. Chromatogr. 585: 233–238.

Kong, X., H. Corke and E. Bertoft. 2009. Fine structure characterization of amylopectins from grain amaranth starch. Carbohydr. Res. 344: 1701–1708.

Kooistra, E. 1962. On the differences between smooth and three types of wrinkled peas. Euphytica 11: 357–373.

Koroteeva, D.A., V.I. Kiseleva, A.V. Krivandin, O.V. Shatalova, W. Blaszczak, E. Bertoft et al. 2007. Structural and thermodynamic properties of rice starches with different genetic background. Part 2. Defectiveness of different supramolecular structures in starch granules. Int. J. Biol. Macromol. 41: 534–547.

Kozlov, S.S., A. Blennow, A.V. Krivandin and V.P. Yuryev. 2007a. Structural and thermodynamic properties of starches extracted from GBSS and GWD suppressed potato lines. Int. J. Biol. Macromol. 40: 449–460.

Kozlov, S.S., A.V. Krivandin, O.V. Shatalova, T. Noda, E. Bertoft, J. Fornal et al. 2007b. Structure of starches extracted from near-isogenic wheat lines. Part II. Molecular organization of amylopectin clusters. J. Therm. Anal. Cal. 87: 575–584.

Laohaphatanaleart, K., K. Piyachomkwan, K. Sriroth, V. Santisopasri and E. Bertoft. 2009. A study of the internal structure in cassava and rice amylopectin. Starch/Stärke 61: 557–569.

Laohaphatanaleart, K., K. Piyachomkwan, K. Sriroth and E. Bertoft. 2010. The fine structure of cassava amylopectin. Part 1. Organization of clusters. Int. J. Biol. Macromol. 47: 317–324.

LeBail, P., H. Bizot, M. Ollivon, G. Keller, C. Bourgaux and A. Buléon. 1999. Monitoring the crystallization of amylose-lipid complexes during maize starch melting by synchrotron X-ray diffraction. Biopolymers 50: 99–110.

Lee, E.Y.C., C. Mercier and W.J. Whelan. 1968. A method for the investigation of the fine structure of amylopectin. Arch. Biochem. Biophys. 125: 1028–1030.

Lelievre, J., J.A. Lewis and K. Marsden. 1986. The size and shape of amylopectin: a study using analytical ultracentrifugation. Carbohydr. Res. 153: 195–203.

Li, J.-Y. and A.-I. Yeh. 2001. Relationships between thermal, rheological characteristics and swelling power for various starches. J. Food Eng. 50: 141–148.

Li, J.H., T. Vasanthan, R. Hoover and B.G. Rossnagel. 2003. Starch from hull-less barley: Ultrastructure and distribution of granule-bound proteins. Cereal Chem. 80: 524–532.

Li, L., H. Jiang, M. Campbell, M. Blanco and J.-l. Jane. 2008. Characterization of maize amylose-extender (ae) mutant starches. Part I: Relationship between resistant starch contents and molecular structures. Carbohydr. Polym. 74: 396–404.

Linsberger-Martin, G., B. Lukasch and E. Berghofer. 2012. Effects of high hydrostatic pressure on the RS content of amaranth, quinoa and wheat starch. Starch/Stärke 64: 157–165.

Lloyd, J.R., C.L. Hedley, V.J. Bull and S.G. Ring. 1996. Determination of the effect of r and rb mutations on the structure of amylose and amylopectin in pea (*Pisum sativum* L.). Carbohydr. Polym. 29: 45–49.

Lourdin, D., J.-L. Putaux, G. Potocki-Véronèse, C. Chevigny, A. Roland-Sabaté and A. Buléon. 2015. Crystalline structure in starch. pp. 61–90. *In*: Nakamura, Y. (ed.). Starch: Metabolism and Structure. Springer Japan, Tokyo.

Manelius, R. and E. Bertoft. 1996. The effect of Ca2+-ions on the α-amylolysis of granular starches from oats and waxy-maize. J. Cereal Sci. 24: 139–150.

Manion, B., M. Ye, B.E. Holbein and K. Seetharaman. 2011. Quantification of total iodine in intact granular starches of different botanical origin exposed to iodine vapor at various water activities. Carbohydr. Res. 346: 2482–2490.

Manners, D.J. 1989. Recent developments in our understanding of amylopectin structure. Carbohydr. Polym. 11: 87–112.

Matheson, N.K. and L.A. Welsh. 1988. Estimation and fractionation of the essentially unbranched (amylose) and branched (amylopectin) components of starches with concanavalin A. Carbohydr. Res. 180: 301–313.

Matheson, N.K. 1990. A comparison of the structures of the fractions of normal and high-amylose pea-seed starches prepared by precipitation with concanavalin A. Carbohydr. Res. 199: 195–205.

McIntyre, A.P., R. Mukerjea and J.F. Robyt. 2013. Reducing values: dinitrosalicylate gives over-oxidation and invalid results whereas copper bicinchoninate gives no over-oxidation and valid results. Carbohydr. Res. 380: 118–123.

Mercier, C. 1973. The fine structure of corn starches of various amylose-percentage: waxy, normal and amylomaize. Stärke 25: 78–83.

Morell, M.K., M.S. Samuel and M.G. O'Shea. 1998. Analysis of starch structure using fluorophore-assisted carbohydrate electrophoresis. Electrophoresis 19: 2603–2611.

Morrison, W.R. and B. Laignelet. 1983. An improved colorimetric procedure for determining apparent and total amylose in cereal and other starches. J. Cereal Sci. 1: 9–20.

Morrison, W.R., T.P. Milligan and M.N. Azudin. 1984. A relationship between the amylose and lipid contents of starches from diploid cereals. J. Cereal Sci. 2: 257–271.

Morrison, W.R., D.C. Scott and J. Karkalas. 1986. Variation in the composition and physical properties of barley starches. Stärke 38: 374–379.

Morrison, W.R., R.V. Law and C.E. Snape. 1993. Evidence for inclusion complexes of lipids with V-amylose in maize, rice and oat starches. J. Cereal Sci. 18: 107–109.

Mouille, G., M.-L. Maddelein, N. Libessart, P. Talaga, A. Decq, B. Delrue et al. 1996. Preamylopectin processing: A mandatory step for starch biosynthesis in plants. Plant Cell 8: 1353–1366.

Naguleswaran, S., J. Li, T. Vasanthan and D. Bressler. 2011. Distribution of granule channels, protein, and phospholipid in triticale and corn starches as revealed by confocal laser scanning microscopy. Cereal Chem. 88: 87–94.

Nakamura, Y. 2018. Rice starch biotechnology: Rice endosperm as a model of cereal endosperms. Starch/Stärke 70: 1600375.

Nakazawa, Y. and Y.-J. Wang. 2003. Acid hydrolysis of native and annealed starches and branch-structure of their Naegeli dextrins. Carbohydr. Res. 338: 2871–2882.

Nikuni, Z. 1969. Starch and cooking (in Japanese). Science of Cookery 2: 6–14.

Nuessli, J., B. Sigg, B. Conde-Petit and F. Escher. 1997. Characterization of amylose–flavour complexes by DSC and X-ray diffraction. Food Hydrocoll. 11: 27–34.

O'Shea, M.G., M.S. Samuel, C.M. Konik and M.K. Morell. 1998. Fluorophore-assisted carbohydrate electrophoresis (FACE) of oligosaccharides: efficiency of labelling and high-resolution separation. Carbohydr. Res. 307: 1–12.

O'Sullivan, A.C. and S. Pérez. 1999. The relationship between internal chain length of amylopectin and crystallinity in starch. Biopolymers 50: 381–390.

Obiro, W.C., S.S. Ray and M.N. Emmambux. 2012. V-amylose structural characteristics, methods of preparation, significance, and potential applications. Food Rev. Int. 28: 412–438.

Ohtani, T., T. Yoshino, S. Hagiwara and T. Maekawa. 2000. High-resolution imaging of starch granule structure using atomic force microscopy. Starch/Stärke 52: 150–153.

Orford, P.O., S.G. Ring, V. Carroll, M.J. Miles and V.J. Morris. 1987. The effect of concentration and botanical source on the gelation and retrogradation of starch. J. Sci. Food Agric. 39: 169–177.

Palmer, T.N., L.E. Macaskie and K.K. Grewel. 1983. The unit chain distribution chain profiles of branched (1→4)-α-D-glucans. Carbohydr. Res. 114: 338–342.

Park, H., S. Xu and K. Seetharaman. 2011. A novel *in situ* atomic force microscopy imaging technique to probe surface morphological features of starch granules. Carbohydr. Res. 346: 847–853.

Paton, D. 1979. Oat starch: Some recent developments. Starch/Stärke 31: 184–187.

Peat, S., W.J. Whelan and G.J. Thomas. 1952. Evidence of multiple branching in waxy maize starch. J. Chem. Soc. 4546–4548.

Perera, C., Z. Lu, J. Sell and J. Jane. 2001. Comparison of physicochemical properties and structures of sugary-2 cornstarch with normal and waxy cultivars. Cereal Chem. 78: 249–256.

Pérez, S. and E. Bertoft. 2010. The molecular structures of starch components and their contribution to the architecture of starch granules: A comprehensive review. Starch/Stärke 62: 389–420.

Perez-Herrera, M., T. Vasanthan and R. Hoover. 2017. Characterization of maize starch nanoparticles prepared by acid hydrolysis. Cereal Chem. 93: 323–330.

Perez-Rea, D., B. Bergenståhl and L. Nilsson. 2015. Development and evaluation of methods for starch dissolution using asymmetrical flow field-flow fractionation. Part I: Dissolution of amylopectin. Anal. Bioanal. Chem. 407: 4315–4326.

Peymanpour, G., M. Marcone, S. Ragaee, I. Tetlow, C.C. Lane, K. Seetharaman et al. 2016. On the molecular structure of the amylopectin fraction isolated from "high-amylose" ae maize starches. Int. J. Biol. Macromol. 91: 768–777.

Pilling, E. and A.M. Smith. 2003. Growth ring formation in the starch granules of potato tubers. Plant Physiol. 132: 365–371.

Popov, D., A. Buléon, M. Burghammer, H. Chanzy, N. Montesanti, J.-L. Putaux et al. 2009. Crystal structure of A-amylose: A revisit from synchrotron microdiffraction analysis of single crystals. Macromolecules 42: 1167–1174.

Potter, A.L., V. Silveira, R.M. McCready and H.S. Owens. 1953. Fractionation of starches from smooth and wrinkled seeded peas. Molecular weights, end-group assays and iodine affinities of the fractions. J. Am. Chem. Soc. 75: 1335–1338.

Putaux, J.-L., A. Buléon, R. Borsali and H. Chanzy. 1999. Ultrastructural aspects of phytoglycogen from cryo-transmission electron-microscopy and quasi-elastic light scattering data. Int. J. Biol. Macromol. 26: 145–150.

Putaux, J.-L., A. Buléon and H. Chanzy. 2000. Network formation in dilute amylose and amylopectin studied by TEM. Macromolecules 33: 6416–6422.

Putaux, J.-L., S. Molina-Boisseau, T. Momaur and A. Dufresne. 2003. Platelet nanocrystals resulting from the disruption of waxy maize starch granules by acid hydrolysis. Biomacromolecules 4: 1198–1202.

Putseys, J.A., L. Lamberts and J.A. Delcour. 2010. Amylose-inclusion complexes: Formation, identity and physico-chemical properties. J. Cereal Sci. 51: 238–247.

Rappenecker, G. and P. Zugenmaier. 1981. Detailed refinement of the crystal structure of Vh-amylose. Carbohydr. Res. 89: 11–19.

Ratnayake, W.S. and D.S. Jackson. 2007. A new insight into the gelatinization process of native starches. Carbohydr. Polym. 67: 511–529.

Ren, S. 2017. Comparative analysis of some physicochemical properties of 19 kinds of native starches. Starch/Stärke 68: 1600367.

Ridout, M.J., A.P. Gunning, M.L. Parker, R.H. Wilson and V.J. Morris. 2002. Using AFM to image the internal structure of starch granules. Carbohydr. Polym. 50: 123–132.

Ridout, M.J., M.L. Parker, C.L. Hedley, T.Y. Bogracheva and V.J. Morris. 2006. Atomic force microscopy of pea starch: Granule architecture of the rug3-a, rug4-b, rug5-a and lam-c mutants. Carbohydr. Polym. 65: 64–74.

Robin, J.P., C. Mercier, R. Charbonnière and A. Guilbot. 1974. Lintnerized starches. Gel filtration and enzymatic studies of insoluble residues from prolonged acid treatment of potato starch. Cereal Chem. 51: 389–406.

Robyt, J. and D. French. 1963. Action pattern and specificity of an amylase from *Bacillus subtilis*. Arch. Biochem. Biophys. 100: 451–467.

Rundle, R.E. and D. French. 1943. The configuration of starch and the starch-iodine complex. II. Optical properties of crystalline starch fractions. J. Am. Chem. Soc. 65: 558–561.

Sande-Bakhuyzen, H.L.v.d. 1926. The structure of starch grains from wheat grown under constant conditions. Proc. Soc. Exp. Biol. and Med. 24: 302–305.

Sanderson, J.S., R.D. Daniels, A.M. Donald, A. Blennow and S.B. Engelsen. 2006. Exploratory SAXS and HPAEC-PAD studies of starches from diverse plant genotypes. Carbohydr. Polym. 64: 433–443.

Santacesaria, E., F. Trulli, G.F. Brussani, D. Gelosa and M. DiSerio. 1994. Oxidized glucosidic oligomers: a new class of sequestering agents—preparation and properties. Carbohydr. Polym. 23: 35–46.

Scheffler, S.L., X. Wang, L. Huang, F.S.-M. Gonzalez and Y. Yao. 2010. Phytoglygogen octenyl succinate, an amphiphilic carbohydrate nanoparticle, and ε-polylysine to improve lipid oxidative stability of emulsions. J. Agric. Food Chem. 58: 660–667.

Schulman, A.H., S. Tomooka, A. Suzuki, P. Myllärinen and S. Hizukuri. 1995. Structural analysis of starch from normal and shx (shrunken endosperm) barley (*Hordeum vulgare* L.). Carbohydr. Res. 275: 361–369.

Schwall, G.P., R. Safford, R.J. Westcott, R. Jeffcoat, A. Tayal, Y.-C. Shi et al. 2000. Production of very-high-amylose potato starch by inhibition of SBE A and B. Nat. Biotechnol. 18: 551–554.

Shamekh, S., P. Myllärinen, K. Poutanen and P. Forssell. 2002. Film formation properties of potato starch hydrolysates. Starch/Stärke 54: 20–24.

Shen, X., E. Bertoft, G. Zhang and B.R. Hamaker. 2013. Iodine binding to explore the conformational state of internal chains of amylopectin. Carbohydr. Polym. 98: 778–783.

Shi, Y.-C. and P.A. Seib. 1995. Fine structure of maize starches from four wx-containing genotypes of the W64A inbred line in relation to gelatinization and retrogradation. Carbohydr. Polym. 26: 141–147.

Shibanuma, K., Y. Takeda, S. Hizukuri and S. Shibata. 1994. Molecular structures of some wheat starches. Carbohydr. Polym. 25: 111–116.

Shin, J.-E., S. Simsek, B.L. Reuhs and Y. Yao. 2008. Glucose release of water-soluble starch-related α-glucans by pancreatin and amyloglucosidase is affected by the abundance of α-1,6-glucosidic linkages. J. Agric. Food Chem. 56: 10879–10886.

Smith, A.M., M. Bettey and I.D. Bedford. 1989. Evidence that the rb locus alters the starch content of developing pea embryos through an effect on ADP glucose pyrophosphorylase. Plant Physiol. 89: 1279–1284.

Srichuwong, S., T.C. Sunarti, T. Mishima, N. Isono and M. Hisamatsu. 2005. Starches from different botanical sources I: Contribution of amylopectin fine structure to thermal properties and enzyme digestibility. Carbohydr. Polym. 60: 529–538.

Stacy, C.J. and J.F. Foster. 1957. Molecular weight heterogeneity in starch amylopectins. J. Polym. Sci., Part A 25: 39–50.

Steffolani, M.E., A.E. León and G.T. Pérez. 2013. Study of the physicochemical and functional characterization of quinoa and kaniwa starches. Starch/Stärke 65: 976–983.

Sterling, C. 1962. A low angle spacing in starch. J. Polym. Sci. 56: S10–S12.

Summer, R. and D. French. 1956. Action of β-amylase on branched oligosaccharides. J. Biol. Chem. 222: 469–477.

Takeda, Y., K. Shirasaka and S. Hizukuri. 1984. Examination of the purity and structure of amylose by gel-permeation chromatography. Carbohydr. Res. 132: 83–92.

Takeda, Y., S. Hizukuri and B.O. Juliano. 1987a. Structures of rice amylopectins with low and high affinities for iodine. Carbohydr. Res. 168: 79–88.

Takeda, Y., S. Hizukuri, C. Takeda and A. Suzuki. 1987b. Structures of branched molecules of amyloses of various origins, and molecular fractions of branched and unbranched molecules. Carbohydr. Res. 165: 139–145.

Takeda, Y., T. Shitaozono and S. Hizukuri. 1988. Molecular structure of corn starch. Starch/Stärke 40: 51–54.

Takeda, Y., S. Hizukuri and B.O. Juliano. 1989a. Structures and amounts of branched molecules in rice amyloses. Carbohydr. Res. 186: 163–166.

Takeda, Y., N. Maruta, S. Hizukuri and B.O. Juliano. 1989b. Structures of indica rice starches (IR48 and IR64) having intermediate affinities for iodine. Carbohydr. Res. 187: 287–294.

Takeda, Y., T. Shitaozono and S. Hizukuri. 1990. Structures of sub-fractions of corn amylose. Carbohydr. Res. 199: 207–214.

Takeda, Y., N. Maruta and S. Hizukuri. 1992a. Examination of the structure of amylose by tritium labelling of the reducing terminal. Carbohydr. Res. 227: 113–120.

Takeda, Y., N. Maruta and S. Hizukuri. 1992b. Structures of amylose subfractions with different molecular sizes. Carbohydr. Res. 226: 279–285.

Takeda, Y. and J. Preiss. 1993. Structures of B90 (sugary) and W64A (normal) maize starches. Carbohydr. Res. 240: 265–275.

Takeda, Y., S. Tomooka and S. Hizukuri. 1993. Structures of branched and linear molecules of rice amylose. Carbohydr. Res. 246: 267–272.

Takeda, Y., C. Takeda, H. Mizukami and I. Hanashiro. 1999. Structures of large, medium and small starch granules of barley grain. Carbohydr. Polym. 38: 109–114.

Takeda, Y., S. Shibahara and I. Hanashiro. 2003. Examination of the structure of amylopectin molecules by fluorescent labeling. Carbohydr. Res. 338: 471–475.

Tang, H., T. Mitsunaga and Y. Kawamura. 2006. Molecular arrangement in blocklets and starch granules architecture. Carbohydr. Polym. 63: 555–560.

Tetlow, I.J. 2011. Starch biosynthesis in developing seeds. Seed Sci. Res. 21: 5–32.

Tetlow, I.J. and M.J. Emes. 2017. Starch biosynthesis in the developing endosperms of grasses and cereals. Agronomy 7: 81; doi: 10.3390/agronomy7040081.

Thompson, D.B. 2000. On the non-random nature of amylopectin branching. Carbohydr. Polym. 43: 223–239.

Umeki, K. and T. Yamamoto. 1972a. Enzymatic determination of structure of singly branched hexaose dextrins formed by liquefying α-amylase of *Bacillus subtilis*. J. Biochem. 72: 101–109.

Umeki, K. and T. Yamamoto. 1972b. Structures of branched dextrins produced by saccharifying α-amylase of *Bacillus subtilis*. J. Biochem. 72: 1219–1226.

Umeki, K. and T. Yamamoto. 1975a. Structures of singly branched heptaoses produced by bacterial liquefying α-amylase. J. Biochem. 78: 889–896.

Umeki, K. and T. Yamamoto. 1975b. Structures of multi-branched dextrins produced by saccharifying α-amylase from starch. J. Biochem. 78: 897–903.

Umeki, K. and K. Kainuma. 1981. Fine structure of nägeli amylodextrin obtained by acid treatment of defatted waxy-maize starch—structural evidence to support the double-helix hypothesis. Carbohydr. Res. 96: 143–159.

Vamadevan, V., E. Bertoft and K. Seetharaman. 2013a. On the importance of organization of glucan chains on thermal properties of starch. Carbohydr. Polym. 92: 1653–1659.

Vamadevan, V., E. Bertoft, D.V. Soldatov and K. Seetharaman. 2013b. Impact on molecular organization of amylopectin in starch granules upon annealing. Carbohydr. Polym. 98: 1045–1055.

Vamadevan, V. and E. Bertoft. 2018. Impact of different structural types of amylopectin on retrogradation. Food Hydrocoll. 80: 88–96.

Vamadevan, V., A. Blennow, A. Buléon, A. Goldstein and E. Bertoft. 2018. Distinct properties and structures among B-crystalline starch granules. Starch/Stärke 70: 1700240.

Vamadevan, V. and E. Bertoft. 2020. Observations on the impact of amylopectin and amylose structure on the swelling of starch granules. Food Hydrocoll. 103: 105663.

Vermeylen, R., B. Goderis, H. Reynaers and J.A. Delcour. 2004. Amylopectin molecular structure reflected in macromolecular organization of granular starch. Biomacromolecules 5: 1775–1786.

Waduge, R.N., D.N. Kalinga, E. Bertoft and K. Seetharaman. 2014. Molecular structure and organization of starch granules from developing wheat endosperm. Cereal Chem. 91: 578–586.

Waigh, T.A., M.J. Gidley, B.U. Komanshek and A.M. Donald. 2000. The phase transformations in starch during gelatinisation: a liquid crystalline approach. Carbohydr. Res. 328: 165–176.

Walker, G.J. and W.J. Whelan. 1960. The mechanism of carbohydrase action. 8. Structures of the muscle-phosphorylase limit dextrins of glycogen and amylopectin. Biochem. J. 76: 264-268.

Wang, L.Z. and P.J. White. 1994. Structure and properties of amylose, amylopectin, and intermediate material of oat starches. Cereal Chem. 71: 263–268.

Wang, S., J. Blazek, E. Gilbert and L. Copeland. 2012. New insights on the mechanism of acid degradation of pea starch. Carbohydr. Polym. 87: 1941–1949.

Wang, S., M.E. Hassani, B. Crossett and L. Copeland. 2013. Extraction and identification of internal granule proteins from waxy wheat starch. Starch/Stärke 65: 186–190.

Wang, S. and L. Copeland. 2015. Effect of acid hydrolysis on starch structure and functionality: A review. Crit. Rev. Food Sci. Nutr. 55: 1081–1097.

Wang, Y.-J., P. White, L. Pollak and J. Jane. 1993. Amylopectin and intermediate materials in starches from mutant genotypes of the Oh43 inbred line. Cereal Chem. 70: 521–525.

Wattebled, F., V. Planchot, Y. Dong, N. Szydlowski, B. Pontoire, A. Devin et al. 2008. Further evidence for the mandatory nature of polysaccharide debranching for the aggregation of semicrystalline starch and for overlapping functions of debranching enzymes in *Arabidopsis* leaves. Plant Physiol. 148: 1309–1323.

Whistler, R.L. and W.M. Doane. 1961. Characterization of intermediary fractions of high-amylose corn starches. Cereal Chem. 38: 251–255.

Wikman, J., F.H. Larsen, M.S. Motawia, A. Blennow and E. Bertoft. 2011. Phosphate esters in amylopectin clusters of potato tuber starch. Int. J. Biol. Macromol. 48: 639–649.

Wikman, J., A. Blennow, A. Buléon, J.-L. Putaux, S. Pérez, K. Seetharaman et al. 2014. Influence of amylopectin structure and degree of phosphorylation on the molecular composition of potato starch lintners. Biopolymers 101: 257–271.

Wilhelm, E., T. Aberle, W. Burchard and R. Landers. 2002. Pecularities of aqueous amaranth starch suspensions. Biomacromol. 3: 17–26.

Witt, T., J. Doutch, E.P. Gilbert and R.G. Gilbert. 2012. Relations beetween molecular, crystalline, and lamellar structures of amylopectin. Biomacromol. 13: 4273–4282.

Witt, T. and R.G. Gilbert. 2014. Causal relations between structural features of amylopectin, a semicrystalline hyperbranched polymer. Biomacromol. 15: 2501–2511.

Wong, K.S. and J. Jane. 1997. Quantitative analysis of debranched amylopectin with a postcolumn enzyme reactor. J. Liq. Chromatogr. 20: 297–310.

Wong, K.-S., A. Kubo, J.-l. Jane, K. Harada, H. Satoh and Y. Nakamura. 2003. Structure and properties of amylopectin and phytoglycogen in the endosperm of sugary-1 mutants of rice. J. Cereal Sci. 37: 139–149.

Yano, M., K. Okuno, J. Kawakami, H. Satoh and T. Omura. 1985. High amylose mutants of rice, *Oryza sativa* L. Theor. Appl. Genet. 69: 253–257.

Yao, Y., M.J. Guiltinan and D.B. Thompson. 2005. High-performance size-exclusion chromatography (HPSEC) and fluorephore-assisted carbohydrate electrophoresis (FACE) to describe the chain-length distribution of debranched starch. Carbohydr. Res. 340: 701–710.

Yoon, J.-W. and S.-T. Lim. 2003. Molecular fractionation of starch by density-gradient ultracentrifugation. Carbohydr. Res. 338: 611–617.

Yoshimoto, Y., J. Tashiro, T. Takenouchi and Y. Takeda. 2000. Molecular structure and some physicochemical properties of high-amylose barley starches. Cereal Chem. 77: 279–285.

Yoshimoto, Y., T. Takenoushi and Y. Takeda. 2002. Molecular structure and some physicochemical properties of waxy and low-amylose barley starches. Carbohydr. Polym. 47: 159–167.

Yun, S.-H. and N.K. Matheson. 1993. Structures of the amylopectins of waxy, normal, amylose-extender, and wx:ae genotypes and of the phytoglycogen of maize. Carbohydr. Res. 243: 307–321.

Zeeman, S.C., J. Kossman and A.M. Smith. 2010. Starch: Its metabolism, evolution, and biotechnological modification in plants. Annu. Rev. Plant Biology 61: 209–234.

Zhang, T. and C.G. Oates. 1999. Relationship between α-amylase degradation and physico-chemical properties of sweet potato starches. Food Chem. 65: 157–163.

Zhao, J., M.A. Madson and R.L. Whistler. 1996. Cavities in porous corn starch provide a large storage space. Cereal Chem. 73: 379–380.

Zhu, F., H. Corke and E. Bertoft. 2011a. Amylopectin internal molecular structure in relation to physical properties of sweetpotato starch. Carbohydr. Polym. 84: 907–918.

Zhu, F., H. Corke, P. Åman and E. Bertoft. 2011b. Structures of building blocks in clusters of sweetpotato amylopectin. Carbohydr. Res. 346: 2913–2925.

Zhu, F., H. Corke, P. Åman and E. Bertoft. 2011c. Structures of clusters in sweetpotato amylopectin. Carbohydr. Res. 346: 1112–1121.

Zhu, F., E. Bertoft, A. Källman, A.M. Myers and K. Seetharaman. 2013a. Molecular structure of starches from maize mutants deficient in starch synthase III. J. Agric. Food Chem. 61: 9899–9907.

Zhu, F., E. Bertoft and K. Seetharaman. 2013b. Composition of clusters and building blocks in amylopectins of starch mutants deficient in starch synthase III. J. Agric. Food Chem. 61: 12345–12355.

Zhu, F. 2015. Isolation, composition, structure, properties, modifications, and uses of yam starch. Compr. Rev. Food Sci. Food Safety 14: 357–386.

Zhu, F., E. Bertoft, Y. Wang, M. Emes, I. Tetlow and K. Seetharaman. 2015. Structure of Arabidopsis leaf starch is markedly altered following nocturnal degradation. Carbohydr. Polym. 117: 1002–1013.

Zhu, T., D.S. Jackson, R.L. Wehling and B. Geera. 2008. Comparison of amylose determination methods and the development of a dual wavelength iodine binding technique. Cereal Chem. 85: 51–58.

Zobel, H.F., A.D. French and M.E. Hinkle. 1967. X-ray diffraction of oriented amylose fibers. II. Structure of V amyloses. Biopolymers 5: 837–845.

CHAPTER 9

Regulation of Assimilatory Starch Metabolism by Cellular Carbohydrate Status

Maria Grazia Annunziata and *John Edward Lunn**

9.1 Introduction

In their natural environment, plants are dependent on sunlight to supply energy for photosynthesis, and they are faced with a daily challenge to survive through the night when photosynthesis is no longer possible. To meet this challenge, plants invest some of the photoassimilates produced during the day in transitory carbon reserves that are stored in the leaves and remobilized at night to provide carbon and energy for maintenance respiration and growth (Stitt and Zeeman 2012). By far, the most common reserves are starch and soluble sugars, especially sucrose, glucose and fructose. Depending on the species, these can be supplemented by fructans, sugar alcohols (e.g., sorbitol and mannitol), raffinose-family oligosaccharides (e.g., raffinose, stachyose, verbascose), mannans, or organic acids (e.g., malate and citrate). Starch is an insoluble, semi-crystalline material comprised of two types of glucan polymer—amylose and amylopectin—both constructed from α-1,4-linked chains of glucose. In amylose, the glucans are present as simple linear chains with few branches. Amylopectin has a more complex branched structure, with short α-1,4-linked chains of similar length attached via α-1,6 glycosidic bonds to a core α-1,4-linked glucan in evenly spaced clusters. The even chain length and regular spacing of these clusters allows amylopectin to form a semi-crystalline structure, with regular layers within the starch granule. This structure is an efficient way of storing a large amount of carbohydrate and accumulation of insoluble starch granules has little or no impact on the osmoregulation of the cell.

The amount of starch accumulated in leaves during the day varies between species, and both the synthesis and degradation of starch are influenced by environmental conditions, especially photoperiod, irradiance and temperature (Britz et al. 1985a,b, 1987, Chatterton and Silvius 1979, 1980a,b, 1981, Fondy et al. 1989, Geiger et al. 1991, Servaites et al. 1989, Pyl et al. 2012). Some species, for example wheat (*Triticum aestivum*) and rice (*Oryza sativa*), store very little starch in their leaves and accumulate mostly sucrose during the day, but others, for example garden pea (*Pisum sativum*), are heavily dependent on their leaf starch reserves (Lunn 2016). One of the most intensively studied model plants, Arabidopsis (*Arabidopsis thaliana*), is another starch-storing species, retaining up to 50% of its photoassimilate in leaf starch reserves (Stitt and Zeeman 2012, Streb and Zeeman 2012). This represents a huge investment of resources for the plant, and these valuable reserves

Max Planck Institute of Molecular Plant Physiology, 14476 Potsdam-Golm, Germany.
Email: annunziata@mpimp-golm.mpg.de
Corresponding author: lunn@mpimp-golm.mpg.de

must be carefully managed if the plant is to thrive and achieve reproductive success (Sulpice et al. 2010). The pathways of transitory starch synthesis and degradation in leaves have been reasonably well defined by studies of various mutants and biochemical analysis of enzymes and intermediates in the pathways (Stitt et al. 2010), and attention is now focussed on understanding the regulation of these processes (Stitt and Zeeman 2012). In this chapter, we shall outline the pathways of transitory starch synthesis and degradation, and then describe our current understanding of their regulation. In particular, we shall focus on how these processes are linked to the carbon status of the plant and regulated accordingly, allowing the plant to adapt its starch metabolism to an ever-changing environment (Annunziata et al. 2017, 2018).

9.2 Pathways of Transitory Starch Metabolism

9.2.1 Starch Synthesis

Atmospheric CO_2 enters leaves through the stomata and is fixed via the Calvin-Benson cycle in the chloroplasts. Net carbon fixation allows phosphorylated intermediates to be withdrawn from the Calvin-Benson cycle for synthesis of various products, of which sucrose and starch are usually the most abundant. Sucrose is synthesized in the cytosol using triose-phosphates exported from the chloroplast via the triose-phosphate translocator, while starch is made in the chloroplasts. Most of the sucrose is exported from the leaf, via the phloem, to heterotrophic sink organs, providing carbon and energy for growth and accumulation of long-term storage reserves, for example in seeds and tubers, while the starch is retained in the leaf for remobilization at night.

Starch synthesis in the chloroplasts uses fructose 6-phosphate (Fru6P) drawn directly from the Calvin–Benson cycle (Fig. 9.1). This is converted into glucose 6-phosphate (Glc6P) and then glucose 1-phosphate (Glc1P) by the sequential action of the plastidial phosphoglucose isomerase (PGI; EC 5.3.1.9) and phosphoglucomutase (PGM; EC 5.4.2.2) (Caspar 1985, Streb and Zeeman 2012). Up to this point, the reactions are readily reversible. The first committed step in the pathway of starch synthesis is the production of adenosine-5´-diphosphoglucose (ADPG) from Glc1P by ADPG pyrophosphorylase (AGPase; EC 2.7.7.27). This reaction is rendered irreversible by hydrolysis of the inorganic pyrophosphate (PPi) by-product by plastidial pyrophosphatase (PPase; EC 3.6.1.1).

ADPG is the substrate for starch synthesis by a combination of starch synthases (Streb and Zeeman 2012). Soluble starch synthases (SS; EC 2.4.1.21) transfer the glucosyl moiety from ADPG to the non-reducing end of an acceptor glucan, with the SSI, SSII and SSIII type isoforms showing preferences for short (≤ 10 glucose residues), medium (13–18 glucose residues) and long glucan chains, respectively (Streb and Zeeman 2012, Nougué et al. 2014, Pfister et al. 2014, Pfister and Zeeman 2016, Lu et al. 2018). SSIV plays a major role in the initiation of starch granule formation, with SSIII also apparently contributing to this function in the absence of SSIV (Roldán et al. 2007, Szydlowski et al. 2009, D'Hulst and Merida 2012, Crumpton-Taylor et al. 2012, 2013, Ragel et al. 2013, Malinova et al. 2017, Seung and Smith 2018). SSV-type isoforms, which are not present in all species, have not yet been functionally characterized in detail and may represent a sub-class of SSIV (Liu et al. 2015). In addition to the soluble enzymes, there is also a granule-bound starch synthase (GBSS), which is responsible for synthesis of amylose (Zeeman et al. 2002, Pfister and Zeeman 2016). Binding of GBSS to starch granules was shown to involve the PROTEIN TARGETING TO STARCH (PTST) protein (Seung et al. 2015). Two other members of this protein family, PTST2 and PTST3, have recently been implicated in targeting of SSIV and initiation of starch granules (Seung et al. 2017).

Starch branching enzymes (SBE; EC 2.4.1.18) are responsible for the introduction of branches in amylopectin. They cleave fragments from linear α-1,4-linked glucan chains and then attach these via α-1,6-linkages to either the parental glucan (intramolecular) or a different glucan (intermolecular) (Wychowski et al. 2017). Type 1 SBEs (also known as B-family SBEs) act preferentially on amylose, transferring longer glucan chains with a degree of polymerisation up to 40 (DP ≤ 40), while the type 2 (A-family) SBEs transfer shorter chains (DP 6–15) and preferentially use amylopectin as substrate.

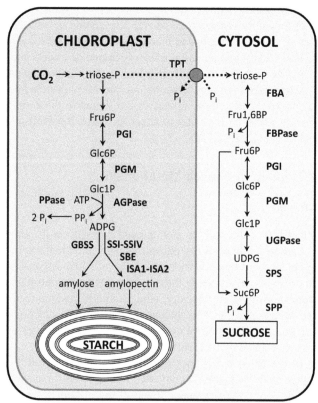

Figure 9.1. Pathway of transitory starch synthesis. Starch is synthesised in the chloroplast during the day, using Fru6P from the Calvin-Benson cycle. AGPase, ADP-glucose pyrophosphorylase; FBA, fructose-1,6-bisphosphate aldolase; FBPase, fructose-1,6-bisphosphatase; GBSS, granule-bound starch synthase; ISA, isoamylase; PGI, phosphoglucose isomerase; PGM, phosphoglucomutase; PPase, inorganic pyrophosphatase; SBE, starch branching enzyme; SS, soluble starch synthase; SPP, sucrose-phosphate phosphatase; SPS, sucrose-phosphate synthase; TPT, triose phosphate translocator; UGPase, UDP-glucose pyrophosphorylase.

Two types of starch debranching enzyme (DBE)—isoamylase 1 (ISA1; EC 3.2.1.68) and isoamylase 2 (ISA2)—form a heteromeric complex, and are important for amylopectin synthesis. It is thought that they have an "editing" function, trimming or removing any excess or incorrectly positioned branches that disrupt the regular clusters and prevent the amylopectin forming a semi-crystalline structure (Delatte et al. 2005). Mutants that lack either or both ISA1 and ISA2 proteins accumulate large amounts of soluble, highly-branched polymers, known as phytoglycogen, which resembles the glycogen synthesized in animal cells (Wattebled et al. 2005).

9.2.2 Starch Degradation

The degradation of starch at night is a complex process involving many different enzymes and ancillary proteins (Fig. 9.2), many of which have been identified by study of Arabidopsis *starch excess* (*sex*) mutants that are unable to fully degrade their starch at night (Stitt and Zeeman 2012). The first stage in the process is phosphorylation of glucan chains at the granule surface by α-glucan, water dikinase (GWD1; EC 2.7.9.4), which adds phosphate at the C6 position of glucose residues (Caspar et al. 1991, Yu et al. 2001, Ritte et al. 2002, 2006). This initial phosphorylation is followed by addition of more phosphate in the C3 position by phosphoglucan, water dikinase (PWD, also known as GWD3; EC 2.7.9.5) (Baunsgaard et al. 2005, Kötting et al. 2005). It is thought that phosphorylation disrupts the intertwined α-helical structure of the glucan chains, making them more accessible to starch degrading

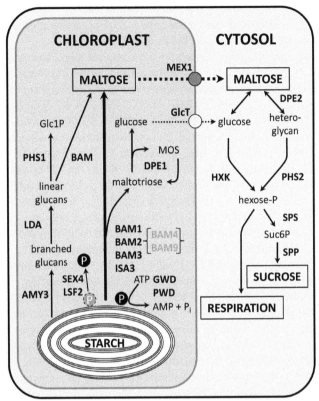

Figure 9.2. Pathway of transitory starch degradation in Arabidopsis leaves. Starch is mainly degraded by β-amylase (BAM) releasing maltose as the major product. BAM4 and BAM9 are regulatory non-catalytic isoforms. Maltose is exported to the cytosol, where it is catabolized to provide substrates for respiration and sucrose synthesis. AMY3, α-amylase; DPE1, plastidic disproportionating enzyme; DPE2, cytosolic disproportionating enzyme; GlcT, glucose transporter; GWD, glucan, water dikinase; HXK, hexokinase; ISA, isoamylase (debranching enzyme); LDA, limit dextrinase (debranching enzyme); MEX1, maltose transporter; MOS, malto-oligosaccharides; PHS1, plastidic starch phosphorylase; PHS2, cytosolic glucan phosphorylase; PWD, phosphoglucan, water dikinase; SEX4/LSF2 phosphoglucan phosphatase; SPP, sucrose-phosphate phosphatase; SPS, sucrose-phosphate synthase.

enzymes. There is growing evidence that GWD also plays a role in starch synthesis, but this function is not yet fully understood (Mahlow et al. 2014, Hejazi et al. 2012, 2014). It has been speculated that C6 phosphorylated glucans generated by GWD contribute in some way to starch granule initiation or the control of starch granule number in the chloroplast (Malinova and Fettke 2017).

Starch can be degraded hydrolytically by endo- and exo-amylases to produce malto-oligosaccharides of various sizes, or phosphorolytically by starch phosphorylase (PHS or PHO1; EC 2.4.1.1) to produce Glc1P. In germinating seeds and sprouting tubers, endo-amylases (α-amylases) often play a major role in starch breakdown. Arabidopsis has three α-amylase isoforms—AMY1, AMY2 and AMY3—but only AMY3 is located in chloroplasts. Neither AMY3 nor PHS appears to play essential roles in transitory starch degradation in Arabidopsis leaves (Zeeman et al. 2004, Yu et al. 2005). Instead, mutant analyses have shown that the predominant pathway of starch degradation in Arabidopsis leaves is mediated by β-amylase (BAM; EC 3.2.1.2), which is an exo-amylase, and the major product of starch degradation is maltose (Weise et al. 2004, Niittylä et al. 2004). In Arabidopsis, there are nine *BAM* genes, five of which (*BAM1-BAM4* and *BAM9*) encode proteins that are targeted to the chloroplast (Lao et al. 1999, Sparla et al. 2006, Fulton et al. 2008). BAM3 is the predominant isoform in mesophyll cells (Fulton et al. 2008). BAM1 is important for degradation of starch in guard cells, and in mesophyll cells under water-limiting conditions (Valerio et al. 2011, Prasch et al. 2015, Horrer et al. 2016). BAM2 has a very low specific activity and its function is unclear. BAM4 and BAM9 lack catalytic activity. Nevertheless, BAM4 appears to have an important role, possibly as a regulatory protein, as starch degradation is severely disrupted in *bam4* mutants (Fulton et al. 2008). Isoamylase 3 (ISA3) and limit-dextrinase (LDA; EC 3.2.1.142) are responsible for debranching of starch by hydrolysis of the α-1,6 glycosidic bonds in amylopectin.

Several additional enzymes are required for the complete degradation of glucan chains. The progressive removal of maltose by exo-amylolytic activity of BAM is blocked when the enzyme encounters phosphorylated glucose residues, so these are removed by two phosphoglucan phosphatases, STARCH EXCESS4 (SEX4) and LIKE-STARCH-EXCESS-FOUR2 (LSF2) as the glucan chains are being degraded (Kötting et al. 2009, Santelia et al. 2011). SEX4 can act on phosphate groups in either the C6 or C3 positions and is essential for complete starch degradation (Hejazi et al. 2010, Streb and Zeeman 2014). LSF2 has a strong preference for the C3 position. It appears not to be essential, probably due to functional redundancy with SEX4, as the *lsf2* mutant is indistinguishable from wild-type in its ability to degrade starch, although the *sex4 lsf2* double mutant has a more severe starch excess phenotype than *sex4* (Santelia et al. 2011). A related isoform, LSF1, appears not to have phosphoglucan phosphatase activity but might have a regulatory function (Comparot-Moss et al. 2010). The smallest malto-oligosaccharide that can be hydrolyzed efficiently by BAM is maltopentaose (DP5). Any residual maltotriose (DP3) or maltotetraose (DP4) is metabolized by the plastidial disproportionating enzyme (DPE1, EC 2.4.1.25). This glucotransferase randomly transfers one or more glucose residue(s) from one malto-oligosaccharide to another, generating higher DP oligosaccharides that can be degraded by BAM (Critchley et al. 2001). Together, DPE1 and BAM fully degrade small malto-oligosaccharides to produce maltose and a lesser amount of glucose.

The main products of starch degradation, maltose and glucose, are exported from the chloroplasts via the MALTOSE EXCESS1 (MEX1) transporter and the plastidial glucose transporter (GlcT), respectively (Weber et al. 2000, Niittylä et al. 2004, Weise et al. 2004, 2006, Cho et al. 2011). Both transporters are located in the inner chloroplast envelope membrane. Disruption of maltose export in *mex1* mutants leads to accumulation of maltose in the chloroplasts (up to 40-fold higher than in wild-type plants), inhibition of starch degradation and severely stunted growth (Niittylä et al. 2004). In contrast, mutants lacking GlcT have a wild-type phenotype (Cho et al. 2011), consistent with glucose being only a minor product of starch breakdown. The severity of the *mex1* phenotype might be alleviated to some extent by diverting more malto-oligosaccharides to glucose production and export, via DPE1 and GlcT, as loss of both transporters in the *mex1 glct* double mutant leads to severe disruption of chloroplast development and function (Cho et al. 2011).

In the cytosol, the exported maltose is metabolized by the cytosolic disproportionating enzyme (DPE2), which transfers one glucosyl moiety to a cytosolic heteroglycan and releases the other as free glucose (Chia et al. 2004, Lu and Sharkey 2004, Fettke et al. 2006). The glucose, along with that exported from the chloroplasts by GlcT, is phosphorylated by hexokinase (HXK; EC 2.7.1.1) to give Glc6P, which enters the cytosolic pool of hexose-phosphates. The glucosyl moiety from maltose that is attached to the cytosolic heteroglycan is cleaved by cytosolic phosphorylase (PHO2) to give Glc1P, which also enters the cytosolic pool of hexose phosphates. This pool of hexose-phosphates provides the substrates for maintenance respiration in the leaf (via glycolysis, the oxidative pentose phosphate pathway and the tricarboxylic acid cycle), and for sucrose synthesis and export to heterotrophic sink organs at night.

9.3 Regulation of Transitory Starch Synthesis

9.3.1 Why is Starch Synthesis Regulated?

The regulation of transitory starch synthesis is a complex balancing act that involves multiple molecular mechanisms to get the balance right. First, starch synthesis must be coordinated with the photosynthetic CO_2 fixation, as either too much or too little starch synthesis can limit photosynthetic capacity. Second, the partitioning of photoassimilates between sucrose and starch must offset demand for sucrose for immediate growth against the need to accumulate enough starch to survive the night.

Third, starch synthesis must be attuned to environmental conditions, as starch reserves provide an important buffer against day-to-day variation in irradiance, temperature and water supply, etc., which can limit photosynthesis (i.e., carbon supply) in plants growing in the wild or in the field. We have a fair understanding of how starch synthesis is regulated to meet the first two requirements, and are starting to elucidate the mechanisms involved in matching starch synthesis to environmental conditions.

9.3.2 Coordination of Starch Synthesis with Photosynthesis and Sucrose Synthesis

Photosynthetic carbon fixation is dependent on the availability of ribulose-1,5-bisphosphate (RuBP), the CO_2 acceptor in the ribulose-1,5-bisphosphate carboxylase-oxygenase (RubisCO; EC 4.1.1.39) reaction, and supplies of NADPH and ATP from the photosynthetic electron transport chain and photophosphorylation. A crucial feature of the Calvin-Benson cycle is the autocatalytic build-up of phosphorylated intermediates, including RuBP, that enables flux to be increased when the leaf is illuminated until the cycle is operating at full capacity. During this autocatalytic phase, excessive withdrawal of intermediates from the cycle for starch and sucrose synthesis would restrict RuBP regeneration and so limit net CO_2 fixation. Conversely, as the Calvin-Benson cycle approaches its maximum capacity, end product synthesis is needed to release orthophosphate (Pi) for photophosphorylation, providing ATP for CO_2 fixation to continue. In addition to balancing the total synthesis of end products with CO_2 fixation, the plant also needs to regulate the partitioning of photoassimilates between different end products, maximizing the production of sucrose for immediate growth while ensuring that it keeps enough carbon in reserve for the coming night.

During the 1970s and 1980s, a model was developed (Fig. 9.3) to explain how end product synthesis is coordinated with CO_2 fixation, and how the plant regulates partitioning of photoassimilates between starch and sucrose (reviewed in MacRae and Lunn 2006, Stitt et al. 2010). The model was based on comparison of maximal enzyme activities, measurements of pathway intermediates, including their subcellular compartmentation and dynamics in responses to changes in irradiance or CO_2 concentrations, and *in vitro* characterization of purified enzymes. In this model, the rate of starch synthesis is controlled primarily by regulation of AGPase. This enzyme is allosterically activated by 3-phosphoglycerate (3PGA) and inhibited by Pi (Preiss 1988). During the autocatalytic phase, newly fixed carbon is retained in the chloroplasts to raise the levels of Calvin-Benson cycle intermediates, and there is little export of triose-phosphates to the cytosol for sucrose synthesis. Under these conditions, the activities of two key enzymes in the pathway of sucrose synthesis, the cytosolic fructose-1,6-bisphosphatase (cytFBPase; EC 3.1.3.11) and sucrose-phosphate synthase (SPS; EC 2.4.1.14), are substrate limited. Furthermore, the cytFBPase is competitively inhibited by the relatively high concentrations of fructose-2,6-bisphosphate (Fru2,6BP), and the low cytosolic Glc6P:Pi ratio allosterically inhibits SPS (Stitt et al. 1987, Huber and Huber 1996). Once the stromal concentration of triose-phosphates has built up sufficiently to maintain high rates of CO_2 fixation, more triose-phosphate is exported to the cytosol. This not only provides more substrates for sucrose synthesis, but also leads to a decrease in Fru2,6BP levels, lifting inhibition of the cytFBPase, and an increase in the cytosolic Glc6P:Pi ratio, which activates SPS. This coordinated feed-forward activation rapidly increases the rate of sucrose synthesis once a threshold level of triose-phosphate export has been reached, thereby releasing Pi in the cytosol, which can then be transported back to the chloroplast via the triose-phosphate translocator in exchange for more triose-phosphates. However, if the rate of sucrose synthesis exceeds the capacity of the leaf to export sucrose or store it in the vacuole, the accumulation of sucrose leads to inhibition of SPS, which in turn leads to a rise in Fru2,6BP levels and inhibition of the cytFBPase. This coordinated feedback inhibition of sucrose synthesis decreases the release of Pi in the cytosol, thereby restricting export of triose-phosphates from the chloroplasts. This results in accumulation of triose-phosphates and other phosphorylated intermediates, including 3PGA, in the chloroplasts, and a concomitant fall in stromal Pi levels. The resulting increase in the stromal 3PGA:Pi ratio allosterically activates AGPase, diverting photoassimilates into starch synthesis.

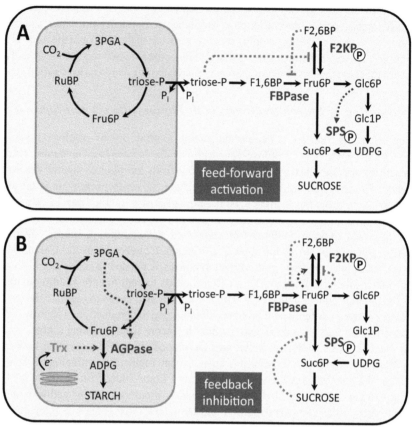

Figure 9.3. Regulation of photoassimilate partitioning between sucrose and starch. **A)** Feed-forward activation. Upon illumination, photosynthetic CO2 fixation leads to production and export of triose-P from the chloroplasts to the cytosol, thereby inhibiting synthesis of fructose-2,6-bisphosphate (F2,6BP) by fructose-6-phosphate 2-kinase (F2KP). This lifts inhibition of the cytosolic fructose-1,6-bisphosphatase (FBPase), leading to production of hexose-phosphates, activating sucrose-phosphate synthase (SPS) and sucrose synthesis. F2KP and SPS are also regulated by protein phosphorylation. **B)** Feedback inhibition. Accumulation of sucrose leads to inactivation of SPS, resulting in accumulation of hexose-P and F2,6BP, which inhibits FBPase. Less Pi is released by sucrose synthesis in the cytosol, restricting export of triose-P from the chloroplasts. Phosphorylated intermediates accumulate in the chloroplast stroma, and the high 3-phosphoglycerate (3PGA):Pi ratio allosterically activates ADP-glucose pyrophosphorylase (AGPase) and starch synthesis. AGPase is also activated by reduction of an intermolecular disulphide bridge, mediated by thioredoxin (Trx). Black dashed arrows show activation, grey dashed lines show inhibition.

This qualitative model considers starch to be an "overflow" product that is produced when the rate of photosynthesis exceeds the capacity of the leaf to export and store sucrose.

Studies of various mutants and transgenic plants with altered enzymes activities allowed the application of metabolic control theory to quantify the contribution made by individual enzymes to control of flux through the pathway (Neuhaus et al. 1989, 1990, Neuhaus and Stitt 1990, Stitt and Sonnewald 1995, Scott et al. 2000, Strand et al. 2000, Hädrich et al. 2011). This approach confirmed that AGPase exerts considerable control over the rate of starch synthesis, while control of sucrose synthesis is largely shared between the cytFBPase and SPS (Stitt et al. 2010). It is important to note, however, that none of these enzymes exerts complete control over their respective pathways, and so should not be considered as "rate-limiting" steps in the classical sense. It also became clear that the distribution of control in a pathway is not fixed, being influenced by environmental conditions and shifting as fluxes increase or decrease.

The regulatory model evolved over the years as new discoveries revealed further layers of regulation, in particular the post-translational modulation of SPS by reversible protein phosphorylation (reviewed in Huber and Huber 1996). Similarly, the bifunctional fructose-6-phosphate 2-kinase/fructose-2,6-bisphosphatase (F2KP; EC 2.7.1.105/EC 3.1.3.46) enzyme responsible for synthesis and breakdown of Fru2,6BP is also regulated by protein phosphorylation (Nielsen et al. 2004). Surveys of the Arabidopsis phosphoproteome have identified phosphorylation sites in other enzymes of sucrose and starch synthesis, but in most cases it is not yet known if these have regulatory potential (reviewed in Kötting et al. 2010). For many years, it has been known that several enzymes from the Calvin-Benson cycle in the chloroplasts are light/dark regulated by changes in their redox status, mediated by thioredoxins (Buchanan and Balmer 2005). More recently, it was discovered that AGPase is also sensitive to redox modulation. AGPase is a heterotetrameric enzyme consisting of two small subunits (APS) and two large subunits (APL) (Crevillén et al. 2003, 2005). In the dark, the two small subunits are linked by an intermolecular disulphide bridge between conserved cysteine residues (Cys81 in Arabidopsis APS1, Hädrich et al. 2012), and in this oxidized (dimeric APS1) form the enzyme has low activity (Ballicora et al. 1999, Tiessen et al. 2003, Hendriks et al. 2003). In the light, the disulphide bridge is reduced to the dithiol form, via thioredoxin, and this reduced (monomeric APS1) form of the enzyme has higher activity.

In heterotrophic organs, for example developing potato (*Solanum tuberosum*) tubers, the redox status of AGPase is regulated via NADP-thioredoxin reductase C and is dependent on sucrose availability, with the enzyme, and thus starch synthesis, being activated when sucrose is abundant (Tiessen et al. 2003, Michalska et al. 2009, Thormählen et al. 2013). The redox status of AGPase in leaves is also positively correlated with sucrose (Hendriks et al. 2003, Lunn et al. 2006). Trehalose 6-phosphate (Tre6P), the intermediate of trehalose biosynthesis, has been proposed as an intermediary linking the redox status of AGPase to sucrose (Kolbe et al. 2005). Tre6P is made by Tre6P synthase (TPS; EC 2.4.1.15) and dephosphorylated to trehalose by Tre6P phosphatase (EC 3.1.3.12). Kolbe et al. (2005) observed that constitutive over-expression of TPS in Arabidopsis plants led to reductive activation of AGPase and higher starch levels, while over-expression of TPP had the opposite effects. Although not measured, these effects were attributed to the TPS and TPP over-expressing lines having higher and lower Tre6P contents, respectively. In addition, *in vitro* experiments showed that addition of Tre6P to isolated chloroplasts, in the presence of dithiothreitol (DTT), led to reductive activation of AGPase (Kolbe et al. 2005). The hypothesis that Tre6P links the redox status of AGPase to sucrose availability received further support from measurements of Tre6P using a newly developed mass-spectrometry based assay for this metabolite (Lunn et al. 2006). It was observed that the level of Tre6P in Arabidopsis rosettes (and also seedlings) is correlated with both sucrose content and the redox status of AGPase.

Sucrose-dependent activation of AGPase via Tre6P is a simple and attractive model, and is widely referenced in the literature. However, later work cast doubt on the validity, or at least the physiological importance, of this mechanism. Short-term increases in Tre6P, brought about by ethanol-inducible over-expression of TPS in Arabidopsis, did not lead to any significant increase in the reductive activation of AGPase (Figueroa et al. 2016). This suggested that the changes in redox status of AGPase in constitutive TPS (and TPP) over-expressers observed by Kolbe et al. (2005) were only indirectly linked to higher (or lower) Tre6P levels in the plants, which have highly pleiotropic growth phenotypes (Schluepmann et al. 2003). The commercial supplies of Tre6P available at that time were relatively impure, with enzymatic analysis indicating only 60–70% purity (Lunn et al. 2006). They were later found to contain at least 40 contaminants, including several detergent-like compounds (Yadav et al. 2014), which might have increased the permeability of the chloroplast envelope to the reducing agent, DTT, when isolated chloroplasts were incubated with Tre6P and DTT (Kolbe et al. 2005). In retrospect, an important control was missing from the isolated chloroplast experiment in Kolbe et al. (2005), namely incubation of chloroplasts with Tre6P that had been pre-incubated with TPP to destroy Tre6P and thereby test for the effect of the contaminants in the commercial Tre6P supply.

In parallel, several studies have also questioned the importance of redox modulation of AGPase in regulation of starch synthesis in Arabidopsis leaves. Site-directed mutagenesis of the *APS1* coding region was used to change the Cys81 residue of the APS1 protein to Ser. Constructs containing the mutated *APS1(C81S)* gene under the control of its endogenous promoter (Hädrich et al. 2012) or a constitutive 35S promoter (Li et al. 2012) were used to complement the *adg1* mutant that lacks APS1. The substitution of Cys81 by Ser prevented formation of an intermolecular disulphide bridge between the two APS1 subunits, effectively abolishing the enzyme's redox modulation. This had surprisingly little effect on the rate of starch synthesis, except in plants growing under moderate light (150 µmol m^{-2} s^{-1}) and short-day (8-hr photoperiod) conditions (i.e., C-limiting conditions), where starch accumulation was increased by about 40% (Hädrich et al. 2012). In contrast, abolishing the allosteric regulation of AGPase, by complementation of *adg1* with a triple-mutated bacterial AGPase (GlcC), showed that allosteric regulation of AGPase is crucial for the plant to adjust its rate of starch accumulation to changes in day length (Mugford et al. 2014).

In conclusion, there is conflicting evidence for and against the concept that Tre6P acts as an intermediary linking the redox status of AGPase to sucrose levels in leaves. Furthermore, redox modulation of AGPase appears to be relatively unimportant, compared to allosteric regulation of the enzyme, in determining the net rate of transitory starch accumulation, except under C-limiting conditions.

9.3.3 Adaptation of Starch Synthesis to Environmental Conditions

Environmental conditions have a significant impact on starch synthesis. In several species, it has been observed that the amount of starch accumulated during the day is particularly dependent on photoperiod (Britz et al. 1985a,b, Chatterton and Silvius 1979, 1980a,b, 1981, Geiger and Servaites 1994, Mullen and Koller 1988, Lorenzen and Ewing 1992), and is adjusted rapidly in response to a change in day length (Gibon et al. 2004b, Mengin et al. 2017). Various mechanisms have been proposed to explain the higher rate of starch accumulation in short days. In Arabidopsis, the maximum catalytic activity of AGPase, relative to SPS, is higher in short-day than long-day grown plants, indicating that differences in gene and protein expression contribute to long-term adaptation of photoassimilate partitioning to photoperiod (Gibon et al. 2009, Sulpice et al. 2014). Redox regulation of AGPase might also be involved in long-term adaptation (Hädrich et al. 2012), but appears to be unimportant for short-term responses to a change in day length (Mugford et al. 2014). In contrast, allosteric regulation of AGPase is necessary for photoperiod adjustment of starch synthesis (Mugford et al. 2014).

The circadian clock plays a central role in many responses to day length, such as flowering time (McClung 2006, 2008). Changes in photoperiod lead to shifts in the timing of expression of core clock genes (Flis et al. 2016). Several clock mutants, including *early flowering 3* (*elf3*) and *prr7prr9* (defective in two PSEUDO-RESPONSE REGULATOR proteins), accumulate starch more slowly than wild-type plants (Flis et al. 2018), and the *gigantea* mutant is unable to adjust starch synthesis in response to a change in day length (Mugford et al. 2014). The mechanisms linking net starch synthesis to the circadian clock are not yet understood. However, several studies point to a restriction in growth in the first hours of the day following a shift to short days (or extension of the night), which leads to accumulation of sugars in the leaf and activation of starch synthesis (Gibon et al. 2004b, 2009, Sulpice et al. 2014, Mugford et al. 2014, Mengin et al. 2017).

In natural environments, plants are not exposed to sudden changes in day length, as this fluctuates in a gradual and predictable way with the seasons. However, light levels and light quality can vary over much shorter timescales, including predictable changes from the diurnal progression of the sun through the sky, as well as unpredictable changes due to sun flecks and shading in lower levels of the canopy and variation in cloud cover (Vialet-Chabrand et al. 2016, 2017, Annunziata et al. 2017, 2018). Although temperature tends to fluctuate in a somewhat predictable manner during the diurnal light-dark cycle, plants are also exposed to unpredictable, weather-dependent changes in temperature. Water availability is another factor that has a major impact on a plant's ability to capture carbon by photosynthesis. Seasonal and unseasonal periods of drought usually impose a progressive stress on plants, giving them time to adjust their metabolism and growth (Hummel et al. 2010), but flooding can be unpredictable. Simultaneous variation in irradiance, temperature and water availability pose a huge challenge for plants in the wild, as they try to balance total starch and sucrose synthesis with a constantly fluctuating supply of photoassimilates, and at the same time meet the competing demands of carbon for immediate growth and accumulation of carbon reserves to survive the night. To better understand how plants meet this challenge, in the future we need to investigate gene expression, enzyme activities, metabolite levels and metabolic fluxes in plants growing in natural conditions, or in controlled environment facilities that simulate the multifactorial fluctuations that occur in the wild (Vialet-Chabrand et al. 2016, 2017, Annunziata et al. 2017, 2018).

9.4 Regulation of Transitory Starch Degradation

9.4.1 Why is Starch Degradation Regulated?

As discussed above, plants growing in the wild are at the mercy of the elements during the day when starch reserves are being accumulated, and starch synthesis is regulated by multiple mechanisms to ensure that some carbon reserves are set aside for the coming night. The mobilization of these reserves in the dark must also be carefully regulated if the plant is to survive through the night and make the best use of its investment. Compared to the huge fluctuations in irradiance that complicated regulation of photosynthetic metabolism during the day, night-time conditions are relatively stable—it is essentially dark all night. Furthermore, the length of the night does not change very much from one day to the next and is therefore reasonably predictable. These features of the dark period potentially allow plants to have more control over remobilization of their starch reserves at night than they have over starch accumulation during the day, which is more dependent on the vagaries of the weather.

Over the last decade or so, it has become clear that the circadian clock plays a major role in regulation of transitory starch degradation, especially under C-limiting (e.g., short-day) conditions, helping the plant to anticipate how long the night will last and use this information to set an appropriate rate of starch degradation at dusk (Weise et al. 2006, Graf et al. 2010, Scialdone et al. 2013). The plant can potentially also use this timekeeper to adjust the rate of starch degradation during the night if changes in temperature lead to starch being degraded too quickly or too slowly (Pilkington et al. 2015). In Arabidopsis, transitory starch is degraded in a fairly linear manner throughout the night, and under C-limiting conditions the rate of degradation is such that most, but not quite all, of the starch is remobilized by the end of the night (Gibon et al. 2004b, 2009, Graf et al. 2010). In this way, the plant maximizes use of its starch reserves for growth at night, while ensuring that it has enough reserves to see it through to dawn. This is important because premature exhaustion of reserves before the end of the night would leave the plant short of carbon, triggering C-starvation responses (e.g., autophagy) that lead to catabolism of proteins and other cellular components that are metabolically expensive to replace. While such responses enable the plant to survive a short-term C-deficit, if repeated over the longer term they have a negative impact on growth, and ultimately compromise the survival and reproductive success of the plant.

Our understanding of the regulation of transitory starch degradation is fragmented. At a biochemical level, we know of some potential mechanisms for regulation of individual enzymes in the pathway, but have little idea of how the control of flux is distributed between different enzymes. Several models have been proposed to explain how plants might predict the length of the night using outputs from the circadian clock, and integrate this with information about the amount of starch at dusk, but we do not yet understand how this is translated to regulation of starch degrading enzymes at a biochemical level. In the following sections, we outline what we do know about regulation of individual enzymes, describe the various models for how the clock helps to determine the maximum permissible rate of starch degradation, and how the rate of starch degradation might be fine-tuned by Tre6P to match the demand for carbon for maintenance respiration and growth at night.

9.4.2 Transcriptional, Translational and Post-Translational Regulation of Starch Degrading Enzymes

Transcripts for many of the starch degrading enzymes show pronounced changes in expression during the diurnal light-dark cycle, with many being under the control of the circadian clock and rising in a coordinated manner during the first hours of the day (Smith et al. 2004). However, despite the pronounced diurnal fluctuations in their transcript levels, the abundance and/or activities of the GWD, AMY3 and DPE2 proteins remain fairly constant through the light-dark cycle (Lu et al. 2005, Yu et al. 2005, Skeffington et al. 2014). Similar discrepancies between transcript and protein levels or activities have been observed for AGPase and many other enzymes of primary metabolism (Gibon et al. 2004a), and can be explained by the relatively high abundance of the enzyme proteins and their slow turnover rates, with half-lives often measured in days (Piques et al. 2009, Baerenfaller et al. 2012, Nelson et al. 2014). One exception to this pattern is BAM3 which, with a half-life of about 10 hr, has one of the most rapid turnovers of all the measured proteins in Arabidopsis rosettes (Li et al. 2017). From these observations, it seems likely that transcriptional and translational regulation of starch-degrading enzymes, except possibly BAM3, is relatively unimportant in regulation of starch degradation on a daily basis, although it probably does contribute to adjustments of starch degrading capacity over the longer term.

Several starch-degrading enzymes are potentially regulated at a post-translational level. GWD contains conserved Cys residues that can form an intramolecular disulphide bridge (Mikkelsen et al. 2005, Sokolov et al. 2006). As observed for several redox-regulated enzymes from the Calvin-Benson cycle, the *in vitro* activity of the reduced (dithiol) form of GWD is higher than that of the oxidized (disulphide) form. Paradoxically, this suggests that GWD should be more active in the light, when there is a more reducing environment in the chloroplasts, than in the dark when the enzyme is needed for starch degradation. However, expression of a mutated, redox-insensitive version of GWD in a null *gwd* (*sex1*) mutant fully complemented the phenotype of the mutant, restoring a wild-type pattern of diurnal starch turnover, showing that redox regulation of GWD is not essential for its function (Skeffington et al. 2014).

There are nine BAM proteins in Arabidopsis but only one, BAM1, has been shown to have potential for redox modulation, with the reduced (dithiol) form of the enzyme being more active than the oxidized (disulphide) form (Sparla et al. 2006). As for GWD, this suggests that BAM1 should be more active in the light. This apparent paradox is explained by the localization of BAM1 primarily in guard cell chloroplasts, where the diurnal pattern of starch degradation is very different from that in mesophyll cells (Santelia and Lunn 2017). The starch reserves in guard cells are rapidly degraded in the light to generate osmolytes that drive stomatal opening (Horrer et al. 2016), so it makes sense for BAM1 to be activated by reduction in the light. Under water-limiting conditions, generation of reactive oxygen species could lead to oxidation, i.e., inactivation, of BAM1, thereby inhibiting starch degradation and stomatal opening (Valerio et al. 2011). BAM1 is also present in mesophyll cells of drought-stressed plants, which have higher starch levels than well-watered control plants (Valerio et al. 2011, Monroe et al. 2014). The accumulation of starch probably reflects a combination of factors. The capacity for starch degradation in mesophyll cells could be more constrained if it is dependent on the redox-sensitive activity of BAM1 than the redox-insensitive BAM3, which is the predominant starch hydrolyzing enzyme in non-stress conditions. Growth may also be actively suppressed in part or all of the plant under stress conditions (Hummel et al. 2010, Paparelli et al. 2013), lowering demand for carbon from starch turnover at night, allowing the plant to eke out its starch reserves until conditions improve.

Phosphoproteomic studies indicate that several starch degrading enzymes, for example BAM1 and AMY3, are phosphorylated *in vivo* (Kötting et al. 2010), but it is unknown whether protein phosphorylation modulates their activities or not (Thalmann et al. 2016). In yeast two-hybrid assays, the SEX4 phosphoglucan phosphatase has been shown to interact with SUCROSE-NON-FERMENTING1-RELATED-KINASE1 (SnRK1) (Fordham-Skelton et al. 2002). SnRK1 is a heterotrimeric protein kinase consisting of a catalytic (α) subunit and two regulatory subunits (β and γ, or β and βγ) that plays a central role in energy homeostasis and stress responses in plants (Baena-Gonzalez et al. 2007). The interaction with SnRK1 suggests that SEX4 could be regulated by reversible protein phosphorylation, but so far this possibility has not been explored in detail. It has been reported that *snrk1* loss-of-function mutants of Arabidopsis and the moss *Physcomitrella patens* accumulate more starch than wild-type plants, and remobilization of starch at night is impaired under Pi-limiting conditions (Thelander et al. 2004, Baena-Gonzalez et al. 2007, Fragoso et al. 2009). Conversely, Arabidopsis plants that over-express the catalytic SnRK1α1 (KIN10) subunit of SnRK1 have less starch under some growth conditions (Baena-Gonzalez et al. 2007, Jossier et al. 2009).

SnRK1 regulates many metabolic and growth-related processes in plants, including sucrose synthesis via phosphorylation of SPS (Huber and Huber 1996), so SnRK1 mutants typically display highly pleiotropic phenotypes. Therefore, it was unclear if their starch phenotypes were due to direct regulation of starch metabolizing enzymes by SnRK1, or were indirect effects of perturbations in other metabolic pathways and growth. However, a recent report that SnRK1α, β and βγ-type subunits are localized in chloroplasts, and that the SnRK1β3 and SnRK1βγ subunits are associated with starch granules, is consistent with SnRK1 playing a direct role in regulation of transitory starch metabolism (Ruíz-Gayosso et al. 2018). It was also discovered that: (i) SnRK1 complexes from Arabidopsis leaves are activated by maltose, but only when these were isolated from leaves harvested around dusk, (ii) reconstituted SnRK1α1/β3/βγ complexes were also activated by maltose *in vitro*, while complexes containing the SnRK1β1 or β2 subunit were not, and (iii) sensitivity to maltose was dependent on a carbohydrate binding module in the SnRK1βγ subunit (Ruíz-Gayosso et al. 2018). These observations suggested that regulation and/or changes in the subunit composition of SnRK1 complexes around dusk lead to SnRK1 becoming sensitive to regulation by maltose, and that SnRK1 is activated by rising maltose levels as starch breakdown begins. It has been speculated that regulation of SnRK1 by maltose might be a way to activate the pathway of maltose catabolism (Ruíz-Gayosso et al. 2018).

9.4.3 *Regulation of Starch Degrading Enzymes by Metabolites*

Unlike photosynthesis, there has been almost no application of metabolic control theory to transitory starch degradation, to map the distribution of control between different enzymes in the pathway. The most common approach is to generate an allelic series of plants with a range of activities of the targeted enzyme, for example using antisense (Stitt et al. 1991) or TILLING (Hädrich et al. 2011), and determine the sensitivity of flux to progressive loss of enzyme activity. Using this approach, it was found that GWD exerts little or no control over starch degradation in Arabidopsis leaves, with loss of up to 70% of the protein having no detectable impact on rates of starch turnover (Skeffington et al. 2014). None of the other enzymes in the pathway has so far been investigated in this way, with most studies focussing on null mutants. Although these have been crucial for elucidating the pathway of starch degradation, they usually tell us little about the control of flux. Nevertheless, a few potential regulatory mechanisms have been revealed by analysis of null mutants. For example, *mex1* and *dpe1* mutants accumulate maltose and maltotriose, respectively, in the chloroplasts, and both have a *sex* phenotype (Niittylä et al. 2004, Critchley et al. 2001, Lu and Sharkey 2004), suggesting that high stromal levels of maltose or maltotriose inhibit starch degradation. Here it is worth noting that BAM3, the predominant starch hydrolyzing enzyme in Arabidopsis mesophyll cells, was recently reported to be inhibited by maltotriose (Li et al. 2017). The *dpe2* mutant also accumulates maltose, this time in the cytosol, and has a *sex* phenotype, suggesting that cytosolic maltose can also exert feedback inhibition on starch degrading enzymes in the chloroplasts in some way (Chia et al. 2004). The high levels of maltose in the *mex1* and *dpe2* mutants might also be expected to lead to activation of SnRK1 (Ruíz-Gayosso et al. 2018), but this seems unlikely to lead to inhibition of starch degradation and a starch excess because plants with increased SnRK1 activity tend to have less starch than wild-type plants (Baena-Gonzalez et al. 2007, Jossier et al. 2009). Thus, the mechanisms linking maltose and maltotriose accumulation to inhibition of starch degradation are uncertain. It is also questionable if such mechanisms are physiologically significant in wild-type plants, where the levels of maltose and maltotriose are much lower than in the *mex1*, *dpe2* and *dpe1* mutants, for example wild-type Arabidopsis plants have 40 times less maltose than *mex1* (Niittylä et al. 2004).

The majority of the maltose produced by starch turnover is destined for synthesis of sucrose, which is then exported to growing sink organs, and several recent studies have implicated Tre6P in feedback regulation of starch degradation by sucrose. Tre6P is thought to function as both a signal and regulator of sucrose levels in plants (Lunn et al. 2006, 2014, Figueroa and Lunn 2016). It was reported that a short-term increase in Tre6P levels in Arabidopsis rosettes, from ethanol-induced over-expression of TPS, led to inhibition of starch degradation at night (Martins et al. 2013). Here it is worth noting that the high starch levels seen in *35S:TPS* plants with constitutively high Tre6P levels (Kolbe et al. 2005) can probably be explained by inhibition of starch degradation, rather than stimulation of starch synthesis. In response to a short-term increase in Tre6P, plants with elevated levels of Tre6P had less maltose than non-induced control plants (Martins et al. 2013). This indicated that an early step in the pathway of starch degradation in the chloroplasts had been inhibited, rather than the export of maltose from the chloroplasts or its catabolism in the cytosol, which would lead to maltose accumulation. Starch granules from induced plants with high Tre6P contained more phosphate than those from non-induced control plants. This suggested that high Tre6P perturbed starch phosphorylation by GWD and PWD, or dephosphorylation by SEX4 and LSF2, or both. However, there were no obvious differences between induced and control plants in the abundance of the GWD, PWD or SEX4 proteins, and *in vitro* assays with recombinant enzymes revealed no obvious sensitivity to Tre6P (Martins et al. 2013), so the target and mode of inhibition by Tre6P remain unknown. As Tre6P is primarily localized in the cytosol (Martins et al. 2013), it is possible that Tre6P itself does not act directly in the chloroplasts, but via an intermediary.

Tre6P interacts with SnRK1 in several ways that can differ between source leaves and sink organs. SnRK1 complexes from developing tissues are inhibited by Tre6P, and this inhibition is dependent on a so far unidentified protein factor (Zhang et al. 2009); however, the catalytic activity of the SnRK1α1 (KIN10) subunit on its own is activated by Tre6P (Zhai et al. 2018). SnRK1 is activated via phosphorylation by the GEMINIVIRUS REP-INTERACTING KINASE1/2 (GRIK1/2) protein kinase (also known as SnRK1-ACTIVATING KINASE1/2), but association of SnRK1 with this kinase in inhibited by binding of Tre6P to SnRK1α1, thereby hindering activation of SnRK1 (Zhai et al. 2018). Changes in Tre6P levels might also affect SnRK1 indirectly, via concomitant changes in sucrose and other metabolites that trigger other mechanisms of regulation of SnRK1 (Figueroa and Lunn 2016). Further work is needed to resolve the complex, and sometimes contradictory, interactions between Tre6P and SnRK1. Given the defective starch turnover in *snrk1* mutants (Thelander et al. 2004, Baena-Gonzalez et al. 2007, Fragoso et al. 2009), this is an avenue worth exploring to see if this might explain how Tre6P inhibits starch degradation. The physiological significance of inhibition of starch degradation by Tre6P will be discussed below, in the context of regulation of starch turnover by the circadian clock.

9.4.4 Regulation of Starch Degradation by the Circadian Clock

The circadian clock is an endogenous timekeeper that enables plants to anticipate and adapt to rhythmic changes in their environment, especially the 24-hr light-dark cycle (Harmer et al. 2000, Dodd et al. 2005, Locke et al. 2005, McClung 2006, Song et al. 2010). The clock consists of a series of transcriptional relays that generates periodic fluctuations in expression of core clock genes and output signals that are used to time many metabolic and developmental processes. The core clock genes can be broadly divided into four groups based on the timing of the peak in their transcript abundance: (i) dawn genes—*LHY, CCA1*, (ii) day genes—*PRR9, PRR7*, (iii) dusk genes—*PRR5, TOC1, GIGANTEA*, and (iv) evening complex genes—*LUX, ELF4* and *ELF3* (Pokhilko et al. 2012, 2013). When plants are growing in light-dark cycles, the clock is reset, or entrained, every day, primarily by light signalling around dawn. Experiments involving shifts in the light-dark cycle (i.e., timing of dawn and dusk) led to the recognition of the circadian clock as a key factor in regulation of transitory starch degradation (Weise et al. 2006, Graf et al. 2010). It was found that Arabidopsis plants adjust their rate of starch degradation to shifts in the length of the night, with degradation being timed so that their starch reserves last until dawn as predicted by the clock, and that the timing of starch degradation is compromised in some clock mutants. Plants are also able to adjust the rate of starch degradation to compensate for a shift in night-time temperature (Pilkington et al. 2015).

Multiple models have been put forward to explain how the circadian clock might regulate starch turnover. Several models involve dynamic sensing of C-status in some form and continuous adjustment of starch degradation via modulation of the core clock or the output from the clock (Feugier and Satake 2013, 2014, Pokhilko et al. 2014, Mishra and Panigrahi 2015, Seki et al. 2017 Shin et al. 2017). In support of such models, there is evidence that sugars are able to entrain the clock (Haydon et al. 2013). SnRK1 is another potential factor, acting as a sensor of energy or C-status and modulating either the core clock, via GIGANTEA (Mishra and Panigrahi 2015) or TIME FOR COFFEE (Shin et al. 2017, Sánchez-Villarreal et al. 2013, 2017), or clock outputs, via an osmo-sensitive kinase (Pokhilko and Ebenhoh 2015).

An alternative, "arithmetic division" model, postulates that plants sense the amount of starch present in the leaves at dusk (proportional to an entity "*S*") and anticipate the length of the night (proportional to an entity "*T*"), using information from the clock, and then perform a division calculation ("*S/T*") to set an appropriate rate of starch degradation (Scialdone et al. 2013). A variant of this model has the *T* entity representing the reciprocal of the expected time until dawn, so that setting the rate of starch degradation is done via multiplication of *S* and *T*, rather than division (Scialdone and Howard 2015). Using some similar concepts, Seaton et al. (2014) proposed a model in which starch levels and "time to dawn" are continuously monitored and the rate of starch degradation is determined by competition between activating and inhibitory factors. Three variants of this model were developed to explore its behaviour in different scenarios, involving either continuous resetting of starch degradation by the clock or light-gated control of starch degradation via different components (Seaton et al. 2014). Predictions from all of these arithmetic models fitted well with experimental observations. In particular, they appear to be better than models, based on modulation of the clock by C-inputs, at predicting the effects of unexpected shifts in the timing of dawn or dusk on starch breakdown.

The identity of the hypothetical entities postulated in the arithmetic models remains a matter of speculation. From shift experiments on several mutants, only the *pwd* mutant was found to be unable to adjust its rate of starch degradation, implicating PWD and the phosphorylation status of the starch granules as factors in regulation of starch degradation by the clock (Scialdone et al. 2013). Using a forward genetics approach, a 50-kDa starch-binding protein named EARLY STARVATION1 (ESV1) has been identified as an important factor in regulation of starch degradation (Feike et al. 2016, Malinova et al. 2018). There is potential for such a protein to act as a signal of starch content. For example, if the total amount of the protein is fixed and the protein becomes sequestered in or on the starch granules in a constant ratio with starch content, then the amount of the protein that remains in an unbound form in the chloroplast stroma would be inversely related to the total starch content in a quantitative manner.

Although the molecular mechanisms involved in regulation of starch degradation by the clock are not yet known, the importance of this regulation is clear, especially under C-limiting conditions when efficient management of transitory starch reserves is most critical. An emerging concept is that the clock sets the maximum permissible rate of starch degradation, ensuring that reserves last through to dawn, but the actual rate of degradation can be varied below this maximum via the inhibitory action of Tre6P, thereby linking the rate of starch degradation to demand for sucrose (Fig. 9.4) (Martins et al. 2013). This concept was explored further in a recent study to determine the sensitivity of starch degradation to Tre6P under different conditions, and whether this is dependent on the clock. Ethanol-inducible-TPS lines of Arabidopsis (Martins et al. 2013) were grown in continuous light and constant temperature from the time of germination, so that the clock was not entrained, then TPS expression

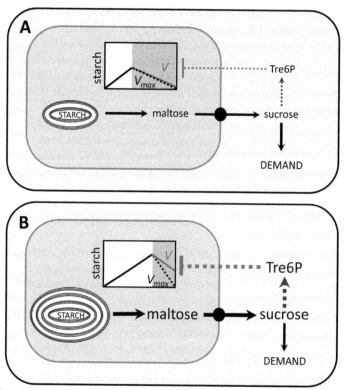

Figure 9.4. Regulation of transitory starch degradation by the circadian clock and trehalose 6-phosphate (Tre6P). **A)** Under carbon-limiting conditions, for example short days, the circadian clock sets the maximum permissible rate of starch synthesis (Vmax) to ensure that starch reserves are not exhausted before dawn. Sucrose supply is just sufficient to meet demand, so sucrose does not accumulate in the leaf and Tre6P levels are low, exerting little or no inhibition of starch breakdown. Therefore, the actual rate of starch degradation (v) is close to the Vmax set by the clock. **B)** Under carbon-replete conditions, for example long days, the circadian clock sets the maximum permissible rate of starch degradation (Vmax) to ensure that starch reserves are not exhausted before dawn. Sucrose supply exceeds demand, so sucrose accumulates in the leaf and Tre6P levels are high, inhibiting starch breakdown. Therefore, the actual rate of starch degradation (v) is below the Vmax set by the clock. (Figure modified from dos Anjos et al. 2018).

was induced to increase Tre6P shortly before darkening the plants. The rate of starch degradation was significantly lower in the induced plants with elevated Tre6P levels than in control plants, indicating that the inhibition of starch breakdown by Tre6P is not dependent on the clock being entrained (dos Anjos et al. 2018).

Starch degradation was also compared in wild-type Arabidopsis plants and the *sweet11;12* mutant, which is defective in phloem loading and accumulates sucrose in its leaves (Chen et al. 2012), and has elevated Tre6P levels throughout the day-night cycle (dos Anjos et al. 2018). Plants were grown under standard laboratory conditions (12-h photoperiod and moderate irradiance of 160 μmol m^{-2} s^{-1}) for 3 weeks and then shifted to a longer day (15-h photoperiod), or high light (320 μmol m^{-2} s^{-1}), or both, for 1 day to increase the amount of starch accumulated by dusk. In each condition, the *sweet11;12* mutants had higher levels of starch at dusk than the corresponding wild-type plants, and starch was degraded in a linear manner through the following night. The wild-type plants remobilized almost all of their starch by the end of the night, whereas the *sweet11;12* mutants retained a substantial amount of starch at dawn (dos Anjos et al. 2018). The starch excess in these plants was comparable to that observed in inducible-TPS lines with similarly high night-time levels of Tre6P (Martins et al. 2013). Thus, the starch excess in these plants was consistently associated with elevated Tre6P at night, rather than sucrose or other sugars, which were high in *sweet11;12* but low in the inducible-TPS plants. Despite differences in starch content between the genotypes, the absolute rates of starch degradation in the *sweet11;12* mutants were the same or very similar to those in the corresponding wild-type plants (dos Anjos et al. 2018). However, when related to the maximum potential rate of starch degradation, derived by dividing starch content at dusk by the length of the night, the relative rate of degradation was significantly lower in the *sweet11;12* mutants than in wild-type plants. The degree to which starch degradation was reduced below the potential maximum rate was highly correlated with the average night-time level of Tre6P, not only across the wild-type and *sweet11;12* plants, but also in the inducible-TPS plants (dos Anjos et al. 2018). From these observations, it was concluded that the action of Tre6P is superimposed on the regulation of starch degradation by the circadian clock, lowering the actual rate of degradation below the maximum permissible rate set by the clock.

The relative influence of the clock and Tre6P probably varies with the C-status of the plants (Fig. 9.4). In C-limiting conditions, such as short days, the slow rate of starch degradation through the long night means that sucrose and Tre6P levels remain low. Therefore, there is little inhibition of starch breakdown by Tre6P, and the rate of degradation approaches the maximum allowable rate set by the clock, such that the plant remobilizes almost all of its starch by dawn. The dominance of the clock under C-limiting conditions ensures that the plant survives the night and makes the best use of its limited starch reserves. In C-replete conditions, such as long days, sucrose and Tre6P levels are higher during the night, and Tre6P exerts significant inhibition on starch degradation, so that not all of the starch is remobilized by the end of the night. The sensitivity of starch degradation to Tre6P means that the remobilization of starch to sucrose can be adjusted according to the demand for sucrose from growing tissues. If growth is restricted by unfavourable conditions, for example low night-time temperature, the starch reserves will be preserved until conditions become more conducive for growth. Under long-day conditions in the wild, starch can also act as a buffer during the final hours of the photoperiod when sunlight levels are falling. As the light fades, sucrose and Tre6P levels decline (Annunziata et al. 2017) and starch becomes more susceptible to degradation via a clock-gated mechanism (Fernandez et al. 2017). Together, these changes allow starch degradation to occur during the last hours of the day when light levels are falling (Fernandez et al. 2017), supplementing the declining supply of sucrose coming directly from photosynthesis. This smoothes the plant's transition from light to dark, avoiding an acute C-crisis at dusk and disruption of growth (Pal et al. 2013, Annunziata et al. 2017, 2018).

9.5 Conclusion

Transitory starch reserves play a central role in the metabolism of many plants, providing carbon and energy for survival through the night, and buffering metabolism against unavoidable fluctuations in their environment. A complex network of regulatory mechanisms balances the rate of starch synthesis in leaves with photosynthetic CO_2 fixation and with the synthesis and export of sucrose for immediate growth. At night, the remobilization of starch is tightly regulated by the circadian clock to ensure that reserves last through the night, with Tre6P superimposing a further layer of regulation to match starch turnover with carbon demand for growth.

References

Annunziata, M.G., F. Apelt, P. Carillo, U. Krause, R. Feil, K. Koehl et al. 2018. Response of Arabidopsis primary metabolism and circadian clock to low night temperature in a natural light environment. Journal of Experimental Botany 69: 4881–4895.

Annunziata, M.G., F. Apelt, P. Carillo, U. Krause, R. Feil, V. Mengin et al. 2017. Getting back to nature: a reality check for experiments in controlled environments. Journal of Experimental Botany 68: 4463–4477.

Baena-Gonzalez, E., F. Rolland, J.M. Thevelein and J. Sheen. 2007. A central integrator of transcription networks in plant stress and energy signalling. Nature 448: 938–942.

Baerenfaller, K., C. Massonnet, S. Walsh, S. Baginsky, P. Bühlmann, L. Lars Hennig et al. 2012. Systems-based analysis of Arabidopsis leaf growth reveals adaptation to water deficit. Molecular System Biology 8: 606.

Ballicora, M.A., Y. Fu, J.B. Frueauf and J. Preiss. 1999. Heat stability of the potato tuber ADP-glucose pyrophosphorylase: role of Cys residue 12 in the small subunit. Biochemical and Biophysical Research Communications 257: 782–786.

Baunsgaard, L., H. Lütken, R. Mikkelsen, M.A. Glaring, T.T. Pham and A. Blennow. 2005. A novel isoform of glucan, water dikinase phosphorylates pre-phosphorylated α-glucans and is involved in starch degradation in Arabidopsis. The Plant Journal 41: 595–605.

Britz, S.J., W.E. Hungerford and D.R. Lee. 1985a. Photoperiodic regulation of photosynthate partitioning in leaves of *Digitaria decumbens* Stent. Plant Physiology 78: 710–714.

Britz, S.J., W.E. Hungerford and D.R. Lee. 1985b. Photosynthate partitioning into *Digitaria decumbens* leaf starch varies rhythmically with respect to the duration of prior incubation in continuous dim light. Photochemical Photobiology 42: 741–744.

Britz, S.J., W.E. Hungerford and D.R. Lee. 1987. Rhythms during extended dark periods determine rates of net photosynthesis and accumulation of starch and soluble sugars in subsequent light periods in leaves of Sorghum. Planta 171: 339–345.

Buchanan, B.B. and Y. Balmer. 2005. Redox regulation: a broadening horizon. Annual Review of Plant Biology 56: 187–220.

Caspar, T., S. Huber and C. Somerville. 1985. Alterations in growth, photosynthesis, and respiration in a starchless mutant of *Arabidopsis thaliana* (L.) deficient in chloroplast phosphoglucomutase activity. Plant Physiology 79: 11–17.

Caspar, T., T.P. Lin, G. Kakefuda, L. Benbow, J. Preiss and C. Somerville. 1991. Mutants of Arabidopsis with altered regulation of starch degradation. Plant Physiology 95: 1181–1188.

Chatterton, N.J. and J.E. Silvius. 1979. Photosynthate partitioning into starch in soybean leaves. I. Effects of photoperiod versus photosynthetic period duration. Plant Physiology 64: 749–753.

Chatterton, N.J. and J.E. Silvius. 1980a. Acclimation of photosynthate partitioning and photosynthetic rates to changes in length of daily photosynthetic period. Annals of Botany 46: 739–745.

Chatterton, N.J. and J.E. Silvius. 1980b. Photosynthate partitioning into leaf starch as affected by daily photosynthetic period duration in six species. Physiologia Plantarum 49: 141–144.

Chatterton, N.J. and J.E. Silvius. J.E. 1981. Photosynthate partitioning into starch in soybean leaves. II. Irradiance level and daily photosynthetic period duration effects. Plant Physiology 67: 257–260.

Chen, L.Q., X.Q. Qu, B.H. Hou, D. Sosso, S. Osorio, A.R. Fernie et al. 2012. Sucrose efflux mediated by SWEET proteins as a key step for phloem transport. Science 335: 207–211.

Chia, T., D. Thorneycroft, A. Chapple, G. Messerli, J. Chen, S.C. Zeeman et al. 2004. A cytosolic glucosyltransferase is required for conversion of starch to sucrose in Arabidopsis leaves at night. The Plant Journal 37: 853–863.

Cho, M.H., H. Lim, D.H. Shin, J.S. Jeon, S.H. Bhoo, Y.I. Park et al. 2011. Role of the plastidic glucose translocator in the export of starch degradation products from the chloroplasts in *Arabidopsis thaliana*. New Phytologist 190: 101–112.

Comparot-Moss, S., O. Kötting, M. Stettler, C. Edner, A. Graf, S.E. Weise et al. 2010. A putative phosphatase, LSF1, is required for normal starch turnover in Arabidopsis leaves. Plant Physiology 152: 685–697.

Crevillén, P., M.A. Ballicora, A. Mérida, J. Preiss and J.M. Romero. 2003. The different large subunit isoforms of *Arabidopsis thaliana* ADP-glucose pyrophosphorylase confer distinct kinetic and regulatory properties to the heterotetrameric enzyme. Journal of Biological Chemistry 31: 28508–28515.

Crevillen, P., T. Ventriglia, F. Pinto, A. Orea, A. Merida and J.M. Romero. 2005. Differential pattern of expression and sugar regulation of *Arabidopsis thaliana* ADP-glucose pyrophosphorylase-encoding genes. Journal of Biological Chemistry 280: 8143–8149.

Critchley, J.H., S.C. Zeeman, T. Takaha, A.M. Smith and S.M. Smith. 2001. A critical role for disproportionating enzyme in starch breakdown is revealed by a knock-out mutation in Arabidopsis. The Plant Journal 26: 89–100.

Crumpton-Taylor, M., S. Grandison, K.M.Y. Png, A.J. Bushby and A.M. Smith. 2012. Control of starch granule numbers in Arabidopsis chloroplasts. Plant Physiology 158: 905–916.

Crumpton-Taylor, M., M. Pike, K.J. Lu, C.M. Hylton, R. Feil, S. Eicke et al. 2013. Starch synthase 4 is essential for coordination of starch granule formation with chloroplast division during Arabidopsis leaf expansion. New Phytologist 200: 1064–1075.

Delatte, T., M. Trevisan, M.L. Parker and S.C. Zeeman. 2005. Arabidopsis mutants Atisa1 and Atisa2 have identical phenotypes and lack the same multimeric isoamylase, which influences the branch point distribution of amylopectin during starch synthesis. The Plant Journal 41: 815–30.

D'Hulst, C. and A. Mérida. 2012. The synthesis and breakdown of starch pp. 55–76. *In*: Tetlow, I.J. (eds.). Essential Reviews in Experimental Biology, volume 5. Society for Experimental Biology, London.

Dodd, A.N., N. Salathia, A. Hall, E. Kévei, R. Tóth, F. Nagy et al. 2005. Plant circadian clocks increase photosynthesis, growth, survival, and competitive advantage. Science 309: 630–633.

dos Anjos, L., P.K. Pandey, T.A. Moraes, R. Feil, J.E. Lunn and M. Stitt. 2018. Feedback regulation by trehalose 6-phosphate slows down starch mobilization below the rate that would exhaust starch reserves at dawn in Arabidopsis leaves. Plant Direct 2: 1–16.

Farre, E.M. and S.E. Weise. 2012. The interactions between the circadian clock and primary metabolism. Current Opinion in Plant Biology 15: 293–300.

Fernandez, O., H. Ishihara, G.M. George, V. Mengin, A. Flis, D. Sumner et al. 2017. Leaf starch turnover occurs in long days and in falling light at the end of the day. Plant Physiology 174: 2199–2212.

Feike, D., D. Seung, A. Graf, S. Bischof, T. Ellick, M. Coiro et al. 2016. The starch granule-associated protein EARLY STARVATION1 (ESV1) is required for the control of starch degradation in *Arabidopsis thaliana* leaves. The Plant Cell 28: 1472–1489.

Fettke, J., T. Chia, N. Eckermann, A. Smith and M. Steup. 2006. A transglucosidase necessary for starch degradation and maltose metabolism in leaves at night acts on cytosolic heteroglycans (SHG). The Plant Journal 46: 668–684.

Feugier, F.G. and A. Satake. 2013. Dynamical feedback between circadian clock and sucrose availability explains adaptive response of starch metabolism to various photoperiods. Frontiers in Plant Science 3: 305. doi: 10.3389/fpls.2012.00305.

Feugier, F.G. and A. Satake. 2014. Hyperbolic features of the circadian clock oscillations can explain linearity in leaf starch dynamics and adaptation of plants to diverse light and dark cycles. Ecological Modelling 290: 110–120.

Figueroa, C.M. and J.E. Lunn. 2016. A tale of two sugars: trehalose 6-phosphate and sucrose. Plant Physiology 172: 7–27.

Figueroa, C.M., R. Feil, H. Ishihara, M. Watanabe, K. Kölling, U. Krause et al. 2016. Trehalose 6-phosphate coordinates organic and amino acid metabolism with carbon availability. The Plant Journal 85: 410–423.

Flis, A., V. Mengin, A.A. Ivakov, S.T. Mugford, H.M. Hubberten, B. Encke et al. 2018. Multiple circadian clock outputs regulate diel turnover of carbon and nitrogen reserves. Plant, Cell and Environment. doi: 10.1111/pce.13440.

Flis, A., R. Sulpice, D.D. Seaton, A.A. Ivakov, M. Liput, C. Abel et al. 2016. Photoperiod-dependent changes in the phase of core clock transcripts and global transcriptional outputs at dawn and dusk in Arabidopsis. Plant, Cell and Environment 39: 1955–1981.

Fondy, B.R., D.R. Geiger and J.C. Servaites. 1989. Photosynthesis, carbohydrate metabolism, and export in *Beta vulgaris* L. and *Phaseolus vulgaris* L. during square and sinusoidal light regimes. Plant Physiology 89: 396–402.

Fordham-Skelton, A.P., P. Chilley, V. Lumbreras, S. Reignoux, T.R. Fenton, C.C. Dahm et al. 2002. A novel higher plant protein tyrosine phosphatase interacts with SNF1-related protein kinases via a KIS (kinase interaction sequence) domain. The Plant Journal 29: 705–715.

Fragoso, S., L. Espíndola, J. Páez-Valencia, A. Gamboa, Y. Camacho, E. Martínez-Barajas et al. 2009. SnRK1 isoforms AKIN10 and AKIN11 are differentially regulated in Arabidopsis plants under phosphate starvation. Plant Physiology 149: 1906–1916.

Fulton, D.C., M. Stettler, T. Mettler, C.K. Vaughan, J. Li, P. Francisco et al. 2008. β-AMYLASE4, a noncatalytic protein required for starch breakdown, acts upstream of three active β-amylases in Arabidopsis chloroplasts. The Plant Cell 20: 1040–1058.

Geiger, D.R. and J.C. Servaites. 1994. Diurnal regulation of photosynthetic carbon metabolism in C3 plants. Annual Review of Plant Physiology and Plant Molecular Biology 45: 235–256.

Geiger, D.R., W.-J. Shieh, L. Su Lu and J.C. Servaites. 1991. Carbon assimilation and leaf water status in sugar beet leaves during a simulated natural light regimen. Plant Physiology 97: 1103–1108.

Gibon, Y., O.E. Blaesing, J. Hannemann, P. Carillo, M. Höhne, J.H. Hendriks et al. 2004a. A robot-based platform to measure multiple enzyme activities in Arabidopsis using a set of cycling assays: comparison of changes of enzyme activities and transcript levels during diurnal cycles and in prolonged darkness. The Plant Cell 16: 3304–3325.

Gibon, Y., O.E. Bläsing, N. Palacios-Rojas, D. Pankovic, J.H.M. Hendriks, J. Fisahn et al. 2004b. Adjustment of diurnal starch turnover to short days: depletion of sugar during the night leads to a temporary inhibition of carbohydrate utilization, accumulation of sugars and post-translational activation of ADP-glucosepyrophosphorylase in the following light period. The Plant Journal 39: 847–862.

Gibon, Y., E.T. Pyl, R. Sulpice, J.E. Lunn, M. Höhne, M. Günther et al. 2009. Adjustment of growth, starch turnover, protein content and central metabolism to a decrease of the carbon supply when Arabidopsis is grown in very short photoperiods. Plant, Cell and Environment 32: 859–874.

Graf, A., A. Schlereth, M. Stitt and A.M. Smith. 2010. Circadian control of carbohydrate availability for growth in Arabidopsis plants at night. Proceedings of the National Academy of Sciences USA 107: 9458–9463.

Hädrich, N., Y. Gibon, C. Schudoma, T. Altmann, J.E. Lunn and M. Stitt. 2011. Use of TILLING and robotised enzyme assays to generate an allelic series of *Arabidopsis thaliana* mutants with altered ADP-glucose pyrophosphorylase activity. Journal of Plant Physiology 168: 1395–1405.

Hädrich, N., J.H.M. Hendriks, O. Kötting, S. Arrivault, R. Feil, S.C. Zeeman et al. 2012. Mutagenesis of cysteine 81 prevents dimerization of the APS1 subunit of ADP-glucose pyrophosphorylase and alters diurnal starch turnover in *Arabidopsis thaliana* leaves. The Plant Journal 70: 231–242.

Harmer, S.L., J.B. Hogenesch, M. Straume, H.S. Chang, B. Han, T. Zhu et al. 2000. Orchestrated transcription of key pathways in Arabidopsis by the circadian clock. Science 290: 2110–2113.

Haydon, M.J., O. Mielczarek, F.C. Robertson, K.E. Hubbard and A.A. Webb. 2013. Photosynthetic entrainment of the *Arabidopsis thaliana* circadian clock. Nature 502: 689–692.

Hendriks, J.H.M., A. Kolbe, Y. Gibon, M. Stitt and P. Geigenberger. 2003. ADP-glucose pyrophosphorylase is activated by posttranslational redox modification in response to light and to sugars in leaves of Arabidopsis and other plant species. Plant Physiology 133: 838–849.

Hejazi, M., J. Fettke, O. Kötting, S.C. Zeeman and M. Steup. 2010. The Laforin-like dual-specificity phosphatase SEX4 from Arabidopsis hydrolyzes both C6- and C3-phosphate esters introduced by starch related dikinases and thereby affects phase transition of α-glucans. Plant Physiology 152: 711–722.

Hejazi, M., J. Fettke and M. Steup. 2012. Starch phosphorylation and dephosphorylation: The consecutive action of starch-related dikinases and phosphatases. *In*: Tetlow I.J. (ed.). Essential Reviews in Experimental Biology: Starch: Origins, Structure and Metabolism. SEB, London 5: 279–308.

Hejazi, M., S. Mahlow and J. Fettke. 2014. The glucan phosphorylation mediated by α-glucan, water dikinase (GWD) is also essential in the light phase for a functional transitory starch turn-over. Plant Signalling and Behavior 9(7): e28892.

Horrer, D., S. Flütsch, D. Pazmino, J.S.A. Matthews, M. Thalmann, A. Nigro et al. 2016. Blue light induces a distinct starch degradation pathway in guard cells for stomatal opening. Current Biology 26: 362–370.

Huber, S.C. and J.L. Huber. 1996. Role and regulation of sucrose-phosphate synthase in higher plants. Annual Review of Plant Physiolgy and Plant Molecular Biology 47: 431–444.

Hummel, I., F. Pantin, R. Sulpice, M. Piques, G. Rolland, M. Dauzat et al. 2010. Arabidopsis plants acclimate to water deficit at low cost through changes of carbon usage: an integrated perspective using growth, metabolite, enzyme, and gene expression analysis. Plant Physiology 154: 357–72.

Jossier, M., J.P. Bouly, P. Meimoun, A. Arjmand, P. Lessard, S. Hawley et al. 2009. SnRK1 (SNF1-related kinase 1) has a central role in sugar and ABA signalling in *Arabidopsis thaliana*. The Plant Journal 59: 316–328.

Kim, J.A., H.-S. Kim, S.-H., J.-Y. Jang, M.-J. Jeong and S.I. Lee. 2017. The importance of the circadian clock in regulating plant metabolism. International Journal of Molecular Sciences 18: 2680, doi:10.3390/ijms18122680.

Kolbe, A., A. Tiessen, H. Schluepmann, M. Paul, S. Ulrich and P. Geigenberger. 2005. Trehalose 6-phosphate regulates starch synthesis via posttranslational redox activation of ADPglucose pyrophosphorylase. Proceedings of the National Academy of Sciences USA 102: 11118–11123.

Kötting, O., J. Kossman, S.C. Zeeman and J.R. Lloyd. 2010. Regulation of starch metabolism: the age of enlightment. Current Opinion of Plant Biology 13: 320–328.

Kötting, O., K. Pusch, A. Tiessen, P. Geigenberger, M. Steup and G. Ritte. 2005. Identification of a novel enzyme required for starch metabolism in Arabidopsis leaves. The phosphoglucan, water dikinase. Plant Physiology 137: 242–252.

Kötting, O., D. Santelia, C. Edner, S. Eicke, T. Marthaler, M.S. Gentry et al. 2009. STARCH-EXCESS4 is a laforin-like phosphoglucan phosphatase required for starch degradation in *Arabidopsis thaliana*. The Plant Cell 21: 334–346.

Lao, N.T., O. Schoneveld, R.M. Mould, J.M. Hibberd, J.C. Gray and T.A. Kavanagh. 1999. An Arabidopsis gene encoding a chloroplast targeted β-amylase. The Plant Journal 20: 519–527.

Li, J., G. Almagro, F.J. Muñoz, E. Baroja-Fernández, A. Bahaji, M. Montero et al. 2012. Post-translational redox modification of ADP-glucose pyrophosphorylase in response to light is not a major determinant of fine regulation of transitory starch accumulation in Arabidopsis leaves. Plant Cell Physiology 53: 433–44.

Li, J., W. Zhou, P. Francisco, R. Wong, D. Zhang and S.M. Smith. 2017. Inhibition of Arabidopsis chloroplast β-amylase BAM3 by maltotriose suggests a mechanism for the control of transitory leaf starch mobilisation. PLoS ONE 12(2): e0172504.

Liu, H., G. Yu, B. Wei, Y. Wang, J. Zhang, Y. Hu et al. 2015. Identification and phylogenetic analysis of a novel starch synthase in maize. Frontiers in Plant Science 6: 1013. doi: 10.3389/fpls.2015.01013.

Locke, J.C.W., M.M. Southern, L. Kozma-Bognar, V. Hibberd, P.E. Brown, M.S. Turner et al. 2005. Extension of genetic network model by iterative experimentation and mathematical analysis. Molecular Systems Biology 1: 2005–2013.

Lorenzen, J.H. and E.E. Ewing. 1992. Starch accumulation in leaves of potato (*Solanum tuberosum* L.) during the first 18 days of photoperiod treatment. Annals of Botany 69: 481–485.

Lu, K.-J., B. Pfister, C. Jenny, S. Eicke and S.C. Zeeman. 2018. Distinct functions of starch synthase 4 domains in starch granule formation. Plant Physiology 176: 566–581.

Lu, Y., J.P. Gehan and T.D. Sharkey. 2005. Daylength and circadian effects on starch degradation and maltose metabolism. Plant Physiology 138: 2280–2291.

Lu, Y. and T.D. Sharkey. 2004. The role of amylomaltase in maltose metabolism in the cytosol of photosynthetic cells. Planta 218: 466–473.

Lunn, J.E. 2016. Sucrose metabolism. *In*: Encyclopedia of Life Science (ELS). John Wiley & Sons, Ltd, Chichester. doi: 10.1002/9780470015902.a0021259.pub2.

Lunn, J.E., I. Delorge, C.M. Figueroa, P. Van Dijck and M. Stitt. 2014. Trehalose metabolism in plants. The Plant Journal 79: 544–567.

Lunn, J.E., R. Feil, J.H.M. Hendriks, Y. Gibon, R. Morcuende, D. Osuna et al. 2006. Sugar induced increases in trehalose-6-phosphate are correlated with redox activation of ADPglucose pyrophosphorylase and higher rates of starch synthesis in *Arabidopsis thaliana*. Biochemical Journal 397: 139–148.

MacRae, E.A. and J.E. Lunn. 2006. Control of sucrose biosynthesis. pp. 234–257. *In*: Plaxton, W.C. and M.T. McManus (eds.). Advances in Plant Research: Control of Primary Metabolism in Plants, volume 22. Blackwell, Oxford.

Mahlow, S., M. Hejazi, F. Kuhnert, A. Garz, H. Brust, O. Baumann et al. 2014. Phosphorylation of transitory starch by α-glucan, water dikinase during starch turnover affects the surface properties and morphology of starch granules. New Phytologist 203: 495–507.

Malinova, I., S. Alseekh, R. Feil, A.R. Fernie, O. Baumann, M.A. Schöttler et al. 2017. Starch synthase 4 and plastidal phosphorylase differentially affect starch granule number and morphology. Plant Physiology 174: 73–85.

Malinova, I., H. Mahto, F. Brandt, S. Al-Rawi, H. Qasim, H. Brust et al. 2018. Early starvation1 specifically affects the phosphorylation action of starch-related dikinases. Plant Journal. doi: 10.1111/tpj.13937.

Malinova, I. and J. Fettke. 2017. Reduced starch granule number per chloroplast in the *dpe2/phs1* mutant is dependent on initiation of starch degradation. PLoS ONE 12(11): e0187985.

Martins, M.C.M., M. Hejazi, J. Fettke, M. Steup, R. Feil, U. Krause et al. 2013. Feedback inhibition of starch degradation in Arabidopsis leaves mediated by trehalose 6-phosphate. Plant Physiology 163: 1142–1163.

McClung, C.R. 2006. Plant circadian rhythms. The Plant Cell 1: 792–803.

McClung, C.R. 2008. Comes a time. Current Opinion in Plant Biology 11: 514–520.

Mengin, V., E.-T. Pyl, T.A. Moraes, R. Sulpice, N. Krohn, B. Encke et al. 2017. Photosynthate partitioning to starch in *Arabidopsis thaliana* is insensitive to light intensity but sensitive to photoperiod due to a restriction on growth in the light in short photoperiods. Plant, Cell and Environment 40: 2608–2627.

Michalska, J., H. Zauber, B.B. Buchanan, F.J. Cejudo and P. Geigenberger. 2009. NTRC links built-in thioredoxin to light and sucrose in regulating starch synthesis in chloroplasts and amyloplasts. Proceedings of the National Academy of Sciences USA 106: 9908–9913.

Mikkelsen, R., K.E. Mutenda, A. Mant, P. Schürmann and A. Blennow. 2005. α-Glucan, water dikinase (GWD): A plastidic enzyme with redox-regulated and coordinated catalytic activity and binding affinity. Proceedings of the National Academy of Sciences USA 102: 1785–1790.

Mishra, P. and K.C. Panigrahi. 2015. GIGANTEA–an emerging story. Frontiers in Plants Science 6. doi: 10.3389/fpls.2015.00008.

Monroe, J.D., A.R. Storm, E.M. Badley, M.D. Lehman, S.M. Platt, L.K. Saunders et al. 2014. β-amylase1 and β-amylase3 are plastidic starch hydrolases in Arabidopsis that seem to be adapted for different thermal, pH, and stress conditions. Plant Physiology 166: 1748–1763.

Mugford, S.T., O. Fernandez, J. Brinton, A. Flis, N. Krohn, B. Encke et al. 2014. Regulatory properties of ADP glucose pyrophosphorylase are required for adjustment of leaf starch synthesis in different photoperiods. Plant Physiology 166: 1733–1747.

Mullen, J.A. and H.R. Koller. 1988. Daytime and nighttime carbon balance and assimilate export in soybean leaves at different photon flux densities. Plant Physiology 86: 880–884.

Nelson, C.J., R. Alexova, R.P. Jacoby and A.H. Millar. 2014. Proteins with high turnover rate in barley leaves estimated by proteome analysis combined with in planta isotope labeling. Plant Physiology 166: 91–108.

Neuhaus, H.E. and M. Stitt. 1990. Control analysis of photosynthate partitioning: impact of reduced activity of ADP-glucose pyrophosphorylase or plastid phosphoglucomutase on the fluxes to starch and sucrose in *Arabidopsis thaliana* L. Heynh. Planta 182: 445–454.

Neuhaus, H.E., A.L. Kruckeberg, R. Feil, L. Gottlieb and M. Stitt. 1989. Dosage mutants of phosphoglucose isomerase in the cytosol and chloroplasts of *Clarkia xantiana*. II. Study of the mechanisms which regulate photosynthate partitioning. Planta 178: 110–122.

Neuhaus, H.E., W.P. Quick, G. Siegl and M. Stitt. 1990. Control of photosynthate partitioning in spinach leaves: analysis of the interaction between feedforward and feedback regulation of sucrose synthesis. Planta 181: 583–592.

Nielsen, T.H., J.H. Rung and D. Villadsen. 2004. Fructose-2,6-bisphosphate: a traffic signal in plant metabolism. Trends Plant Science 9: 556–563.

Niittylä, T., G. Messerli, M. Trevisan, J. Chen, A.M. Smith and S.C. Zeeman. 2004. A previously unknown maltose transporter essential for starch degradation in leaves. Science 303: 87–89.

Nougué, O., J. Corbi, S.G. Ball, D. Manicacci and M.I. Tenaillon. 2014. Molecular evolution accompanying functional divergence of duplicated genes along the plant starch biosynthesis pathway. BMC Evolutionary Biology 14: 103.

Pal, S.K., M. Liput, M. Piques, H. Ishihara, M.C.M. Martins, R. Sulpice et al. 2013. Diurnal changes of polysome loading track sucrose content in the rosette of wildtype Arabidopsis and the starchless pgm mutant. Plant Physiology 162: 1246–1265.

Paparelli, E., S. Parlanti, S. Gonzali, G. Novi, L. Mariotti, N. Ceccarelli et al. 2013. Nighttime sugar starvation orchestrates gibberellin biosynthesis and plant growth in Arabidopsis. The Plant Cell 25: 3760–3769.

Pfister, B. and S.C. Zeeman. 2016. Formation of starch in plant cells. Cellular and Molecular Life Sciences 73: 2781–2807.

Pfister, B., K.-J. Lu, S. Eicke, R. Feil, J.E. Lunn, S. Streb et al. 2014. Genetic evidence that chain length and branch point distributions are linked determinants of starch granule formation in Arabidopsis. Plant Physiology 165: 1457–1474.

Pilkington, S.M., B. Enke, N. Krohn, M. Höhne, M. Stitt, E.-T. Pyl. 2015. Relationship between starch degradation and carbon demand for maintenance and growth in *Arabidopsis thaliana* in different irradiance and temperature regimes. Plant, Cell and Environment 38: 157–171.

Piques, M., W.X. Schulze, M. Höhne, B. Usadel, Y. Gibon, J. Rohwer et al. 2009. Ribosome and transcript copy numbers, polysome occupancy and enzyme dynamics in Arabidopsis. Molecular Systems Biology 5: 314. doi: 10.1038/msb.2009.68.

Pokhilko, A. and O. Ebenhoh. 2015. Mathematical modelling of diurnal regulation of carbohydrate allocation by osmo-related processes in plants. Journal of the Royal Society Interface 12. doi: 10.1098/rsif.2014.1357.

Pokhilko, A., A.P. Fernandez, K.D. Edwards, M.M. Southern, K.J. Halliday and A.J. Millar. 2012. The clock gene circuit in Arabidopsis includes a repressilator with additional feedback loops. Molecular Systems Biology 8: 574.

Pokhilko, A., A. Flis, R. Sulpice, M. Stitt and O. Ebenhoh. 2014. Adjustment of carbon fluxes to light conditions regulates the daily turnover of starch in plants: A computational model. Molecular BioSystem 10: 613–627.

Pokhilko, A., P. Mas and A.J. Millar. 2013. Modelling the widespread effects of TOC1 signalling on the plant circadian clock and its outputs. BMC Systems Biology 7: 23.

Prasch, C.M., K.V. Ott, H. Bauer, P. Ache, R. Hedrich and U. Sonnewald. 2015. ß-amylase1 mutant Arabidopsis plants show improved drought tolerance due to reduced starch breakdown in guard cells. Journal of Experimental Botany 66: 6059–6067.

Preiss, J. 1988. Biosynthesis of starch and its regulation. pp. 181–254. *In*: Preiss, J. (ed.). The Biochemistry of Plants, volume 14. Academic Press, San Diego.

Pyl, E.-T., M. Piques, A. Ivakov, W.X. Schulze, M. Stitt and R. Sulpice. 2012. Metabolism and growth in Arabidopsis depend on the daytime temperature but are temperature-compensated against cool nights. The Plant Cell 24: 2443–2469.

Ragel, P., S. Streb, R. Feil, M. Sahrawy, M.G. Annunziata, J.E. Lunn et al. 2013. Loss of starch granule initiation has a deleterious effect on the growth of Arabidopsis plants due to an accumulation of ADP-glucose. Plant Physiology 163: 75–85.

Ritte, G., M. Heydenreich, S. Mahlow, S. Haebel, O. Kötting and M. Steup. 2006. Phosphorylation of C6- and C3-positions of glucosyl residues in starch is catalysed by distinct dikinases. FEBS Letter 580: 4872–4876.

Ritte, G., J.R. Lloyd, N. Eckermann, A. Rottmann, J. Kossmann and M. Steup. 2002. The starch-related R1 protein is an α-glucan, water dikinase. Proceedings of the National Academy of Sciences USA 99: 7166–7171.

Roldán, I., F. Wattebled, M. Mercedes Lucas, D. Delvallé, V. Planchot, S. Jiménez et al. 2007. The phenotype of soluble starch synthase IV defective mutants of *Arabidopsis thaliana* suggests a novel function of elongation enzymes in the control of starch granule formation. The Plant Journal 49: 492–504.

Ruiz-Gayosso, A., R. Rodríguez-Sotres, E. Martínez-Barajas and P. Coello. 2018. A role for the carbohydrate-binding module (CBM) in regulatory SnRK1 subunits: the effect of maltose on SnRK1 activity. The Plant Journal 96: 163–175.

Sánchez-Villarreal, A., J. Shin, N. Bujdoso, T. Obata, U. Neumann, S.-X. Du et al. 2013. Time for coffee is an essential component in the maintenance of metabolic homeostasis in *Arabidopsis thaliana*. The Plant Journal 76: 188–200.

Sánchez-Villarreal, A., A.M. Davis and S.J. Davis. 2017. AKIN10 Activity as a cellular link between metabolism and circadian-clock entrainment in *Arabidopsis thaliana*. Plant Signaling and Behavior doi: 10.1080/15592324.2017.1411448.

Santelia, D. and J.E. Lunn. 2017. Transitory starch metabolism in guard cells: unique features for a unique function. Plant Physiology 174: 539–549.

Santelia, D., O. Kötting, D. Seung, M. Schubert, M. Thalmann, S. Bischof et al. 2011. The phosphoglucan phosphatase Like Sex Four2 dephosphorylates starch at the C3-position in Arabidopsis. The Plant Cell 23: 4096–4111.

Schluepmann, H., T. Pellny, van Dijken, A., S. Smeekens, and M. Paul. 2003. Trehalose 6-phosphate is indispensable for carbohydrate utilization and growth in Arabidopsis thaliana. Proceedings of the National Academy of Sciences USA 100: 6849–6854.

Scialdone, A. and M. Howard. 2015. How plants manage food reserves at night: quantitative models and open questions. Frontiers in Plant Science 6: 204. doi: 10.3389/fpls.2015.00204.

Scialdone, A., S.T. Mugford, D. Feike, A. Skeffington, P. Borrill, A. Graf et al. 2013. Arabidopsis plants perform arithmetic division to prevent starvation at night. Elife 2:e00669.doi:10.7554/eLife.00669.

Scott, P., A.J. Lange and N.J. Kruger. 2000. Photosynthetic carbon metabolism in leaves of transgenic tobacco (*Nicotiana tabacum* L.) containing decreased amounts of fructose 2,6-bisphosphate. Planta 211: 864–873.

Seaton, D.D., O. Ebenhöh, A.J. Millar and A. Pokhilko. 2014. Regulatory principles and experimental approaches to the circadian control of starch turnover. Journal of the Royal Society Interface 11:20130979.doi:10.1098/rsif.2013.0979.

Seki, M., T. Ohara, T.J. Hearn, A. Frank, V.C.H. da Silva, C. Caldana et al. 2017. Adjustment of the Arabidopsis circadian oscillator by sugar signalling dictates the regulation of starch metabolism. Scientific Reports 7, article number 8305.

Servaites, J.C., D.R. Geiger, M.A. Tucci and B.R. Fondy. 1989. Leaf carbon metabolism and metabolite levels during period of sinusoidal light. Plant Physiology 89: 403–408.

Seung, D. and A. Smith. 2018. Starch granule initiation and morphogenesis—progress in Arabidopsis and cereals. Journal of Experimental Botany. doi: 10.1093/jxb/ery412.

Seung, D., J. Boudet, J. Monroe, T.B. Schreier, L.C. David, M. Abt et al. 2017. Homologs of protein targeting to starch control starch granule initiation in Arabidopsis leaves. The Plant Cell 29: 1657–1677.

Seung, D., S. Soyk, M. Coiro, B.A. Maier, S. Eicke and S.C. Zeeman. 2015. Protein targeting to starch is required for localising granule-bound starch synthase to starch granules and for normal amylose synthesis in Arabidopsis. PLoS Biology 13: e1002080.

Shin, J., S. Sánchez-Villarreal, A.M. Davis, Sx. Du, K.W. Berendzen, C. Koncz et al. 2017. The metabolic sensor AKIN10 modulates the Arabidopsis circadian clock in a light-dependent manner. Plant, Cell and Environment 40: 997–1008.

Skeffington, A.W., A. Graf, Z. Duxbury, W. Gruissem and A.M. Smith. 2014. Gucan, water dikinase exerts little control over starch degradation in Arabidopsis leaves at night. Plant Physiology 165: 866–879.

Smith, S.M., D.C. Fulton, T. Chia, D. Thorneycroft, A. Chapple, H. Dunstan et al. 2004. Diurnal changes in the transcriptome encoding enzymes of starch metabolism provide evidence for both transcriptional and post-transcriptional regulation of starch metabolism in Arabidopsis leaves. Plant Physiology 136: 2687–2699.

Sokolov, L.N., J.R. Dominguez-Solis, A.L. Allary, B.B. Buchanan and S. Luan. 2006. A redox-regulated chloroplast protein phosphatase binds to starch diurnally and functions in its accumulation. Proceedings of the National Academy of Sciences USA 103: 9732–9737.

Song, Y.H., S. Ito and T. Imaizumi. 2010. Similarities in the circadian clock and photoperiodism in plants. Current Opinion in Plant Biology 13: 594–603.

Sparla, F., A. Costa, F. Lo Schiavo, P. Pupillo and P. Trost. 2006. Redox regulation of a novel plastid-targeted β-amylase of Arabidopsis. Plant Physiology 141: 840–850.

Stitt, M. and U. Sonnewald. 1995. Regulation of metabolism in transgenic plants. Annual Review of Plant Physiology and Plant Molecular Biology 46: 341–368.

Stitt, M. and S.C. Zeeman. 2012. Starch turnover: pathways, regulation and role in growth. Current Opinion in Plant Biology 15: 282–292.

Stitt, M., S. Huber and P. Kerr. 1987. Control of photosynthetic sucrose synthesis. pp. 327–409. *In*: Hatch, M.D. and N.K. Boardman (eds.). The Biochemistry of Plants, volume 10. New York: Academic Press.

Stitt, M., J.E. Lunn and B. Usadel. 2010. Arabidopsis and primary photosynthetic metabolism—more than the icing on the cake. The Plant Journal 61: 1067–1091.

Stitt, M.,, W.P. Quick, U. Schurr, E.D. Schulze, S.R. Rodermel and L. Bogorad. 1991. Decreased ribulose-1,5-bisphosphate carboxylase-oxygenase in transgenic tobacco transformed with 'antisense' *rbcS* : II. Flux-control coefficients for photosynthesis in varying light, CO_2, and air humidity. Planta 183: 555–566.

Strand, A., R. Zrenner, M. Stitt and P. Gardestrom. 2000. Antisense inhibition of two key enzymes in the sucrose biosynthesis pathway, cytosolic fructose bisphosphatase and sucrose phosphate synthase, has different consequences for photosynthetic carbon metabolism in transgenic Arabidopsis thaliana. The Plant Journal 23: 751–770.

Streb, S. and S.C. Zeeman. 2012. Starch metabolism in Arabidopsis. The Arabidopsis Book. doi: 10.1199/tab.0160.

Streb, S. and S.C. Zeeman. 2014. Replacement of the endogenous starch debranching enzymes ISA1 and ISA2 of Arabidopsis with the rice orthologs reveals a degree of functional conservation during starch synthesis. PLoS One 9: e92174.

Sulpice, R., A. Flis, A.A. Ivakov, F. Apelt, H. Krohn, B. Encke et al. 2014. Arabidopsis coordinates the diurnal regulation of carbon allocation and growth across a wide range of photoperiods. Molecular Plant 7: 137–155.

Sulpice, R., S. Trenkamp, M. Steinfath, B. Usadel, Y. Gibon, H. Witucka-Wall et al. 2010. Network analysis of enzyme activities and metabolite levels and their relationship to biomass in a large panel of Arabidopsis accessions. The Plant Cell 22: 2872–2893.

Szydlowski, N., P. Ragel, S. Raynaud, M.M. Lucas, I. Roldán, M. Montero et al. 2009. Starch granule initiation in Arabidopsis requires the presence of either class IV or class III starch synthases. The Plant Cell 21: 2443–2457.

Thalmann, M., D. Pazmino, D. Seung, D. Horrer, A. Nigro, T. Meier et al. 2016. Regulation of leaf starch degradation by abscisic acid is important for osmotic stress tolerance in plants. The Plant Cell 28: 1860–1878.

Thelander, M., T. Olsson and H. Ronne. 2004. Snf1-related protein kinase 1 is needed for growth in a normal day-night light cycle. Embo Journal 23: 1900–1910.

Thormählen, I., J. Ruber, E. von Roepenack-Lahaye, S.M. Ehrlich, V. Massot, C. Hümmer et al. 2013. Inactivation of thioredoxin f1 leads to decreased light activation of ADP-glucose pyrophosphorylase and altered diurnal starch turnover in leaves of Arabidopsis plants. Plant Cell and Environment 36: 16–29.

Tiessen, A., K. Prescha, A. Branscheid, N. Palacios, R. McKibbin, N.G. Halford et al. 2003. Evidence that SNF1-related kinase and hexokinase are involved in separate sugar-signalling pathways modulating post-translational redox activation of ADP-glucose pyrophosphorylase in potato tubers. The Plant Journal 35: 490–500.

Valerio, C., A. Costa, L. Marri, E. Issakidis-Bourguet, P. Pupillo, P. Trost et al. 2011. Thioredoxin-regulated beta-amylase (BAM1) triggers diurnal starch degradation in guard cells, and in mesophyll cells under osmotic stress. Journal of Experimental Botany 62: 545–555.

Vialet-Chabrand, S., J.S. Matthews, O. Brendel, M.R. Blatt, Y. Wang, A. Hills et al. 2016. Modelling water use efficiency in a dynamic environment: An example using *Arabidopsis thaliana*. Plant Science 251: 65–74.

Vialet-Chabrand, S., J.S.A. Matthews, A.J. Simkin, C.A. Raines and T. Lawson. 2017. Importance of fluctuations in light on plant photosynthetic acclimation. Plant Physiology 173: 2163–2179.

Wattebled, F., Y. Dong, S. Dumez, D. Delvallé, V. Planchot, P. Berbezy et al. 2005. Mutants of Arabidopsis lacking a chloroplastic isoamylase accumulate phytoglycogen and an abnormal form of amylopectin. Plant Physiology 138: 184–195.

Weber, A., J.C. Servaites, D.R. Geiger, H. Kofler, D. Hille, F. Gröner et al. 2000. Identification, purification, and molecular cloning of a putative plastidic glucose translocator. The Plant Cell 12: 787–802.

Weise, S.E., S.M. Schrader, K.R. Kleinbeck and T.D. Sharkey. 2006. Carbon balance and circadian regulation of hydrolytic and phosphorolytic breakdown of transitory starch. Plant Physiology 141: 879–886.

Weise, S.E., A.P.M. Weber and T.D. Sharkey. 2004. Maltose is the major form of carbon exported from the chloroplast at night. Planta 218: 474–482.

Wychowski, A., C. Bompard, F. Grimaud, G. Potocki-Veronese, C. D'Hulst, F. Wattebled and X. Roussel. 2017. Biochemical characterization of *Arabidopsis thaliana* starch branching enzyme 2.2 reveals an enzymatic positive cooperativity. Biochimie 140: 146–158.

Yadav, U.P., A. Ivakov, R. Feil, G.Y. Duan, D. Walther, P. Giavalisco et al. 2014. The sucrose-trehalose 6-phosphate (Tre6P) nexus: specificity and mechanisms of sucrose signalling by Tre6P. Journal of Experimental Botany 65: 1051–1068.

Yu, T.S., H. Kofler, R.E. Häusler, D. Hille, U.I. Flügge, S.C. Zeeman et al. 2001. The Arabidopsis *sex1* mutant is defective in the R1 protein, a general regulator of starch degradation in plants, and not in the chloroplast hexose transporter. The Plant Cell 13: 1907–1918.

Yu, T.S., S.C. Zeeman, D. Thorneycroft, D.C. Fulton, H. Dunstan, W.L. Lue et al. 2005. α-Amylase is not required for breakdown of transitory starch in Arabidopsis leaves. Journal of Biological Chemistry 280: 9773–9779.

Zeeman, S.C., S.M. Smith and A.M. Smith. 2002. The priming of amylose synthesis in Arabidopsis leaves. Plant Physiology 128: 1069–1076.

Zeeman, S.C., D. Thorneycroft, N. Schupp, A. Chapple, M. Weck, H. Dunstan et al. 2004. Plastidial α-glucan phosphorylase is not required for starch degradation in Arabidopsis leaves but has a role in the tolerance of abiotic stress. Plant Physiology 135: 849–858.

Zhai, Z., J. Keereetaweep, H. Liu, R. Feil, J.E. Lunn and J. Shanklin. 2018. Trehalose 6-phosphate positively regulates fatty acid synthesis by stabilizing WRINKLED1. The Plant Cell. doi: 10.1105/tpc.18.00521.

Zhang, Y., L.F. Primavesi, D. Jhurreea, P.J. Andralojc, R.A.C. Mitchell, S.J. Powers et al. 2009. Inhibition of SNF1-related protein kinase1 activity and regulation of metabolic pathways by trehalose-6-phosphate. Plant Physiology 149: 1860–1871.

CHAPTER 10

Reserve Starch Metabolism

Yasunori Nakamura

10.1 Introduction

The annual production of cereal such as maize, rice, and wheat accounts for 260–280 million tons (FAO-STATS 2015), whereas the demand for starch is approximately 38 million tons worldwide per year. Reserve starch production is the main reason for the agricultural production of cereals. Reserve starch supports the world population by supplying people with staple food and more than 3,000 food-related products. In addition, it is used for non-food materials, such as papers, adhesives, textiles, and cosmetics.

The oldest evolutionary function of starch is probably immediately linked to photosynthesis performed by unicellular or multicellular organisms. This type of starch (usually designated as assimilatory starch) serves as a reservoir of energy and reduced carbon which is accumulated in chloroplasts during photosynthesis. It is degraded when photosynthesis is not possible or incapable to sustain metabolic activities. Typically, assimilatory starch is turned over once during the light-dark period. Many higher plant species have developed photosynthetically incompetent reserve organs or tissues that accumulate large quantities of starch in non-green amyloplasts. This type of starch (often designated as reserve starch) can be stored for a prolonged period. Once degradation is initiated it permits, for a limited period of time, to maintain the plants to survive, develop, and germinate independent of photosynthesis. Turnover of the reserve starch is usually much slower and far less frequent than that of assimilatory starch. Biosynthesis of reserve starch is coupled only indirectly to photosynthesis and in many cases it proceeds for weeks. Likewise, reserve starch is often degraded over an extended period of time. Reserve starch biosynthesis requires an intact cellular structure and, therefore, it is strictly linked to living tissue. By contrast, degradation happens either in living cells or is associated with the lysis of the entire tissue. Therefore, reserve starch degradation is a diverse process depending on the plant species under investigation.

Starch has mainly two components, amylopectin, a highly branched glucan, and amylose, a basically linear but occasionally few branched glucan. The structural features of amylopectin are explained by the presence of a unit structure called "cluster" (Nikuni 1969, Kainuma and French 1972, Hizukuri 1986) (Fig. 10.1), which is formed by the rhythmical and asymmetrical nature of the branch distribution (Thompson 2000). Most importantly, the external segments of neighboring chains form right-turn double helices when their lengths exceed the degree of glucose polymerization (DP) of 10 (Gidley and Bulpin 1987). By massively forming double helices, hydrophobicity of amylopectin is increased and semi-crystalline properties of the starch are enhanced. Starch particles are densely

Starch Technologies Co., Ltd., Akita Prefectural University, Akita, 010-0195 Japan, Email: nakayn@silver.plala.or.jp

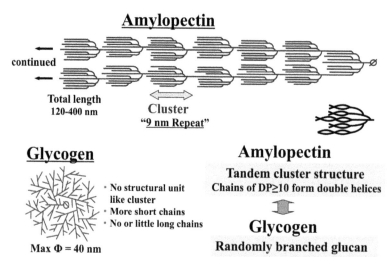

Figure 10.1. Schematic representation of structural features of amylopectin in comparison with glycogen. A section of amylopectin is shown. The fine structure of amylopectin is characterized by the structural unit called cluster, which has the uniform size of approximately 9 nm. Thus, this is referred to as the "9 nm repeat". Longer α-glucan chains connect several clusters. Basically, the orientation of these longer chains relative to that of the cluster is not known. In this figure, these chains and the clusters have a parallel orientation; there is, however, another model according to which the clusters are placed perpendicularly to the long chains (see Chapter 8 of this volume). When the linear segments of neighboring side chains within a cluster exceed the length of degree of polymerization (DP) of 10, they form double helices which cause the hydrophobic nature of the amylopectin. Wild-type glycogen does not have this structural unit "cluster" because branches are more equally distributed throughout the entire molecule. Since the average length of side chains of glycogen is significantly shorter than that of amylopectin, double helices are not detectable in wild-type glycogen. The maximum diameter of glycogen granules is approximately 40 nm.

packed and highly ordered. Their semi-crystallinity is often visible as birefringence observed with polarized light and as distinct patterns using various approaches such as wide-angle (Katz and Itallie 1930, Jenkins and Donald 1998) and small-angle X-ray diffraction patterns (Sterling 1962, Jenkins et al. 1993), NMR spectroscopy (Gidley and Bociiek 1985, Flanagan et al. 2015), SHG (Mizutani et al. 2000, Cisek et al. 2018; see Chapter 2 of this volume), and SFG (Miyauchi et al. 2006). The cereal starch granules exhibit the A-type X-ray diffraction pattern, while tuber starch granules show the B-type pattern (Katz and Itallie 1930). The C-type allomorph found in starch granules from rhizome and bean is known to be a mixture of A-type and B-type starches (Hizukuri and Nikuni 1957, Imberty and Pérez 1988, Imberty et al. 1988, 1991, Bogracheva et al. 1998). It is particularly interesting that the A-type pattern can be changed into the B-type pattern when the chains of amylopectin are elongated, i.e., the ratio of branching to glucosyl transfer is diminished during amylopectin biosynthesis (Hizukuri et al. 1983).

The morphology and size of reserve starch granules are entirely specific for the plant species and organs. Various granular shapes have been documented, such as polygonal, ellipsoidal, oval, spherical, and disc-shaped (Jane et al. 1994, Jane 2009). Their sizes range from 0.1 to over 100 μm, and it seems that there is no clear upper size limit. The cereal endosperm generally contains polygonal starch granules. Starch granules exhibit a wide variety of physicochemical properties including thermo-gelatinization properties, pasting properties, retrogradation properties, film-forming properties and elasticities but frequently it is difficult to correlate these properties with a distinct (physico)chemical feature.

Since the discovery of nucleotide-sugar, such as UDPglucose and ADPglucose, as a glucosyl donor for α-glucan biosynthesis by Leloir et al. (1961), numerous biochemical and genetic approaches have been attempted to resolve the sophisticated system of starch biosynthesis in various plant tissues from many plant species (Preiss and Levi 1980). During the last three decades, our understanding of starch's biosynthetic and degradation processes in both photosynthetic and heterotrophic tissues of a wide range of plant species has increased dramatically in accordance with the development of whole genome analysis and various molecular biology approaches, especially generation and identification of starch-related mutants and transformants (Ball and Morell 2003, Nakamura 2015a, Tetlow and Emes 2017). Currently, the knowledge on regulation of assimilatory starch metabolism is more advanced. Nevertheless, it is realistic (but still challenging) to overview the whole scope of starch metabolism and to apply the knowledge for the industrial use of starches applying the biotechnological methodology. Although the list of starch-related proteins is not yet complete, at present the fundamental schemes of the metabolic process as to how and to what extent individual enzymes contribute to the structures and physicochemical properties of starch molecules have largely been disclosed. Thus, we often can predict changes of starch quality and quantities per cell or even per plant when the activity of a target enzyme is modified. This enables us to design the molecular starch biotechnology.

The starting material for reserve starch biosynthesis is usually sucrose, a representative product of photosynthesis that is translocated from green leaves by long distance transport through sieve tubes. In the reserve starch storing tissue, sucrose is metabolized through several enzymatic reactions to finally form ADPglucose. It is widely accepted that in the cereal endosperm, most of ADPglycose is synthesized by a cytosolic form of ADPglucosepyrophosphorylase (AGPase; Denyer et al. 1996, Comparat-Moss and Denyer 2009). According to this, the cytosolic formation of ADPglucose is unique to cereals. This also requires the ADPglucose/ADP antiporter to be functional and they relate this type of metabolism to genome duplication. In endosperm from maize (Shannon et al. 1998) and rice (Cakir et al. 2016, Li et al. 2017), this transporter has been identified and mutants lacking this transporter have a reduced starch content with different physicochemical properties. In other reserve organs, ADPglucose is formed by a plastidial AGPase inside the amyloplasts. ADPglucose acts as the principal glucosyl donor for the biosynthesis of reserve starch. ATP, which is needed as another substrate for the AGPase reaction, is supplied from the oxidative phosphorylation system in mitochondria. Therefore, reserve starch biosynthesis requires the normal cellular organization as present in living tissues.

Chains of amylopectin are elongated by repetitively transferring a glucosyl residue from ADPglucose to the non-reducing terminals of the α-glucan as mediated by one of the various soluble starch synthase (SS) isozymes. The growth of chains is, therefore, strictly unidirectional. Chain elongation is closely associated with the action of starch branching enzyme (BE) and debranching enzyme (DBE) to finally reach the correctly branched amylopectin (Fig. 10.1). Amylose is synthesized by the granule-bound starch synthase I (GBSSI) using ADPglucose molecules as glucosyl donors.

In this chapter, the present status of our understanding of starch biosynthetic process in reserve organs and tissues is summarized, particularly focusing on the process in cereal endosperm. Recently, Nakamura (2015a) and Tetlow and Emes (2017) comprehensively reviewed the regulation of reserve starch biosynthesis, citing many papers so far available from studies with various plant species. Thus, here we focus on selected topics on the mechanisms of metabolic regulation of starch biosynthesis, the relationship between enzyme activity levels, the fine structure of amylopectin, and, finally, properties of starch granules.

10.2 Reserve Starch Biosynthesis

In higher plants, starch metabolism is based on a close collaboration of more than 30 gene products not including distinct heteromeric protein complexes (Ball et al. 2011). The relative activities of each isozyme in SS, BE, and DBE and the expression of subunit isoforms of AGPase largely vary between cereal endosperm and green leaves. Thus, the reserve starch biosynthesis and its regulation largely differ from those in assimilatory starch biosynthesis. For example, BEIIb, GBSSI, and a small subunit 2b of AGPase are selectively expressed in cereal endosperm (Ohdan et al. 2005). They are, therefore, considered to play important roles in the structure and the high rate of reserve starch biosynthesis in the endosperm.

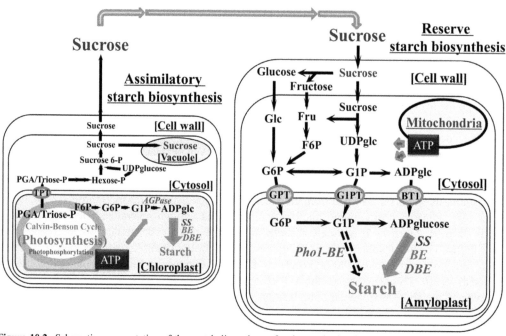

Figure 10.2. Schematic representation of the metabolic pathway for the synthesis of reserve starch in cereal endosperm (right) and transitory starch in green leaves (left). The figures are drawn based on the experimental results mainly conducted using rice endosperm and/or rice enzymes, although these conclusions are mostly applicable to other cereal endosperms. For more details see text. The situation is the same in Figs. 10.2–10.5.

In cereal endosperm, the major AGPase activity is located in the cytosol (Denyer et al. 1996). Subsequently, ADPglucose is transported into the amyloplast in exchange of ADP by the ADPglucose translocator (Fig. 10.2; Shannon et al. 1998). Amylose and amylopectin are synthesized in the amyloplasts and form the main constituents of the starch granules. Both size and number of the starch granules increase with the increase in volume and number of amyloplasts. At the end of maturing stage, most of the endosperm is occupied by amyloplasts including large starch granules. In contrast, at the early stage of endosperm development, vacuoles account for the bulk of the cellular volume. Valuoles transiently store large amounts of sugars, such as sucrose, organic acids, amino acids, and various inorganic ions at the early developmental stage. In later states, however, these compounds almost disappeared. In fact, at the dehydration or desiccation stage, about 80–90% (w/v) of the kernel consists of starch.

In cereal reserve organs, amylopectin is synthesized by coordinate actions of SS, BE, and DBE. The biochemical contributions of the major isozymes are summarized in Table 10.1 (Nakamura 2018). In developing rice endosperm, the major enzymes responsible for amylopectin biosynthesis are three branching isozymes (BEI, BEIIa, and BEIIb), three soluble starch synthases (SSI, SSIIa, and SSIIIa), and two types of debranching enzymes (isoamylase, ISA and pullulanase, PUL), which both follow the mode of direct debranching (ISA1 and PUL). During starch biosynthesis, the debranching enzymes are thought to remove those branches that do not fit to the branching clusters (James et al. 1995, Ball et al. 1996, Nakamura 2002). Amylose is synthesized essentially by a starch synthase isozyme that exclusively occurs inside the starch granule and, therefore, is named as granule-bound starch synthase isoform I (GBSSI). Please note that the enzymatic reactions and the isozymes involved are common among cereal endosperm from various plant species, pointing to a significant similarity of the biosynthetic routes of cereal reserve starch.

It is particularly important to clarify the relationship between the activity level of each isozyme and the fine structure of amylopectin and properties of starch granules. To discuss explicitly the contributions of individual isozymes to the fine structure of α-glucans including amylopectin, data

Table 10.1. The Contributions and Specificities of Major Amylopectin Biosynthetic Isozymes in Rice Endosperm.

		Reactivity/Specificity to chain-length	Contribution to amylopectin structure	Specificity of function	Major References
Branching enzyme (BE)	BEI	Formation of intermediate/ long chains (peak, DP12)	Synthesis of intermediate/ long chains	Moderate	Satoh et al. 2003, Nakamura et al. 2010
	BEIIa	Formation of short chains of DP6 to ca. 10	Synthesis of short chains	Low	Nakamura 2002, Satoh et al. 2003, Nakamura et al. 2010
	BEIIb	Act on DP12-14 to form DP7 and 6 chains	Synthesis of very short A chains	Very high	Nishi et al. 2001, Nakamura et al. 2010
Starch synthase (SS)	SSI	Act on DP6&7 to form DP8-12 (mainly DP8)	Elongation of very short chains	High	Fujita et al. 2006, Nakamura et al. 2014
	SSIIa	Synthesis of intermediate chains of DP≤24	Elongation of cluster chains	Very high	Nakamura et al. 2005
	SSIIIa	Synthesis of long B1 and B2-3 chains	Synthesis of long B1 and B2-3 chains	Moderate	Fujita et al. 2017
	SSIVb	(Not directly involved in starch synthesis)	(Contribution to granule morphology)	(Possibly high)	Toyosawa et al. 2016
Debranching enzyme (DBE)	ISA1	Act on a wide range of chains	Trimming by clearing ill-positioned chains	Very high	Utsumi et al. 2011, Kobayashi et al. 2016
	PUL	Act on a wide range of chains	Support of ISA function	Low	Fujita et al. 2009, Kobayashi et al. 2016

on chain-length distribution of glucans are needed that include not only short and intermediate side chains but also long chains interconnecting clusters. For this purpose, we used experimental results obtained from rice plants because these data are comprehensive and have been obtained most recently.

10.2.1 Metabolic Functions of BEI, BEIIa, and BEIIb

Biochemical analysis of BE-deficient mutants of rice revealed the distinct functions of BEI (Satoh et al. 2003) and BEIIb (Nishi et al. 2001). Specifically, the function of BEIIb could not be exerted by BEI and BEIIa. The starch from the so-called *amylose-extender* (*ae*) mutant defective in BEIIb had an increased hydrophobicity due to a high proportion of long chains in amylopectin, as found in maize *ae* starch (Yuan et al. 1993, Klucinec and Thompson 2002). Detailed comparative analysis of the *ae* and wild-type amylopectins showed that BEIIb plays a specific role in the transfer of short chains of amylopectin and determines the number of the side chains (A1 + B1) per cluster (see Chapter 8 of this volume). Based on the results with BE-related mutants and transformants, it was concluded that there are two types of branches in amylopectin cluster: branches forming the basal part of the cluster and those residing in the region between crystal lamellae and amorphous lamellae (Nakamura 2002, 2015a, 2018). The former branches are mainly synthesized by BEI, while the latter branches are specifically formed by BEIIb (see Fig. 10.3).

10.2.2 Metabolic Functions of SSI, SSIIa, and SSIIIa

Analysis of the chain-length distribution of amylopectin from the SSI-defective mutant (*ss1* mutant) indicated that this enzyme plays an important role in the synthesis of short chains having a degree of polymerization (DP) from 8 to 12 (Fujita et al. 2006). These results are consistent with the *in vitro* studies indicating that SSI certainly plays an important role in the synthesis of short chains having a degree of polymerization (DP) from 8 to 12 (Fujita et al. 2006). Actually, it is the only enzyme that

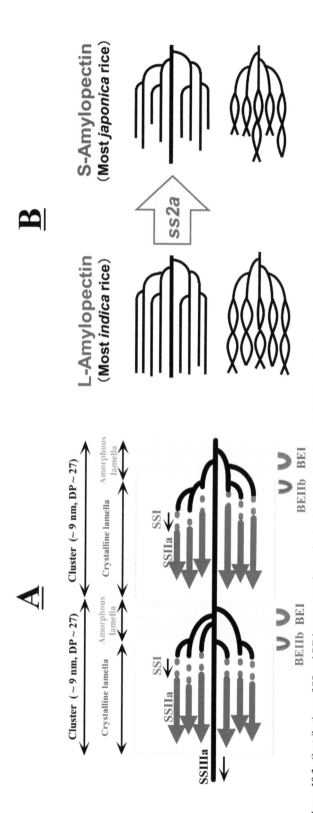

Figure 10.3. Contributions of SS and BE isozymes to the synthesis of amylopectin cluster chains in cereal endosperm. (**A**) SSI and SSIIa elongate short and intermediate chains within the cluster, respectively, while SSIIIa plays an important role in the long chains such as the cluster-interconnecting chains (B2 and B3 chains) as well as long B1 chains. See details in the text. (**B**) When the SSIIa activity is lost as in *japonica*-type rice varieties, cluster chains (A and B1 chains) are not fully elongated, and therefore, the cluster has more short chains (S-amylopectin) compared to L-amylopectin synthesized under high SSIIa condition as in *indica*-type rice varieties. As a result, S-amylopectin has fewer and weaker double helices than L-amylopectin.

mainly synthesizes these chains. In the *in vitro* studies, recombinant SS1 from maize (Commuri and Keeling 2001) and rice (Nakamura et al. 2014) has been used.

Rice cultivars are divided into the *japonica*-type and the *indica*-type cultivars, and the gelatinization temperature of the former starch is significantly lower than that of the latter one. Interestingly, the *japonica* rice is defective in the *ss2a* gene. This mutation is strictly associated with lower levels of intermediate chains of amylopectin. Therefore, in the rice SSIIa has an indispensable role in the elongation of A and B1 chains (Nakamura et al. 2003, 2005, Umemoto et al. 2002). Fujita et al. (2007) isolated the *ss3a* mutant lines of rice. With respect to the kernel phenotype, these mutants are similar to the maize *dull-1* mutants in showing an opaque to tarnished kernel appearance and significantly less weight. The *in vivo* results showed that SSIIIa plays an important part in the biosynthesis of long chains, particularly B2-3 chains (Fujita et al. 2006, Crofts et al. 2017b).

In summary, the soluble starch synthase isozymes, SSI, SSIIa, and SSIIIa, consecutively act on the side chains of amylopectin and thereby they mediate a series of glucosyl transfers covering a wide range of chain lengths. SSI synthesizes short chains having a DP of 8 up to approximately 12, SSIIa elongates these chains until they reach a DP of about 10 to 24, and, finally, SSIIIa synthesizes side chains with a DP of approximately 40 and occasionally up to the size of B3-/B4-chains to a lesser extent (Nakamura 2002, 2015a), although it is most likely that SSs recognize the length of external segments, rather than that of total length of chains including internal segments between branches. These results are basically consistent with *in vitro* analytical data obtained for SS isozymes from other cereal endosperms such as maize (Imparl-Radosevich et al. 1998, 1999, Commuri and Keeling 2001) and barley (Cuesta-Seiyo et al. 2016).

Recently, Toyosawa et al. (2016) found that a fourth soluble starch synthase isozyme, SSIVb, has a pivotal role in determining the starch morphology in rice endosperm. In the *ss4b/ss3a* double mutant, but not in the single *ss4b* mutant, the polygonal morphology of starch granules in the wild-type was replaced by a spherical one, suggesting that the contribution of SSIVb to starch granule morphology is partially overlapped with that of SSIIIa. This observation shows the novel role of SS in forming starch granule morphology.

10.2.3 *Metabolic Function of Starch Debranching Isozymes during Reserve Starch Biosynthesis: Isoamylase (ISA) and Pullulanase (PUL)*

James et al. (1995) verified that the *Sugary-1* gene of maize encodes ISA1, showing that deficiency of ISA activity resulted in accumulation of phytoglycogen instead of starch in the endosperm. Subsequently, the essential role of ISA-type DBE in reserve and assimilatory starch biosynthesis has been confirmed in rice (Kubo et al. 1999), potato (Bustos et al. 2004), barley (Burton et al. 2002), *Arabidopsis* (Zeeman et al. 1998), *Chlamydomonas* (Mouille et al. 1996), and a Cyanobacterium CLg1 (Cenci et al. 2013) (for review see Hennen-Bierwagen et al. 2012). Nakamura's group examined starch-related phenotypes of *sugary-1* mutants and transformants from rice in which the ISA activities were suppressed using biochemical, genetic, and molecular approaches, concluding that ISA1 is directly involved in normal amylopectin biosynthesis in developing rice endosperm (Kubo et al. 1999, Fujita et al. 2003, Utsumi et al. 2011). They also showed that PUL significantly mimics ISA1 function especially when ISA1 activity is lost or greatly inhibited (Fujita et al. 2009). Of interest is that there is a variation in the active form of ISA1 activities among plant species and tissues. For example, the ISA1-homomer is an active form in endosperm from rice (Utsumi et al. 2011) and maize (Kubo et al. 2010), whereas in potato tubers the ISA-ISA2 heteromer is involved in the synthesis of amylopectin (Ishizaki et al. 1983, Bustos et al. 2004).

10.2.4 *Fundamental Properties of Reserve Starch Biosynthesizing Metabolism*

Several functional interactions of starch-related isozymes appear to play essential roles in the formation of new clusters in amylopectin (Fig. 10.4). Functional interaction between BEI and SSIIIa coordinates

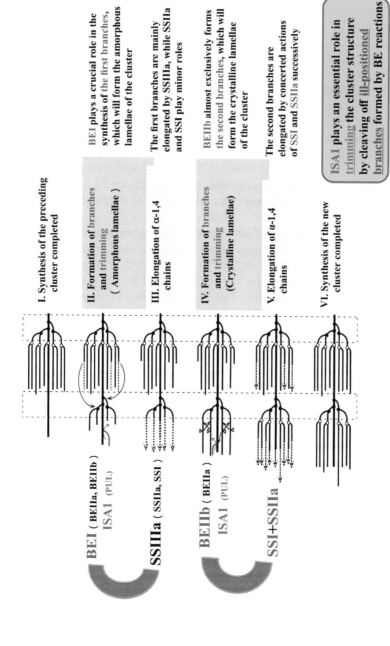

I. Synthesis of the preceding cluster completed

II. Formation of branches and trimming (Amorphous lamellae)

III. Elongation of α-1,4 chains

IV. Formation of branches and trimming (Crystalline lamellae)

V. Elongation of α-1,4 chains

VI. Synthesis of the new cluster completed

BEI plays a crucial role in the synthesis of the first branches, which will form the amorphous lamellae of the cluster

The first branches are mainly elongated by SSIIIa, while SSIIa and SSI play minor roles

BEIIb almost exclusively forms the second branches, which will form the crystalline lamellae of the cluster

The second branches are elongated by concerted actions of SSI and SSIIa successively

ISA1 plays an essential role in trimming the cluster structure by cleaving off ill-positioned branches formed by BE reactions

BEI (BEIIa, BEIIb)
ISA1 (PUL)

SSIIIa (SSIIa, SSI)

BEIIb (BEIIa)
ISA1 (PUL)

SSI+SSIIa

Figure 10.4. Schematic representation of the synthesis of amylopectin in cereal endosperm. The figure shows a model proposed by Nakamura (2002) with a slight modification. This model assumes that the synthesis of amylopectin is composed of the process in which the preceding cluster is reproduced. In this process, two combinations of BE and SS reactions play key roles: firstly, the branching reaction by BEI is followed by the elongation reaction by SSIIIa; secondly, the second branches are formed by BEIIb, and the resulting short chains of mainly DP 7 and 6 are elongated by SSI to chains of DP8 and a little bit longer, which are further elongated by SSIIa. ISA1 plays an essential role to remove the wrongly positioned branches which otherwise would interfere with the formation of double helices. See details in the text.

chain elongation and branching and, therefore, is expected to have an important function in the synthesis of branched chains at the beginning of a new cluster, which eventually forms the amorphous lamella of the cluster. Both enzymes react to longer chains as compared to BEIIa or BEIIb and SSI or SSIIa (Nakamura 2002, 2015a). The chain-length specificity of BEIIb is very selective because it almost exclusively forms the DP7 and DP6 chains by attacking the DP12-14 linear chains (external segments; Nakamura et al. 2010). The formed DP6-7 chains are elongated by SSI to form mainly DP8 chains as well as slightly longer chains of DP approximately 12 (Fujita et al. 2006, Nakamura et al. 2014). Subsequently, these short chains are further elongated by SSIIa until their lengths reach the end of the unit cluster equivalent to approximately DP24 (Nakamura et al. 2005, 2015a, 2018). In summary, BEIIb, SSI, and SSIIa functionally interact very closely in terms of chain-lengths of branched chains of the amylopectin clusters.

To accelerate such functional interaction, some isozymes are known to physically interact by forming heteromeric enzyme complexes (Tetlow et al. 2015; see also Chapter 11 this volume). These protein complexes, composed of several starch biosynthetic isozymes, have been isolated from the cereal endosperm of wheat (Tetlow et al. 2004, 2008), maize (Hennen-Bierwagen et al. 2008, 2009, Liu et al. 2009, 2012a,b), rice (Crofts et al. 2015), and barley (Ahmed et al. 2015). In these complexes, the enzymatic activities of the protein compounds are mutually favored (see also Tetlow et al. 2015).

10.2.5 A Model for Cluster Biosynthesis in Amylopectin

Based on biochemical, genetic, and molecular studies, a model to explain the amylopectin biosynthesis has been proposed (Fig. 10.4; Nakamura 2002, 2015a). The model claims that two functional interactions between branching reactions by BE and chain-elongation reactions by SS play crucial roles in creating the new cluster of amylopectin using the preceding cluster as primer. The first and second combined reactions consist of BEI/SSIIIa and BEIIb/SSI+SSIIa reactions, respectively: (a) BEI forms the first branches to the longest chain(s) in the preceding cluster, followed by elongation of the branches by SSIIIa. The elongated chains are further branched by BEI and elongated by SSIIIa. Thereby, elongated chains increase in number; (b) when the length of chains reaches DP12-14, BEIIb attacks them to form DP7 and DP6 chains, as described in 10.2.4. The resulting DP7 and 6 chains are elongated by SSI to the length of at least DP8. These short chains are further elongated to form intermediate chains of DP up to about 24. The second process can be repeated until space for the formation of new chains is occupied. In fact, both processes will operate simultaneously in the outward direction of the granules. In addition, the role of DBEs such as ISA1 is indispensable so that the wrongly-positioned branches formed accidentally by BEs reactions might be removed. Through DBEs, the structure of the new cluster can be trimmed during biosynthesis.

10.2.6 The Initiation and Amplification Process

Most of past studies with reserve starch biosynthesis have been performed to clarify the process in which the pre-existing clusters of precursor amylopectin (pre-amylopectin or immature amylopectin) and/or glucans are reproduced to form finally the mature amylopectin molecules by revealing the contributions of individual starch biosynthetic isozymes such as SSs, BEs, and DBEs to the fine structure of amylopectin (Ball et al. 1996, Myers et al. 2000, Nakamura 2002, 2015a, Ball and Morell 2003). This pathway of the amylopectin biosynthesis does not fully describe the formation of clusters, because at least two cases of cluster formation are distinguished: (a) an enlargement of an existing pre-amylopectin molecule occurs by forming additional clusters, (b) *de novo* synthesis of glucan primers having cluster structure from simple sugars and malto-oligosaccharides.

The latter case requires formation of additional linear α-glucans and then the formation of branches to yield a cluster. Reserve starch biosynthesis covers both processes that are mechanistically different as it requires both an increase in the number of clusters in a given number of amylopectin molecules and the total number of amylopectin molecules (Jeon et al. 2010, Nakamura 2015a, Nakamura et al. 2017). Therefore, two questions arise. First, how is the primer synthesized from simple sugars such as glucose and maltose? Second, how is the glucan primer used in the process of amplification of cluster numbers during rapid starch biosynthesis? In this chapter, the former and the latter processes are referred to as the "initiation process" and the "amplification process", respectively (Jeon et al. 2010, Nakamura 2015b, Crofts et al. 2017a, Nakamura et al. 2017).

Recent studies strongly suggested the involvement of plastidial phosphorylase (Pho1) in the initiation of amylopectin biosynthesis in rice endosperm (Satoh et al. 2008, Nakamura 2015b). Pho1 is capable of the synthesis of glucans under usual physiological conditions in spite of a low concentration of glucose 1-P and a high ratio of Pi/G1P (Hwang et al. 2010). It was found that rice Pho1 can use maltose and maltotriose as primer for glucan synthesis. The synthetic rate is, however, markedly lower than using maltotetraose, which has been so far considered to be a minimum primer (Nakamura et al. 2017). Pho1 has some specific properties in terms of an efficient glucan synthesis. First, Pho1 can synthesize glucans under a very low concentration of glucan primers by closely interacting with any BE isozymes (Nakamura et al. 2012, 2017), although the physical interaction between them seems to be less stable than other heteromeric protein complexes (Nakamura et al. 2017). In the functional interaction between Pho1 and BE, the apparent affinity of BE for glucans significantly increases (Nakamura et al. 2017). Second, Pho1 reacts differently with branched glucans and linear glucans. Pho1 functionally interacts with BE only when both enzymes use branched glucans as primer and the products are also branched ones (Nakamura et al. 2012, 2017). Pho1 is, however, also capable of the synthesis of linear glucans from malto-oligosaccharides such as maltose and maltotriose as well as longer linear glucans. It is possible that plastidial disproportionating enzyme (DPE1) is involved in the linear glucan synthesis with Pho1 by forming a protein complex of Pho1-DPE1 (Hwang et al. 2016a). In this way, Pho1 is considered to play a critical role in the glucan initiation process. In this connection, it is noted that long linear maltodextrins of DP up to at least 23 are present in rice endosperm at the very early developmental stage such as at 3 days after anthesis (Nakamura et al. 2020), suggesting that these maltodextrins as well as malto-oligosaccharides play some role in the initiation process of starch synthesis.

In addition, rice SSI can also synthesize branched glucans by close interaction with any BEs, whereas such interaction between SSIIa or SSIIIa with BEs is very low (Nakamura et al. 2014). A similar trend was also found between *Arabidopsis* SSI, SSII, and SSIII with BEs (Brust et al. 2014). Considering a chain-length specificity of SSI, it is highly possible that the SSI-BE interaction is involved in later steps than the Pho1-BE interaction. Thereby, a broad range of chain lengths would be involved in the initiation process. These results strongly suggest that the initiation process of starch biosynthesis in cereal endosperm is plant-specific and the self-glycosylation protein, such as glycogenin, is not needed which is known to play a pivotal role in glycogen initiation in animal cells (Roach et al. 2012). Notably, it is highly possible that Pho1-BE and SSI-BE are involved only in the initiation process, but not in the reproductive/amplification process in starch biosynthesis. The temperature dependency is different for Pho1 ad SS isoforms as Pho1 can act at lower temperatures almost completely whereas the synthases do not; therefore, what appears to be some redundancy is true only for a certain temperature range (Hwang et al. 2016b).

What is the physiological meaning for the separation of two processes? The fact that no detectable amounts of intermediate glucans lacking the cluster structure and lower molecular sizes compared with amylopectin are found shows that the carbon flow of the amplification process is much higher than that of the initiation process. This way must be an efficient way for the starch synthesis. It is highly possible that the amplification process includes rather simple steps in comparison with the initiation process because the former process does not require the steps for construction of the cluster structure from the cluster-lacking primer(s). The amplification process must be a more rational way to keep the same structural basis as the amylopectin cluster, maintaining the low value of entropy of the glucan structure.

10.2.7 Comparison with Assimilatory Starch Metabolism

The reserve starch metabolic process is likely to differ from the assimilatory starch metabolic process in several ways and aspects (see Chapter 9 of this volume). Here, some features of differences between both processes are compared between starch biosynthesis in chloroplasts of green leaves and that in amyloplasts of cereal endosperm. Most of cereal endosperm starches exhibit the A-type X-ray diffraction pattern, in which amylopectin double helices are most densely packed with a specific gravity up to about 1.5 (Imberty et al. 1991). The size and morphology of starch granules greatly vary depending on cereal species, whereas all leaf starch granules are likely to show ellipsoidal bodies. The assimilatory starch is usually synthesized during photosynthesis in chloroplasts under the light condition, and degraded to support the energy for numerous biological reactions under the dark condition. By contrast, the reserve starch synthesized in non-photosynthetic reserve organs is kept stored for long period for several months or years. To make starch granules more degradable, assimilatory starch is more phosphorylated by two glucan-related dikinases (GWD) and (PWD) so that starch can be easily attacked by hydrolytic enzymes (Hejazi et al. 2012). It has been elaborately examined that the efficient process for starch mobilization is equipped with leaf cells. For example, glucan chains are hydrolyzed by ß-amylase, DPE1, and DBE to maltose, and the resulting maltose is translocated into cytosol by the maltose transporter located in chloroplast envelope. Then in the cytosol, maltose is used for further metabolism with the help of DPE2 and Pho2 (see details by Hejazi et al. 2012). No report is available regarding such process in the endosperm.

The isozymes of SS, GBSS, BE, and DBE and subunits of AGPase are greatly different between endosperm and leaf of rice plants (Ohdan et al. 2005). For example, SSIIa, GBSSI, BEIIb, ISA1, AGPL1, and AGPL2 are preferentially expressed in the endosperm, while SSIIb, SSIIIb, GBSSII, BEIIa, ISA3, and AGPS2a are more highly expressed in the leaf.

10.3 Integration of Reserve Starch Biosynthesis into Cell Biology

Various reserve organs or tissues develop in plant species-specific manners. For example, there are several distinct stages in cereal seed development: (1) the early development stage including double fertilization, syncytium formation, and cellularization; (2) the differentiation stage, which includes the formation of transfer cells, aleurone, starchy endosperm, and embryo-surrounding cells, mitosis and endo-reduplication, and the production of storage compounds; and (3) the maturation stage, comprising the rapid production and storage of starch granules accompanied by programmed cell death, dormancy, and desiccation (Sabelli and Larkins 2009, Dante et al. 2014). During these processes, the function of cereal endosperm is specified to synthesize and store a huge amount of starch very efficiently. Proliferation of the endosperm nuclei leads to several dozen copies in a single cell, which might enable the high expression of genes responsible for starch biosynthesis (Sabelli and Larkins 2009).

It is known that the patterns of expression of active starch biosynthetic isozymes dramatically differ in accordance with the endosperm development. For example, these patterns can be classified into 3 groups in developing rice endosperm (Ohdan et al. 2005). The first group is expressed only at a very early stage of endosperm development. For example, *BEIIa* belongs to this group. The transcript level reaches a peak at 3 to 5 days after flowering (DAF), but rapidly declines thereafter. The second group including SSI and Pho1 is characterized by a medium expression at the very early developmental stages, and rises to reach peak at 5 DAF. Then it gradually declines to the end of maturation stage, although the transcript levels are significantly high at 10–20 DAF. The third group, composed of *SSIIa, SSIIIa, GBSSI, BEI, BEIIb, ISA1*, and *PUL*, is very low at the very early stage of endosperm development, but rapidly increases from 5 or 7 DAF, and the higher level is maintained until seed maturation. It is noted that this group includes the genes of which products play key roles in starch biosynthesis in rice endosperm, and that these transcript levels are roughly in accordance with the rate of starch production. This trend is mostly consistent with the changes in starch biosynthetic enzyme activities (Nakamura and Yuki 1992). In summary, these results show that the expression of different sets of isozymes is regulated in accordance with the developmental stages of the endosperm, playing distinct roles in starch biosynthesis.

As stated above, most cereal starches show the A-type X-ray diffraction pattern, but some tuber starches give the B-type pattern, while legume starches generally exhibit the C-type pattern. In terms of relative isozyme activities, SSI is a major SS form in cereal endosperms (Boyer and Preiss 1978, Fujita et al. 2006), whereas SSIII and SSII account for the major SS activities in potato tuber (Marshall et al. 1996) and pea embryo (Denyer and Smith 1992), respectively. The observations that the amount of DP6 and 7 chains of amylopectin is higher in the wild-type potato tuber (Edwards et al. 1999) than that in the wild-type rice endosperm (Fujita et al. 2005) is considered to be due to the lower SSI activity in the former than in the latter, because the role of SSI is responsible for the elongation of very short amylopectin chains of DP6 and 7 chains.

On the other hand, BEIIb, expressed in endosperm, plays a distinct role in forming short A and B1 chains, and this is responsible for the synthesis of the A-type starch granules, while loss of BEIIb such as *amylose-extender* mutant lines results in the formation of the B-type starch, being similar to starches in dicotyledonous reserve organs and tissues which all lack the BEIIb-type isozyme. The facts would reflect the number of chains per cluster (A + B1-chains/B2-3 chains), which is higher in cereal amylopectin than in tuber amylopectin (Hizukuri 1996, Bertoft et al. 2008).

What is the reason why reserve organs/tissues can store large amounts of starch at the high velocity in amyloplasts? It is known that the enzymatic activities (Nakamura et al. 1989), the expression levels of genes encoding starch metabolizing enzymes (Ohdan et al. 2005), and levels of metabolites involved in starch biosynthesis-related metabolites are by far higher in cereal reserve tissues than those in green leaves. Great amounts of simple sugars, such as sucrose, glucose, and fructose, can be stored in vacuoles, and this might support the continuous supply of a glucosyl donor (ADPglucose) with the starch biosynthetic machinery and in addition, might play a role in regulating the expression of genes for starch biosynthesis-related proteins. The steady-state metabolic flux has been studied using ^{13}C-labeling of sugars as substrate with NMR and/or GC-MS analysis of labeled metabolites aided by computer-modeling (Alonso et al. 2007, 2011). The results indicate an efficient partitioning of sugars such, as glucose, into starch and are consistent with the other results suggesting the close interaction of starch biosynthetic processes and the other metabolism (Yamakawa et al. 2007, Yamakawa and Hakata 2010, Ishimaru et al. 2015, Yu and Wang 2016).

The relationship between the starch biosynthetic process and other metabolic processes in cereal endosperm is not necessarily fully understood. Some literature has reported that loss of enzymes involved in the intermediate carbon metabolism such as pyruvate orthophosphate dikinase (PPDK) results in the formation of floury-type endosperm compared with the translucent appearance of the endosperm (Kang et al. 2005). The result suggests that the PPDK activity affects the smooth carbon flow to the production of ATP in mitochondria to support the AGPase reaction and/or the hexokinase reaction in the cytosol. The defect of the other enzymes such as pyruvate kinase (Cai et al. 2018) involved in the intermediary metabolism is known to cause serious effects on starch biosynthesis in cereal endosperm (see also review by Tetlow and Emes 2018, Nakamura 2018). Recently, several studies have indicated that the enhanced α-amylase protein level induced by higher temperature during endosperm development of rice is responsible for the floury phenotypes, which are considered to happen by degradation of starch granules with elevated α-amylase activity (Kaneko et al. 2016, Nakata et al. 2017). It is also interesting that the floury phenotype in rice endosperm appears in different sites caused by different amylase isozymes (Nakata et al. 2017).

It is widely known that the morphology, size, and composition of starch granules are largely different among Poaceae species and are classified into four types (Tateoka 1962, Matsushima et al. 2013). The first group is characterized by a compound type such as rice, which possesses only compound starch granules in the endosperm. The first group having only a single starch granule in each amyloplast is further divided into two-types: the uniform type and the bimodal type. The uniform type species such as maize is characterized by having a single starch granule with the similar size in amyloplast, whereas the bimodal type species such as barley and wheat develop two populations of starch granules, the small and large single granules in amyloplasts and both of them reside in the same cells. The fourth type such as a number of species of *Miscanthus*, *Perotis*, *Gymnopogon*, and *Thuarea* includes a mixture of compound and single starch granules in the same endosperm cells. However, the phylogenetic analysis shows vast variations in morphology of starch granules developed in endosperms among Poaceae species (Matsushima et al. 2013).

The structural development of amyloplasts in cereal endosperm is dynamic and varies depending on plant species. Kawagoe and his colleagues (Yun and Kawagoe 2010, Kawagoe 2013) claim that the septum-like structure (SLS) is synthesized beneath the outer envelope of amyloplast in developing rice endosperm, and SLS plays a role as a partition-wall in the stroma, which results in the formation of compound starch granules with hexagonal morphology having sharp edges. The *ss3a/ss4b* double mutation changed the polyhedral granules into the spherical granules with similar size (Toyosawa et al. 2016). Since the major amount of SSIVb protein was observed to be present between starch granules, possibly locating at the SLS membrane component, the authors propose that both SSIVb and SSIIIa play an important role in determining the polygonal starch morphology by maintaining the amyloplast membrane. The observation seems to be consistent with the result showing the essential role of SSIV in coordination of starch granule formation with chloroplast division during development of *Arabidopsis* leaf (Crumpton-Taylor et al. 2013).

10.4 Reserve Starch Degradation

In the same tissue, reserve starch degradation occurs far less frequently as compared to assimilatory starch. In cereals, typically it happens once and results in the elimination of the total starch storing tissue. In reserve starch storing dicotyledons, such as in potato tubers and storage roots, reserve starch can, in principle, be degraded several times, each time followed by a period of resynthesis, but there are other complications (see below).

Thus, depending on the plant system under study, the process of degradation strongly varies. In cereals, reserve starch storing tissue undergoes the senescence program before starch mobilization is initiated. This process is associated with the hydrolysis of other macromolecular compounds such as cell wall and protein hydrolysis. Apparently, thereby two purposes are followed: (1) by degrading the entire tissue, the accessibility of starch is ensured for starch degrading enzymes that are produced and exported by surrounding living cells. (2) Hydrolysis of cellular compounds other than starch yields additional low molecular compounds that also can be used by living cells for biosynthetic purposes provided they are imported by these cells. In other plant systems, such as the well-known potato tuber, reserve starch is degraded in living tissue and, therefore, degradation of reserve starch appears to be more selective as other cellular compounds appear to remain intact. In these systems, starch degradation is an intracellular process.

There are some reports pointing to a correlation of starch-related enzymes and/or the central carbon metabolism on the one side and the intensity of potato tuber sprouting on the other side (Duwenig et al. 1997, Hajirezaei et al. 1999) but it is difficult to define any biochemical link. Furthermore, it is difficult to define the area around the initiation of sprouting and the reserve starch degradation that is needed to sustain sprouting. As organs such as potato tuber have been selected for high starch during domestication, the capacity of storing starch exceeds by far the need of starch-derived degradation products for sprouting and most of the starch will remain unaffected by these processes. This unclearness certainly is a handicap for more detailed studies.

It should be mentioned that cotyledons of leguminosae often undergo a rapid senescence process during seed germination. Therefore, reserve starch degradation seems to be initiated in living tissue which, however, dies as soon as germination proceeds. This complexity prevents more detailed studies on reserve starch degradation and, therefore, we will not consider these cases in more detail.

10.4.1 Reserve Starch Degradation in Dead Tissues

Reserve starch degradation in dead tissue is usually preceded by a period of dormancy in which seed germination is not induced (Finkelstein et al. 2008). Dormancy is not restricted to cereal seeds but has been studied in more detail using cereal seeds (and Arabidopsis seeds as well).

10.4.1.1 Dormancy

Dormancy usually prevents germination in short periods of time that, in principle, would allow growth of the young plant but are placed in a longer season unfavorable for plant development. Dormancy is an unwanted trait as crops are supposed to rapidly germinate after sowing and, therefore, it has certainly been diminished during domestication of most crop species (Nakamura et al. 2016, Torada et al. 2016, Née et al. 2017). If, however, dormancy is inappropriately lost it may lead to pre-harvest sprouting and a significant diminishment of yield and/or quality of agricultural products (Simsek et al. 2014).

Several types of dormancy have been classified; the most prevalent form is the physiological dormancy (Baskin and Baskin 2004). Interestingly, in Arabidopsis several environmental conditions (such as temperature, light, and nitrate concentration in the soil) experienced by the mother plant affect the level of physiological dormancy (He et al. 2014). These results suggest the existence of a memory mechanism (Chen et al. 2014). Seeds harvested from the same individual plant usually differ, to some extent, in physiological dormancy. As a consequence, these seeds do not germinate at the same time and, therefore, the risk is minimized that, under unfavorable environmental conditions, the entire progeny of this plant is lost following germination.

Mechanistically, it appears that chromatin remodeling plays a central role in dormancy but also in germination. Therefore, it appears to be possible to retain several germination-related genes in an inactive state. Although activating transcription factors for these genes are available, due to steric hindrance, they cannot bind to and activate these genes. Abscisic acid (ABA) is the plant hormone known since decades as dormancy favoring factor appears to be involved in chromatin restructuring. When dormancy is released, the chromatin structure is altered leading to an activation of these genes (Shu et al. 2017).

10.4.1.2 Germination and Reserve Starch Mobilization

Following dormancy, germination and starch degradation are initiated (Rajjou et al. 2012). First, seeds take up massive water (phase I according to Bewley 1997). Water uptake during this phase is purely driven by the very low matrix potential of the seed material and is possible even in dead tissues. In phase II, water uptake is strongly diminished but in living cells that surround the dead tissue many processes occur, such as translation of newly synthesized mRNAs. Phase III is characterized by massive mobilization of storage products, such as cell wall, starch, and storage proteins, in the dead endosperm (Tan-Wilson and Wilson 2012). Degradation products are imported by living cells that are located in the surrounding of the dead endosperm. They are used for biosynthetic processes including cell division.

Degradation of storage products appears to be controlled, in principle, by the ratio of ABA and gibberellins (GA), as affected by various other phytohormones. Furthermore, several downstream-acting proteins are involved which act as transcription factors as well as the DELLA proteins which are GA-depending inhibitory regulators (Shu et al. 2016).

10.4.1.3 Potential Paths of Cereal Reserve Starch Degradation

The paths of reserve starch degradation in dead tissue appear to be more heterogeneous than the biosynthetic routes but none of the routes has been definitively established.

The aleuron layer that surrounds the starchy endosperm consists of living cells. These cells synthesize several α- and β-amylase isozymes massively under the influence of GA. Following secretion, both the α- and β-amylases can, in principle, hydrolyze α-1,4, interglucose linkages at the starch granules.

However, the *in vivo* function of the β-amylases is not clear as in several cereal species, reserve starch degradation is reported to be similar in mutants that essentially or completely lack β-amylases and in the wild type (Daussant et al. 1981, Kreis et al. 1987, Kiraha et al. 1999). By contrast, the degradation of assimilatory starch in chloroplasts includes, as essential steps, the plastidial starch-derived formation of this disaccharide and its export, by a specific transporter, into the cytosol (see Chapter 9 of this volume).

Multiplicity of α-amylase encoding genes is common (Ji et al. 2012, Mieog et al. 2017, Nakata et al. 2017). In germinating rice plants, at least 10 genes encoding for α-amylase are induced (Orchiai et al. 2014, Mitsui and Itoh 2014). The pattern of the induced α-amylase isozymes can be changed if the environmental conditions are altered, suggesting that the various isoforms do not have the same *in vivo* function (Hwang et al. 1999).

This induction is, however, far from being selective. In cereals like barley and rice, up to 1,300 genes are up-regulated under the influence of GA that encode many hydrolases but also proteins having a high functional diversity (Chen and An 2006, Tsuji et al. 2006). Different induction kinetics and different transcript levels have been reported for the α-amylase isoforms from germinating wheat seeds (Mieog et al. 2017). It is reasonable to assume that α-amylase forms are involved in cereal reserve starch degradation but it is not possible to describe the precise role(s) they exert. Therefore, a detailed metabolic path of reserve starch degradation cannot be presented. In fact, there may exist different paths depending on the biological system under consideration.

Under the influence of GA, several α-glucosidase isoforms are also strongly induced in the aleuron cells. There is evidence that, at least in some cereals, a distinct α-glucosidase has a dual function in reserve starch degradation. All isoforms of the α-glucosidase are traditionally assumed to convert oliglucans, such as the disaccharide maltose, into glucose. Following the induction by GA, the total α-glucosidase activity and the total α-amylase activity in the medium of isolated aleuron cells increase essentially coincidentally (Sun and Henson 1990). The formation of glucose is usually localized downstream of the processes taking place at the granule (Stanley et al. 2011).

Based on *in vitro* studies and scanning electron microscopy of isolated starch granules, at least two α-glucosidase forms were able to undergo a strong synergism with α-amylases. When starch granules were incubated with α-amylases alone, the sites of the enzymatic attack were exclusively located at equatorial grooves. The addition of both, α-amylase and α-glucosidase, however, resulted in larger and more holes at the entire surface of the granules (Sun and Henson 1990). This effect is explained by the assumption that the α-glucosidase increases the accessibility of starch for the α-amylases. Obviously, a more detailed knowledge of the starch granule surface is needed.

In any case, it is likely that the biochemical path(s) of cereal reserve starch degradation differ(s) from that of assimilatory starch mobilization.

10.4.1.4 Potential Control Mechanisms in Cereal Reserve Starch Degradation

As, at present, the whole reaction sequence(s) of cereal reserve starch breakdown cannot be given, the control mechanisms that regulate the entire process remain obscure. Only a few potentially relevant control elements are to be mentioned:

(a) A proteinaceous inhibitor of the activity of a pullulanase-type I debranching enzyme has been described (Møller et al. 2015b). This α-1,6- glucan hydrolysing enzyme activity appears to be the only debranching enzyme that exists for reserve starch degradation in barley. It has a high activity using pullulan as substrate but low activity towards amylopectin (Møller et al. 2015a). Presumably, *in vivo* the enzyme debranches oligo-glucans downstream from the particulate starch and the amylolytic degradation of starch granules on soluble branched α-glucans that are formed by the action of α-amylase (plus β-amylase). Inhibition of this type of debranching enzyme appears to be widely distributed in plants (Møller et al. 2015b and references therein).

(b) Some starch-related carbohydrate-active enzymes, such as an α-glucosidase isozyme, are strongly inhibited by various iminosugars (Stanley et al. 2011, Andriotis et al. 2016). In principle, this allows to study *in vivo* the effect of iminosugars on the mobilization of cereal reserve starch and, possibly, to identify as-yet unknown enzymes of the cereal starch breakdown.

(c) The phosphorylation of starch appears to have only a minor effect on cereal reserve starch degradation. Mobilization of barley reserve starch did not differ largely from that of the wild type (Shaik et al. 2014) when seeds from transgenic plants were germinated permanently expressing the potato glucan, water dikinase (StGWD1) (Carciofi et al. 2011). This concurs with a very low phosphate content of cereal starch. By contrast, the degradation of assimilatory starch is strongly inhibited (and growth of the plants is clearly reduced) when the content of C6 phosphate esters is largely diminished (Ritte et al. 2002, van Dijk et al. 2014).

Using a similar approach, Chen et al. (2017) generated transgenic rice that constitutively expressed the same potato GWD gene. The transgenic rice starch had a glucosyl 6-phosphate content approximately 9-fold higher as compared to the wild type control. Gelatinization temperatures of the transgenic rice starch were significantly lower than that of the control. This supports the notion that when overexpressing the starch phosphorylating enzyme, GWD, actually a different type of starch is formed which, indirectly, may affect several starch-related processes.

Likewise, for assimilatory starch it is known that when suppressing GWD activity, the starch that is formed differs in its morphology and in surface properties from the wild type control (Mahlow et al. 2014). Transitory starch that has been formed in the absence of GDW differs in morphology and surface properties from that of the wild type control (Mahlow et al. 2014).

The starch-phosphorylating enzymes certainly do not belong to the group of 'starch-degrading enzymes' as they are involved in both synthesis and degradation of starch.

10.4.2 Reserve Starch Degradation in Living Tissues

Reserve starch degradation in living tissues is highly heterogeneous. It includes the well-studied potato tuber (Hajirezaei et al. 2003), but also the rhizome of curcuma (Leonel et al. 2003, Hansdah et al. 2015) as well as fruits from tomatoes (Luengwilai and Beclies 2009, Maria et al. 2016), banana (Xiao et al. 2018, Peroni-Okita et al. 2013, Nascimento et al. 2005) and mango (Simao et al. 2008, Peroni et al. 2008). In part, studies were performed for practical reasons.

For two reasons, investigations are not as successful as those with Arabidopsis leaves or chloroplasts: (a) for some of these plants, no complete genome data are available; (b) for tissues with a high starch content, it is difficult (if not impossible) to obtain intact subcellular fractions and, therefore, the intact compartmentation of starch storing tissues is often assumed rather than experimentally proven.

Despite these uncertainties, it seems that starch degradation is more similar to that of chloroplasts (for details see Smirnova et al. 2015). But the entire path of mobilization remains unclear. It might even be that, depending on the species, more than one path exists.

As revealed by recent proteome analysis, many proteins that have been identified in Arabidopsis leaves (such as Protein Targeting to Starch, Early Starvation Protein 1, Like Early Starvation Protein 1) were found associated to starch granules. Likewise, almost all the proteins for phosphorylation/dephosphorylation of starch (see Chapter 7 of this volume) have been identified in the starch granule fraction (Helle et al. 2018; see also Stensballe et al. 2008). It is known that potato starch contains approximately 0.06% proteins based on a weight basis (Jobling 2004), the majority of which is granule-bound starch synthase (De Fekete et al. 1960).

In potato starch, a relatively high degree of phosphorylation occurs as approximately 0.5% of the starch-related glucosyl residues are monoesterified, the majority of phosphate being linked to the C6 (Hizukuri et al. 1970). Interestingly, using potato tuber discs and glucose-driven starch biosynthesis, incorporation of ^{32}P into starch was observed first during starch formation. Labeling of starch was proportional to the amount of starch synthesized (Nielsen et al. 1994). Although in the so-called high-phosphorylated starches only far less than 1% of the total starch-related glucosyl residues are monoesterified, the low degree of phosphorylation certainly affects significant physicochemical properties of starch, such as gel-forming capacity, swelling power and paste stability (Visko-Nielsen et al. 2001, Jobling 2004). The covalent modification of starch is, therefore, quite efficient.

In the autotetraploid potato, reserve starch phosphorylation is affected by allelic variation in the genes encoding the glucan, water dikinase, but also two branching enzymes, and starch synthase III (Carpenter et al. 2015). This indicates that the site of starch phosphorylation is rather precisely defined.

10.5 Conclusions and Future Perspectives

Recent studies proved that SS elongates glucosyl chains of glucans by adding a glucose moiety from ADPglucose to their non-reducing ends (Cuesta-Seijo et al. 2016, Larson et al. 2016), that is consistent with the chain elongation reaction by Pho1 using glucose 1-phosphate as glucosyl donor (Cuesta-Seijo et al. 2017, Nakamura et al. 2017). Comparative *in vitro* analyses of major isozymes of SS (Nakamura et al. 2014, Cuesta-Seijo et al. 2016, Huang et al. 2017), BE (Nakamura et al. 2010, Sawada et al. 2014), DBE (Kobayashi et al. 2016), and Pho (Cuesta-Seijo et al. 2017, Nakamura et al. 2017) clarified the fundamental functional properties of these isozymes to the fine structures of glucans. These *in vitro* experimental results and *in vivo* studies using starch-related mutants and transformants provide a concrete basis for strategy of generation of new cultivars of cereals and other plants. At the present time, we know to what extent the starch properties can be modified by altering activities and properties of enzymes involved in the starch biosynthetic process in reserve tissues. We also know that the double and triple mutants produce starches having much more marked properties than expected from those by simply summing the single mutation (Shannon et al. 2009, Nakamura 2018). Most of the various combinations of individual mutations have not been examined yet, and thus the mutant lines with simultaneous modification of multiple gene functions will be possibly the most promising cultivars in the next generation breeding. Since some mutants have lower yields than their wild-types, it is important to choose some measures to circumvent such disadvantage. In fact, the kernel weight of the rice backcrossed *ss3a/be2b* double mutant line increased by about 50%, the level being significantly higher than its wild-type, by crossing the original mutant with an elite high yield-type cultivar Akita63 (Fujita 2015). The backcrossed line contains the same functional properties of the starch as the original double mutant, and it will be cultivated under domestic climate conditions and used as a new cultivar producing high-resistant starch.

The past numerous investigations have accumulated evidence from a variety of approaches toward a better understanding of the mechanisms for biochemical processes of starch biosynthesis using various plant species. At present, it might be the appropriate timing to conduct comparative analysis to clarify both the common and different properties underlying the biochemical processes between different plant species. In addition, the relationship between the fine structure of starch molecules and the internal structure of starch granules is needed to be elaborated.

Acknowledgement

I would like to express my great gratitude to Dr. Martin Steup for stimulating discussion, critical reviewing, and helpful advice with a variety of topics, particularly related to starch degradation.

References

Alonso, A.P., F.G. Goffman, J.B. Ohlrogge and Y. Shachar-Hill. 2007. Carbon conversion efficiency and central metabolic fluxes in developing sunflower (*Helianthus annuus* L.) embryos. Plant J. 52: 296–308.

Alonso, A.P., D.L. Val and Y. Shachar-Hill. 2011. Central metabolic fluxes in the endosperm of developing maize seeds and their implications for metabolic engineering. Metabolic Engineer 13: 96–107.

Andriotus, V.M.E., M. Rejzek, M.D. Rugen, B. Sventsson, A.M. Smith and R.A. Field. 2016. Iminosugar inhibitors of carbohydrate-active enzymes that underpin cereal grain germination and endosperm metabolism. Biochem. Soc. Trans. 44: 159–165.

Ball, S., H. Guan, M. James, A. Myers, P. Keeling, G. Mouille et al. 1996. From glycogen to amylopectin: a model for the biosynthesis of the plant starch granule. Cell 86: 349–352.

Ball, S.G. and M.K. Morell. 2003. From bacterial glycogen to starch: Understanding the biogenesis of the plant starch granule. Ann. Rev. Plant Biol. 54: 207–233.

Ball, S., C. Colleoni, U. Cenci, J.N. Raj and C. Tirtiaux. 2011. The evolution of glycogen and starch metabolism in eukaryotes gives molecular clues to understand the establishment of plastid endosymbiosis. J. Exp. Bot. 62: 1775–1801.

Baskin, J.M. and C.C. Baskin. 2004. A classification system for seed dormancy. Seed Sci. Res. 14: 1–16.

Bertoft, E., K. Piyachomkwan, P. Chatakanonda and K. Sriroth. 2008. Internal unit chain in amylopectin. Carbohydr. Polym. 74: 527–543.

Bewley, D.J. 1997. Seed germination and dormancy. Plant Cell 9: 1400–1415.

Bogracheva, T.Y., v.J. Morris, S.G. Ring and C.L. Hedley. 1998. The granular structure of C-type pea starch and its role in gelatinization. Biopolym. 45: 323–332.

Boyer, C.D. and J. Preiss. 1978. Multiple forms of (1→4)-α-D-glucan, (1→4)-α-D-glucan-6-glycosyl transferase from developing *Zea mays* L. kernels. Carbohydr. Res. 61: 321–334.

Brust, H., T. Lehman, C. D'Hulst and J. Fettke. 2014. Analysis of the functional interaction of Arabidopsis starch synthase and branching enzyme isoforms reveals that the cooperative action of SSI and BEs results in glucans with polymodal chain length distribution similar to amylopectin. PLoS ONE 9: e102364.

Burton, R.A., H. Jenner, L. Carrangis, B. Fahy, G.B. Fincher, C. Hylton et al. 2002. Starch granule initiation and growth are altered in barley mutants that lack isoamylase activity. Plant J. 31: 97–112.

Bustos, R., B. Fahy, C.M. Hylton, R. Seale, N.M. Nebane, A. Edwards et al. 2004. Starch granule initiation is controlled by a heteromultimeric isoamylase in potato tubers. Proc. Nat. Acad. Sci. U.S.A. 101: 2215–2220.

Cai, Y., S. Li, G. Jiao, Z. Sheng, Y. Wu, G. Shao et al. 2018. *OsPK2* encodes a plastidic pyruvate kinase involved in rice endosperm starch synthesis, compound formation and grain filling. Plant Biotech. J. in press.

Cakir, B., S. Shiraishi, A. Tuncel, H. Matsusaka, R. Satoh, S. Singh et al. 2016. Analysis of the rice ADP-glucose transporter (OsBT1) indicates the presence of regulatory processes in the amyloplast stroma that control ADP-glucose flux into starch. Plant Physiol. 170: 1271–1283.

Carciofi, M., S.S. Shaik, S.L. Jenson, A. Blennow, J.T. Svensson, É. Vincze et al. 2011. Hyperphosphorylation of cereal starch. J. Cereal Sci. 54: 339–346.

Carperter, M.A., N.I. Joyce, R.A. Genet, R.D. Cooper, S.R. Murray, A.D. Noble et al. 2015. Starch phosphorylation is influenced by allelic variation in the genes encoding glucan water dikinase, starch branching enzymes I and II, and starch synthase. Frontiers in Plant Science 6: 143.

Cenci, U., M. Chabi, M. Ducatez, K. Tirtiaux, J.N. Raji, Y. Utsumi et al. 2013. Convergent evolution of polysaccharide debranching defines a common mechanism for starch accumulation in cyanobacteria and plants. Plant Cell 25: 3961-3975.

Chen, K. and Y-Q.C. An 2006. Transcriptional responses to gibberelin and abscisic acid in barley aleuron. J. Integr. Plant Biol. 48: 591–612.

Chen, M., D.R. MacGregor, A. Dave, H. Florance, H. Moore, K. Paszkiewicz et al. 2014. Maternal temperature history activates Flowering Locus T in fruits to control progeny dormancy according to time of year. Proc. Natl. Acad. Sci. USA 111: 18787–18792.

Chen, Y., X. Sun, X. Zhou, K.H. Hebelstrup, A. Blennow and J. Bao. 2017. Highly phosphorylated functionalized rice starch produced by transgenic rice expressing the potato GWD1 gene. Sci. Rep. 7: 3339.

Cisek, R., D. Tokarz, L. Kontenis, V. Barzda and M. Steup. 2018. Polarimetric second harmonic generation microscopy: An analytical tool for starch bioengineering. Starch 70: 1700031.

Commuri, P.D. and P.L. Keeling. 2001. Chain-length specificities of maize starch synthase I enzyme: studies of glucan affinity and catalytic properties. Plant J. 25: 475–486.

Crofts, N., N. Abe, N.F. Oitome, R. Matsushima, M. Hayashi, I.J. Tetlow et al. 2015. Amylopectin biosynthetic enzymes from developing rice seed form enzymatically active protein complexes. J. Exp. Bot. 66: 4469–4482.

Comparat-Moss, S. and K. Denyer. 2009. The evolution of the starch biosynthetic pathway in cereals and other grasses. J. Exp. Bot. 60: 2481–2492.

Crofts, N., Y. Nakamura and N. Fujita. 2017a. Critical and speculative review of the roles of multi-protein complexes in starch biosynthesis in cereals. Plant Sci. 262: 1–8.

Crofts, N., K. Sugimoto, N.F. Oitome, Y. Nakamura and N. Fujita. 2017b. Differences in specificity and compensatory functions among three starch synthases determine the structure of amylopectin in rice endosperm. Plant Mol. Biol. 94: 399–417.

Crumpton-Taylor, M., M. Pike, K. Lu, C.M. Hylton, R. Feil, S. Eicke et al. 2013. Starch synthase 4 is essential for coordination of starch granule formation with chloroplast division during Arabidopsis leaf expansion. New Phytol. 200: 1064–1075.

Cuesta-Seijo, J.A., M.M. Nielsen, C. Ruzanski, K. Krucewicz, S.R. Beeren, M.G. Rydhal et al. 2016. *In vitro* biochemical characterization of all barley endosperm starch synthases. Front. Plant Sci. 6: 1265.

Cuesta-Seijo, J.A., C. Ruzanski, K. Krucewicz, S. Meier, P. Hägglund, B. Svensson et al. 2017. Functional and structural characterization of plastidic starch phosphorylase during barley endosperm development. PLoS ONE 12: e0175488.

Daussant, J., J. Zbaszyniak, J. Sadowski and I. Wiatroszak. 1981. Cereal β-amylase: immunochemical study on two enzyme-deficient inbred lines of rye. Planta 151: 176–179.

Dante, R.A., B.A. Larkins and P.A. Sabelli. 2014. Cell cycle control and seed development. Front. Plant Sci. 5: 493.

De Fekete, R.M.A., L.F. Leloir and C.E. Cardini. 1960. Mechanism of starch biosynthesis. Nature 187: 918–919.

De Souza, A.P., J.-C. Cocuron, A.C. Garcia, A.P. Alonso and M.S. Buckeridge. 2015. Changes in whole-plant metabolism during the grain-filling stage in sorghum growth under elevated CO_2 and drought. Plant Physiol. 169: 1755–1765.

Denyer, K. and A.M. Smith. 1992. The purification and characterization of the two forms of soluble starch synthase from developing pea embryos. Planta 186: 609–617.

Denyer, K., F. Dunlap, T. Thorbjørnsen, P. Keeling and A.M. Smith. 1996. The major form of ADP-glucose pyrophosphorylase in maize endosperm is extra-plastidial. Plant Physiol. 112: 779–785.

Duwenig, E., M. Steup, L. Willmitzer and J. Kossmann. 1997. Antisense inhibition of cytosolic phosphorylase in potato plants (*Solanum tuberosum L.*) affects tuber sprouting and flower formation with only little impact on carbohydrate metabolism. Plant J. 12: 323–333.

Edwards, A., D.C. Fulton, C.M. Hylton, S.A. Jobling, M. Gidley, U. Rössner et al. 1999. A combined reduction in activity of starch synthases II and III of potato has novel effects on the starch of tubers. Plant J. 17: 251–261.

Flanagan, B.M., M.J. Gidley and F.J. Warren. 2015. Rapid quantification of starch molecular order through multivariate medelling of ^{13}C CP/MAS NMR spectra. Chem. Commun. 51: 14856–14858.

Finkelstein, R., W. Reevers, T. Arilzumi and C. Steber. 2008. Molecular aspects of seed dormancy. Annu. Rev. Plant Biol. 59: 387–415.

Fujita, N., M. Yoshida, N. Asakura, T. Ohdan, A. Miyao, H. Hirochika et al. 2006. Function and characterization of starch synthase I using mutants in rice. Plant Physiol. 140: 1070–1084.

Fujita, N., M. Yoshida, T. Kondo, K. Saito, Y. Utsumi, T. Tokunaga et al. 2007. Characterization of SSIIIa-deficient mutants of rice: The function of SSIIIa and pleiotropic effects by SSIIIa deficiency in the rice endosperm. Plant Physiol. 144: 2009–2023.

Fujita, N., Y. Toyosawa, Y. Utsumi, T. Higuchi, I. Hanashiro, A. Ikegami et al. 2009. Characterization of pullulanase (PUL)-deficient mutants of rice (*Oryza sativa* L.) and the function of PUL on starch biosynthesis in the developing rice endosperm. J. Exp. Bot. 60: 1009–1023.

Gidley, M.J. and S.M. Bociek. 1985. Molecular organization of starches: a 13C C/MAS NMR study. J. Am. Chem. Soc. 107: 7040–7044.

Gidley, M.J. and P.V. Bulpin. 1987. Crystallization of malto-oligosaccharides as models of the crystalline forms of starch. Carbohydr. Res. 161: 291–300.

Hajirezai, M.R. and U. Sonnewald. 1999. Inhibition of potato tuber sprouting: low levels of cytosolic pyrophosphate lead to non-sprouting harvested from transgenic potato plants. Potato Res. 42: 353–372.

Hajirezaei, M.R., F. Börnke, M. Peisker, Y. Takahata, J. Lerchl, A. Kirakosyan et al. 2003. Decreased sucrose content triggers starch breakdown and respiration in stored tubers (*Solanum tuberosum*). J. Exp. Bot. 54: 477–488.

Hansdah, R., P.K. Prabhakar, P.P. Srivastav and N.H. Mishra. 2015. Physico-chemical characterization of lesser known Palo (*Curcuma leucorrhiza*) starch. Intern. Food Res. J. 22: 1368–1373.

He, H., D.S. Vidigal, L.B. Snoek, S. Schabel, H. Nijveen, H. Hilhorst et al. 2014. Interaction between parental environment and genotype affects plant and seed performance in Arabidopsis. J. Exp. Bot. 65: 6603–6615.

Hejazi, M., J. Fettke and M. Steup. 2012. Starch phosphorylation and dephosphorylation: the consecutive action of starch-related dikinases and phosphatases. pp. 279–309. *In*: Tetlow, I. (ed.). Starch: Origins, Structure and Metabolism, Essential Reviews in Experimental Biology, vol. 5. The Society for Experimental Biology, London.

Helle, S., F. Bray, J. Verbeke, S. Devassine, A. Courseaux, M. Facon et al. 2018. Proteome analysis of potato starch reveals the presence of new starch metabolic proteins as well as multiple protease inhibitors. Front. Plant Sci. 9: 746.

Hennen-Bierwagen, T.A., M.G. James and A.M. Myers. 2012. Involvement of debranching enzymes in starch biosynthesis. pp. 179–215. *In*: Tetlow, I. (ed.). Starch: Origins, Structure and Metabolism, Essential Reviews in Experimental Biology, vol. 5. The Society for Experimental Biology, London.

Hennen-Bierwagen, T.A, F. Liu, R.S. Marsh, S. Kim, Q. Gan, I.J. Tetlow et al. 2008. Starch biosynthetic enzymes from developing maize endosperm associate in multisubunit complexes. Plant Physiol. 146: 1892–1908.

Hennen-Bierwagen, T.A., Q. Lin, F. Grimaud, V. Planchot, P.L. Keeling, M.G. James et al. 2009. Proteins from multiple metabolic pathways associate with starch biosynthetic enzymes in high molecular weight complexes: A model for regulation of carbon allocation in maize endosperm. Plant Physiol. 149: 1541–1559.

Hizukuri, S. 1986. Polymodal distribution of the chain lengths of amylopectins, and its significance. Carbohydr. Res. 147: 342–347.

Hizukuri, S. 1996. Starch: Analytical aspects. pp. 347–428. *In*: Eliasson, A.-C. (ed.). Carbohydrates in Food, Marcel Dekker, Inc., New York, Basel, Hong Kong.

Hizukuri, S. and Z. Nikuni. 1957. X-ray diffractomeric studies on starches. II. Structure of "C"-type crystallite. J. Agric. Chem. Soc. Jpn. 31: 525–527.

Hizukuri, S., T. Kaneko and Y. Takeda. 1983. Measurement of the chain length of amylopectin and its relevance to the origin of crystalline polymorphism of starch granules. Biochim. Biophys. Acta 760: 188–191.

Hizukuri, S., S. Tabata and Z. Nikumi. 1970. Studies on starch phosphate. Part 1. Estimation of glucose-6-phosphate residues in starch and the presence of other bound phosphate(s). Stärke 22: 238–243.

Huang, B., P.L. Keeling, T.A. Hennen-Bierwagen and A.M. Myers. 2016. Comparative *in vitro* analysis of recombinant maize starch synthases SSI, SSIIa, and SSIII reveal direct regulatory interactions of thermosensitivity. Arch. Biochem. Biophys. 596: 63–72.

Hwang, S., A. Nishi, H. Satoh and T.W. Okita. 2010. Rice endosperm-specific plastidial α-glucan phosphorylase is important for synthesis of short-chain malto-oligosaccharides. Arch. Biochem. Biophys. 495: 82–92.

Hwang, S.K., K. Koper, H. Satoh and T.W. Okita. 2016a. Rice endosperm starch phosphorylase (Pho1) assembles with disproportionating enzyme (Dpe1) to form a protein complex that enhances synthesis of malto-oligosaccharides. J. Biol. Chem. 291: 19994–20007.

Hwang, S.K., S. Singh, B. Cakir, H. Satoh and T.W. Okita. 2016b. The plastidial starch phosphorylase from rice endosperm: catalytic properties at low temperature. Planta 243: 999–1009.

Hwang, Y.S., B.R. Thomas and R.L. Rodriguez. 1999. Differential expression of rice α-amylase genes during seedling development under anoxia. Plant Mol. Biol. 40: 911–920.

Imberty, A., H. Chanzy, S. Pérez, A. Buléon and V. Tran. 1988. The double-helical nature of the crystalline part of A-starch. J. Mol. Biol. 201: 365–378.

Imberty, A. and S. Pérez. 1988. A revisit to the three-dimensional structure of B-type starch. Biopolym. 27: 1205–1221.

Imberty, A., A. Buléon, V. Tran and S. Pérez. 1991. Recent advances in knowledge of starch structure. Starch 43: 375–384.

Imparl-Radosevich, J.M., P. Li, L. Zhang, A.L. McKean, P.L. Keeling and H. Guan.1998. Purification and characterization of maize starch synthase I and its truncated forms. Arch. Biochem. Biophys. 353: 64–72.

Imparl-Radosevich, J.M., D.J. Nichols, P. Li, A.L. McKean, P.L. Keeling and H. Guan. 1999. Analysis of purified maize starch synthases IIa and IIb: SS isoforms can be distinguished based on their kinetic properties. Arch. Biochem. Biophys. 362: 131–138.

Ishimaru, T., M. Ida, S. Hirose, S. Shimamura, T. Matsumura, N. Nishizawa et al. 2015. Laser microdissection-based gene expression analysis in the aleurone layer and starchy endosperm of developing rice caryopses in the early storage phase. Rice 8: 22.

Ishizaki, Y., H. Taniguchi, Y. Maruyama and M. Nakamura. 1983. Debranching enzymes of potato tubers (*Solanum tuberosam* L.). I. Purification and some properties of potato isoamylase. Agric. Biol. Chem. 47: 771–779.

James, M.G., D.S. Robertson and A.M. Myers. 1995. Characterization of the maize gene *sugary1*, a determinant of starch composition in kernels. Plant Cell 7: 417–429.

Jane, J. 2009. Structural features of starch granules II. pp. 193–236. *In*: BeMiller, J. and R. Whistler (ed.). Starch: Chemistry and Technology, Third Edition, Academic Press, New York.

Jane, J., T. Kasemsuwan, S. Leas, H. Zobel and J.F. Robyt. 1994. Anthology of starch granule morphology by scanning electron microscopy. Starch 46: 121–129.

Jenkins, P.J., R.E. Cameron and A.M. Donald. 1993. A universal feature in the starch granules from different botanical sources. Starch 45: 417–420.

Jenkins, P.J. and A.M. Donald. 1998. Gelatinisation of starch: a combined SAXS/WAXS/SANS study. Carbohydr. Res. 308: 133–147.

Jeon, J.S., N. Ryoo, T.R. Hahn, H. Walia and Y. Nakamura. 2010. Starch biosynthesis in cereal endosperm. Plant Physiol. Biochem. 48: 383–392.

Jobling, S. 2004. Improving starch for food and industrial applications. Curr. Opin. Plant Biol. 7: 210–218.

Kainuma, K. and D. French. 1972. Naegeli amylodextrin and its relationship to starch granule structures. III. Role of water in crystallization of B-starch. Biopolym 11: 2241–2250.

Kaneko, K., M. Sasaki, N. Kuribayashi, H. Suzuki, Y. Sasuga, T. Shiraya et al. 2016. Proteomic and glycomic characterization of rice chalky grains produced under moderate and high-temperature conditions in field system. Rice 9: 26.

Kang, H., S. Park, M. Matsuoka and G. An. 2005. White-core endosperm *floury endosperm-4* in rice is generated by knockout mutations in the C4-type pyruvate orthophosphate dikinase gene (*OsPPDKB*). Plant J. 42: 901–911.

Kawagoe, Y. 2013. The characteristic polyhedral, sharp-edged shape of compound-type starch granules in rice endosperm is achieved via the septum-like structure of the amyloplast. J. Appl. Glycosci. 60: 29–36.

Katz, J.R. and T.B. van Itallie. 1930. Abhandlungen zur physikalischen Chemie der Stärke und Brotbereitung, V. Alle Stärkearten haben das gleiche Retrogradationsspektrum. Z. Phys. Chem. A150: 90–99.

Kihara, M., T. Kaneko, K. Ito, Y. Aida and K. Takeda. 1999. Geographic variation of β-amylase thermostability among varieties of barley *(Hordeum vulgare)* and β-amylase deficiency. Plant Breed. 118: 453–455.

Klucinec, J.D. and D.B. Thompson. 2002. Structure of amylopectins from ae-containing maize starches. Cereal Chem. 79: 19–23.

Kobayashi, T., S. Sasaki, Y. Utsumi, N. Fujita, K. Umeda, T. Sawada et al. 2016. Comparison of chain-length preferences and glucan specificities of isoamylase-type α-glucan debranching enzymes from rice, cyanobacteria, and bacteria. PLoS ONE 11: e0157020.

Kreis, M., M. Williamson, B. Buxton, J. Pywell, J. Hejgaard and I. Svendsen. 1987. Primary structure and differential expression of β-amylase in normal and mutant barley. Eur. J. Biochem. 169: 517–525.

Kubo, A., N. Fujita, K. Harada, T. Matsuda, H. Satoh and Y. Nakamura. 1999. The starch-debranching enzymes isoamylase and pullulanase are both involved in amylopectin biosynthesis in rice endosperm. Plant Physiol. 121: 399–409.

Kubo, A., C. Colleoni, J.R. Dinges, Q. Lin, R.R. Lappe and J.G. Rivenbark. 2010. Functions of heteromeric and homomeric isoamylase-type starch-debranching enzymes in developing maize endosperm. Plant Physiol. 153: 956–969.

Larson, M.E., D.J. Faiconer, A.M. Myers and A.W. Barb. 2016. Direct characterization of the maize starch synthase IIa product shows maltodextrin elongation occurs at the non-reducing end. J. Biol. Chem. 291: 24951–24960.

Leloir, L.F., M.A. de Fekete and C.E. Cardini. 1961. Starch and oligosaccharide synthesis from uridine diphosphoglucose. J. Biol. Chem. 236: 636–641.

Leonel, M., S.B.S. Samento and M.P. Cereda. 2003. New starches for the fool industry: *Curcuma longa* and *Curcuma zedoaria*. Carbohydr. Polym. 54: 385–388.

Li, S., X. Wei, Y. Ren, J. Qiu, G. Jiao, X. Guo et al. 2017. *OsBT1* encodes an ADP-glucose transporter involved in starch synthesis and compound granule formation in rice endosperm. Sci. Rep. 7: 40124.

Liu, F., A. Makhmoudova, E.A. Lee, R. Wait, M.J. Emes and I.J. Tetlow. 2009. The *amylose extender* mutant of maize conditions novel protein-protein interactions between starch biosynthetic enzymes in amyloplasts. J. Exp. Bot. 60: 4423–4440.

Liu, F., Z. Ahmed, E.A. Lee, E. Donner, Q. Liu, R. Ahmed et al. 2012a. Allelic variants of the amylose extender mutation of maize demonstrate phenotypic variation in starch structure resulting from modified protein-protein interactions. J. Exp. Bot. 63: 1167–1183.

Liu, F., N. Romanova, E.A. Lee, R. Ahmed, E. Martin, E.P. Gilbert et al. 2012b. Glucan affinity of starch synthase IIa determines binding of starch synthase I and starch-branching enzyme IIb to starch granules. Biochem. J. 448: 373–387.

Mahlow, S., M. Hejazi, F. Kuhnert, A. Garz, H. Brust, O. Baumann et al. 2014. Phosphorylation of transitory starch by a-glucan, water dikinase during starch turnover affects the surface properties and morphology of starch granules. New Phytol. 203: 495–507.

Marshall, J., C. Sidebottom, M. Debet, C. Martin, A. Smith and A. Edwards. 1996. Identification of the major starch synthase in the soluble fraction of potato tubers. Plant Cell 8: 1121–1135.

Matsushima, R. 2015. Morphological variations of starch grains. pp. 425–441. *In*: Y. Nakamura (ed.). Starch: Metabolism and Structure, Springer, Tokyo, Japan.

Matsushima, R., J. Yamashita, S. Kariyama, T. Enomoto and W. Sakamoto. 2013. A phylogenetic re-evaluation of morphological variations of starch grains among Poaceae species. J. Appl. Glycosci. 60: 37–44.

Mieog, J.C., Š. Janeček and J.-P. Ral. 2017. New insight in cereal starch degradation: identification and structural characterization of four a-amylases in bread wheat. Amylase 1: 35–49.

Mitsui, T. and K. Itoh. 1997. The α-amylase multigene family. Trends Plant Sci. 2: 255–161.

Miyauchi, Y., H. Sano and G. Mizutani. 2006. Selective observation of starch in a water plant using optical sum-frequency microscopy. J. Opt. Soc. Am. A 23: 1687–1690.

Mizutani, G., Y. Sonoda, H. Sano, M. Sakamoto, T. Takahashi and S. Ushioda. 2000. Detection of starch granules in a living plant by optical second harmonic microscopy. J. Lumin. 87: 824–826.

Møller, M.S., M.S. Windahl, L. Sim, M. Bøjstrup, M. Abou-Hachem, O. Hindsgaul et al. 2015a. Oligosaccharide and substrate binding in the starch debranching enzyme barley limit dextrinase. J. Mol. Biol. 427: 1263–1277.

Møller, M.S., M.B. Vester-Christensen, J.M. Jensen, M. Abou-Hachem, A. Henriksen and B. Svensson. 2015b. Crystal structure of barley limit dextrinase-limit dextrinase inhibitor (LD-LDI) complex reveals insights into mechanism and diversity of cereal type inhibitors. J. Biol. Chem. 290: 12614–12629.

Mouille, G., M.L. Maddelein, N. Libessart, P. Talaga, A. Decq, B. Delrue et al. 1996. Preamylopectin processing: A mandatory step for starch biosynthesis in plants. Plant Cell 8: 1353–1368.

Myers, A.M., M.K. Morell, M.G. James and S.G. Ball. 2000. Recent progress toward understanding biosynthesis of the amylopectin crystal. Plant Physiol. 122: 989–997.

Nakamura, Y. 2002. Towards a better understanding of metabolic system for amylopectin biosynthesis: rice endosperm as a model tissue. Plant Cell Physiol. 43: 718–725.

Nakamura, Y. 2015a. Biosynthesis of reserve starch. pp. 161–209. *In*: Nakamura, Y. (ed.). Starch: Metabolism and Structure, Springer, Tokyo, Japan.

Nakamura, Y. 2015b. Initiation process of starch biosynthesis. pp. 315–332. *In*: Nakamura, Y. (ed.). Starch: Metabolism and Structure, Springer, Tokyo, Japan.

Nakamura, Y. 2018. Rice starch biotechnology: Rice endosperm as a model of cereal endosperms. Starch 70: 1600375.

Nakamura, Y., K. Yuki, S. Park and T. Ohya. 1989. Carbohydrate metabolism in the developing endosperm of rice grains. Plant Cell Physiol. 30: 833–839.

Nakamura, Y. and K. Yuki. 1992. Changes in enzyme activities associated with carbohydrate metabolism during the development of rice endosperm. Plant Sci. 82: 15–20.

Nakamura, Y., A. Sakurai, Y. Inaba, K. Kimura, N. Iwasawa and T. Nagamine. 2002. The fine structure of amylopectin in endosperm from Asian cultivated rice can be largely classified into two classed. Starch. 54: 117–131.

Nakamura, Y., P.B. Francisco, Jr., Y. Hosaka, A. Sato, T. Sawada, A. Kubo et al. 2005. Essential amino acids of starch synthase IIa differentiate amylopectin structure and starch quality between *japonica* and *indica* rice varieties. Plant Mol. Biol. 58: 213–227.

Nakamura, Y., Y. Utsumi, T. Sawada, S. Aihara, C. Utsumi, M. Yoshida et al. 2010. Characterization of the reactions of starch branching enzyme from rice endosperm. Plant Cell Physiol. 51: 776–794.

Nakamura, Y., S. Aihara, N. Crofts, T. Sawada and N. Fujita. 2014. *In vitro* studies of enzymatic properties of starch synthases and interactions between starch synthase I and starch branching enzymes from rice. Plant Sci. 224: 1–8.

Nakamura, Y., M. Ono, T. Sawada, N. Crofts, N. Fujita and M. Steup. 2017. Characterization of the functional interactions of plastidial starch phosphorylase and starch branching enzymes from rice endosperm during reserve starch biosynthesis. Plant Sci. 264: 83–95.

Nakamura, Y., M. Ono, M. Suto and H. Kawashima. 2020. Analysis of malto-oligosaccharides and related metabolites in rice endosperm during development. Planta in press.

Nakamura, S., M. Pourkheirandish, H. Morishige, Y. Kubo, M. Nakamura, K. Ichimura et al. 2016. *Mitogen-activated protein kinase kinase 3* regulates seed dormancy in barley. Curr. Biol. 26: 775–781.

Nakata, M., Y. Fukamatsu, T. Miyashita, M. Hakata, R. Kimura, Y. Nakata et al. 2017. High temperature-induced expression of rice α-amylases in developing endosperm produces chalky grains. Front. Plant Sci. 8: 2089.

Née, G., Y. Xiang and W.J.J. Soppe. 2017. The release of dormancy, a wake-up call for seeds to germinate. Curr. Opin. Plant Biol. 35: 8–14.

Nielsen, T.H., B. Wischmann, K. Enevoldsen and B.L. Møller. 1994. Starch phosphorylation in potato tubers proceeds concurrently with the *de novo* biosynthesis of starch. Plant Physiol. 105: 111–117.

Nikuni, Z. 1969. Starch and cooking (in Japanese). Sci. Cooking 2: 6–14.

Nishi, A., Y. Nakamura and H. Satoh. 2001. Biochemical and genetic analysis of *amylose-extender* mutations of rice endosperm. Plant Physiol. 127: 459–472.

Ochiai, A., H. Sugai, K. Harada, S. Tanaka, Y. Ishiyama, K. Itoh et al. 2014. Crystal structure of α-amylase from *Oryza sativa*: molecular insights into enzyme activity and thermostability. Biosci. Biotechnol. Biochem. 78: 989–997.

Ohdan, T., P.B. Francisco, Jr., T. Sawada, T. Hirose, T. Terao, H. Satoh et al. 2005. Expression profiling of genes involved in starch synthesis in sink and source organs of rice. J. Exp. Bot. 56: 3229–3244.

Pérez, S. and E. Bertoft. 2010. The molecular structures of starch components and their contribution to the architecture of starch granules: A comprehensive review. Starch 62: 389–420.

Preiss, J. and C. Levi. 1980. Starch biosynthesis and degradation. pp. 371–423. *In*: Preiss, J. (ed.). The Biochemistry of Plants, vol. 3, Carbohydrates: Structure and Function, Academic Press, New York, London, Toronto, Sydney, and San Francisco.

Rajjou, L., M. Duval, K. Gallardo, J. Catusse, J. Baily, C. Job et al. 2012. Seed germination and vigor. Annu. Rev. Plant Biol. 63: 507–533.

Ritte, G., J.R. Lloyd, N. Eckermann, A. Rottmann, J. Kossmann and M. Steup. 2002. The starch-related R1 protein is an alpha-glucan, water dikinase. Proc. Natl. Acad. Sci. USA 99: 7166–7171.

Roach, P.J., A.A. Depaoli-Roach, T.D. Hurley and V. Tagliabracci. 2012. Glycogen and its metabolism: some new developments and old themes. Biochem. J. 441: 763–787.

Sabelli, P.A. and B.A. Larkins. 2009. The development of endosperm in grasses. Plant Physiol. 149: 14–26.

Satoh, H., A. Nishi, K. Yamashita, Y. Takemoto, Y. Tanaka, Y. Hosaka et al. 2003. Starch-branching enzyme I-deficient mutation specifically affects the structure and properties of starch in rice endosperm. Plant Physiol. 133: 1111–1121.

Satoh, H., K. Shibahara, T. Tokunaga, A. Nishi, M. Tasaki, S.K. Hwang et al. 2008. Mutation of the plastidial α-glucan phosphorylase gene in rice affects the synthesis of branched maltodextrins. Plant Cell 20: 1833–1849.

Sawada, T., Y. Nakamura, T. Ohdan, A. Saitoh, P.B. Francisco, Jr., E. Suzuki et al. 2014. Diversity of reaction characteristics of glucan branching enzymes and the fine structure of α-glucan from various sources. Arch. Biochem. Biophys. 562: 9–21.

Shannon, J.C., F. Pien, H. Cao and K. Liu. 1998. Brittle-1, an adenylate translocator, facilitates transfer of extraplastidial synthesized ADP-glucose into amyloplasts of maize endosperms. Plant Physiol. 117: 1235–1252.

Shannon, J.C., D.L. Garwood and C.D. Boyer. 2009. Genetics and physiology of starch development. pp. 23–82. *In*: BeMiller, J. and R. Whistler (ed.). Starch: Chemistry and Technology, Third Ed., Academic Press, New York.

Shaik, S.S., M. Carciofi, H.J. Martens, K.H. Hebelstrup and A. Blennow. 2014. Starch bioengineering affects cereal grain germination and seedling establishment. J. Exp. Bot. 65: 2257–2270.

Shu, K., X. Liu, Q. Xie and Z. He. 2016. Two faces of one seed: hormonal regulation of dormancy and germination. Mol. Plant 9: 34–45.

Simsek, S., J.B. Ohm, H. Lu, M. Rugg, W. Berzonsky, M.S. Alamri et al. 2014. Effect of pre-harvest sprouting on physicochemical changes of proteins in wheat. J. Sci. Food Agric. 94: 205–212.

Smirnova, J., A.R. Fernie and M. Steup. 2015. Starch degradation. pp. 239–290. *In:* Nakamura, Y. (ed.). Starch Metabolism and Structure. Springer Japan.

Sonnewald, S. and U. Sonnewald. 2014. Regulation of potato tuber sprouting. Planta 239: 27–38.

Stanley, D., M. Rejcek, H. Naested, M. Smedley, S. Otero, B. Fahy et al. 2011. The role of α-glucosidase in germinating barley grains. Plant Physiol. 155: 932–943.

Stensballe, A., S. Hald, G. Bauw, A. Blennow and K.G. Welinder. 2008. The amyloplast proteome of potato tuber. FEBS J. 275: 1723–1741.

Sterling, C.J. 1962. A low angle spacing in starch. J. Polym. Sci. 56: S10–12.

Sun, Z. and C.A. Henson. 1990. Degradation of native starch granules by barley α-glucosidases. Plant Physiol. 94: 320–327.

Tan-Wilson, A.L. and K.A. Wilson. 2012. Mobilisation of protein reserves. Physiol. Plant. 145: 140–153.

Tanaka, N., N. Fujita, A. Nishi, H. Satoh, Y. Hosaka, M. Ugaki et al. 2004. The structure of starch can be manipulated by changing the expression levels of *starch branching enzyme IIb* in rice endosperm. Plant Biotech. J. 2: 507–516.

Tetlow, I.J., R. Wait, Z. Lu, C.G. Akkasaeng, S. Bowsher, B. Esposito et al. 2004. Protein phosphorylation in amyloplasts regulates starch branching enzyme activity and protein-protein interactions. Plant Cell 16: 694–708.

Tetlow, I.J., K.G. Beisel, S. Cameron, A. Makhmoudova, F. Liu, N.S. Bresolin et al. 2008. Analysis of protein complexes in amyloplasts reveals functional interactions among starch biosynthetic enzymes. Plant Physiol. 146: 1878–1891.

Tetlow, I.J., F. Liu and M.J. Emes. 2015. Biosynthesis of reserve starch. pp. 291–313. *In*: Nakamura, Y. (ed.). Starch: Metabolism and Structure, Springer, Tokyo, Japan.

Tetlow, I.J. and M.J. Emes. 2017. Starch biosynthesis in the developing endosperms of grasses and cereals. Agronomy 7: 81.

Thompson, D.B. 2000. On the non-random nature of amylopectin branching. Carbohydr. Polym. 43: 223–239.

Torada, A., M. Koike, T. Ogawa, Y. Takenouchi, K. Tadamura, J. Wu et al. 2016. A causal gene for seed dormancy on wheat chromosome 4A encodes a MAP kinase kinase. Curr. Biol. 26: 782–787.

Toyosawa, Y., Y. Kawagoe, R. Matsushima, N. Crofts, M. Ogawa, M. Fukuda et al. 2016. Deficiency of starch synthase IIIa and IVb alters starch granule morphology from polyhedral to spherical in rice endosperm. Plant Physiol. 170: 1255–1270.

Tsuji, H., K. Aya, M. Ueguchi-Tanaka, Y. Shimida, M. Nakazono, R. Watanabe et al. 2006. GAMYB controls different sets of genes and is differently regulated by microRNA in aleuron cells and anthers. Plant J. 47: 427–444.

Umemoto, T., M. Yano, H. Satoh, A. Shomura and Y. Nakamura. 2002. Mapping of a gene responsible for the difference in amylopectin structure between *japonica*-type and *indica*-type rice varieties. Theor. Appl. Genet. 104: 1–8.

Utsumi,Y., C. Utsumi, T. Sawada, N. Fujita and Y. Nakamura. 2011. Functional diversity of isoamylase (ISA) oligomers: The ISA1 homo-oligomer is essential for amylopectin biosynthesis in rice endosperm. Plant Physiol. 156: 61–77.

Vikso-Nielsen, A., A. Blennow, K. Jorgensen, K.H. Kristensen, A. Jensen and B.L. Moller. 2001. Structural, physicochemical and pasting properties of starches from potato plants with repressed r1-gene. Biomacromol. 2: 836–843.

Vriet, C., WT. Wellham, A. Brachmann, M. Pike, J. Perry, M. Parsinke et al. 2010. A suite of *Lotus japonicus* starch mutants reveals both conserved and novel features of starch metabolism. Plant Physiol. 154: 643–655.

Xiao, Y., J. Kuang, X. Qi, Y. Ye, Z. Wu, J. Chen et al. 2018. A comprehensive investigation of starch degradation process and identification of a transcriptional activator MabHLH6 during banana fruit ripening. Plant Biotechnol. J. 16: 151–164.

Yamakawa, H., T. Hirose, M. Kuroda and T. Yamaguchi. 2007. Comprehensive expression profiling of rice grain filling-related genes under high temperature using DNA microarray. Plant Physiol. 144: 258–277.

Yamakawa, H. and M. Hakata. 2010. Atlas of rice grain filling-related metabolism under high temperature: joint analysis of metabolome and transcriptome demonstrated inhibition of starch accumulation and induction of amino acid accumulation. Plant Cell Physiol. 51: 795–809.

Yoon, M. and Y. Kawagoe. 2010. Septum formation in amyloplasts produces compound granules in the rice endosperm and is regulated by plastid division proteins. Plant Cell Physiol. 51: 1469–1479.

Yu, H. and T. Wang. 2016. Proteomic dissection of endosperm starch granule associated proteins reveals a network coordinating starch biosynthesis and amino acid metabolism and glycolysis in rice endosperm. Front. Plant Sci. 7: 707.

Yuan, R.C., D.B. Thompson and C.D. Boyer. 1993. Fine-structure of amylopectin in relation to gelatinization and retrogradation behavior of maize starches from 3 *wx*-containing genotypes in 2 inbred lines. Cereal Chem. 70: 81–89.

Zeeman, S.C., T. Umemoto, W. Lue, P. Au-Yeung, C. Martin, A.M. Smith et al. 1998. A mutant of *Arabidopsis* lacking a chloroplastic isoamylase accumulates both starch and phytoglycogen. Plant Cell 10: 1699–1711.

CHAPTER 11

Heteromeric Protein Interactions in Starch Synthesis

Michael J. Emes, Gregory J. MacNeill and *Ian J. Tetlow**

11.1 Introduction

Glucose (Glc, for list of abbreviations see 11.8) storage in the form of polyglucans appears to be a universal feature of carbon storage in the natural world, offering a compact, osmotically inert reserve of energy. In plants, as well as various freshwater and marine algae, starch is a major form of carbon storage, whilst in many prokaryotes and eukaryotes such as yeasts and mammals, the major storage polymer is glycogen. The ability to synthesize polyglucans as a carbon store confers important fitness advantages to organisms. In plants and green algae, the biological significance of starch accumulation is emphasized by studies of mutants lacking the ability to produce starch (Caspar et al. 1985, Dumez et al. 2006, Krishnan et al. 2015). Starchless plants may show reduced growth rates and fitness, reduced seed set, reduced pollen fertility, loss of gravitropic response, and increased susceptibility to oxidative stress (Brauner et al. 2014, Caspar and Pickard 1989, Kiss et al. 1997, Lee et al. 2018). Similarly, in bacteria and yeast the ability to store glycogen strongly impacts fitness and survival (McMeechan et al. 2005, Jones et al. 2008, Anderson and Tatchell 2001, Silljé et al. 1999).

At the most basic level, starch and glycogen are structurally similar, both being comprised of α-1,4-oxygen (O)-linked glucosyl chains and α-1,6-O-linked branch linkages, each susceptible to hydrolytic cleavage by amylases and debranching enzymes, respectively. Both glucan polymers contain variable amounts of phosphate covalently linked to Glc residues (Hizukuri et al. 1970, Nitschke et al. 2013), although the biological function and mechanism of phosphate deposition in the two polymers likely differs (see Chapters 3 and 7 of this volume). However, the higher structural levels of organization of the two polymers reveal major differences in the degree of complexity of their architecture and assembly. Glycogen is relatively highly and evenly branched, leading to a water-soluble particle whose growth is restricted to the nanometer level (Roach et al. 2016). Starch, on the other hand, is a highly organized polymer, made up of a minor, essentially linear, amylose component, and a major, highly branched component of amylopectin, the latter largely defining the structure of the granule (Jenkins et al. 1993). In amylopectin, clustered branch points lead to localized crystallinity of the polymer, resulting in a water-insoluble granule, whose size varies, but is in the order of micrometers, although there is no fundamental restriction in the growth of the granule (for details of the structure of starch, see Chapter 8 of this volume). The structure of the granule is such that very high densities of glycosyl residues (up to 1.5 g/cm^3) can be stored (Buléon et al. 1982).

University of Guelph, College of Biological Sciences, Department of Molecular and Cellular Biology, Guelph, ON, N1G 2W1, Canada, Emails: memes@uoguelph.ca, macneilg@uoguelph.ca
* Corresponding author: itetlow@uoguelph.ca

There are broadly two types of starch found in plants: transitory starches produced in chloroplasts of photosynthetic tissues, and storage starches found in non-green plastids of developing seeds and tubers. Transitory starch can make up as much as 30% of daytime fixed carbon (Stitt and Zeeman 2012) and as it is degraded at night, plays an essential role in nocturnal growth and survival, while storage starches provide a long term store of carbon for growth of the next generation (in seeds) or vegetative growth under favorable environmental conditions (in stolons, roots and tubers). The fine structures of transitory and storage starches differ, and is related to their biological functionality, and the biochemical pathways of the two starch types differ in terms of tissue specific isoforms of the various enzyme classes (Lloyd and Kossmann 2015, Tetlow and Emes 2017) which generally have overlapping functions (Nougué et al. 2014). The biochemistry of starch granule biosynthesis is complex, and begins with the formation of the soluble precursor, ADP-glucose (ADP-Glc) by the enzyme ADP-Glc pyrophosphorylase (AGPase). Conversion of this simple water-soluble precursor into the highly ordered starch granule structure requires the careful coordination of many enzymes: three major classes of enzymes, starch synthases (SS), starch branching enzymes (SBE), and starch debranching enzymes (DBE), with multiple isoforms of each class operating inside the plastid, as well as other enzyme activities. For recent reviews on the enzymes of the starch biosynthetic pathway, see reviews by Pfister and Zeeman (2016), Tetlow and Emes (2017), Goren et al. (2018), and in relation to modelling of the pathway see Pfister et al. (2020). By contrast, the biochemical pathway responsible for the synthesis of glycogen is relatively simple; in eukaryotes, two isoforms of glycogen synthase (GS) (in multicellular organisms, these GS isoforms often show organ-specific expression) form linear α-1,4-linked glucosyl chains, although regulation of GS is extremely complex and includes both covalent modification and allosteric factors (Wilson et al. 2010). A single isoform of branching enzyme introduces α-1,6-linked branch points in a relatively short distance (on average with a degree of polymerization (DP) of 13) (Meléndez-Hevia et al. 1993).

The complexity in the structural organization of the starch granule appears also to be reflected in the regulation of the enzymes involved in its biosynthesis. All of the major classes of enzymes of starch biosynthesis have been shown to form heteromeric protein complexes in a variety of plant species, contrasting with the situation in glycogen biosynthesis where, currently, there is much less evidence for interactions between the biosynthetic enzymes. However, recently a physical interaction between glycogenin and GS has been reported (Zeqiraj et al. 2014).

This chapter deals with the various heteromeric protein complexes involved in the biosynthesis of amylose and amylopectin, and includes details of recent developments in our understanding of protein complex assemblies involved in starch granule initiation. The chapter will focus on those protein-protein interactions amongst enzymes of the starch biosynthetic pathway that have been identified by a number of direct approaches, and discuss their possible role in the pathway. Details of experimental approaches and genetic evidence for the existence of various protein-protein interactions in starch synthesis has been covered in detail in recent reviews, and will only be briefly mentioned here (Tetlow et al. 2015, Crofts et al. 2017). Many of the components of protein complexes are themselves subject to regulation by protein phosphorylation, and this will also be discussed, as well as the roles of various non-catalytic proteins in regulation of these multi-protein assemblies.

11.2 Distinct Steps within the Starch Biosynthetic Paths and their Regulation

11.2.1 ADP–Glc Formation through AGPase and its Regulation

AGPase catalyses a reversible reaction which, under physiological conditions, synthesizes ADP-Glc, the substrate of SSs, and inorganic pyrophosphate (PPi) from ATP and glucose 1-phosphate (Glc1P) (Ghosh and Preiss 1966). Although not directly involved in starch granule formation, AGPase represents the first committed step of starch biosynthesis via provision of the immediate soluble substrate ADP-Glc, and plays a pivotal role in regulating carbon flux, from triose phosphates to transient starch in chloroplasts, and sucrose to starch in storage organs, and is therefore a highly regulated enzyme. AGPase is subject to allosteric regulation, and in different plant tissues its catalytic activity is variously sensitive to activation by 3-phosphoglyceric acid (3PGA) and inhibition by inorganic orthophosphate (Pi). In photosynthetic tissues, the chloroplast AGPase is highly sensitive to 3PGA/Pi ratios, making starch biosynthesis responsive to carbon availability (Ghosh and Preiss 1966, Kleczkowski 1999). In grasses (Poaceae), the bulk of the AGPase in the developing endosperm is extra-plastidial (Denyer et al. 1996a, Thorbjørnsen et al. 1996, Tetlow et al. 2003), and it is this cytosolic isoform which is responsible for the majority of the storage starch synthesized in these tissues (Lee et al. 2007, Huang et al. 2014). The extra-plastidial AGPase is largely insensitive to allosteric modulation (Kleczkowski et al. 1993, Tetlow et al. 2003). AGPase is a heterotetrameric enzyme consisting of two large subunits (AGP-L) and two small subunits (AGP-S) encoded by separate genes, and tissue specific expression of AGP-L and AGP-S can modulate AGPase activities in different tissues at different stages of plant development (for reviews of AGPase regulation see Geigenberger (2011) and Tuncel and Okita (2013). The catalytic activity of plastidial AGPase is also regulated by redox modulation. Reduction of an intermolecular disulfide bridge between AGP-S subunits at conserved cysteine residues activates the enzyme in relation to sugar signals (Fu et al. 1998, Hädrich et al. 2012, Tiessen et al. 2002). Reductive activation of plastidial AGPase requires interaction with thioredoxin, a 10 kDa protein which is a member of a large multi-gene family and responsible for redox modulation of many other plastidial metabolic enzymes (Buchanan and Balmer 2005). In chloroplasts, AGPase is inactivated by oxidative conditions during the night, and the inactive (oxidized) AGP-S dimer is activated during daylight hours by reduction through photoreduction and transfer of electrons from photosystem I to thioredoxin, via ferredoxin-thioredoxin reductase (Hendriks 2003, Ballicora et al. 2004)). In non-photosynthetic dicot tissues, such as potato (*Solanum tuberosum* L.) tuber, reductive activation may operate through a novel NADPH-thioredoxin reductase in the amyloplasts (Michalska et al. 2009). By contrast, the cytosolic AGPase in the endosperm of Poaceae is most likely insensitive to redox modulation because it lacks the conserved cysteine (Cys-82) residues of the plastidial enzyme required for reductive activation (Geigenberger 2011). Interestingly, not all chloroplastic AGPases appear to possess the cysteine residue required for redox modulation which, for example, is absent in the plastidial AGPase from the green alga *Chlamydomonas reinhardtii* (Zabawinski et al. 2001).

The regulation of AGPase activity is also linked to whole plant carbon availability through complex sugar signaling pathways involving sucrose, signal metabolite trehalose 6-phosphate (T6P), and the regulatory protein kinase, sucrose-nonfermenting1 (SNF1)-related protein kinase 1 (SnRK1). T6P is thought of as a surrogate for sucrose, and studies in Arabidopsis suggest a role for T6P in promoting SnRK1-mediated redox activation of chloroplast AGPase (Kolbe et al. 2005). Both SnRK1 and hexokinase can independently increase starch formation through redox activation of AGPase in response to elevated sucrose and Glc, respectively (Tiessen et al. 2003, McKibbin et al. 2006). The sugar signalling pathway linking SnRK1 and AGPase is also complicated by the fact that AGPase is also activated, indirectly, by T6P, which is an inhibitor of SnRK1 in both leaf and storage tissues, without affecting hexokinase (Debast et al. 2011, Martinez-Barajas et al. 2011, Gonzali et al. 2002). The SnRK1-mediated control of AGPase redox state has been studied in plants with plastidial AGPase (leaf chloroplasts and amyloplasts of dicots); it is not clear if this pathway operates in cereal endosperms. The specific interactions between AGPase and the various components of the sugar signaling cascades described above are unknown.

Proteomic studies in Arabidopsis (*Arabidopsis thaliana* L.) leaves as well as developing wheat and maize endosperms indicate that both AGP-L and AGP-S are phosphorylated *in vivo* (Heazlewood et al. 2008, Lohrig et al. 2009, Zhen et al. 2017, Walley et al. 2013) and suggest both plastidial and cytosolic forms are regulated by protein phosphorylation. The precise role of phosphorylation of AGPase in plants is not clear, and the protein kinase(s) responsible have not been identified. Maize endosperm amyloplast AGPase was recently shown to form a protein complex with SSIII and pyruvate phosphate dikinase (PPDK1) (Lappe et al. 2017). The precise function of this complex is not understood, and it is not known whether a putative 14-3-3 domain on PPDK1 (see section 11.6.1 below) is required for this interaction. However, a model was proposed whereby PPi generated by PPDK1 may be channelled towards AGPase for Glc1P formation (rather than that for ADP-Glc), generating ATP, and thus regulating starch formation in favour of energy generation for other biochemical pathways (Hennen-Bierwagen et al. 2009).

11.2.2 Starch Granule Initiation

We have only recently begun to identify and understand the interplay between the enzymatic components involved in the initiation of starch granules in plant cells, and this is in sharp contrast to our much firmer understanding of the subsequent steps of starch granule assembly. Unlike glycogen particle initiation, which mainly requires a dedicated priming protein termed glycogenin to interact with GS (Pitcher et al. 1987, Smythe and Cohen 1991, Zeqiraj et al. 2014), starch granule initiation appears to utilize a single SS isoform (SSIV) (Lu et al. 2017). However, it appears that other enzymes of starch metabolism (for example SSIII and SP) may also be involved in granule initiation, indicating a degree of flexibility in the pathway adopted, depending on tissue type or environmental conditions, which is perhaps not surprising given the importance of starch metabolism in plants. Recent evidence suggests protein-protein interactions play a critical role in starch granule initiation, and in some plants and tissues the sub-organellar localization of granule initiation and formation are controlled through interactions of α-glucan-acting enzymes with a variety of other plastidial proteins. For a recent summary of starch granule initiation, the reader is referred to a review by Malinova et al. (Malinova et al. 2018b).

Mutant studies in Arabidopsis established the importance of two phylogenetically related SS isoforms, SSIII and SSIV in starch granule initiation, and it appears there is a degree of overlap in their actions (Leterrier et al. 2008, Szydlowski et al. 2009). However, studies in Arabidopsis suggest that SSIV plays a specialized, and perhaps unique role in starch granule initiation (Roldán et al. 2007, Crumpton-Taylor et al. 2012). In a recent study by Seung et al. (Seung et al. 2017), it was shown that SSIV interacts with a class of non-catalytic proteins previously identified as potential regulatory scaffold proteins (Lohmeier-Vogel et al. 2008). The protein, PTST2, is a member of the PROTEIN TARGETING TO STARCH (PTST) group (see section 11.6 on non-catalytic proteins). PTSTs contain a coiled-coil domain, structures known to be involved in protein-protein interactions (Mason and Arndt 2004), and a family 48 carbohydrate binding (CBM48) domain at the C-terminus which facilitates binding to α-glucans (Janeček et al. 2011). Seung et al. (2017) also showed that another isoform of PTST, PTST3, is implicated in the pathway of granule initiation, although it appears to play a less prominent role than PTST2, and apparently does not bind to SSIV (Seung et al. 2017). PTST2 and PTST3 interact, but the functional significance of this interaction is not clear (Seung et al. 2018). SSIV forms homodimers whose interactions are mediated through a region of the protein between one of two coiled-coil domains, and the glycosyltransferase region containing a conserved tyrosine (putative phospho-tyrosine) residue (Raynaud et al. 2016). Dimerization is important for catalytic activity, and may also regulate its interactions with other proteins (Raynaud et al. 2016). SSIV truncation experiments suggest that the N-terminus may be important in determining granule morphology, whilst the C-terminal catalytic region governs granule number (Lu et al. 2017). In the model put forward by Seung et al. (Seung et al. 2017), PTST2 recognizes maltooligosaccharides (MOS) of a specific three-dimensional (helical) shape via its CBM48; such conformations arise spontaneously when linear MOS reach a critical DP (Gidley and Bulpin 1987, Puteaux et al. 2006). The PTST2/MOS complex then interacts with a SSIV dimer allowing SSIV to elongate the glucan, possibly releasing PTST2 and allowing it to bind to other helical MOS structures for further interaction with SSIV (see Fig. 11.1). In this scheme, aggregated polyglucan structures are formed that escape the degradative activities of α- and β-amylases and may be acted on by other SSs and SP, and branched by SBEs. SSIII has been shown to be an important enzyme in granule initiation, and it is possible that it plays a similar role to SSIV. However, there is no evidence for it interacting with any of the PTSTs despite the fact that SSIII also possesses coiled-coil domains and 14-3-3 binding sites (Hennen-Bierwagen et al. 2009, Wayllace et al. 2010). It is possible that SSIII does not require the aid of PTST2 to

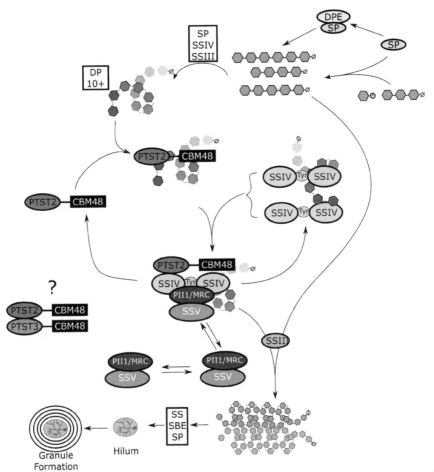

Figure 11.1. Protein-protein interactions involved in starch granule initiation. Formation of the hilum. An emerging model of the starch granule initiation pathway in plants. Unprimed α-glucan synthesis is possible via SSIII and SP, followed by further elongation and/or metabolism by SP or the SP/DPE protein complex, and SS isoforms. MOS of a critical shape and minimum DP (>10) are recognized by the CBM48 domain of PTST2, which then forms a protein complex with a SSIV homodimer to allow catalysis (in chloroplasts, this may also involve interaction of PTST2 with MFP1 for localization of granule initiation machinery to be targeted to the thylakoid membranes, but is not shown here). PTST2 may then dissociate from the SSIV dimer and collect more MOS for further SSIV/glucan interaction. SSIV may require interaction with the PII1 (MRC) protein for catalysis. In addition, PII1 (MRC) interacts with SSV, a non-catalytic protein structurally related to SSIV. SSV may serve a similar role to PTST in that its function appears to be that of guiding other proteins to MOS using its α-glucan binding domains. PTST2 and PTST3 interact, but the role of this protein complex is not known. It is hypothesized by Seung et al. (2017) that long, helical α-glucans produced by the SSIV/PTST2 complex form dense structures which are less susceptible to attack by hydrolytic amylases, and form the initiation point for the hilum and subsequent granule growth. It is possible that other enzymes can fulfill this function without PTST2; for example, SP or SSIII may produce similar glucan structures that can form a granule initiation point. Further growth of the initiation point forming the hilum of the starch granule at some stage will involve α-glucan branching and debranching, but the role of these enzymes at this early stage of granule formation is not clear.

recognize and bind to helical MOS structures, as it is known to readily elongate longer α-glucans as mentioned above (Delrue et al. 1992). SSIII may therefore produce α-glucan chains of sufficient length (DP ≥ 30) to form self-assembled, three-dimensional structures that evade hydrolytic enzymes, possibly explaining the overlapping functions of SSIII and SSIV in the granule initiation process. SSIV requires pre-existing α-glucan chains for elongation and is particularly active with maltotriose (Szydlowski et al. 2009), which begs the question of the source of glucan primer for starch granule initiation. In addition to the SSIII and SSIV mediated pathway of starch granule synthesis, there is emerging evidence that starch phosphorylase (SP) plays a role in granule initiation (Malinova et al.

2014, 2017), but it is not clear whether the two pathways operate independently or cooperatively under different circumstances or in different plant tissues. It has been suggested by Nakamura and coworkers that functional and physical interactions between SP (Pho1) and SBE isoforms modify MOS that can then be used for starch granule initiation (Nakamura et al. 2017). SSIII has been shown to be capable of unprimed α-glucan formation in the presence of ADP-Glc (Szydlowski et al. 2009). The plastidial SP (Pho1) in barley and rice endosperm is capable of producing and extending MOS in the absence of α-glucan primer using Glc1P (Cuesta-Seijo et al. 2017). In addition to its possible priming role, SP may work in conjunction with other enzymes such as SSs and SBEs to produce an initiation point for continued granule growth (the hilum) from the structures arising from the SSIV/PTST2 interaction (Nakamura et al. 2012) (see Fig. 11.1). SP has been shown to form a protein complex with disproportionating enzyme (DPE) (Hwang et al. 2016). However, it is possible that the actions of the SP/DPE complex are involved in provision of MOS for the process of granule initiation. DPE is involved in modification of MOS chain length, and may well play a role in both the granule initiation process as well as subsequent steps in starch synthesis. DPE has also been shown to form a functional homodimer in Arabidopsis (O'Neill et al. 2015). In addition to SP and DPE, other activities, for example amylases and the DBEs, ISA3 and limit dextrinase (pullulanase), probably maintain turnover of the MOS pool. Indeed, studies in Arabidopsis indicate that the MOS pool turns over rapidly, and it has been proposed to play an important role in regulating potential granule initiation sites for further elongation by SSs (Critchley et al. 2001, Chia et al. 2004).

Studies with chloroplasts of Arabidopsis indicate that SSIV interacts with fibrillin during the process of starch granule initiation (Gámez-Arjona et al. 2014a, Raynaud et al. 2016). Fibrillins are a class of hydrophobic proteins localized to plastoglobules, specialized sub-plastidial lipid bodies, and implicated in pigment accumulation, thylakoid function, and protection of the photosynthetic apparatus from photo-oxidative damage (Eugeni Piller et al. 2012, Bréhélin et al. 2007). SSIV interacts with fibrillin 1a (FBN1a) and fibrillin 1b (FBN1b), and the interaction has been suggested to control localization of SSIV during starch granule initiation (Gámez-Arjona et al. 2014a). FBN1a and FBN1b form a protein complex via a head-to tail-like interaction (Gámez-Arjona et al. 2014b), and it is thought that the FBN1a/FBN1b complex interacts with SSIV via its N-terminal coiled-coil domain. FBN1a and FBN1b are phosphorylated at their N-termini, possibly by a plastidial casein kinase IIα activity (Schönberg et al. 2014, Lohscheider et al. 2016), but it is not known whether phosphorylation controls the interaction with SSIV, or FBN1a/FBN1b protein complex formation. An attractive hypothesis for granule initiation being localized to plastoglobules is that such highly hydrophobic micro-environments may be conducive to the formation of the hilum and subsequent growth of the nascent starch granule, since these preliminary structures may be protected from the actions of hydrolytic enzymes such as α- and β-amylases. It is not clear whether the interactions between SSIV and fibrillins during starch granule initiation represent just one of a number of possible mechanisms of granule initiation. Recent proteomic analyses in Arabidopsis chloroplasts have failed to detect interactions between SSIV and fibrillin (Lundquist et al. 2017, Ytterberg et al. 2006). Nevertheless, there is a clear link between starch formation in chloroplasts and the metabolic activities of plastoglobules. Studies with the ABC-like kinase (ABC1K) family, a group of highly conserved atypical kinases localized to plastoglobules (Vidi et al. 2006), suggests that these proteins provide a link between photosynthetic activity and carbon metabolism, as *abc1k1* mutants lack starch (Martinis et al. 2014). Although plastoglobules are found in most plastid types (Bréhélin et al. 2007), fibrillins are associated with maintenance of the photosynthetic machinery in chloroplasts and stabilization of pigments and antioxidants in chromoplasts (Singh and McNellis 2011), and may therefore not be present in all plastid types. It is not yet known if similar sub-organellar localization of SSIV and starch granule initiation machinery is present in non-photosynthetic amyloplasts which produce storage starch. Recent work by Vandromme et al. (2018) further illustrates the complex pathway of starch granule initiation by the discovery of a novel chloroplast protein in Arabidopsis known as PROTEIN INVOLVED IN STARCH INITIATION1 (PII1, At4g32190) which forms a protein complex with SSIV (Vandromme et al. 2018). It is suggested that the interaction between

SSIV and PII1 is required for catalytic activity of SSIV, and mutant *pii1* lines lacking PII1 show a single, large starch granule in the chloroplast (as opposed to the usual 5–7), suggestive of interference with starch priming and the normal pathway of granule initiation (Vandromme et al. 2018). PII1 is also referred to as MYOSIN-RESEMBLING CHLOROPLAST PROTEIN (MRC), which was identified as one of two proteins interacting with PTST2 in a recent study by Seung et al. (Seung et al. 2018). This recent study by Seung et al. showed that PTST2 interacts with both MRC (PII1) and thylakoid-associated MAR-BINDING FILAMENT-LIKE PROTEIN (MFP1), both coiled-coil containing proteins. The *mrc/mfp1* double mutants produce a single large starch granule as opposed to multiple granules in WT, and the proposed function of MFP1 is to target the starch granule initiation machinery to the thylakoid membranes in chloroplasts (Seung et al. 2018). It is not clear if similar non-catalytic proteins are required for localizing starch granule formation within non-photosynthetic plastids such as amyloplasts, which lack thylakoids. Most recently, another component of the granule initiation machinery has been discovered in Arabidopsis. SSV, originally detected in potato starch granules (Helle et al. 2018) has been studied in Arabidopsis, and appears to regulate the number of starch granules formed inside plastids acting through a glucan binding domain and interacting with PII1/MRC. Interestingly, SSV lacks glycosyltransferase activity and is closely related to SSIV; *ssv* mutants show reduced numbers of starch granules per chloroplast but the starch granules produced by the mutant are larger than those of wild-type plants, indicating that other components of the granule initiation machinery can compensate for loss of SSV to some degree (Abt et al. 2020).

As the granule grows from the central hilum, microtubules have been observed to radiate from this central structure forming proteinaceous channels, terminating at the surface of some starch granules (Glaring et al. 2006, Fannon et al. 2004). Actin-like and tubulin-like (FtsZ) proteins are associated with these channels, including biosynthetic enzymes such as AGPase and SS isoforms (Benmoussa et al. 2010). The biological role of these protein-protein interactions is not understood. FtsZ is a highly conserved plastid division protein (Osteryoung and Pyke 2014) and it has been implicated in the formation of compound starch granules in species such as rice (*Oryza sativa* L.) and bamboo (Bambusoideae). During formation of compound starch granules, membranous cross walls are formed which contain FtsZ and other plastid division proteins such as Min and PDV2 (Yun and Kawagoe 2010). It has been suggested that SSIV localizes to these cross walls, possibly interacting with FtsZ during compound granule initiation (Toyosawa et al. 2016). A recent study by Matsushima et al. identified a gene, *SSG6*, by map-based cloning and microarray analysis involved in determining starch granule size in rice endosperm (Matsushima et al. 2016). It appears that SSG6 is an amyloplast membrane protein, sharing structural similarity to aminotransferases, and in some way interacts with the starch biosynthetic machinery to control starch granule size (Matsushima et al. 2016). Interactions between starch granule initiation proteins, plastid division machinery and membrane proteins within amyloplasts require further study to determine the generality of the mechanisms identified in chloroplasts.

11.2.3 Amylose Biosynthesis

Amylose is an essentially linear, α-1,4-linked, glucosyl polymer contributing a minor, but significant, component of starch granules, accounting for approximately 25% of the total granule mass in most storage starches (Bertoft 2017). Amylose is synthesized within an existing matrix of amylopectin, and thought to be more abundant near the surface of the starch granule (Pan and Jane 2000). Although amylose plays no direct role in contributing to amylopectin structure, its presence in the granule influences physicochemical properties and granule integrity (Swinkels 1985). Mutant studies have long established that a single glucosyltransferase activity, GBSSI, encoded by the *Waxy* locus, is responsible for the synthesis of amylose (Shure et al. 1983, Ball et al. 1998). A tissue-specific isoform of GBSS (GBSSII), encoded by a separate gene, is responsible for amylose synthesis in the chloroplasts of leaves and photosynthetic tissues, which accumulate transient starch (Vrinten and Nakamura 2000). GBSS isoforms are found almost exclusively in the starch granule matrix amongst

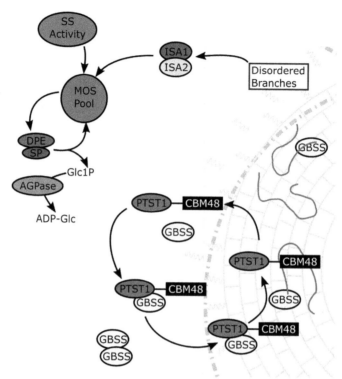

Figure 11.2. Amylose biosynthesis and MOS metabolism in plastids showing protein-protein interactions. Amylose biosynthesis is dependent on interactions between GBSS (GBSSI in non-photosynthetic plastids, and GBSSII in chloroplasts) and PTST1, the latter being required to target GBSS to the nascent starch granule via its CBM48 domain, which binds α-glucans. PTST is probably released from the granule to continue trafficking available stromal GBSS to the granule. GBSSI is known to form homo-dimers, but their role is not understood, and it is not known if dimers of GBSSI interact with PTST1. GBSS synthesizes linear amylose within the semi-crystalline amylopectin matrix (shown as brick-like structures in diagram), and is stimulated by MOS. In plastids, the MOS pool arises from the actions of soluble SSs, SP, and the actions of ISA during the trimming of α-1,6-branch linkages during amylopectin synthesis and is probably subject to rapid turnover by amylolytic activities. In most plants, the catalytically active ISA1 isoform forms a heteromeric complex with the catalytically inactive ISA2 isoform *in vivo*. A heteromeric protein complex comprising of SP and DPE is involved in modification of MOS glucan chain lengths, and may also provide Glc1P for the synthesis of ADP-Glc by AGPase.

the starch granule-associated proteins (Denyer et al. 1993, Mu-Forster et al. 1995) and their catalytic activity is stimulated by the presence of MOS (Denyer et al. 1996b). The various enzymes involved in MOS metabolism (e.g., SP, DPE and SS isoforms, in particular SSIV) appear to play a role in granule initiation (see above) as well as aspects of later granule formation. A study in rice endosperm showed that GBSSI is able to form oligomers, and that oligomerization is promoted by increasing the physiological concentrations of ADP-Glc in the plastid stroma, as well as by post-translational phosphorylation of GBSS (Liu et al. 2013). It is not clear how widespread oligomerization of GBSS is in other plant species and tissues. Studies in other cereal endosperms such as maize (*Zea mays* L.), wheat (*Triticum aestivum* L.) and barley (*Hordeum vulgare* L.) have shown that GBSSI in the granule is phosphorylated (Grimaud et al. 2008, Liu et al. 2009, Chen et al. 2016, Ahmed et al. 2015), but the precise role of protein phosphorylation in the regulation of GBSS is not understood. However, recent studies by Seung et al. (Seung et al. 2015) indicate that the synthesis of amylose requires a non-catalytic protein termed PROTEIN TARGETING TO STARCH (PTST). At least three isoforms of PTST exist in Arabidopsis and other plant species (see section 11.6 on non-catalytic proteins below), and PTST1, formerly identified as scaffolding protein At5g39790 in Arabidopsis (Lohmeier-Vogel et al. 2008), is the isoform which interacts with GBSSI (Seung et al. 2015). Interestingly, in addition to coiled-coil protein interaction domains, PTST1 also possesses a family-48 carbohydrate-binding domain (CBM48), and it is likely that the coiled-coil domains interact with GBSS, while the CBM48 domain targets it to the starch granule. Notably, GBSS does not have any apparent carbohydrate binding module (Nielsen et al. 2018) (see Fig. 11.2). Loss of PTST1 is associated with GBSSI-free starch granules and the lack of amylose (Seung et al. 2015). Conservation of PTST1 throughout the plant kingdoms argues for a critical role of this protein in amylose biosynthesis. Previous attempts at increasing amylose content by over-expression of GBSS have not met with success (Sestili et al. 2012), perhaps because delivery of GBSS to the granule may be limited by PTST1 availability.

11.2.4 Amylopectin Biosynthesis

The structure of the starch granule, from basic clusters to higher orders of organization, is determined by the amylopectin component of starch. The complex structure of amylopectin is thought to emanate from a central zone termed the hilum, a region of the nascent granule resulting from the initiation processes described above. Little is known of the structure of the hilum other than it appears to be comprised of dis-organized polyglucan (Ziegler et al. 2005), and that it appears to be essential for the subsequent build-up of more organized α-glucan polymers (Puteaux et al. 2006). The basic units of amylopectin are termed clusters, made up of short and intermediate sized glucan chains with clustered branch points showing a periodicity of ~ 9 nm (Bertoft 2013; see Chapter 8 of this volume). Linear α-glucan chains within clusters form left-handed helical arrangements via self-assembly mechanisms (Donald 2001), excluding water, and thus enabling granule crystallinity. Clusters are arranged and joined via long, inter-cluster spanning α-glucan chains (termed C-chains). Structural and genetic studies have determined the importance of the positioning and frequency of α-1,6-branch linkages in defining the final structure of starch granules (Sullivan and Perez 1999, Li and Gilbert 2016). Discrete units of these structures are called blocklets, which arrange into regions of tighter packing and crystallinity to form large scale, crystalline growth rings interspersed with regions of apparent disorder (amorphous regions) possibly arising through the interactions of amylopectin super-helices (Gallant et al. 1997). A large suite of enzymes, with multiple isoforms in each class, is responsible for amylopectin biosynthesis, and most plants possess tissue-specific isoforms involved in transient starch synthesis in the chloroplast, or storage starch synthesis in amyloplasts or other heterotrophic plastid types. The enzymes of amylopectin biosynthesis are described in the following section in relation to the regulation and coordination of their particular catalytic activities. Specific details of each of the enzymes in the pathway are beyond the scope of this chapter and are only briefly mentioned in relation to the discussion of protein-protein interactions. The following section summarizes our current knowledge of protein-protein interactions amongst the enzymes of amylopectin biosynthesis, and includes a model describing how the different protein complexes may build and help organize the basic structures. Various biochemical approaches, including gel permeation chromatography, immunoprecipitation, cross linking and blue-native polyacrylamide gel electrophoresis (BN-PAGE) have been employed to investigate protein interactions, in addition to genetic approaches, which will be discussed briefly in relation to mutations in the starch pathway. Much of what is known of the biochemistry of protein complexes involved in amylopectin biosynthesis has been derived from studies with developing cereal endosperm tissue. By contrast, relatively little is known of protein-protein interactions between these enzyme classes in chloroplasts or dicotyledonous storage tissues such as potato. What becomes apparent is that understanding the coordinated activities of the different classes of amylopectin synthesizing enzymes facilitates a higher resolution model of the synthesis of clusters.

11.2.5 Starch Granule–Associated Proteins

All of the enzymes of amylopectin biosynthesis are "soluble"; in other words, they are also found in the plastid stroma, unlike the amylose-synthesizing enzymes GBSSI and GBSSII, which are observed exclusively bound to the starch granule matrix as so-called granule-associated proteins. In addition, α-glucan, water dikinase (ATP: α-1,4-glucan, water phosphotransferase (GWD), and phospho-α-glucan, water dikinase (ATP: phospho-α-1,4-glucan, water phosphotransferase (PWD)) enzymes involved in modification of α-glucans by covalent phosphate addition are also found associated with starch granules (Ritte et al. 2002, Hejazi et al. 2008). However, specific enzymes of amylopectin biosynthesis are consistently found both as stromal and granule-associated proteins in normal (i.e., wild-type (WT)) starches. Granule-associated proteins are proteins which remain attached to the starch granules following extensive washing with detergents and acetone, as well as protease treatments (Denyer et al. 1995, 1993, Rahman et al. 1995, Mu-Forster et al. 1995, Borén et al. 2004) and are routinely found in higher plants and green algae. It is not clear whether granule-associated proteins play a functional role within the granule matrix, or are present as a result of entrapment during polymer biosynthesis. The mechanism by which enzymes become granule-associated is not clear, and may vary. The individual components of enzyme complexes thought to be involved in amylopectin cluster synthesis (see below) are consistently found in WT starch as granule-associated proteins, and there is compelling evidence that protein complex formation is in some way responsible for their association with the starch granule matrix. Interestingly, other enzymes and enzyme complexes are not found as granule-associated proteins, and these are either involved in MOS metabolism, or in the synthesis of long cluster-connecting chains and branched chains found in the less ordered, amorphous regions of the starch granule.

11.2.6 Specialized Functions of GBSS and SSIV

Of the five characterized SS isoforms in plants, two, GBSS and SSIV, appear to have uniquely specialized functions as described above. Although GBSS may provide extra-long chains of amylopectin (Delrue et al. 1992), it is essential for amylose synthesis. SSIV may produce MOS for other enzymatic activities during the later stages of granule development, but, based on the genetic evidence described above, it is clear that it is essential for granule initiation. The remaining three SS isoforms play an important role in amylopectin synthesis; SSI, SSII and SSIII are successively involved in the elongation and the production of increasingly longer α-1,4-linked glucan chains (Cuesta-Seijo et al. 2016) by adding a series of glucosyl residues onto the non-reducing end of a pre-existing chain. In all these reactions, ADP-Glc serves as glucosyl donor (Larson et al. 2016). As stated above, it appears that SSIII also plays a role in starch granule initiation. Studies in all of the major cereals (wheat, maize, rice and barley) indicate that SSI, SSII and SSIII are found in a variety of heteromeric complexes with other enzymes of starch synthesis in a bewildering number of permutations (see Crofts et al. (2017) for a recent review).

11.3 Interactions of Starch Synthase Isoforms and those of Branching Enzymes

11.3.1 Trimeric Protein Complexes

One of the better characterized protein-protein interactions in the starch pathway was studied by biochemical analyses of wheat and maize endosperm amyloplasts. The heteromeric protein complex includes SSI, SSIIa (as opposed to the chloroplast-specific SSIIb isoform in cereals), and SBEII class enzymes (Tetlow et al. 2008, Hennen-Bierwagen et al. 2008). All three proteins are partitioned between the plastid stroma and granules as they are found both as soluble and starch-associated proteins. Formation of the trimeric protein complex appears to be controlled by protein phosphorylation (see Section 11.5 below on protein phosphorylation). The trimeric protein complex consisting of SSI, SSIIa and SBEII has been detected in other cereals using GPC and immunoprecipitation techniques (Ahmed et al. 2015, Crofts et al. 2015). This protein complex is particularly interesting, as it contains all the enzyme activities needed for cluster formation and, therefore, its function may be more easily understood compared to many other protein complexes in the starch pathway. Both genetic evidence

(Nakamura et al. 2010, Abe et al. 2014) and biochemical evidence with immuno-purified protein complexes (Liu et al. 2012a) suggest that the branched products formed by SBEII are elongated by SSI, since SSI generally utilizes small α-glucan chains (of approximately DP 6-7) to produce intermediate-sized chains of DP 8-12, the substrates for SSIIa, which in turn synthesizes longer glucan chains of DP 12–30. The glucan products formed by the enzymes in this trimeric protein complex closely match the structures of amylopectin clusters observed *in vivo* (Bertoft 2013; see also Chapter 8 of this volume). The precise function of the trimeric complex, beyond that arising from the actions of the individual, monomeric enzymes found in this complex, is not clear. The trimeric protein complex may be a more efficient means of performing repeated actions of cluster biosynthesis. Interestingly, the activities of SBEs have been shown to stimulate SS activity (Hawker et al. 1974, Brust et al. 2014), and *in vitro* experiments with protein complexes comprised of SSs and SBEs indicated increased catalytic activities of SSs in the complex (Liu et al. 2012a). The trimeric SSI/SSIIa/SBEII protein complex may position itself on the α-glucan substrate such that particular branching configurations are promoted. For example, SBEs perform two types of branch-linkage formation following α-1,4-bond cleavage: the cleaved glucan can be transferred to an acceptor chain which is either part of the original α-glucan chain (intra-chain transfer), or part of an adjacent glucan chain (inter-chain transfer). The particular chain transfer mechanism adopted appears to depend on localized concentrations of linear α-glucan chains, with the inter-chain transfer mechanism predominating when chains are closely associated (Borovsky et al. 1979). The binding characteristics of both the SSs and SBEs may therefore play an important role in positioning the protein complex within the α-glucan substrate and thus influence the relative distances between neighboring branch linkages.

All of the trimeric protein complexes characterized from cereal endosperms contain SBEII. In cereals, the SBEII class is sub-divided into two isoforms, SBEIIa and SBEIIb isoforms, discrete gene products, which each display specific tissue expression patterns which varies between species (Gao et al. 1996, 1997, Rahman 2001). The SBEII form found in the trimeric complex appears to be the isoform which is most highly expressed in this tissue; SBEIIb is expressed specifically in the endosperm tissue, whilst SBEIIa appears to be expressed ubiquitously (Mizuno et al. 1993). The relative proportions of SBEIIa and SBEIIb is a function of the tissue and species in which they are expressed. In developing wheat endosperm, SBEIIb is expressed at much lower levels than the IIa isoform (Regina et al. 2006). However, in maize endosperm SBEIIb is the predominant form, and is expressed at around fifty times the level of SBEIIa (Blauth et al. 2002). In barley, both SBEII isoforms are similarly expressed in the endosperm (Regina et al. 2010). Such variations in expression of the SBEII isoforms is reflected in their relative presence in protein complexes in these tissues (Hennen-Bierwagen et al. 2008, Ahmed et al. 2015, Crofts et al. 2015, Tetlow et al. 2008) and may have implications in the fine structure of amylopectin in different species since the catalytic properties of the two SBEII isoforms are different; SBEIIb appears to have a more narrow range of glucans preferentially transferred (DP 6-7) as compared to SBEIIa (Nakamura et al. 2010).

11.3.2 Mutations of the Starch Synthase Isoforms II in Cereal Endosperm

Analysis of mutants of starch synthesis in the endosperms of various cereals has shed light on the functioning and importance of the trimeric protein complex in starch biosynthesis. Mutations of the gene encoding SSIIa, leading to a non-functional or absent protein in cereals such as wheat, barley (*sex6*) and rice, result in reduced starch contents, misshapen starch granules and altered physical properties of starch, and the granules are devoid of SSI and the SBEII class protein (Yamamori et al. 2000, Morell et al. 2003, Umemoto and Aoki 2005, Ahmed et al. 2015, Miura et al. 2018). Indeed, loss of SSII activity in cereals leads to one of the most striking phenotypes in comparison to that of other starch biosynthetic enzymes (Pfister and Zeeman 2016, Tetlow and Emes 2017). In maize mutation of SSIIa, the endosperm-specific SSII isoform is known as the *sugary 2* (*su2*) mutation, and leads to similar phenotypic effects as other cereals, with a marked decrease, if not loss, of SSI and SBEIIb from the starch granules (Grimaud et al. 2008). In addition to the important catalytic role of SSIIa, these studies suggest that the presence of a functional SSIIa is required for the association of SSI and class II SBEs with starch granules. A study by Liu et al. examined the effects of a loss-of-function mutation in maize SSIIa on the formation of the trimeric protein complex (Liu et al. 2012a). The *su2* mutant examined was a result of a single nucleotide polymorphism giving rise to substitution of arginine (R) for glycine (G) at residue 522 (R522G) and a catalytically inactive protein which was unable to bind to starch. Interestingly, the R522G SSIIa mutant protein was able to form a protein complex with SSI and SBEIIb in the amyloplast stroma as in the WT plants, and both SSI and SBEIIb were shown to be catalytically active, both in a monomeric state and in the complex. However, as a consequence of the inability of the R522G SSIIa mutant protein to bind to starch, the SSI/SSIIa/SBEIIb protein complex was not bound to starch granules, and analysis of starch granule-associated proteins from *su2* starch showed that SSI, SSIIa, and SBEIIb are undetectable in this fraction (Liu et al. 2012a). In addition, analysis of both SSI and SBEIIb from the *su2* mutant showed that their respective α-glucan-binding characteristics were unaltered from the corresponding WT enzymes. Similar results were obtained with respect to protein complex formation and alteration in starch granule proteins in a *sex6* barley mutant lacking SSIIa (Ahmed et al. 2015). These experiments provide compelling evidence that the presence of SSI, SSIIa and class II SBEs as starch granule-associated proteins results from their association in a trimeric complex, and that SSIIa plays a crucial role, both catalytically, and in transporting SSI and SBEII to starch granules. It is proposed that the SSI/SSIIa/SBEII protein complex is primarily responsible for the synthesis of branched glucan clusters in amylopectin (see Fig. 11.3).

11.3.3 Mutations of the Starch Branching Isoenzymes in Cereal Endosperm

In maize, the *amylose extender* (*ae*) mutation results in the loss of SBEIIb, the major SBEII isoform in the endosperm with a characteristic starch structure phenotype (low branching frequency and increased long α-glucan chains in the amylopectin clusters) (Banks et al. 1974). Studies with allelic variants of the *ae* mutation, in which SBEIIb protein is either absent (*ae1.1*) or where a catalytically inactive SBEIIb is expressed (*ae 1.2*), have provided insight into the biological function of protein complexes (Liu et al. 2009, 2012b). Analysis of protein complexes in the null *ae1.1* mutant revealed that in the absence of SBEIIb, a new set of protein-protein interactions replace the WT trimeric SSI/SSIIa/SBEIIb protein complex, and SSI and SSIIa could be detected in protein complexes with SBEI, SBEIIa, and starch phosphorylase (SP) (Liu et al. 2009). The precise stoichiometry of the various enzymes in the protein complexes detected in *ae1.1* amyloplasts is not clear, although gel permeation chromatography suggested that multiple permutations of different heteromeric protein complexes is more likely than a single "super-complex". Interestingly, SBEI, SBEIIa, and SP, normally confined to the stroma in WT amyloplasts, were found as starch granule-associated proteins in *ae1.1* endosperm (Grimaud et al. 2008, Liu et al. 2009), which is consistent with the hypothesis (above, and Fig. 11.3) that components of protein complexes involved in amylopectin cluster biosynthesis become entrapped as granule-associated proteins. The detection of SBEI, SBEIIa, and SP, in association with SSI and SSIIa in the stroma, and as starch granule-associated proteins, suggests that these proteins have in some way functionally substituted for the loss of SBEIIb. Similar complementation of protein complexes has been observed in a recent study of rice lines with various SS isoform mutations (Hayashi et al. 2018). Analysis of the starch in *ae1.1* supports the idea that novel protein complexes are responsible for cluster biosynthesis in this *ae* mutant, as branching frequency of the *ae* starch is reduced, and amylopectin chain length is increased (Klucinec and Thompson 2002). Studies with maize SBEI indicate that the enzyme has low branching frequency and a preference for branching long chains (Guan and Preiss 1993), indicating that the starch phenotype of the *ae1.1* is not merely a result of loss of SBEIIb, but a result of the activities of new protein complexes. SBEIIa is expressed at low levels in maize, so its contribution to the *ae1.1* starch structure may be expected to be small.

The role of SP in the protein complexes in the *ae1.1* mutant is not clear. In rice, mutation of SBEIIb leads to appearance of SBEI in starch granules (Abe et al. 2014), and studies in barley endosperm

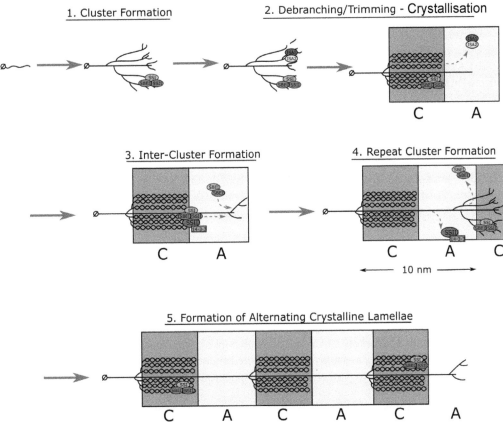

Figure 11.3. Model of the role of protein complexes in the formation of amylopectin clusters in storage starch of cereal endosperm. **1.** The combined catalytic activities of SBEII, SSI and SSIIa in the trimeric protein complex synthesizes amylopectin clusters. **2.** Self-association of short and intermediate-sized α-glucan chains within the clusters is promoted by trimming of inappropriately positioned branch points by the ISA1/ISA2 complex, which facilitates crystallinity within the polymer (in cereals, this is also a ISA1/ISA1 homomeric complex which is not shown here). The properties of the ISA1/ISA2 complex are such that it does not readily associate with the trimmed, ordered cluster structure, and are therefore never found as granule-associated proteins. **3.** The inter-cluster connecting chains and longer branch points forming the base of the clusters form the amorphous, less dense regions in amylopectin. SSIII may associate with SBEII and SSs already on the clusters, possibly through association with 14-3-3 proteins and/or its coiled-coil domains, and extend α-glucan chains to form long inter-cluster connecting chains. SBEI and SBEII form a complex which may form long branch points which begin a new cluster. **4.** As the new cluster is synthesized by the actions of the trimeric complex and actions of ISA1/ISA2, SSIII and the SBEI/SBEII complex dissociate from the amorphous region. **5.** Repeated actions of the various protein complexes form repeat semi-crystalline structures (growth rings). Such action may explain the retention of relatively large amounts of specific enzymes as granule-associated proteins, i.e. those proteins and protein complexes involved in the formation of crystalline regions in amylopectin, whilst enzymes involved in the formation of less densely packed α-glucan (called the amorphous region) are generally not found as granule-associated proteins. The actions of the various protein complexes do not explain the formation of larger-scale structures of the starch granule. The model outlined in Fig. 11.3 is compatible with both the cluster models of Hizukuri (1986) and the backbone model of Bertoft (2013).

using lines in which SBEIIa or SBEIIb were down-regulated using RNAi showed that SBEI and SP formed protein complexes with SSI and SSIIa, and SBEI and SP became granule associated (Ahmed et al. 2015). This suggests a general mechanism operating in cereals whereby loss of the major SBEII isoform causes complementation with SBEI and SP. Interestingly, Ahmed et al. (2015) separated the two starch granule populations present in the barley endosperm (as well as other Festucoid grasses; Tetlow and Emes 2017): the large A-type and the small B-type starch granules. In the SBEII down-regulated lines, the protein complexes containing SBEI and SP were found only in the large A-type starch granules, suggesting a distinct mechanism involving different protein complexes for the synthesis of the two granule types.

Interestingly, studies with a maize *ae* mutant expressing an inactive SBEIIb (*ae1.2*) showed that SBEI, but not SP, interacted with the trimeric complex containing the inactive SBEIIb (Liu et al. 2012b). As with the other mutations described above, the composition of the new protein complex was reflected in the starch granule proteome. Analysis of the different starches from *ae1.1* and *ae 1.2*, which are derived from near-isogenic maize lines, revealed marked differences in starch characteristics such as the apparent amylose content, starch content, granule size and a small difference in amylopectin chain length distribution (Liu et al. 2012b). The phenotypic differences between the two allelic variants of *ae* are significant, as both lines lack a functional SBEIIb, but structural differences in starch must be a result of the activities of different functional protein complexes involved in amylopectin cluster formation. It is not understood how alternate protein complexes arise from the loss of SBEIIb, or when an inactive SBEIIb is expressed. Interestingly, SBEI, SBEIIb, and SP are known to interact (Tetlow et al. 2004) and perhaps SBEIIb has preference for binding with SSI/SSIIa over SBEI.

11.3.4 Protein Complexes Containing Starch Synthase and Branching Enzyme Isoforms in Leaves and Non−Green Tissues from Dicotyledonous Plants

In contrast with cereal endosperm tissue, little is known of the protein-protein interactions between SS and SBE classes in leaf tissue, or indeed in non-photosynthetic tissues of dicotyledonous plants. Biochemical studies in Arabidopsis indicate protein-protein interactions between *At*SS2, and *At*SBE2.2, one of two SBEII class isozymes in Arabidopsis (Patterson et al. 2018). In common with the trimeric protein complex assembly in cereals noted above, this interaction is ATP-dependent, and probably involves one or both partner(s) being phosphorylated by plastidial protein kinase activities (see section 11.5 below on protein phosphorylation) (Patterson et al. 2018). The use of recombinant proteins of various Arabidopsis SS and SBE isoforms demonstrated some form of interaction affecting function (Brust et al. 2014), including between recombinant *At*SS1 and both recombinant *At*SBE isoforms. The difficulty in demonstrating these interactions directly may be due to a difference in the heteromeric complex in Arabidopsis *in planta* compared to cereal endosperm. However, as functional changes have been observed (Brust et al. 2014), the difficulty may be due to lower protein abundances in chloroplasts making detection of some proteins more difficult. Analysis of WT Arabidopsis starch granules indicates that the major granule-associated proteins, in addition to GBSSI, are SS2 and SBE2.2 (Feike et al. 2016), and consistent with the notion that enzyme complexes involved in amylopectin synthesis become entrapped within the starch granule. Other than the *At*SS2/*At*SBE2.2 protein complex identified by Patterson et al. (2018), there is no reported evidence for interactions between *At*SS1 or *At*SBE2.1, either with each other, or with other starch biosynthetic enzymes.

11.3.5 Protein Complexes Containing Starch Synthase III

SSIII is known to interact with a number of starch metabolic proteins, and may possess regulatory properties beyond its direct catalytic activity in amylopectin biosynthesis (Zhang et al. 2005, 2008). SSIII probably associates with other proteins through an unusually long N-terminus termed the homology domain (SSIIIHD) and in maize is found as a high order multimeric protein complex eluting from gel permeation columns at approximately 670 kDa (Hennen-Bierwagen et al. 2009, 2008). Biochemical experiments showed that the SSIIIHD could interact with SSIIa, SBEIIa and SBEIIb (Hennen-Bierwagen et al. 2009). The SSIIIHD contains starch-binding domains, two coiled-coil domains and a consensus motif for binding 14-3-3 proteins (see section 11.6.1 below) (Hennen-Bierwagen et al. 2008, 2009, Valdez et al. 2008, Wayllace et al. 2010). It is thought that these partial sequences allow for interaction with a number of enzymes, acting as both a metabolic enzyme, and a scaffold for large protein complexes. Genetic studies suggest functional interactions between

SSIII and ISA1 (Lin et al. 2012). One model suggests that SSIII may interact with SBEII isoforms located at amylopectin clusters, and components of the trimeric protein complex mentioned above. This would position SSIII for synthesis of the long, inter-cluster spanning chains in the amorphous region of the granule, and perhaps explaining why SSIII is not detected as a major starch granule-associated protein (Fig. 11.3). Functional interactions between SSIII and ISA isoforms have been demonstrated, which may explain how glucan elongation and amylopectin (glucan) trimming (see above) are coordinated (Lin et al. 2012).

11.3.6 Functional Interactions between SBE Isoforms

SBEs are components of many protein complexes identified in all cereal endosperms studied to date (Tetlow et al. 2015, Crofts et al. 2017). In addition to interactions with SS isoforms (above), there is evidence that SBEI and SBEIIb interact in wheat, maize and rice (Tetlow et al. 2004, Liu et al. 2012b, Crofts et al. 2015). In common with the trimeric SSI/SSIIa/SBEII protein complex described above, assembly of the SBEI/SBEIIb protein complex in wheat and maize endosperm seems to be dependent on protein phosphorylation (Tetlow et al. 2004, Liu et al. 2009) (see section 11.5 below). The precise role of this protein complex is not clear, but it may be involved in regulating the availability of SBEIIb either as a free monomer, dimer (Tetlow et al. 2008), or assembly of the trimeric protein complex described above. Regulation of SBE catalytic activity in relation to SS activity is probably important during starch granule formation, as several studies have indicated the importance of the correct/optimal ratio of SS: SBE activity to promote granule crystallinity (Boyer et al. 2016, Tanaka et al. 2004). Genetic studies with maize endosperm, where loss of SBEI from a SBEIIb-deficient background causes increased branching, are suggestive of a regulatory role for SBEI in influencing other SBEs (Yao et al. 2004). SBEI is expressed in the endosperm of all grasses and cereals, and its loss can lead to alterations in starch structure which can reduce plant fitness (Xia et al. 2011). The SBEI/SBEIIb protein complex appears to be confined to the plastid stroma, since in WT plants SBEI (and therefore SBEIIb complexed with SBEI) is not found in starch granules. It is possible that the long branched chains found at the base of amylopectin clusters, and found in the amorphous regions of amylopectin, are produced by SBEI and/or the SBEI/SBEIIb protein complex (see Fig. 11.3) and therefore do not become granule associated. In photosynthetic tissues, and plants such as Arabidopsis and canola (*Brassica napus* L.) which lack SBEI (and produce no storage starch), the role of SBEI, for example forming branches at the base of clusters, is served by SBEII isoforms, which may account, in part, for some of the structural differences between transient and storage starches.

11.4 Other Starch–Related Heteromeric Protein Complexes

11.4.1 Heteromeric ISA Complexes

Debranching enzymes (DBE) catalyze the hydrolysis of α-1, 6-branch linkages in starch and other branched polyglucans (direct debranching). During evolution of the starch biosynthetic pathway, the catabolic function of specific DBE isoforms has been adapted, such that they play an essential role in the synthesis of amylopectin and contribute to the clustered branching (Cenci et al. 2013, 2014). Two groups of DBEs are found in plants: isoamylase-type (ISA), which debranches α-1, 6-linkages in amylopectin and other polyglucans, and one pullulanase-type (or limit-dextrinase, LDA), which, in addition to debranching polyglucans, is distinct from ISA in that it can debranch the fungal polymer pullulan (Møller et al. 2016). Three isoforms of ISA exist in plants, two of which (ISA1 and ISA2) are involved in amylopectin biosynthesis and the third (ISA3) appears to work together with LDA during starch degradation and turnover by hydrolyzing and removing α-1, 6-linkages in MOS that have been released by the preceding action of amylases (Delatte et al. 2006, Yun et al. 2011, Wattebled et al. 2005, Dinges et al. 2003). In rice endosperm, a stable interaction between LDA and SBEI was identified, although the function of this protein complex is not clear, and it has not yet been identified in other species (Crofts et al. 2015). The likely role of ISA1 and ISA2 in starch synthesis is the removal of inappropriately positioned α-1, 6-linkages, thus promoting formation of the semi-crystalline structures found in the amylopectin clusters. When these key ISA isoforms are absent or inactive, a water-soluble α-polyglucan, termed phytoglycogen, is formed in addition to or replacing starch. In phytoglycogen, branch linkages are more evenly distributed as compared to amylopectin (see Fig. 11.3).

Evidence for the involvement of ISA1 and ISA2 in starch biosynthesis comes from a wealth of genetic and biochemical data (Ball et al. 1996), and the reader is referred to recent reviews (Pfister and Zeeman 2016, Tetlow and Emes 2017). In addition, structural analysis of the ISA1 protein supports a role for ISA1 in "glucan trimming" (Sim et al. 2014). In plants and green algae, ISA1 and ISA2 form a functional heteromeric protein complex: ISA1 is catalytically active, whereas ISA2 lacks key amino acid residues essential for debranching and is catalytically inactive. There is a suggestion that the ISA1/ISA2 protein complex in rice may form a tetrameric assembly (Sundberg et al. 2013, Crofts et al. 2015) similar to that found in algae and some archaea (Woo et al. 2008, Sim et al. 2014). In dicotyledonous species, both isoforms of the heteromeric ISA1/ISA2 complex are required for catalytic action (Facon et al. 2013, Sundberg et al. 2013, Delatte et al. 2005, Wattebled et al. 2005). The presence of ISA2 appears to stabilize the protein complex and to modify the catalytic properties of ISA1. Thus, in the heteromeric complex ISA2 appears to have a regulatory role (Sundberg et al. 2013, Hussain et al. 2003). In storage tissues of cereals, and likely other grass species, a homomeric ISA1 complex is found in addition to the heteromeric ISA1/ISA2 complex (Lin et al. 2013, Kubo et al. 2010, Utsumi et al. 2011). Genetic studies with maize and rice mutants demonstrated that the ISA1 homomer is capable of supporting normal rates of starch synthesis (Utsumi et al. 2011, Kubo et al. 2010, Lin et al. 2013). Despite the presence of a functional ISA1 homomer in cereal endosperms alongside the heteromeric ISA1/ISA2 complex, the heteromeric ISA1/ISA2 complex may provide plants with survival advantages, as studies with the rice endosperm ISA1/ISA2 heteromer indicate increased thermostability and affinity for phytoglycogen compared with the ISA1 homomer (Utsumi and Nakamura 2006). In addition, the ISA1/ISA2 may provide specific kinetic characteristics allowing for specialized (storage starch-specific) amylopectin structures to be synthesized. The monocot-specific heteromeric ISA1/ISA2 protein complex appears to be yet another example in the starch biosynthetic pathway of a catalytically inactive protein (ISA2) assisting in targeting a catalytically active enzyme (ISA1) to α-glucans. Studies in rice have shown that ISA1 interacts with a CBM48-containing protein encoded by the *FLOURY ENDOSPERM 6* (*FLO6*) gene (Peng et al. 2014). A recent study in barley shows that mutation of the FLO6 gene is the cause of the well-known "fra" mutation leading to fractured starch granules and an opaque phenotype (Saito et al. 2018). Recent evidence from studies in Arabidopsis suggests that light and sugar-responsive transcription factors, FHY3 and FAR1, positively regulate leaf starch synthesis via transcriptional regulation of ISA2 (Ma et al. 2017). Control over the rate of starch synthesis in leaves may therefore be exercised through transcriptional regulation of ISA2 by regulation of the availability of ISA2 protein for inclusion in the ISA1/ISA2 protein complex.

In accordance with the "glucan trimming" model, ISA1 and ISA2 operate at the periphery of the growing starch granule, promoting amylopectin clusters to form water-insoluble structures. Enzymes, and enzyme complexes operating in the amorphous region of amylopectin, are not retained in the starch granule as granule-associated proteins. Neither of the ISA isoforms involved in amylopectin biosynthesis has been shown to interact with other enzymes involved in starch metabolism (Crofts et al. 2017, Tetlow et al. 2008), although clearly, there needs to be careful coordination of debranching by ISAs with the branching and elongation activities of the trimeric protein complex and other enzymes during polyglucan synthesis. It is perhaps worth considering that many of the experimental methods and techniques used to probe protein-protein interactions in this pathway and others (Crofts et al. 2017, Kudla and Bock 2016) tend to select for strong, stable interactions. There may well be many other more "loose" and transient protein-protein interactions crucial to starch biosynthesis.

11.4.2 Functional Interactions of the Plastidial Starch Phosphorylase Isozyme

In addition to possible roles in starch granule initiation and MOS metabolism, SP may also function in amylopectin biosynthesis throughout the period of starch granule construction, as the enzyme has been detected in protein complexes with a number of starch biosynthetic enzymes and its expression in developing endosperm tissue of a number of cereals closely correlates with periods of active starch accumulation (Yu et al. 2001, Higgins et al. 2013). SP catalyzes a reversible reaction in either a biosynthetic direction, transferring glucosyl units from Glc1P to the non-reducing end of α-1,4-linked glucan chains producing Pi, or a degradative (phosphorolytic) reaction, whereby Pi is utilized to produce Glc1P from the removal of a terminal glucosyl residue from an α-1,4-linked glucan chain. The plastidial isoform (Pho1) of SP involved in starch metabolism is characterized by high affinity for MOS and amylopectin, and inhibited by ADP-Glc (Dauvillée et al. 2006, Burr et al. 1975). Pho1 from a number of cereal endosperms has been shown to form protein complexes with SBE isoforms, supporting a biosynthetic role for SP in amylopectin synthesis (Tetlow et al. 2004, Crofts et al. 2015, Subasinghe et al. 2014).

Pho1 is characterized by high affinity for MOS and amylopectin, and inhibited by ADP-Glc (Dauvillée et al. 2006, Burr et al. 1975). Pho1 from a number of cereal endosperms has been shown to form protein complexes with SBE isoforms, supporting a biosynthetic role for SP in amylopectin synthesis (Tetlow et al. 2004, Crofts et al. 2015, Subasinghe et al. 2014). Of the SSs and SBEs found in heteromeric protein complexes, SP is not normally found to be associated with the starch granule, except in mutants lacking the major SBEII isoform, for example *ae* maize (see above). In WT plants, SP-containing heteromeric protein complexes, such as the SBEIIa/SP complex of rice endosperm, are confined to the plastid stroma (Crofts et al. 2015). Phosphorylases from a number of plant tissues occur as homotetrameric or homodimeric assemblies (Cuesta-Seijo et al. 2017, Buchbinder et al. 2001). Elution profiles of maize Pho1 from gel permeation chromatography is consistent with a homotetrameric form of SP, although evidence suggests SP may exist as monomers or lower complexity multimers when associating with other enzymes (Liu et al. 2009, Subasinghe et al. 2014).

11.5 Function of Protein Phosphorylation in Starch−Related Protein Complexes

Many enzymes of the starch biosynthetic pathway have been shown to be phosphorylated, through direct biochemical analysis or via large scale proteomic screens (Tetlow et al. 2004, Grimaud et al. 2008, Heazlewood et al. 2008, Walley et al. 2013, van Wijk et al. 2014), including AGPase, GBSSI, SSIIa, SBEIIa, SBEIIb, SBEI, ISA2 and SP (Pho1), although in many cases the precise role of phosphorylation is not clear. However, the assembly of some of the protein complexes described above appears to be regulated by protein phosphorylation. *In vitro* studies with isolated amyloplasts from cereal endosperms have shown that formation of a number of heteromeric protein complexes is enhanced by and, in some cases, depends on ATP. By contrast, treatment with alkaline phosphatase, a non-specific dephosphorylating enzyme, causes disassembly of some protein complexes (Tetlow et al. 2004, 2008, Hennen-Bierwagen et al. 2008, Liu et al. 2009, Ahmed et al. 2015).

In terms of understanding its regulation by protein phosphorylation, one of the best characterized protein complexes is the cereal endosperm trimeric SSI/SSIIa/SBEII protein complex (Tetlow et al. 2008, Liu et al. 2009). Studies in wheat and maize endosperm amyloplasts indicate that SBEII and SSIIa are phosphorylated in the trimeric complex, and that phosphorylated forms of SBEII and SSIIa are found as starch granule-associated proteins. This is consistent with the notion that protein phosphorylation in some way controls complex assembly, and that the components of the complex are entrapped in the granule in a phosphorylated state (Tetlow et al. 2004, 2008, Grimaud et al. 2008, Liu et al. 2009). Monomeric and dimeric forms of each of the phosphoproteins in the trimeric complex exist, and are also phosphorylated. Thus, the identification of the function of each phosphorylation site is complex. Protein phosphorylation may affect the catalytic activities of individual (monomeric) proteins as well as when they are components of protein complexes. In wheat, endosperm treatment with alkaline phosphatase reduced both of the SBEII activities, but had no effect on SBEI activity (Tetlow et al. 2004). Limited information is available as to how the ATP-dependent protein complex formation affects the individual enzyme activities in the complex. Data from wheat endosperm indicates that the trimeric SSI/SSIIa/SBEII protein complex has a higher affinity for α-glucan than each of the individual monomeric proteins (Tetlow et al. 2008). The major maize SBEII isoform, SBEIIb, is phosphorylated at three serine (Ser) residues (Makhmoudova et al. 2014). In this study with isolated amyloplasts and recombinant SBEIIb, two plastidial Ca^{2+}-dependent protein kinase (CDPK) activities were identified as responsible for phosphorylation of the three sites on SBEIIb. Ser^{286} is conserved within the cereal SBEII class, Ser^{297} is highly conserved in all classes of SBE enzymes, and Ser^{649} is present in some, but not all SBEIIb homologues (Makhmoudova et al. 2014). The precise function of each of the phosphorylation sites is not yet clear, but modeling studies suggest that phosphorylation could stabilize enzyme structure through salt bridges. Of particular interest was the potential formation of a stabilizing salt bridge between phosphorylated Ser^{297} and arginine (R665) which is far removed from Ser^{297} in terms of primary sequence, but forms part of a disorganized polypeptide loop. This is part of a putative protein-protein interaction domain containing the sequence $KCRR^{665}R$, the two being close in the three dimensional model. It was argued by Makhmoudova et al. that phosphorylation at Ser^{297} and subsequent formation of a salt bridge with R665 of the KCRRR putative protein-protein interaction domain may facilitate interactions with other proteins (e.g., SSIIa) (Makhmoudova et al. 2014). The highly conserved Ser^{297} phosphorylation site also happens to be located at the opposite end of the central catalytic β-barrel from Ser^{286}, and phosphorylation of these sites may also regulate interaction of SBEII enzymes with their α-glucan substrates. The Ser^{649} site is interesting, in that it appears to be confined to some SBEIIb homologues and found in a twelve amino acid region absent in the SBEI class, suggesting a highly specialized function. Amongst dicotyledonous species, Ser^{649} is substituted by aspartic acid (D), a phospho-mimic. In the *ae1.2* mutation noted in the above section, the expressed mutant SBEIIb lacks a region of the enzyme important for catalysis which includes both Ser^{286} and Ser^{297} phosphorylation sites, and becomes hyperphosphorylated, presumably at Ser^{649}, the only other identified phosphorylation site (Liu et al. 2012b). Whether phosphorylation of one of the three phosphorylation sites on SBEIIb regulates phosphorylation of the others remains to be established, but analysis of the properties of the two plastidial CDPK activities from maize endosperm amyloplasts showed that both protein kinases could phosphorylate Ser^{649}, but discriminated between Ser^{286} and Ser^{297}, suggesting distinctive regulatory roles for the different phosphorylation sites (Makhmoudova et al. 2014).

There is direct evidence for phosphorylation of SS isoforms, as mentioned above, and early studies by Drier et al. (1992) showed that the catalytic activity of SS could be stimulated by ATP via a calcium-dependent protein kinase activity, although it was not determined which isoforms of SS were affected by phosphorylation (Drier et al. 1992). It will be interesting to determine whether the CDPK involved in phosphorylation of SBEIIb is also involved in phosphorylation of SS isoforms. A recent study in Arabidopsis found that SS2 is phosphorylated in chloroplasts at two sites (Ser^{63} and Ser^{65}), in a structurally disordered region of the N-terminus of the protein (Patterson et al. 2018). Analysis of the phosphorylation sites indicated motifs likely recognized by a casein kinase II (CKII)-like protein kinase, and biochemical analysis with recombinant CKII indeed showed that it could phosphorylate SS2 at these sites. Patterson et al. showed that SS2 forms homo-dimers, which have low catalytic activity relative to the monomeric protein, but that phosphorylation at either Ser^{63} or Ser^{65} has no effect on protein dimerization, or the ability of the protein to form a protein complex with SBE2.2 (Patterson et al. 2018).

In Pho1, the plastidial isoform of SP involved in starch metabolism, six phosphorylation sites have been identified, five of which are clustered together in a highly acidic region which forms part of a long polypeptide known as L78 (Walley et al. 2013). L78 is specific for Pho1 in storage tissues and can be proteolytically cleaved by a plastidial protease to increase phosphorolytic activity of the enzyme (Chen et al. 2002). Interestingly, five of the identified phosphorylation sites within L78 appear to be specific to cereals and not found in dicotyledonous plants such as sweet potato where other phosphorylation sites were identified (Young et al. 2006), whereas the sixth site is conserved amongst all species homologues of Pho1 (Tetlow et al. 2015). It is not yet known if any of the phosphorylation sites identified in cereal Pho1 are responsible for any of the protein-protein interactions identified in WT or mutant plants.

Identification of the plastidial protein kinases and phosphatases involved in regulating protein-protein interactions and the activities of starch metabolic enzymes in plastids is limited and information, other than that concerning CDPK regulation of SBEIIb mentioned above, is largely indirect. Bioinformatics-based surveys of Arabidopsis protein kinases and phosphatases suggest that the number of protein kinases in chloroplasts might be as few as 15 (Bayer et al. 2012). A recent review by White-Gloria et al. has identified potential protein kinases and protein phosphatases involved in starch metabolism based on phospho-proteomics data (White-Gloria et al. 2018). Much work is still required to identify the protein kinases and protein phosphatases involved in starch metabolism in plastids. Indeed, few plastidic protein kinases and their substrates have been characterized with any certainty for any aspect of plastid metabolism (Bayer et al. 2012).

11.6 Non-Catalytic Regulatory Proteins in Starch Biosynthesis

The roles of some non-catalytic proteins in starch metabolism and in relation to known protein-protein interactions involved in starch synthesis will be discussed here.

11.6.1 14-3-3 Proteins

14-3-3 proteins are a highly conserved group of non-catalytic eukaryotic proteins of approximately 30 kDa, playing a wide variety of roles in cellular regulation (Aitken et al. 1992). Plants possess an unusually large number of 14-3-3 isoforms (for example, Arabidopsis possesses 13 isoforms), and those with known function operate in a diverse number of biological functions, including flowering, hormone signalling and carbon metabolism (Kaneko-Suzuki et al. 2018, Camoni et al. 2018, Huber et al. 2002). The various isoforms of 14-3-3 proteins are broadly split into two groups; epsilon (ε), and non-ε types, all of which exist as homo- and hetero-dimers, allowing a broad specificity of potential interactions (Chevalier et al. 2009). Different isoforms of 14-3-3 are designated by letters of the Greek alphabet. 14-3-3 proteins act through their interactions with phosphorylated target (client) proteins via recognition of specific phosphorylation site motifs. Three consensus motifs for the recognition of client proteins by 14-3-3 isoforms have been established, and these are termed mode I (R/K)XX(pS/pT)XP, mode II (R/K)XXX(pS/pT)XP, and mode III located at the C-terminus, (pS/pT)X1-2-COOH, where X represents any amino acid and pS and pT represent phosphoserine and phosphothreonine, respectively (Yaffe et al. 1997, Muslin et al. 1996, Coblitz et al. 2006).

A number of studies have implicated 14-3-3 proteins in the starch biosynthetic pathway. Indeed, a significant number of enzymes in the starch biosynthetic pathway in maize possess mode I or mode II consensus motifs for binding 14-3-3 proteins (see Table 11.1). Maize SSIII possesses at least two mode I consensus 14-3-3 binding motifs KLFTYP (amino acids 780–785) and RYGTIP (amino acids 1578–1583) (Wayllace et al. 2010, Valdez et al. 2008). Antisense plants lacking expression of the GF14ε isoform showed a 2-4-fold increase in leaf starch, whilst the rate of nocturnal starch degradation was unaffected (Sehnke et al. 2001). Analysis of the starch in the antisense plants suggested that plants lacking GF14ε produced a starch with a higher branching frequency. Biochemical analyses

Table 11.1. Putative 14-3-3 Binding Motifs on Starch Biosynthetic Enzymes from Maize.

Zea mays Starch Biosynthetic Enzyme	Motif Sequence(s)	Mode	Accession No.
SSI	-	-	AAB99957.2
SSIIa	RVGSSP (91-96)	I	AFW76788.1
SSIIb	KTATSAP (105-111)	II	XP_008680930.1
	KAEPSAP (141-147)	II	
SSIII	KLFTYP (780-785)	I	AAC14014.1
	RYGTIP (1568-1573)	I	
	KSVVSVP (489-495)	II	
	SKL-COOH (1672-1674)	III	
SSIV	RYGSVP (812-817)	I	AFW82566.1
GBSSI	KVVGTP* (544-549)	II*	ABW95927.3
SBEI	KVLSPP (750-755)	I	AAC36471.1
	TK-COOH (822-823)	III	
SBEIIa	KTSSSP (63-68)	I	AAB67316.1
SBEIIb	-	-	AAC33764.1
SP	RNDVSYP (252-258)	II	NP_001296783.1
Iso-I	-	-	ACG43008.1
Iso-II	RGCPSP* (145-150)	II*	AAO17048.3
Iso-III	-		NP_001105198.2
PPDK1**	RAETSP (502-507)	I	NP_001105738.2
	KQPLSPP (583-589)	II	
	RLGISYP (745-751)	II	
GWD1	RPAASSP (13-19)	II	NP_001348353.1
	RAIHSEP (476-482)	II	

The motif sequence for each enzyme is given, followed by the amino acid number/position of the motif from the full length protein sequence in parentheses. The type of consensus motif (mode) is given, based on (R/K)XX(pS/pT)XP for mode I, (R/K)XXX(pS/pT)XP for mode II, and (pS/pT)X1-2-COOH for mode III located at the C-terminus, where X represents any amino acid and pS/pT represent phosphoserine or phosphothreonine. *Variant of mode II site ** PPDK plays a variety of roles in metabolism, but has been implicated in the regulation of starch synthesis in developing maize endosperm through protein-protein interactions with SSIII.

showed that 14-3-3 protein was present in the starch fraction in WT Arabidopsis plants as well as maize endosperm starch, and that a recombinant maize endosperm 14-3-3 (*Zm*GF14-12) physically interacted with maize SSIII (Sehnke et al. 2001). Other studies have detected 14-3-3 proteins in the starch granules of immature maize pollen (Datta et al. 2002). Given the potential interaction between SSIII and 14-3-3 proteins, and the results of the antisense GF14ε experiments in Arabidopsis noted above, it is interesting to speculate on the results obtained by Zhang et al. in which loss of SSIII in Arabidopsis caused an increase in leaf starch content and enhanced catalytic activities of other SS isoforms (Zhang et al. 2005). SSIII may exert control over the rate of starch synthesis in the chloroplast by regulation of branching and elongation activities through association with 14-3-3 proteins. Using a recombinant 14-3-3 from developing barley endosperm, Alexander and Morris showed that a number of enzymes involved in starch metabolism physically interacted with the 14-3-3 protein, including GBSSI, SSI, SSIIa, and SBEIIa, as well as α- and β-amylases (Alexander and Morris 2006). A later study in Arabidopsis used tandem affinity chromatography and a recombinant 14-3-3ϕ isoform, and noted that the 14-3-3 interacts with GBSSI and α-glucan water dikinase, amongst other proteins (Chang et al. 2009). These interesting *in vitro* experiments provide evidence of potential 14-3-3 protein interactions with enzymes of starch biosynthesis, although the precise effects of such interactions on specific enzyme function remains unclear. A recent study by Dou et al. (2014) analyzed the biological function of two out of the twelve isoforms of 14-3-3 protein in maize, the *Zm*GF-14-6 and *Zm*GF14-4 isoforms chosen for their high expression in the endosperm during early development and grain filling (6–37 days after pollination) (Dou et al. 2014). Recombinant forms of *Zm*GF-14-6 and *Zm*GF14-4 were used in affinity chromatography experiments with maize endosperm whole cell extracts, and these *in vitro* experiments indicated interaction of both endosperm-specific 14-3-3 isoforms with GBSSII, SBEIIb and ISA (Dou et al. 2014).

The above studies provide evidence supporting a role for 14-3-3 proteins in plastidial starch metabolism. However, evidence for direct involvement of 14-3-3 isoforms in regulating the starch biochemical pathway remains elusive. Finally, direct regulation of starch metabolism by 14-3-3 isoforms necessitates their sub-cellular localization within the plastid. Amino acid sequence analysis suggests no obvious canonical plastidial transit peptide motifs, and direct visual evidence of 14-3-3 proteins in plastids is also lacking.

11.6.2 PTST Family

The PTST family is a group of non-catalytic plastidial proteins with a scaffolding function, interacting with client proteins through their coiled-coil domains (Lohmeier-Vogel et al. 2008). The coiled-coil domains consist of repeated patterns of hydrophobic and hydrophilic residues which facilitates formation of an alpha-helical supercoil between the PTST and its partner protein (Lupas and Gruber 2005). Three isoforms of PTST are known in Arabidopsis, PTST1 (At5g39790), PTST2 (At1g27070), and PTST3 (At5g03420), and have been discussed above in relation to amylose biosynthesis (PTST1) and starch granule initiation (PTST2 and 3). However, other regulatory functions for the PTST family cannot be ruled out. For example, PTST1 exhibits strong co-expression with many other starch metabolic enzymes (Lohmeier-Vogel et al. 2008), and may therefore regulate other enzymes in the pathway of starch metabolism besides GBSS. A recent study in barley shows that PTST1 plays a crucial role in grain filling in the developing endosperm (Zhong et al. 2019). PTST3, unlike PTST2 which interacts with SSIV, did not appear to interact with SSIV in the study described by Seung et al. (2017), although immunoprecipitation experiments showed interactions between PTST2 and PTST3. The function of this interaction is not clear. PTST2 interacts with two non-catalytic coiled-coil containing proteins, PII1 (MRC) and MFP1, as mentioned in the section above, and these interactions are likely important in targeting PTST2 and starch initiation machinery to the chloroplast thylakoid membranes during transient starch synthesis (Seung et al. 2018). Further studies on this class of regulatory protein are needed.

11.6.3 ESV and LESV Proteins

In a series of elegant experiments to determine the factors responsible for adjusting the rate of nocturnal starch degradation in Arabidopsis, Feike et al. (2016) discovered two related non-catalytic starch binding proteins, early starvation 1 (ESV1) and like early starvation 1 (LESV1) (Feike et al. 2016). ESV1 and LESV1 are chloroplast-localized proteins found in the plastid stroma and bound to starch granules as granule-associated proteins via tryptophan-rich carbohydrate binding regions, and are found in all plants and green algae. It is thought that ESV1 and LESV1 regulate starch degradation by modifying glucans at the granule surface. A recent paper by Malinova et al. has shown that ESV1 binds highly ordered α-glucans and affects glucan phosphorylation by reducing phosphorylation by GWD, and increasing the action of PWD at the granule surface (Malinova et al. 2018a). No known interactions between these regulatory proteins are known to date, but it seems improbable that such important regulators of starch metabolism are acting in isolation.

11.7 Conclusions and Future Directions

Protein-protein interactions play an integral role in the regulation of all aspects of starch biosynthesis in plastids of higher plants, from precursor (ADP-Glc) formation, to granule initiation, amylose synthesis, and assembly of clusters, the basic building unit of amylopectin. Many combinatorial interactions between the core enzymes of the starch biosynthetic pathway have been discovered in non-photosynthetic plastids of cereal endosperms, which are responsible for storage starch synthesis, but it is becoming apparent that similar protein-protein interaction networks also operate in leaf chloroplasts responsible for transitory starch synthesis. Recent studies in Arabidopsis have discovered groups of non-catalytic interacting proteins responsible for the regulation of amylose synthesis, and the complex process of starch granule initiation, adding a fascinating new dimension to our understanding of the pathway as a whole. Non catalytic proteins such as the PTSTs, and other components of plastid machinery such as the organelle division machinery and membrane proteins, appear to be involved in the correct localization of granule initiation sites within thylakoid membranes through the use of coiled-coil domains and α-glucan binding sites. It will be interesting to see if comparable regulatory systems are in operation in heterotrophic plastids responsible for storage starch formation. Protein phosphorylation regulates the assembly of some, but not all, protein complexes involved in the starch pathway, and may also regulate catalytic activity of individual enzymes. Identifying the protein kinases and protein phosphatases, and the interaction domains involved in regulating the various protein interactions involved in starch synthesis will be an important area of future research, ultimately allowing for manipulation of specific protein-protein interactions and a greater understanding of their role in the pathway, and a more informed approach to modifying starch structure.

11.8 List of Abbreviations

ADP-Glc, ADP-glucose; AGPase, ADP-Glc pyrophosphorylase; AGP-L, large subunit AGPase; AGP-S, small subunit AGPase; *ae⁻, amylose extender*; CBM, carbohydrate binding module; CDPK, Ca²⁺-dependent protein kinase; CKII, casein kinase II; DBE, debranching enzyme; DP, degree of polymerization; DPE, disproportionating enzyme; ESV, early starvation; FBN, fibrillin; GBSSI, granule-bound starch synthase; GH, glycoside hydrolase family; Glc, glucose; Glc1P, α-D-glucose 1-phosphate; GWD, glucan, water dikinase; ISA, isoamylase; LDA, limit dextrinase; LSF, like Sex4; MFP1, thylakoid-associated MAR-BINDING FILAMENT-LIKE PROTEIN; MOS, malto-oligosaccharides; MRC, MYOSIN-RESEMBLING CHLOROPLAST PROTEIN; 3PGA, 3-phosphoglyceric acid; PII1, PROTEIN INVOLVED IN STARCH INITIATION1; Pi, inorganic orthophosphate; PPDK, pyruvate, phosphate dikinase; PPi, inorganic pyrophosphate; PTST, protein targeting to starch; PWD, phospho-glucan, water dikinase; SBE, starch branching enzyme; SDS-PAGE, SP, starch phosphorylase; SnRK1, sucrose-nonfermenting1 (SNF1)-related protein kinase 1; SS, starch synthase; T6P, trehalose 6-phosphate; WT, wild-type

Acknowledgements

MJE and IJT are grateful for funding from the Natural Sciences and Engineering Research Council (NSERC) for a Team Discovery Grant [no. 435781]. We thank Mr. Matthew Carswell for preparing Table 11.1.

References

Abe, N., H. Asai, H. Yago, N.F. Oitome, R. Itoh, N. Crofts et al. 2014. Relationships between starch synthase I and branching enzyme isozymes determined using double mutant rice lines. BMC Plant Biol. 14: 1–12.

Abt, M., B. Pfister, M. Sharma, S. Eicke, L., Bürgy, I. Neale et al. 2020. Starch Synthase 5: A Non-Canonical Starch Synthase-Like Protein Involved in Starch Granule Initiation in Arabidopsis. Plant Cell. 32: 2543–2565.

Ahmed, Z., I.J. Tetlow, R. Ahmed, M.K. Morell and M.J. Emes. 2015. Protein-protein interactions among enzymes of starch biosynthesis in high-amylose barley genotypes reveal differential roles of heteromeric enzyme complexes in the synthesis of A and B granules. Plant Sci. 233: 95–106.

Aitken, A., D.B. Collinge, B.P.H. van Heusden, T. Isobe, P.H. Roseboom, G. Rosenfeld et al. 1992. 14-3-3 Proteins: a highly conserved, widespread family of eukaryotic proteins. Trends Biochem. Sci. 17: 498–501.

Albrecht, T., B. Greve, K. Pusch, J. Kossmann, P. Buchner, U. Wobus et al. 1998. Homodimers and heterodimers of Pho1-type phosphorylase isoforms in *Solanum tuberosum* L. as revealed by sequence-specific antibodies. Eur. J. Biochem. 251: 343–352.

Alexander, R.D. and P.C. Morris. 2006. A proteomic analysis of 14-3-3 binding proteins from developing barley grains. Proteomics 6: 1886–1896.

Anderson, C. and K. Tatchell. 2001. Hyperactive glycogen synthase mutants of Saccharomyces cerevisiae suppress the glc7-1 protein phosphatase mutant. J. Bacteriol. 183: 821–829.

Ball, S., H. Guan, M. James, A. Myers, P. Keeling, G. Mouille et al. 1996. From glycogen to amylopectin: a model for the biogenesis of the plant starch granule. Cell 86: 349–352.

Ball, S.G., M.H.B.J. Van De Wal and R.G.F. Visser. 1998. Progress in understanding the biosynthesis of amylose. Trends Plant Sci. 3: 462–467.

Ballicora, M.A., A.A. Iglesias and J. Preiss. 2004. ADP-glucose pyrophosphorylase: A regulatory enzyme for plant starch synthesis. Photosynth. Res. 79: 1–24.

Banks, W., C.T. Greenwood and D. Muir. 1974. Studies on starches of high amylose-content. Die Stärke 9: 289–328.

Bayer, R.G., S. Stael, A.G. Rocha, A. Mair, U.C. Vothknecht and M. Teige. 2012. Chloroplast-localized protein kinases: A step forward towards a complete inventory. J. Exp. Bot. 63: 1713–1723.

Benmoussa, M., B.R. Hamaker, C.P. Huang, D.M. Sherman, C.F. Weil and J.N. BeMiller. 2010. Elucidation of maize endosperm starch granule channel proteins and evidence for plastoskeletal structures in maize endosperm amyloplasts. J. Cereal Sci. 52: 22–29.

Bertoft, E. 2013. On the building block and backbone concepts of amylopectin structure. Cereal Chem. 90: 294–311.

Bertoft, E. 2017. Understanding Starch Structure: Recent Progress. Agronomy 7: 56.

Blauth, S.L., K.N. Kim, J. Klucinec, J.C. Shannon, D. Thompson and M. Guiltinan. 2002. Identification of *mutator* insertional mutants of starch-branching enzyme 1 (*sbe1*) in *Zea mays* L. Plant Mol. Biol. 48: 287–297.

Borén, M., H. Larsson, A. Falk and C. Jansson. 2004. The barley starch granule proteome—Internalized granule polypeptides of the mature endosperm. Plant Sci. 166: 617–626.

Borovsky, D., E.E. Smith, W.J. Whelan, D. French and S. Kikumoto. 1979. The mechanism of Q-enzyme action and its influence on the structure of amylopectin. Arch. Biochem. Biophys. 198: 627–631.

Boyer, L., X. Roussel, A. Courseaux, O.M. Ndjindji, C. Lancelon-Pin, J.L. Putaux et al. 2016. Expression of *Escherichia coli* glycogen branching enzyme in an Arabidopsis mutant devoid of endogenous starch branching enzymes induces the synthesis of starch-like polyglucans. Plant Cell Environ. 39: 1432–1447.

Brauner, K., I. Hörmiller, T. Nägele and A.G. Heyer. 2014. Exaggerated root respiration accounts for growth retardation in a starchless mutant of Arabidopsis thaliana. Plant J. 79: 82–91.

Bréhélin, C., F. Kessler and K.J. van Wijk. 2007. Plastoglobules: versatile lipoprotein particles in plastids. Trends Plant Sci. 12: 260–266.

Brust, H., T. Lehmann, C. D'Hulst and J. Fettke. 2014. Analysis of the functional interaction of Arabidopsis starch synthase and branching enzyme isoforms reveals that the cooperative action of SSI and BEs results in glucans with polymodal chain length distribution similar to amylopectin. PLoS One 9, Issue 7 e102364.

Buchanan, B.B. and Y. Balmer. 2005. Redox Regulation: A Broadening Horizon. Annu. Rev. Plant Biol. 56: 187–220.

Buchbinder, J.L., V.L. Rath and R.J. Fletterick. 2001. Structural relationships among regulated and unregulated phosphorylases. Annu. Rev. Biophys. Biomol. Struct. 30: 191–209.

Buléon, A., H. Bizot, M.M. Delage and J.L. Multno. 1982. Evolution of crystallinity and specific gravity of potato starch versus water ad- and desorption. Starch/Stärke 34: 361–366.

Burr, B. and O.E. Nelson. 1975. Maize α-Glucan Phosphorylase. Eur. J. Biochem. 56: 539–546.

Camoni, L., S. Visconti, P. Aducci and M. Marra. 2018. 14-3-3 Proteins in plant hormone signaling: doing several things at once. Front. Plant Sci. 9: 1–8.

Caspar, T., S.C. Huber and C. Somerville. 1985. Alterations in growth, photosynthesis, and respiration in a starchless mutant of *Arabidopsis thaliana* (L.) deficient in chloroplast phosphoglucomutase activity. Plant Physiol 79: 11–17.

Caspar, T. and B.G. Pickard. 1989. Gravitropism in a starchless mutant of Arabidopsis—Implications for the starch-statolith theory of gravity sensing. Planta 177: 185–197.

Cenci, U., M. Chabi, M. Ducatez, C. Tirtiaux, J. Nirmal-Raj, Y. Utsumi et al. 2013. Convergent evolution of polysaccharide debranching defines a common mechanism for starch accumulation in cyanobacteria and plants. Plant Cell 25: 3961–3975.

Cenci, U., F. Nitschke, M. Steup, B.A. Minassian, C. Colleoni and S.G. Ball. 2014. Transition from glycogen to starch metabolism in archaeplastida. Trends Plant Sci. 19: 18–28.

Chen, G.-X., J.-W. Zhou, Y.-L. Liu, X.-B. Lu, C.-X. Han, W.-Y. Zhang et al. 2016. Biosynthesis and regulation of wheat amylose and amylopectin from proteomic and phosphoproteomic characterization of granule-binding proteins. Sci. Rep. 6: 33111.

Chen, H.M., S.C. Chang, C.C. Wu, T.S. Cuo, J.S. Wu and R.H. Juang. 2002. Regulation of the catalytic behaviour of L-form starch phosphorylase from sweet potato roots by proteolysis. Physiol. Plant. 114: 506–515.

Chevalier, D., E.R. Morris and J.C. Walker. 2009. 14-3-3 and FHA domains mediate phosphoprotein interactions. Annu. Rev. Plant Biol. 60: 67–91.

Chia, T., D. Thorneycroft, A. Chapple, G. Messerli, J. Chen, S.C. Zeeman et al. 2004. A cytosolic glucosyltransferase is required for conversion of starch to sucrose in Arabidopsis leaves at night. Plant J. 37: 853–863.

Coblitz, B., M. Wu, S. Shikano and M. Li. 2006. C-terminal binding: An expanded repertoire and function of 14-3-3 proteins. FEBS Lett. 580: 1531–1535.

Critchley, J.H., S.C. Zeeman, T. Takaha, A.M. Smith and S.M. Smith. 2001. A critical role for disproportionating enzyme in starch breakdown is revealed by a knock-out mutation in Arabidopsis. Plant J. 26: 89–100.

Crofts, N., N. Abe, N.F. Oitome, R. Matsushima, M. Hayashi, I.J. Tetlow et al. 2015. Amylopectin biosynthetic enzymes from developing rice seed form enzymatically active protein complexes. J. Exp. Bot. 66: 4469–4482.

Crofts, N., Y. Nakamura and N. Fujita. 2017. Critical and speculative review of the roles of multi-protein complexes in starch biosynthesis in cereals. Plant Sci. 262: 1–8.

Crumpton-Taylor, M., S. Grandison, K.M.Y. Png, A.J. Bushby and A.M. Smith. 2012. Control of starch granule numbers in *Arabidopsis chloroplasts*. Plant Physiol. 158: 905–916.

Cuesta-Seijo, J.A., M.M. Nielsen, C. Ruzanski, K. Krucewicz, S.R. Beeren, M.G. Rydhal et al. 2016. *In vitro* biochemical characterization of all barley endosperm starch synthases. Front. Plant Sci. 6: 1–17.

Cuesta-Seijo, J.A., C. Ruzanski, K. Krucewicz, S. Meier, P. Hägglund, B. Svensson et al. 2017. Functional and structural characterization of plastidic starch phosphorylase during barley endosperm development. PLoS One 12: 1–25.

Datta, R., K.C. Chamusco and P.S. Chourey. 2002. Starch biosynthesis during pollen maturation is associated with altered patterns of gene expression in maize 1. Society 130: 1645–1656.

Dauvillée, D., V. Chochois, M. Steup, S. Haebel, N. Eckermann, G. Ritte et al. 2006. Plastidial phosphorylase is required for normal starch synthesis in *Chlamydomonas reinhardtii*. Plant J. 48: 274–285.

Debast, S., A. Nunes-Nesi, M.R. Hajirezaei, J. Hofmann, U. Sonnewald, A.T. Fernie et al. 2011. Altering trehalose-6-phosphate content in transgenic potato tubers affects tuber growth and alters responsiveness to hormones during sprouting. Plant Physiol. 156: 1754–1771.

Delatte, T., M. Trevisan, M.L. Parker and S.C. Zeeman. 2005. Arabidopsis mutants Atisa1 and Atisa2 have identical phenotypes and lack the same multimeric isoamylase, which influences the branch point distribution of amylopectin during starch synthesis. Plant J. 41: 815–830.

Delatte, T., M. Umhang, M. Trevisan, S. Eicke, D. Thorneycroft, S.M. Smith et al. 2006. Evidence for distinct mechanisms of starch granule breakdown in plants. J. Biol. Chem. 281: 12050–12059.

Delrue, B., T. Fontaine, F. Routier, A. Decq, J.M. Wieruszeski, N. Van den Koornhuyse et al. 1992. Waxy Chlamydomonas reinhardtii: Monocellular algal mutants defective in amylose biosynthesis and granule-bound starch synthase activity accumulate a structurally modified amylopectin. J. Bacteriol. 174: 3612–3620.

Denyer, K., C. Sidebottom, C.M. Hylton and A.M. Smith. 1993. Soluble isoforms of starch synthase and starch-branching enzyme also occur within starch granules in developing pea embryos. Plant J. 4: 191–198.

Denyer, K., C.M. Hylton, C.F. Jenner and A.M. Smith. 1995. Identification of multiple isoforms of soluble and granule-bound starch synthase in developing wheat endosperm. Planta 196: 256–265.

Denyer, K., F. Dunlap, T. Thorbjørnsen, P. Keeling and A.M. Smith. 1996a. The major form of ADP-glucose pyrophosphorylase in maize endosperm is extra-plastidial. Plant Physiol. 112: 779–785.

Denyer, K., B. Clarke, C. Hylton, H. Tatge and A.M. Smith. 1996b. The elongation of amylose and amylopectin chains in isolated starch granules. Plant J. 10: 1135–1143.

Dinges, J.R., C. Colleoni, M.G. James and A.M. Myers. 2003. Mutational analysis of the pullulanase-type debranching enzyme of maize indicates multiple functions in starch metabolism. Plant Cell 15: 666–680.

Donald, A.M. 2001. Plasticization and self assembly in the starch granule. Cereal Chem. 78: 307–314.

Dou, Y., X. Liu, Y. Yin, S. Han, Y. Lu, Y. Liu et al. 2014. Affinity chromatography revealed insights into unique functionality of two 14-3-3 protein species in developing maize kernels. J. Proteomics 114: 274–286.

Dumez, S., F. Wattebled, D. Dauvillee, D. Delvalle, V. Planchot, S.G. Ball et al. 2006. Mutants of Arabidopsis lacking starch branching enzyme II substitute plastidial starch synthesis by cytoplasmic maltose accumulation. Plant Cell 18: 2694–2709.

Eugeni Piller, L., M. Abraham, P. Dörmann, F. Kessler and C. Besagni. 2012. Plastid lipid droplets at the crossroads of prenylquinone metabolism. J. Exp. Bot. 63: 1609–1618.

Facon, M., Q. Lin, A.M. Azzaz, T.A. Hennen-Bierwagen, A.M. Myers, J.-L Putaux et al. 2013. Distinct functional properties of isoamylase-type starch debranching enzymes in monocot and dicot leaves. Plant Physiol. 163: 1363–1375.

Fannon, J.E., J.A. Gray, N. Gunawan, K.C. Huber and J.N. BeMiller. 2004. Heterogeneity of starch granules and the effect of granule channelization on starch modification. Cellulose 11: 247–254.

Feike, D., D. Seung, A. Graf, S. Bischof, T. Ellick, M. Coiro et al. 2016. The starch granule-associated protein EARLY STARVATION1 is required for the control of starch degradation in *Arabidopsis thaliana* leaves. Plant Cell 28: 1472–1489.

Fu, Y., M.A. Ballicora, J.F. Leykam and J. Preiss. 1998. Mechanism of reductive activation of potato tuber ADP-glucose pyrophosphorylase. J. Biol. Chem. 273: 25045–25052.

Gallant, D.J., B. Bouchet and P.M. Baldwin. 1997. Microscopy of starch: evidence of a new level of granule organization. Carbohydr. Polym. 32: 177–191.

Gámez-Arjona, F.M., S. Raynaud, P. Ragel and Á. Mérida. 2014a. Starch synthase 4 is located in the thylakoid membrane and interacts with plastoglobule-associated proteins in Arabidopsis. Plant J. 80: 305–316.

Gámez-Arjona, F.M., J.C. De La Concepción, S. Raynaud and Á. Mérida. 2014b. Arabidopsis thaliana plastoglobule-associated fibrillin 1a interacts with fibrillin 1b *in vivo*. FEBS Lett. 588: 2800–2804.

Gao, M., D.K. Fisher, K.-N. Kim, J.C. Shannon and M.J. Guiltinan. 1996. Evolutionary conservation and expression patterns of maize starch branching enzyme I and IIb genes suggests isoform specialization. Plant Mol. Biol. 30: 1223–1232.

Gao, M., D.K. Fisher, K.N. Kim, J.C. Shannon and M.J. Guiltinan. 1997. Independent genetic control of maize starch-branching enzymes IIa and IIb. Isolation and characterization of a Sbe2a cDNA. Plant Physiol. 114: 69–78.

Geigenberger, P. 2011. Regulation of starch biosynthesis in response to a fluctuating environment. Plant Physiol. 155: 1566–1577.

Ghosh, H. and J. Preiss. 1966. Adenosine diphosphate glucose pyrophosphorylase: a regulatory enzyme in the biosynthesis of starch in spinach leaf chloroplasts. J. Biol. Chem. 241: 4491–4504.

Gidley, M. and P. Bulpin. 1987. Crystallisation of malto-oligosaccharides as models of the crystalline forms of starch: minimum chain-length requirement for the formation of double helices. Carbohydr. Res. 161: 291–300.

Glaring, M.A., C.B. Koch and A. Blennow. 2006. Genotype-specific spatial distribution of starch molecules in the starch granule: A combined CLSM and SEM approach. Biomacromolecules 7: 2310–2320.

Gonzali, S., A. Alpi, F. Blando and L. De Bellis. 2002. Arabidopsis (HXK1 and HXK2) and yeast (HXK2) hexokinases overexpressed in transgenic lines are characterized by different catalytic properties. Plant Sci. 163: 943–954.

Goren, A., D. Ashlock and I.J. Tetlow. 2018. Starch formation inside plastids of higher plants. Protoplasma 255: 1855–1876.

Grimaud, F., H. Rogniaux, M.G. James, A.M. Myers and V. Planchot. 2008. Proteome and phosphoproteome analysis of starch granule-associated proteins from normal maize and mutants affected in starch biosynthesis. J. Exp. Bot. 59: 3395–3406.

Guan, H.P. and J. Preiss. 1993. Differentiation of the properties of the branching isozymes from maize (*Zea mays*). Plant Physiol. 102: 1269–1273.

Hädrich, N., J.H.M. Hendriks, O. Kötting, S. Arrivault, R. Feil, S.C. Zeeman et al. 2012. Mutagenesis of cysteine 81 prevents dimerization of the APS1 subunit of ADP-glucose pyrophosphorylase and alters diurnal starch turnover in *Arabidopsis thaliana* leaves. Plant J. 70: 231–242.

Hawker, J.S., J.L. Ozbun, H. Ozaki, E. Greenberg and J. Preiss. 1974. Interaction of spinach leaf adenosine diphosphate glucose α-1,4-glucan α-4-glucosyl transferase and α-1,4-glucan, α-1,4-glucan-6-glycosyl transferase in synthesis of branched α-glucan. Arch. Biochem. Biophys. 160: 530–551.

Hayashi, M., N. Crofts, N.F. Oitome and N. Fujita. 2018. Analysis of starch biosynthetic protein complexes and starch properties from developing mutant rice seeds with minimal starch synthase activities. BMC Plant Biology 18: 59.

Heazlewood, J.I., P. Durek, J. Hummel, J. Selbig, W. Weckwerth, D. Walther et al. 2008. PhosPhAt : A database of phosphorylation sites in *Arabidopsis thaliana* and a plant-specific phosphorylation site predictor. Nucleic Acids Res. 36: D1015–D1021.

Hejazi, M., J. Fettke, S. Haebel, C. Edner, O. Paris, C. Frohberg et al. 2008. Glucan, water dikinase phosphorylates crystalline maltodextrins and thereby initiates solubilization. Plant J. 55: 323–334.

Helle, S., F. Bray, J. Verbeke, S. Devassine, A. Courseaux, M. Facon et al. 2018. Proteome analysis of potato starch reveals the presence of new starch metabolic proteins as well as multiple protease inhibitors. Front. Plant Sci. 9: 746.

Hendriks, J.H.M. 2003. ADP-glucose pyrophosphorylase is activated by posttranslational redox-modification in response to light and to sugars in laves of Arabidopsis and other plant species. Plant Physiol. 133: 838–849.

Hennen-Bierwagen, T.A., F. Liu, R.S. Marsh, S. Kim, Q. Gan, I.J. Tetlow et al. 2008. Starch biosynthetic enzymes from developing maize endosperm associate in multisubunit complexes. Plant Physiol. 146: 1892–1908.

Hennen-Bierwagen, T.A., Q. Lin, F. Grimaud, V. Planchot, P.L. Keeling, M.G. James et al. 2009. Proteins from multiple metabolic pathways associate with starch biosynthetic enzymes in high molecular weight complexes: A model for regulation of carbon allocation in maize amyloplasts. Plant Physiol. 149: 1541–1559.

Higgins, J.E., B. Kosar-Hashemi, Z. Li, C.A. Howitt, O. Larroque, B. Flanagan et al. 2013. Characterization of starch phosphorylases in barley grains. J. Sci. Food Agric. 93: 2137–2145.

Hizukuri, S., S. Tabata, Kagoshima and Z. Nikuni. 1970. Studies on starch phosphate Part 1. Estimation of glucose-6-phosphate residues in starch and the presence of other bound phosphate(s). Starch/Stärke 22: 338–343.

Hizukuri, S. 1986. Polymodal distribution of the chain lengths of amylopectins, and its significance. Carbohydr. Res. 147: 342–347.

Huang, B., T.A. Hennen-Bierwagen and A.M. Myers. 2014. Functions of multiple genes encoding ADP-glucose pyrophosphorylase subunits in maize endosperm, embryo, and leaf. Plant Physiol. 164: 596–611.

Huber, S.C., C. MacKintosh and W.M. Kaiser. 2002. Metabolic enzymes as targets for 14-3-3 proteins. Plant Mol. Biol. 50: 1053–1063.

Hussain, H., A. Mant, R. Seale, S. Zeeman, E. Hinchliffe, A. Edwards et al. 2003. Three isoforms of isoamylase contribute different catalytic properties for the debranching of potato glucans. Plant Cell 15: 133–149.

Hwang, S.K., K. Koper, H. Satoh and T.W. Okita. 2016. Rice endosperm starch phosphorylase (Pho1) assembles with disproportionating enzyme (Dpe1) to form a protein complex that enhances synthesis of malto-oligosaccharides. J. Biol. Chem. 291: 19994–20007.

Janeček, Š., B. Svensson and E.A. MacGregor. 2011. Structural and evolutionary aspects of two families of non-catalytic domains present in starch and glycogen binding proteins from microbes, plants and animals. Enzyme Microb. Technol. 49: 429–440.

Jenkins, P.J., R.E. Cameron and A.M. Donald. 1993. A universal feature in the structure of starch granules from different botanical sources. Starch/Stärke 45: 417–420.

Jones, S.A., M. Jorgensen, F.Z. Chowdhury, R. Rodgers, J. Hartline, M.P. Leatham et al. 2008. Glycogen and maltose utilization by *Escherichia coli* O157:H7 in the mouse intestine. Infect. Immun. 76: 2531–2540.

Kaneko-Suzuki, M., R. Kurihara-Ishikawa, C. Okushita-Terakawa, C. Kojima, M. Nagano-Fujiwara, I. Ohki et al. 2018. TFL1-Like Proteins in rice antagonize rice FT-Like protein in inflorescence development by competition for complex formation with 14-3-3 and FD. Plant Cell Physiol. 59: 458–468.

Kiss, J.Z., M.M. Guisinger, A.J. Miller and K.S. Stackhouse. 1997. Reduced gravitropism in hypocotyls of starch-deficient mutants of Arabidopsis. Plant Cell Physiol. 38: 518–525.

Kleczkowski, L.A., P. Villand, E. Lüthi and O.P.J. Olsen. 1993. Insensitivity of barley endosperm ADP-glucosepyrophosphorylase to 3-phosphoglycerate and orthophosphate regulation. Plant Physiol. 101: 179–186.

Kleczkowski, L.A. 1999. A phosphoglycerate to inorganic phosphate ratio is the major factor in controlling starch levels in chloroplasts via ADP-glucose pyrophosphorylase regulation. FEBS Lett. 448: 153–156.

Klucinec, J.D. and D.B. Thompson. 2002. Structure of amylopectins from ae-containing maize starches. Cereal Chem. 79: 19–23.

Kolbe, A., A. Tiessen, H. Schluepmann, M. Paul, S. Ulrich and P. Geigenberger. 2005. Trehalose 6-phosphate regulates starch synthesis via posttranslational redox activation of ADP-glucose pyrophosphorylase. Proc. Natl. Acad. Sci. U. S. A. 102: 11118–11123.

Krishnan, A., G.K. Kumaraswamy, D.J. Vinyard, H. Gu, G. Ananyev, M.C. Posewitz et al. 2015. Metabolic and photosynthetic consequences of blocking starch biosynthesis in the green alga *Chlamydomonas reinhardtii sta6* mutant. Plant J. 81: 947–960.

Kubo, A., C. Colleoni, J.R. Dinges, Q. Lin, R.R. Lappe, J.G. Rivenbark et al. 2010. Functions of heteromeric and homomeric Isoamylase-type starch-debranching enzymes in developing maize endosperm. Plant Physiol. 153: 956–969.

Kudla, J. and R. Bock. 2016. Lighting the way to protein-protein interactions: Recommendations on best practices for bimolecular fluorescence complementation analyses. Plant Cell 28: 1002–1008.

Lappe, R.R., J.W. Baier, S.K. Boehlein, R. Huffman, Q. Lin, F. Wattebled et al. 2017. Functions of maize genes encoding pyruvate phosphate dikinase in developing endosperm. Proc. Natl. Acad. Sci. 201715668.

Larson, M.E., D.J. Falconer, A.M. Myers and A.W. Barb. 2016. Direct characterization of the maize starch synthase IIa product shows maltodextrin elongation occurs at the non-reducing end. J. Biol. Chem. 291: 24951–24960.

Lee, S., J. Eom, S. Hwang, D. Shin, G. An and T.W. Okita. 2018. Plastidic phosphoglucomutase and ADP-glucose pyrophosphorylase mutants impair starch synthesis in rice pollen grains and cause male sterility. J. Exp. Bot. 67: 5557–5569.

Lee, S.K., S.K. Hwang, M. Han, J.S. Eom, H.G. Kang, Y. Han et al. 2007. Identification of the ADP-glucose pyrophosphorylase isoforms essential for starch synthesis in the leaf and seed endosperm of rice (*Oryza sativa* L.). Plant Mol. Biol. 65: 531–546.

Leterrier, M., L.D. Holappa, K.E. Broglie and D.M. Beckles. 2008. Cloning, characterisation and comparative analysis of a starch synthase IV gene in wheat: functional and evolutionary implications. BMC Plant Biol. 8: 98.

Li, C. and R.G. Gilbert. 2016. Progress in controlling starch structure by modifying starch-branching enzymes. Planta 243: 13–22.

Lin, Q., B. Huang, M. Zhang, X. Zhang, J. Rivenbark, R.L. Lappe et al. 2012. Functional interactions between starch synthase III and isoamylase-type starch-debranching enzyme in maize endosperm. Plant Physiol. 158: 679–692.

Lin, Q., M. Facon, J.-L. Putaux, J.R. Dinges, F. Wattebled, C. D'Hulst et al. 2013. Function of isoamylase-type starch debranching enzymes ISA1 and ISA2 in the Zea mays leaf. New Phytol. 200: 1009–1021.

Liu, D.R., W.X. Huang and X.L. Cai. 2013. Oligomerization of rice granule-bound starch synthase 1 modulates its activity regulation. Plant Sci. 210: 141–150.

Liu, F., A. Makhmoudova, E.A. Lee, R. Wait, M.J. Emes and I.J. Tetlow. 2009. The amylose extender mutant of maize conditions novel protein-protein interactions between starch biosynthetic enzymes in amyloplasts. J. Exp. Bot. 60: 4423–4440.

Liu, F., N. Romanova, E.A. Lee, R. Ahmed, M. Evans, E.P. Gilbert et al. 2012a. Glucan affinity of starch synthase IIa determines binding of starch synthase I and starch-branching enzyme IIb to starch granules. Biochem. J. 448: 373–387.

Liu, F., Z. Ahmed, E.A. Lee, E. Donner, Q. Liu, R. Ahmed et al. 2012b. Allelic variants of the amylose extender mutation of maize demonstrate phenotypic variation in starch structure resulting from modified protein-protein interactions. J. Exp. Bot. 63: 1167–1183.

Lloyd, J.R. and J. Kossmann. 2015. Transitory and storage starch metabolism: Two sides of the same coin? Curr. Opin. Biotechnol. 32: 143–148.

Lohmeier-Vogel, E.M., D. Kerk, M. Nimick, S. Wrobel, L. Vickerman, D.G. Muench et al. 2008. Arabidopsis At5g39790 encodes a chloroplast-localized, carbohydrate- binding, coiled-coil domain-containing putative scaffold protein. BMC Plant Biol. 8: 1–14.

Lohrig, K., B. Müller, J. Davydova, D. Leister and D.A. Wolters. 2009. Phosphorylation site mapping of soluble proteins: Bioinformatical filtering reveals potential plastidic phosphoproteins in *Arabidopsis thaliana*. Planta 229: 1123–1134.

Lohscheider, J.N., G. Friso and K.J. Van Wijk. 2016. Phosphorylation of plastoglobular proteins in *Arabidopsis thaliana*. J. Exp. Bot. 67: 3975–3984.

Lu, K.-J., B. Pfister, C. Jenny, S. Eicke and S.C. Zeeman. 2017. Distinct functions of STARCH SYNTHASE 4 domains in starch granule formation. Plant Physiol. 176: 01008.2017.

Lundquist, P.K., O. Mantegazza, A. Stefanski, K. Stühler and A.P.M. Weber. 2017. Surveying the oligomeric state of Arabidopsis thaliana chloroplasts. Mol. Plant 10: 197–211.

Lupas, A.N. and M. Gruber. 2005. The structure of α-helical coiled coils. Adv. Protein Chem. 70: 37–78.

Ma, L., N. Xue, X. Fu, H. Zhang and G. Li. 2017. *Arabidopsis thaliana* FAR-RED ELONGATED HYPOCOTYLS3 (FHY3) and FAR-RED-IMPAIRED RESPONSE1 (FAR1) modulate starch synthesis in response to light and sugar. New Phytol. 213: 1682–1696.

Makhmoudova, A., D. Williams, D. Brewer, S. Massey, J. Patterson, A. Silva et al. 2014. Identification of multiple phosphorylation sites on maize endosperm starch branching enzyme IIb, a key enzyme in amylopectin biosynthesis. J. Biol. Chem. 289: 9233–9246.

Malinova, I., S. Mahlow, S. Alseekh, T. Orawetz, A.R. Fernie, O. Baumann et al. 2014. Double knockout mutants of Arabidopsis grown under normal conditions reveal that the plastidial phosphorylase isozyme participates in transitory starch metabolism. Plant Physiol. 164: 907–921.

Malinova, I., S. Alseekh, R. Feil, A.R. Fernie, O. Baumann, M.A. Schöttler et al. 2017. Starch synthase 4 and plastidal phosphorylase differentially affect starch granule number and morphology. Plant Physiol. 174: 73–85.

Malinova, I., H. Mahto, F. Brandt, S. Al-Rawi, H. Qasim, H. Brust et al. 2018a. EARLY STARVATION1 specifically affects the phosphorylation action of starch-related dikinases. The Plant Journal 95: 126–137.

Malinova, I., H.M. Qasim, H. Brust and J. Fettke. 2018b. Parameters of starch granule genesis in chloroplasts of *Arabidopsis thaliana*. Front Plant Sci. 9: 761.

Martinez-Barajas, E., T. Delatte, H. Schluepmann, G.J. de Jong, G.W. Somsen, C. Nunes et al. 2011. Wheat grain development is characterized by remarkable trehalose 6-phosphate accumulation pregrain filling: Tissue distribution and relationship to SNF1-Related Protein Kinase1 activity. Plant Physiol. 156: 373–381.

Martinis, J., G. Glauser, S. Valimareanu, M. Stettler, S.C. Zeeman, H. Yamamoto et al. 2014. ABC1K1/PGR6 kinase: A regulatory link between photosynthetic activity and chloroplast metabolism. Plant J. 77: 269–283.

Mason, J.M. and K.M. Arndt. 2004. Coiled coil domains: Stability, specificity, and biological implications. ChemBioChem 5: 170–176.

Matsushima, R., M. Maekawa, M. Kusano, K. Tomita, H. Kondo, H. Nishimura et al. 2016. Amyloplast membrane protein SUBSTANDARD STARCH GRAIN6 controls starch grain size in rice endosperm. Plant Physiol. 170: 1445–1459.

McKibbin, R.S., N. Muttucumaru, M.J. Paul, S.J. Powers, M.M. Burrell, S. Coates et al. 2006. Production of high-starch, low-glucose potatoes through over-expression of the metabolic regulator SnRK1. Plant Biotechnol. J. 4: 409–418.

McMeechan, A., M.A. Lovell, T.A. Cogan, K.L. Marston, T.J. Humphrey and P.A. Barrow. 2005. Glycogen production by different *Salmonella enterica* serotypes: Contribution of functional glgC to virulence, intestinal colonization and environmental survival. Microbiology 151: 3969–3977.

Meléndez-Hevia, E., T.G. Waddell and E.D. Shelton. 1993. Optimization of molecular design in the evolution of metabolism: the glycogen molecule. Biochem. J. 295: 477–483.

Michalska, J., H. Zauber, B.B. Buchanan, F.J. Cejudo and P. Geigenberger. 2009. NTRC links built-in thioredoxin to light and sucrose in regulating starch synthesis in chloroplasts and amyloplasts. Proc. Natl. Acad. Sci. U. S. A. 106: 9908–9913.

Miura, S., N. Crofts, Y. Saito, Y. Hosaka, N.F. Oitome, T. Watanabe et al. 2018. Starch synthase IIa-deficient mutant rice line produces endosperm starch with lower gelatinization temperature than Japonica rice cultivars. Front. Plant Sci. 9: 645.

Mizuno, K., T. Kawasaki, H. Shimada, H. Satoh, E. Kobayashi, S. Okumura et al. 1993. Alteration of the structural properties of starch components by the lack of an isoform of starch branching enzyme in rice seeds. J. Biol. Chem. 268: 19084–19091.

Møller, M.S., A. Henriksen and B. Svensson. 2016. Structure and function of α-glucan debranching enzymes. Cell. Mol. Life Sci. 73: 2619–2641.

Morell, M.K., B. Kosar-Hashemi, M. Cmiel, M.S. Samuel, P. Chandler, S. Rahman et al. 2003. Barley sex6 mutants lack starch synthase IIa activity and contain a starch with novel properties. Plant J. 34: 173–185.

Mu-Forster, C., R. Huang, J.R. Powers, R.W. Harriman, M. Knight, C.W. Singletary et al. 1995. Physical association. Plant Physiol. 111: 821–829.

Muslin, A.J., J.W. Tanner, P.M. Allen and A.S. Shaw. 1996. Interaction of 14-3-3 with signaling proteins is mediated by the recognition of phosphoserine. Cell 84: 889–897.

Nakamura, Y., Y. Utsumi, T. Sawada, S. Aihara, C. Utsumi, M. Yoshida et al. 2010. Characterization of the reactions of starch branching enzymes from rice endosperm. Plant Cell Physiol. 51: 776–794.

Nakamura, Y., M. Ono, C. Utsumi and M. Steup. 2012. Functional interaction between plastidial starch phosphorylase and starch branching enzymes from rice during the synthesis of branched maltodextrins. Plant Cell Physiol. 53: 869–878.

Nakamura, Y., M. Ono, T. Sawada, N. Crofts, N. Fujita and M. Steup. 2017. Characterization of the functional interactions of plastidial starch phosphorylase and starch branching enzymes from rice endosperm during reserve starch biosynthesis. Plant Sci. 264: 83–95.

Nielsen, M.M., C. Ruzanski, K. Krucewicz, A. Striebeck, U. Cenci, S.G. Ball et al. 2018. Crystal Structures of the Catalytic Domain of *Arabidopsis thaliana* Starch Synthase IV, of Granule Bound Starch Synthase From CLg1 and of Granule Bound Starch Synthase I of *Cyanophora paradoxa* Illustrate Substrate Recognition in Starch Synthases. Front. Plant Sci. 9: 1138.

Nitschke, F., P. Wang, P. Schmieder, J.M. Girard, D.E. Awrey, T. Wang et al. 2013. Hyperphosphorylation of glucosyl c6 carbons and altered structure of glycogen in the neurodegenerative epilepsy Lafora disease. Cell Metab. 17: 756–767.

Nougué, O., J. Corbi, S.G. Ball, D. Manicacci and M.I. Tenaillon. 2014. Molecular evolution accompanying functional divergence of duplicated genes along the plant starch biosynthesis pathway. BMC Evol. Biol. 14: 1–16.

O'Neill, E.C., C.E.M. Stevenson, K. Tantanarat, D. Latousakis, M.I. Donaldson, M. Rejzek et al. 2015. Structural dissection of the maltodextrin disproportionation cycle of the *Arabidopsis plastidial* disproportionating enzyme 1 (DPE1). J. Biol. Chem. 290: 29834–29853.

Osteryoung, K.W. and K.A. Pyke. 2014. Division and dynamic morphology of plastids. Annu. Rev. Plant Biol. 65: 443–472.

Pan, D.D. and J. Jane. 2000. Internal structure of normal maize starch granules revealed by chemical surface gelatinization. Biomacromolecules 1: 126–132.

Patterson, J.S., I.J. Tetlow and M.J. Emes. 2018. Bioinformatic and *in vitro* analyses of Arabidopsis starch synthase 2 reveal post-translational regulatory mechanisms. Front. Plant Sci. 9: 1338.

Peng, C., Y. Wang, F. Liu, Y. Ren, K. Zhou, J. Lv, M. Zheng et al.. 2014. *FLOURY ENDOSPERM6* encodes a CBM48 domain-containing protein involved in compound granule formation and starch synthesis in rice endosperm. Plant J. 77: 917–930.

Pfister, B. and S.C. Zeeman. 2016. Formation of starch in plant cells. Cell. Mol. Life Sci. 73: 2781–2807.

Pfister, B., S.C. Zeeman, M.D. Rugen, R.A. Field, O. Ebenhoh, A. Raguin et al. 2020. Theoretical and experimental approaches to understand the biosynthesis of starch granules in a physiological context. Photosynth. Res. 631.

Pitcher, J., C. Smythe, D.G. Campbell and P. Cohen. 1987. Identification of the 38-kDa subunit as glycogenin of rabbit skeletal muscle glycogen synthase as glycogenin. Eur. J. Biochem. 169: 497–502.

Puteaux, J.-L., G. Potocki-Veronese, M. Remaud-Simeon and A. Buléon. 2006. Alpha-D-glucan-based dendritic nanoparticles prepared by *in vitro* enzymatic chain extension of glycogen. Biomacromolecules 7: 1720–1728.

Rahman, S., B. Kosar-Hashemi, M. Samuel, A. Hill, D. Abbott, J. Skerritt et al. 1995. The major proteins of wheat endosperm starch granules. Aust. J. Plant Physiol. 22: 793.

Rahman, S. 2001. Comparison of starch-branching enzyme genes reveals evolutionary relationships among isoforms. Characterization of a gene for starch-branching enzyme IIa from the wheat D genome donor *Aegilops tauschii.* Plant Physiol. 125: 1314–1324.

Raynaud, S., P. Ragel, T. Rojas and Á. Mérida. 2016. The N-terminal part of Arabidopsis thaliana starch synthase 4 determines the localization and activity of the enzyme. J. Biol. Chem. 291: 10759–10771.

Regina, A., A. Bird, D. Topping, S. Bowden, J. Freeman, T. Barsby et al. 2006. High-amylose wheat generated by RNA interference improves indices of large-bowel health in rats. Proc. Natl. Acad. Sci. U. S. A. 103: 3546–3551.

Regina, A., B. Kosar-Hashemi, S. Ling, Z. Li, S. Rahman and M. Morell. 2010. Control of starch branching in barley defined through differential RNAi suppression of starch branching enzyme IIa and IIb. J. Exp. Bot. 61: 1469–1482.

Ritte, G., J.R. Lloyd, N. Eckermann, A. Rottmann, J. Kossmann and M. Steup. 2002. The starch-related R1 protein is an α-glucan, water dikinase. Proc. Natl. Acad. Sci. U. S. A. 99: 7166–7171.

Roach, P.J., A.A. Depaoli-Roach, T.D. Hurley and V.S. Tagliabracci. 2016. Glycogen and its metabolism: some new developments and old themes. Biochem. J. 441: 763–787.

Roldán, I., F. Wattebled, M. Mercedes Lucas, D. Delvallé, V. Planchot, S. Jiménez et al. 2007. The phenotype of soluble starch synthase IV defective mutants of Arabidopsis thaliana suggests a novel function of elongation enzymes in the control of starch granule formation. Plant J. 49: 492–504.

Saito, M., T. Tanaka, K. Saito, P. Vrinten and T. Nakamura. 2018. A single nucleotide polymorphism in the "*Fra*" gene results in fractured starch granules in barley. Theor. Appl. Genet. 131: 353–364.

Schönberg, A., E. Bergner, S. Helm, B. Agne, B. Dünschede, D. Schünemann et al. 2014. The peptide microarray "chlorophos1.0" identifies new phosphorylation targets of plastid casein kinase II (pCKII) in Arabidopsis thaliana. PLoS One 9.

Sehnke, P.C., H.J. Chung, K. Wu and R.J. Ferl. 2001. Regulation of starch accumulation by granule-associated plant 14-3-3 proteins. Proc. Natl. Acad. Sci. U. S. A. 98: 765–770.

Sestili, F., E. Botticella, G. Proietti, M. Janni, R. D'Ovidio and D. Lafiandra. 2012. Amylose content is not affected by overexpression of the Wx-B1 gene in durum wheat. Plant Breed. 131: 700–706.

Seung, D., S. Soyk, M. Coiro, B.A. Maier, S. Eicke and S.C. Zeeman. 2015. Protein targeting to starch is required for localising granule-bound starch synthase to starch granules and for normal amylose synthesis in Arabidopsis. PLoS Biol. 13: 1–29.

Seung, D., J. Boudet, T.B. Schreier, L.C. David, M. Abt, K.-J. Lu et al. 2017. Homologs of protein targeting to starch control starch granule initiation in arabidopsis leaves. Plant Cell 29: 1657–1677.

Seung, D., T. Schreier, L. Bürgy, S. Eicke and S. Zeeman. 2018. Two plastidial coiled-coil proteins are essential for normal starch granule initiation in Arabidopsis. Plant Cell 30: 1523–1542.

Shure, M., S. Wessler and N. Fedoroff. 1983. Molecular identification and isolation of the Waxy locus in maize. Cell 35: 225–233.

Silljé, H.H.W., J.W.G. Paalman, E.G. ter Schure, S.Q.B. Olsthoorn, A.J. Verkleij, J. Boonstra et al. 1999. Function of trehalose and glycogen in cell cycle progression and cell viability in Saccharomyces cerevisiae. J. Bacteriol. 181: 396–400.

Sim, L., S.R. Beeren, J. Findinier, D. Dauville, S.G. Ball, A. Henriksen et al. 2014. Crystal structure of the Chlamydomonas starch debranching enzyme isoamylase ISA1 reveals insights into the mechanism of branch trimming and complex assembly. J. Biol. Chem. 289: 22991–23003.

Singh, D.K. and T.W. McNellis. 2011. Fibrillin protein function: The tip of the iceberg? Trends Plant Sci. 16: 432–441.

Smythe, C. and P. Cohen. 1991. The discovery of glycogenin and the priming mechanism for glycogen biogenesis. Eur. J. Biochem. 200: 625–631.

Stitt, M. and S.C. Zeeman. 2012. Starch turnover: Pathways, regulation and role in growth. Curr. Opin. Plant Biol. 15: 282–292.

Subasinghe, R.M., F. Liu, U.C. Polack, E.A. Lee, M.J. Emes and I.J. Tetlow. 2014. Multimeric states of starch phosphorylase determine protein-protein interactions with starch biosynthetic enzymes in amyloplasts. Plant Physiol. Biochem. 83: 168–179.

Sullivan, A.C.O. and S. Perez. 1999 The relationship between internal chain length of amylopectin and crystallinity in starch. Biopolymers. 50: 381–390.

Sundberg, M., B. Pfister, D. Fulton, S. Bischof, T. Delatte, S. Eicke et al. 2013. The heteromultimeric debranching enzyme involved in starch synthesis in Arabidopsis requires both isoamylase1 and isoamylase2 subunits for complex stability and activity. PLoS One 8.

Swinkels, J.J.M. 1985. Composition and properties of commercial native starches. Starch/Starke. 37: 1–5.

Szydlowski, N., P. Ragel, S. Raynaud, M.M. Lucas, I. Roldan, M. Montero et al. 2009. Starch granule initiation in arabidopsis requires the presence of either class IV or class III starch synthases. Plant Cell Online 21: 2443–2457.

Tanaka, N., N. Fujita, A. Nishi, H. Satoh, Y. Hosaka, M. Ugaki et al. 2004. The structure of starch can be manipulated by changing the expression levels of starch branching enzyme IIb in rice endosperm. Plant Biotechnol. J. 2: 507–516.

Tetlow, I. and M. Emes. 2017. Starch biosynthesis in the developing endosperms of grasses and cereals. Agronomy 7: 81.

Tetlow, I.J., E.J. Davies, K.A. Vardy, C.G. Bowsher, M.M. Burrell and M.J. Emes. 2003. Subcellular localization of ADPglucose pyrophosphorylase in developing wheat endosperm and analysis of the properties of a plastidial isoform. J. Exp. Bot. 54: 715–725.

Tetlow, I.J., R. Wait, Z. Lu, R. Akkasaeng, C.G. Bowsher, S. Esposito et al. 2004. Protein phosphorylation in amyloplasts regulates starch branching enzyme activity and protein–protein interactions. Plant Cell 16: 694–708.

Tetlow, I.J., K.G. Biesel, S. Cameron, A. Makhmoudova, F. Liu, N.S. Bresolin et al. 2008. Analysis of protein complexes in wheat amyloplasts reveals functional interactions among starch biosynthetic enzymes. Plant Physiol. 146: 1878–1891.

Tetlow, I., F. Liu and M. Emes. 2015. Protein-protein interactions during starch biosynthesis. pp. 291–313. *In*: Nakamura, Y. (ed.). Starch Metabolism and Structure (Springer).

Thorbjørnsen, T., P. Villand, K. Denyer, O.A. Olsen and A.M. Smith. 1996. Distinct isoforms of ADPglucose pyrophosphorylase occur inside and outside the amyloplasts in barley endosperm. Plant J. 10: 243–250.

Tiessen, A., J.H.M. Hendriks, M. Stitt, A. Branscheid, Y. Gibon, E.M. Farré et al. 2002. Starch synthesis in potato tubers is regulated by post-translational redox modification of ADP-glucose pyrophosphorylase. Am. Soc. Plant Biol. 14: 2191–2213.

Tiessen, A., K. Prescha, A. Branscheid, N. Palacios, R. McKibbin, N.G. Halford et al. 2003. Evidence that SNF1-related kinase and hexokinase are involved in separate sugar-signalling pathways modulating post-translational redox activation of ADP-glucose pyrophosphorylase in potato tubers. Plant J. 35: 490–500.

Toyosawa, Y., Y. Kawagoe, R. Matsushima, N. Crofts, M. Ogawa, M. Fukuda et al. 2016. Deficiency of starch synthase IIIa and IVb alters starch granule morphology from polyhedral to spherical in rice endosperm. Plant Physiol. 170: 1255–1270.

Tuncel, A. and T.W. Okita. 2013. Improving starch yield in cereals by over-expression of ADPglucose pyrophosphorylase: Expectations and unanticipated outcomes. Plant Sci. 211: 52–60.

Umemoto, T. and N. Aoki. 2005. Single-nucleotide polymorphisms in rice starch synthase IIa that alter starch gelatinisation and starch association of the enzyme. Funct. Plant Biol. 32: 763–768.

Utsumi, Y. and Y. Nakamura. 2006. Structural and enzymatic characterization of the isoamylase1 homo-oligomer and the isoamylase1-isoamylase2 hetero-oligomer from rice endosperm. Planta 225: 75–87.

Utsumi, Y., C. Utsumi, T. Sawada, N. Fujita and Y. Nakamura. 2011. Functional diversity of isoamylase oligomers: the ISA1 homo-oligomer is essential for amylopectin biosynthesis in rice endosperm. Plant Physiol. 156: 61–77.

Valdez, H.A., M.V. Busi, N.Z. Wayllace, G. Parisi, R.A. Ugalde and D.F. Gomez-Casati. 2008. Role of the N-terminal starch-binding domains in the kinetic properties of starch synthase III from *Arabidopsis thaliana*. Biochemistry 47: 3026–3032.

Vandromme, C., C. Spriet, D. Dauville, A. Courseaux, J.-L. Puteaux, A. Wychowski et al. 2018. PII1: a protein involved in starch initiation that determines granule number and size in Arabidopsis chloroplast. New Phytol. 221: 356–370.

van Wijk, K.J., G. Friso, D. Walther and W.X. Schulze. 2014. Meta-analysis of *Arabidopsis thaliana* phospho-proteomics data reveals compartmentalization of phosphorylation motifs. Plant Cell 26: 2367–2389.

Vidi, P.A., M. Kanwischer, S. Baginsky, J.R. Austin, G. Csucs, P. Dörmann et al. 2006. Tocopherol cyclase (VTE1) localization and vitamin E accumulation in chloroplast plastoglobule lipoprotein particles. J. Biol. Chem. 281: 11225–11234.

Vrinten, P.L. and T. Nakamura. 2000. Wheat granule-bound starch synthase I and II are encoded by separate genes that are expressed in different tissues. Plant Physiol. 122: 255–264.

Walley, J.W., Z. Shen, R. Sartor, K.J. Wu, J. Osborn, L.G. Smith et al. 2013. Reconstruction of protein networks from an atlas of maize seed proteotypes. Proc. Natl. Acad. Sci. U. S. A. 110: E4808–E4817.

Wattebled, F., Y. Dong, S. Dumez, D. Delvallé, V. Planchot, P. Berbezy et al. 2005. Mutants of Arabidopsis lacking a chloroplastic isoamylase accumulate phytoglycogen and an abnormal form of amylopectin. Plant Physiol. 138: 184–195.

Wayllace, N.Z., H.A. Valdez, R.A. Ugalde, M.V. Busi and D.F. Gomez-Casati. 2010. The starch-binding capacity of the noncatalytic SBD2 region and the interaction between the N- and C-terminal domains are involved in the modulation of the activity of starch synthase III from Arabidopsis thaliana: Enzymes and catalysis. FEBS J. 277: 428–440.

White-Gloria, C., J.J. Johnson, K. Marritt, A. Kataya, A. Vahab and G.B. Moorhead. 2018. Protein kinases and phosphatases of the plastid and their potential role in starch metablism. Frontiers in Plant Science 9: 1032.

Wilson, W.A., P.J. Roach, M. Montero, E. Baroja-Fernandez, F.J. Munoz, G. Eydallin et al. 2010. Regulation of glycogen metabolism in yeast and bacteria. FEMS Microbiol. Rev. 34: 952–985.

Woo, E.J., S. Lee, H. Cha, J.T. Park, S.M. Yoon, H.N. Song et al. 2008. Structural insight into the bifunctional mechanism of the glycogen-debranching enzyme TreX from the archaeon *Sulfolobus solfataricus*. J. Biol. Chem. 283: 28641–28648.

Xia, H., M. Yandeau-Nelson, D.B. Thompson and M.J. Guiltinan. 2011. Deficiency of maize starch-branching enzyme I results in altered starch fine structure, decreased digestibility and reduced coleoptile growth during germination. BMC Plant Biol. 11: 95.

Yaffe, M., K. Rittinger, S. Volinia, P. Caron, A. Aitken, H. Leffers et al. 1997. The structural basis for 14-3-3: phosphopeptide binding specificity. Cell 91: 961–971.

Yamamori, M., S. Fujita, K. Hayakawa, J. Matsuki and T. Yasui. 2000. Genetic elimination of a starch granule protein, SGP-1, of wheat generates an altered starch with apparent high amylose. Theor. Appl. Genet. 101: 21–29.

Yao, Y., D.B. Thompson and M.J. Guiltinan. 2004. Maize starch-branching enzyme isoforms and amylopectin structure. In the absence of starch-branching enzyme IIb, the further absence of starch-branching enzyme Ia leads to increased branching. Plant Physiol. 136: 3515–3523.

Young, G.H., H.M. Chen, C.T. Lin, K.C. Tseng, J.S. Wu and R.H. Juang. 2006. Site-specific phosphorylation of L-form starch phosphorylase by the protein kinase activity from sweet potato roots. Planta 223: 468–478.

Ytterberg, A.J., J.B. Peltier and K.J. van Wijk. 2006. Protein profiling of plastoglobules in chloroplasts and chromoplasts. A surprising site for differential accumulation of metabolic enzymes. Plant Physiol. 140: 984–997.

Yu, Y., H.H. Mu, B.P. Wasserman and G.M. Carman. 2001. Identification of the maize amyloplast stromal 112-kD protein as a plastidic starch phosphorylase. Plant Physiol. 125: 351–359.

Yun, M.S. and Y. Kawagoe. 2010. Septum formation in amyloplasts produces compound granules in the rice endosperm and is regulated by plastid division proteins. Plant Cell Physiol. 51: 1469–1479.

Yun, M.S., T. Umemoto and Y. Kawagoe. 2011. Rice debranching enzyme isoamylase3 facilitates starch metabolism and affects plastid morphogenesis. Plant Cell Physiol. 52: 1068–1082.

Zabawinski, C., N. Van Den Koornhuyse, C. D'Hulst, R. Schlichting, C. Giersch, B. Delrue et al. 2001. Starchless mutants of *Chlamydomonas reinhardtii* lack the small subunit of a heterotetrameric ADP-glucose pyrophosphorylase. J. Bacteriol. 183: 1069–1077.

Zeqiraj, E., X. Tang, R.W. Hunter, M. Garcia-Rocha, A. Judd, M. Deak et al. 2014. Structural basis for the recruitment of glycogen synthase by glycogenin. Proc. Natl. Acad. Sci. U. S. A. 111: E2831–E2840.

Zhang M., A.M. Myers and M.G. James. 2005. Mutations affecting starch synthase III in Arabidopsis alter leaf starch structure and increase the rate of starch synthesis. Plant Physiol. 138: 663–674.

Zhang, X., N. Szydlowski, D. Delvallé, C. D'Hulst, M.G. James and A.M. Myers. 2008. Overlapping functions of the starch synthases SSII and SSIII in amylopectin biosynthesis in Arabidopsis. BMC Plant Biol. 8: 1–18.

Zhen, S., X. Deng, M. Zhang, G. Zhu, D. Lv, Y. Wang et al. 2017. Comparative phosphoproteomic analysis under high-nitrogen fertilizer reveals central phosphoproteins promoting wheat grain starch and protein synthesis. Front. Plant Sci. 8: 67.

Zhong, Y., A. Blennow, O. Kofoed-Enevoldsen, D. Jiang and K.H. Hebelstrup. 2019. Protein Targeting to Starch 1 is essential for starchy endosperm development in barley. J. Exp. Bot. 70: 485–496.

Ziegler, G.R., J.A. Creek and J. Runt. 2005. Spherulitic crystallization in starch as a model for starch granule initiation. Biomacromolecules 6: 1547–1554.

Index

Milton Keynes UK
Ingram Content Group UK Ltd.
UKHW050450071024
449327UK00014B/313